Eurotherm Conference No.95: Computational Thermal Radiation in Participating Media IV (2012)

Journal of Physics: Conference Series Volume 369

Nancy, France
18-20 April 2012

ISBN: 978-1-62276-803-5
ISSN: 1742-6588

Printed from e-media with permission by:

Curran Associates, Inc.
57 Morehouse Lane
Red Hook, NY 12571

Some format issues inherent in the e-media version may also appear in this print version.

Printed by Curran Associates, Inc. (2013)

For permission requests, please contact the Institute of Physics
at the address below.

Institute of Physics
Dirac House, Temple Back
Bristol BS1 6BE UK

Phone: 44 1 17 929 7481
Fax: 44 1 17 920 0979

techtracking@iop.org

Additional copies of this publication are available from:

Curran Associates, Inc.
57 Morehouse Lane
Red Hook, NY 12571 USA
Phone: 845-758-0400
Fax: 845-758-2634
Email: curran@proceedings.com
Web: www.proceedings.com

Eurotherm Conference No.95: Computational Thermal Radiation in Participating Media IV (2012)

Journal of Physics: Conference Series Volume 369

Nancy, France
18-20 April 2012

ISBN: 978-1-62276-803-5
ISSN: 1742-6588

IOP Institute of Physics

TABLE OF CONTENTS

RADIATIVE TRANSFER IN COMPLEX MEDIA

APPLICATIONS, COMBUSTION AND HIGH TEMPERATURES

INVERSE METHODS

NEW DEVELOPMENTS FOR RTE SOLUTION

GAS RADIATION MODELING

PARTICLES, DROPLETS AND DISPERSED SYSTEMS

MONTE CARLO METHODS

Preface

This volume of *Journal of Physics: Conference Series* is based on papers presented at the Eurotherm Conference 95: Computational Thermal Radiation in Participating Media IV, which was held in Nancy, France, on 18–20 April 2012. This seminar was the fourth in a series after Mons, Belgium (Eurotherm Seminar 73, April 2003), Poitiers, France (Eurotherm Seminar 78, April 2006) and Lisbon, Portugal (Eurotherm Seminar 83, April 2009).

Around 70 contributions were received during the conference preparation, including submissions for oral and poster presentations. A first selection process based on abstracts and a second based on two peer-reviews of the full papers finally resulted in the acceptance of 38 contributions from oral presentations and 11 from poster presentations.

The conference was attended by almost 80 scientists from 17 different countries: Austria, Belgium, Brazil, Canada, Chile, China, France, Germany, India, Poland, Portugal, Russia, Switzerland, The Netherlands, Tunisia, Turkey and USA.

The 30 contributions presented in this volume relate to the following main topics:

- Radiative transfer in complex media
- Applications, combustion and high temperatures
- Inverse methods
- New developments for RTE solution
- Gas radiation modeling
- Particles, droplets and dispersed systems
- Monte Carlo methods

The conference organizers gratefully acknowledge the members of the scientific committee and the experts who carried out the reviews of the papers, and the local organizing committee for their preparation of the conference. Also acknowledged are the support by the LEMTA, the Société Française de Thermique, ICHMT, the sponsorship of CNRS and the Lorraine Region, and the financial contributions of the Université de Lorraine, FLIR, Baccarat and the city of Nancy.

Conference Chairs

P Boulet	LEMTA, Nancy, France
D Lemonnier	Institut P', Poitiers, France
N Selçuk	Middle East Technical University, Ankara, Turkey
P Lybaert	Faculté Polytechnique de Mons, Belgium

Guest Editors

P Boulet
Université de Lorraine, LEMTA, CNRS, France
pascal.boulet@univ-lorraine.fr

D Lacroix
Université de Lorraine, LEMTA, CNRS, France
david.lacroix@univ-lorraine.fr

D Lemonnier
PPrime Institute, CNRS – ENSMA, University of Poitiers, France
denis.lemonnier@ensma.fr

P Lybaert
UMONS, Thermal Engineering and Combustion Laboratory, Belgium
paul.lybaert@umons.ac.be

N Selçuk
Middle East Technical University, Turkey
selcuk@metu.edu.tr

Peer review statement

All papers published in this volume of *Journal of Physics: Conference Series* have been peer reviewed through processes administered by the proceedings Editors. Reviews were conducted by expert referees to the professional and scientific standards expected of a proceedings journal published by IOP Publishing.

Modeling heat transfer within porous multiconstituent materials

Mathieu NIEZGODA[1], Denis ROCHAIS[1], Franck ENGUEHARD[1], Benoit ROUSSEAU[2], Patrick ECHEGUT[3]

[1]Commissariat à l'Énergie Atomique et aux Énergies Alternatives, DAM, CEA Le Ripault, BP 16, 37260 MONTS, France
[2]Laboratoire de Thermocinétique de Nantes (LTN), CNRS UMR 6607, 44306 NANTES, France
[3]Conditions Extrêmes et Matériaux : Hautes Températures et Irradiation (CEMHTI), CNRS UPR 3079, 45071 ORLÉANS, France

denis.rochais@cea.fr

Abstract. The purpose of our work has been to determine the effective thermal properties of materials considered heterogeneous at the microscale but which are regarded as homogenous in the macroscale environment in which they are used. We have developed a calculation code that renders it possible to simulate thermal experiments over complex multiconstituent materials from their numerical microstructural morphology obtained by volume segmentation through tomography. This modeling relies on the transient solving of the coupled conductive and radiative heat transfer in these voxelized structures.

1. Nomenclature

c_p	massic specific heat $(J.kg^{-1}.K^{-1})$	*Greek symbols*	
$\mathscr{D}$	Dirac function (s^{-1})	δ	voxel thickness (m)
$\mathscr{C}$	conductance $(W.m^{-1})$	Δt	step of time (s)
$\mathscr{E}$	total intensity of the black body $(W.m^{-2})$	κ	absorption coefficient (m^{-1})
$\mathcal{E}$	set of voxels	λ_i	conductivity $(W.m^{-1}.K^{-1})$
E	energy (J)	λ	wavelength (m)
f	form ratio	ϕ	flux (W)
F	form ratio with absorption	φ	flux density $(W.m^{-2})$
$\mathbf{F}$	function (vector) of the equations	ρ	density $(kg.m^{-3})$
$\mathbf{F}'$	jacobian matrix of $\mathbf{F}$	ρ_{ij}	reflectivity
$\mathcal{F}_{\text{up}}$	voxels of the superior face	σ_{SB}	Stefan-Boltzmann constant
h	convecto-radiative coefficient $(W.m^{-2}.K^{-1})$	θ	angle
$I, \mathscr{I}$	intensity $(W.m^{-2})$ or $(W.m^{-2}.sr^{-1})$	τ_{ij}	transmitivity

I^0	total intensity of the black body ($W.m^{-2}$)	*Subscripts and superscripts*	
k	extinction index	$'$	directional parameter
M	intensities arrival point	$=$	matrix
n	optical index	amb	ambiante
N	intensities origin point	i, j, k	voxel locations or face
s	curvilign abscisse (m)	cond	conductif
S	voxel face	e	incoming
$\mathcal{S}$	set of edges voxels	λ	spectral parameter
t	time (s)	n	iteration or step time
T	temperature (K)	O	opposite faces
$\mathbf{X}$	intensities and temperature vector	r, rad	radiative
z	axial coordinate	s	outcoming
		V	adjacent faces

2. Introduction

The French Atomic Energy Commission (CEA) works a great deal with porous materials – carbon composites, ceramics – and aims to optimize their microstructural and thermal properties for specific uses. A material of interest can be composed of several constituents and generally has a complex structure. Its pore size range can be quite large (several tens of micrometers). It is used in large-scale systems that are bigger than its own characteristic scale and can be submitted to very high fluxes and temperatures (until 3000°C). In these conditions, the material is considered as equivalent to a homogeneous medium for the simulation of its behavior in its using environment. Its local morphology is thus not taken into account.

We are therefore interested in characterizing the effective thermal properties of materials with heterogeneous microstructural morphologies. The objective is to determine the properties that theses materials would have if they were homogeneous. In order to do so, we have developed a methodology that consists in starting from a 3D-microstructure of the studied materials obtained by X-microtomography or FIB/SEM-tomography. The observations of the representative rebuilt volumes enable an understanding and analysis of the structural organization of these materials. The microstructures constitute the numerical support to compute the numerical experiment simulations.

The results render it possible to estimate the thermal properties of the materials in question. We are especially interested in the effective thermal diffusivity of porous multiconstituent materials that we estimate as a function of temperature with the help of an inverse method applied to experimental thermograms obtained by the flash method [1]. However, the identification of the diffusivity of porous and/or semitransparent materials is made difficult because of the thermal behavior of these media in which a strong conducto-radiative coupling can quickly occur when the temperature increases. For this reason, we have modeled the coupled conductive and radiative heat transfer as a function of the temperature within porous multiconstituent materials from their morphology discretized into a set of homogeneous voxels. The size of the voxel is larger than the wavelength which is the case in practice because of the studied materials and the high temperature. This modeling rendered it possible, on the one hand, to simulate any kind of numerical thermal experiments, especially the flash method, and on the other hand, to reproduce the thermal behavior of our materials in their using conditions. Our modeling is only accurate in the context we have here presented.

Considering the coupled heat transfer in porous media is very difficult to take into account, already in a homogeneous medium and even more in an inhomogeneous one, because it strongly depends on the microstructure of the constituents. Consequently, the media are mainly studied at ambient temperature where the conductive transfer is dominating and the radiation neglected or they are interested in their radiative properties and only the radiation is considered. Some researchers estimated the effective thermal conductivity especially in foams [2–3] often by being based on analytical model of the porous structure. Other works deal with only the radiative heat transfer [4–5]. Thus, very few studies deal with the coupled heat transfer in porous media at high temperatures. But the flash method was modeled in the case of a coupled conductive and radiative heat transfer within a semi-transparent material [1, 6–9].

Moreover, among the works dealing with the conductive or radiative heat transfer, most of them concerns 1D-problems and a weaker number focus on the numerical microstructure of the sample obtained by 3D-X-ray tomography. Numerical methods were developed to estimate the effective thermal conductivity [10–12] and they were applied to 3D-microstructures obtained by X-ray-tomography [13]. At the same time, the radiative properties (extinction and absorption coefficients, scattering phase function, etc.) were investigated starting from the microstructure [14–15] and thanks to the X-ray tomography [16–21].

Finally, in our knowledge, there is no study looking into the modeling of the coupled conductive and radiative heat transfer from the 3D-microstructure of porous media as a function of the temperature.

The purpose of this paper consists in describing the modeling approach and explaining the analytical and numerical resolution of this thermal problem. We have demonstrated how to solve the flash method in such voxelized structures and how to organize the analytical resolution of this problem to compute it in a calculation code giving the transient temperature field in each voxel. Furthermore, we have enumerated all the thermophysical and radiative data introduced in the developed calculation code for each constituent at the voxel scale to carry out our simulations. Finally, we have obtained the first validation results of the code.

3. Solving the flash method in a voxelized structure

3.1. Obtaining the microstructural morphology of materials

First, a numerical volume of the studied sample is required as the support of the numerical thermal experiment simulation [22]. By using the X-microtomography or FIB/SEM-tomography technique over the sample, it is possible to obtain a set of 2D images from which the sample can be built in 3D and then discretized into a set of voxels. These voxels constitute the elementary mesh over which the considered heat transfer problem is solved. Each voxel is therefore assumed to be homogeneous with the properties of the constituent of which it is composed. This discretization is a very fine description of the morphology of the studied samples. We have to make sure that the size of the voxels is larger than the wavelength of the occurring radiation. It is the case in the context of our study and we present what follows in the next parts.

3.2. Analytical solving

Consider a multiconstituent sample, representative of the studied material, in the form of a parallelepiped some millimeters long. It is composed of cubic homogeneous voxels. Each voxel – with a size of a few tens of microns– is attributed the thermophysical and radiative properties of the material that it represents and can be semi-transparent. The upper face of the sample is submitted to a pulsed energy deposition of short duration at $t = 0$, which is assumed to be uniform over the surface. This face and the opposite one are submitted to heat dissipation through a coupled convecto-radiative coefficient. Moreover, these two faces are assumed to be opaque with emission and diffuse reflection. The lateral walls are assumed to be adiabatic and totally reflective.

The physics of this thermal problem are governed by two highly coupled equations: the radiative transfer equation and a flux balance equation. We need to solve these two equations in each voxel to model the heat transfer in the previously presented problem. To simplify a complex mathematical problem, the following assumptions are made. (*i*) For each face of each voxel, the intensities are assumed to be isotropic in each half-space. Two intensities are therefore considered (back and front) for each face. (*ii*) The radiative properties and variables of each constituent are presumed to be independent of the wavelength (grey media). This is done in order to consider mean radiative properties and variables. Such an assumption, although it may seem strong, is nevertheless legitimate because the sought temperature field is the result of energetic variables issued from the integration of all the contributions of each wavelength. And (*iii*) the properties are assumed to be independent of the temperature since the increase in temperature range is small in the case of the flash method. When it comes to real materials, they can be subjected to greater thermal stresses. This is accounted for by making their properties vary as a function of the temperature.

Thus, we have thirteen unknowns in each voxel (and we especially seek to determine the time-resolved temperature field of each voxel): the temperature and twelve intensities. For each voxel, we solve the radiative transfer equation between each face and the five others (i.e., six equations), as well as the flux balance equation involving the conductive and radiative flux exchanged by the voxel with its six neighbors (i.e., one equation). The mathematical problem is closed with six interface equations – one on each face – relating the incoming intensity on the face of the voxel to the outgoing one of its neighbor at the same face. In order to determine these thirteen unknowns, we need to simultaneously solve these thirteen equations.

3.3. The radiative transfer equation

Consider a voxel as it was previously described. Its faces are arbitrarily numbered as presented in Figure 1. We want an expression of the intensity going outside through one of the faces (for example the face numbered 6) which is obtained by adding the contributions of the intensities going inside through the five other faces. It is likely that the intensity going outside through face 6 is not isotropic.

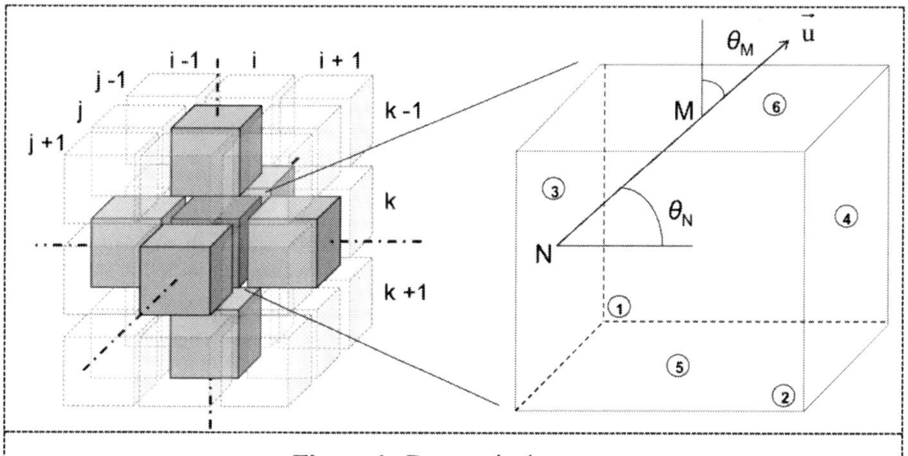

Figure 1. Geometrical aspects

Thus, the radiative flux going outside through this face is sought and the equivalent isotropic intensity going inside is deduced.

$$\phi_6^s = \iint_{M\in S_6} \overrightarrow{\varphi_r^s}(M)\cdot\vec{z}\,\mathrm{d}S_{M_6} \quad \text{where:} \quad \overrightarrow{\varphi_r^s} = \sum_{i=1}^{5}\iint_{N_i\in S_i} I'\left(M;\frac{\overrightarrow{N_iM}}{N_iM}\right)\frac{\overrightarrow{N_iM}}{N_iM}\frac{\cos\theta_{N_i}}{N_iM^2}\,\mathrm{d}S_{N_i} \tag{1}$$

To determine the directional intensity in M in the direction of **u**, the RTE is applied to the voxel between N and M. As the voxel is homogeneous, there is no scattering within each voxel. Thus, we only study materials in which there would not be micro- or nano-porosity and consequently scattering within the voxels. But scattering appears at the sample scale because of its heterogeneity. Indeed, if a voxel has a neighbour of different nature, a part of the outgoing radiative transfer is reflected whereas an other part is transmitted. At a larger scale (the one of the sample), that causes scattering. Under the assumption of grey medium, it is written:

$$\frac{\mathrm{d}I'(s_i)}{\mathrm{d}s_i} + \kappa I'(s_i) = \kappa I^0(T) \quad \text{with:} \quad I^0(T) = \frac{n^2 \sigma_{\mathrm{SB}} T^4}{\pi} = \mathscr{E} \tag{2}$$

It is simply solved, giving:

$$I'(s_i) = (\mathscr{I}_i^{\mathrm{e}} - \mathscr{E})\,\mathrm{e}^{-\kappa \cdot N_i M} + \mathscr{E} \tag{3}$$

where $\mathscr{I}_i^{\mathrm{e}}$ is the (isotropic) intensity going inside through face i. After some easy calculations, it is deduced that:

$$\phi_6^{\mathrm{s}} = \iint\limits_{M\in S_6} \sum_{i=1}^{5} \left[(\mathscr{I}_i^{\mathrm{e}} - \mathscr{E}) \iint\limits_{N_i\in S_i} \mathrm{e}^{-\kappa \cdot N_i M} \frac{\cos\theta_M \cos\theta_{N_i}}{N_i M^2}\, \mathrm{d}S_{N_i} + \mathscr{E} \iint\limits_{N_i\in S_i} \frac{\cos\theta_M \cos\theta_{N_i}}{N_i M^2}\, \mathrm{d}S_{N_i} \right] \mathrm{d}S_M \tag{4}$$

Hence:

$$\phi_6^{\mathrm{s}} = \sum_{i=1}^{5} \left[(\mathscr{I}_i^{\mathrm{e}} - \mathscr{E}) \iint\limits_{M\in S_6} \iint\limits_{N_i\in S_i} \mathrm{e}^{-\kappa \cdot N_i M} \frac{\cos\theta_M \cos\theta_{N_i}}{N_i M^2}\, \mathrm{d}S_{N_i}\, \mathrm{d}S_M \right] + \mathscr{E}\pi\delta^2 \underbrace{\sum_{i=1}^{5} f_{6i}}_{=1} \tag{5}$$

where f_{ij} is the form ratio of which the sum on the five faces gives 1. Similarly to the form ratio, a dimensionless parameter is defined which is called the form ratio with absorption F_{ij}.

$$F_{ij} = \frac{1}{\pi S_i} \iint\limits_{M_i\in S_i} \iint\limits_{N_j\in S_j} \mathrm{e}^{-\kappa \cdot N_j M_i} \frac{\cos\theta_{M_i} \cos\theta_{N_j}}{N_j M_i^2}\, \mathrm{d}S_{N_j}\, \mathrm{d}S_{M_i} \tag{6}$$

By setting an equality between the equivalent isotropic intensity going outside through face 6 and equation 5, we finally obtain:

$$\phi_6^{\mathrm{s}} = \pi\delta^2 \mathscr{I}_6^{\mathrm{s}} = \pi\delta^2 \left[\sum_{i=1}^{5} (\mathscr{I}_i^{\mathrm{e}} - \mathscr{E})\, F_{6i} + \mathscr{E} \right] \tag{7}$$

$$\mathscr{I}_6^{\mathrm{s}} = \sum_{i=1}^{5} (\mathscr{I}_i^{\mathrm{e}} - \mathscr{E})\, F_{6i} + \mathscr{E} \tag{8}$$

We can show by geometrical considerations that the thirty-six form ratios with absorption can be reduced in two. We now have: $F_{61} = F_{62} = F_{63} = F_{64} = F_{\mathrm{V}}$ and $F_{65} = F_{\mathrm{O}}$. These form ratios with absorption depend only on the optical thickness $\kappa\delta$ in the case of a homogeneous cubic voxel and are constant and particular to a material.

Finally, here are the six first equations for a voxel:

$$\begin{aligned}
\mathscr{I}_1^{\mathrm{s}} &= F_{\mathrm{V}} \sum_{i=3}^{6} \mathscr{I}_i^{\mathrm{e}} + F_{\mathrm{O}} \mathscr{I}_2^{\mathrm{e}} + \mathscr{E}(1 - F_{\mathrm{O}} - 4F_{\mathrm{V}}) \\
\mathscr{I}_2^{\mathrm{s}} &= F_{\mathrm{V}} \sum_{i=3}^{6} \mathscr{I}_i^{\mathrm{e}} + F_{\mathrm{O}} \mathscr{I}_1^{\mathrm{e}} + \mathscr{E}(1 - F_{\mathrm{O}} - 4F_{\mathrm{V}}) \\
\mathscr{I}_3^{\mathrm{s}} &= F_{\mathrm{V}}(\mathscr{I}_1^{\mathrm{e}} + \mathscr{I}_2^{\mathrm{e}} + \mathscr{I}_5^{\mathrm{e}} + \mathscr{I}_6^{\mathrm{e}}) + F_{\mathrm{O}} \mathscr{I}_4^{\mathrm{e}} + \mathscr{E}(1 - F_{\mathrm{O}} - 4F_{\mathrm{V}}) \\
\mathscr{I}_4^{\mathrm{s}} &= F_{\mathrm{V}}(\mathscr{I}_1^{\mathrm{e}} + \mathscr{I}_2^{\mathrm{e}} + \mathscr{I}_5^{\mathrm{e}} + \mathscr{I}_6^{\mathrm{e}}) + F_{\mathrm{O}} \mathscr{I}_3^{\mathrm{e}} + \mathscr{E}(1 - F_{\mathrm{O}} - 4F_{\mathrm{V}}) \\
\mathscr{I}_5^{\mathrm{s}} &= F_{\mathrm{V}} \sum_{i=1}^{4} \mathscr{I}_i^{\mathrm{e}} + F_{\mathrm{O}} \mathscr{I}_6^{\mathrm{e}} + \mathscr{E}(1 - F_{\mathrm{O}} - 4F_{\mathrm{V}}) \\
\mathscr{I}_6^{\mathrm{s}} &= F_{\mathrm{V}} \sum_{i=1}^{4} \mathscr{I}_i^{\mathrm{e}} + F_{\mathrm{O}} \mathscr{I}_5^{\mathrm{e}} + \mathscr{E}(1 - F_{\mathrm{O}} - 4F_{\mathrm{V}})
\end{aligned} \tag{9-14}$$

The form ratios with absorption can be written in the following way:

$$F_{\mathrm{O}} = \int_{x=0}^{1}\int_{y=0}^{1}\int_{z=0}^{1}\int_{t=0}^{1} \frac{\mathrm{e}^{-\kappa\delta\sqrt{(x-z)^2+(y-t)^2+1}}}{\left((x-z)^2+(y-t)^2+1\right)^2}\,\mathrm{d}x\,\mathrm{d}y\,\mathrm{d}z\,\mathrm{d}t$$

$$F_{\mathrm{V}} = \int_{x=0}^{1}\int_{y=0}^{1}\int_{z=0}^{1}\int_{t=0}^{1} zt\frac{\mathrm{e}^{-\kappa\delta\sqrt{(y-x)^2+z^2+t^2}}}{\left((y-x)^2+z^2+t^2\right)^2}\,\mathrm{d}x\,\mathrm{d}y\,\mathrm{d}z\,\mathrm{d}t$$

(15-16)

Multiple integral calculations render it possible to simplify these quadruple integrals into a simple one:

$$F_{\mathrm{O}} = \frac{2}{\pi}\int_{0}^{1} \frac{\mathrm{e}^{-\kappa\delta\sqrt{r^2+1}}}{(r^2+1)^2} r(\pi - 4r + r^2)\,\mathrm{d}r + \frac{2}{\pi}\int_{1}^{\sqrt{2}} \frac{\mathrm{e}^{-\kappa\delta\sqrt{r^2+1}}}{(r^2+1)^2} r\left(\pi - 2 - r^2 + 4\sqrt{r^2-1} - 4\arccos\left(\frac{1}{r}\right)\right)\mathrm{d}r$$

$$F_{\mathrm{V}} = \frac{8\kappa\delta - 3 + (3-5\kappa\delta)\mathrm{e}^{-\kappa\delta}}{12\pi(\kappa\delta)^2} + \frac{1}{12\pi}\int_{1}^{\sqrt{2}} \frac{\mathrm{e}^{-\kappa\delta r}}{r^3}\left(6r^4 - 6r^2 + 5 - 16(r^2-1)^{\frac{3}{2}}\right)\mathrm{d}r + \frac{1}{12\pi}\int_{\sqrt{2}}^{\sqrt{3}} \frac{\mathrm{e}^{-\kappa\delta r}}{r^3}\left(-3r^4 + 6r^2 + 1 + 8(r^2-2)^{\frac{3}{2}}\right)\mathrm{d}r$$

(17-18)

We can also write an intensity continuity relation at each face of each voxel (six extra equation for each voxel). For example, the intensity going inside through face 6 is the sum of the transmitted part of the intensity going outside through face 5 of the upper voxel and of the reflected part of the intensity going outside through face 6 of the considered voxel.

$$\begin{aligned}
\mathscr{I}_1^{\mathrm{e}}(i,j,k) &= \rho_{12'}\mathscr{I}_1^{\mathrm{s}}(i,j,k) + \tau_{2'1}\mathscr{I}_2^{\mathrm{s}}(i-1,j,k)\\
\mathscr{I}_2^{\mathrm{e}}(i,j,k) &= \rho_{21'}\mathscr{I}_2^{\mathrm{s}}(i,j,k) + \tau_{1'2}\mathscr{I}_1^{\mathrm{s}}(i+1,j,k)\\
\mathscr{I}_3^{\mathrm{e}}(i,j,k) &= \rho_{34'}\mathscr{I}_3^{\mathrm{s}}(i,j,k) + \tau_{4'3}\mathscr{I}_4^{\mathrm{s}}(i,j-1,k)\\
\mathscr{I}_4^{\mathrm{e}}(i,j,k) &= \rho_{43'}\mathscr{I}_4^{\mathrm{s}}(i,j,k) + \tau_{3'4}\mathscr{I}_3^{\mathrm{s}}(i,j+1,k)\\
\mathscr{I}_5^{\mathrm{e}}(i,j,k) &= \rho_{56'}\mathscr{I}_5^{\mathrm{s}}(i,j,k) + \tau_{6'5}\mathscr{I}_6^{\mathrm{s}}(i,j,k-1)\\
\mathscr{I}_6^{\mathrm{e}}(i,j,k) &= \rho_{65'}\mathscr{I}_6^{\mathrm{s}}(i,j,k) + \tau_{5'6}\mathscr{I}_5^{\mathrm{s}}(i,j,k+1)
\end{aligned}$$

(19-24)

By eliminating intensities going outside in equations (19-24), the equations (9-14) enable to reduce the initial problem with thirteen unknowns for each voxel to a problem with seven unknowns for each voxel.

3.4. The flux balance

The last equation to be formulated is a flux balance: the sum of the conductive and radiative flux entering a voxel *i* contributes to increase its temperature.

$$\forall i \in \mathcal{E}\backslash\mathcal{S}, \quad \sum_{k=1}^{6} \phi^{i}_{\text{cond},k} + \sum_{k=1}^{6} \phi^{i}_{\text{rad},k} = \rho c_p \delta^3 \frac{\partial T_i}{\partial t}$$

$$\forall i \in \mathcal{E}\backslash\mathcal{S}, \quad \sum_{k=1}^{6} \underbrace{\frac{2\delta\lambda_i\lambda_k}{\lambda_i + \lambda_k}}_{\mathscr{C}_{i/k}} \left(T_k^{n+1} - T_i^{n+1}\right) + \pi\delta^2 \sum_{k=1}^{6} \left(\mathscr{I}_k^{e^i} - \mathscr{I}_k^{s^i}\right) - \rho c_p \delta^3 \frac{T_i^{n+1} - T_i^n}{\Delta t} = 0$$

(25-26)

3.5. Boundary conditions

The equations mentioned above are valid for all the problems that can be studied with this modeling approach. Only the equations related to the boundary conditions, concerning the voxels belonging to the six faces of the parallelepiped volume, the twelve sides and the eight corners, can change from one problem to another. In the case of the flash method, in order to take into account the boundary conditions of the lateral walls, the previous equations have to be adjusted for the concerned voxels in the sense that there is no radiation transmitted outside the sample but rather that it is totally reflected on the lateral walls (i.e., the intensities entering and exiting are equal). Moreover, there is neither a conductive flux, nor a radiative one (with the previous consideration) going inside through these walls. In the case of the boundary conditions on the upper and lower walls, we have that at the lower/upper wall of each concerned voxel, the intensity going inside through these face is equal to the black body intensity at the same temperature. Moreover, a flux balance at the wall of each voxel can be written: thus, in the upper face, the sum of the conductive and radiative flux going inside is equal to the flux brought by the flash and exchanged by the convective and radiative dissipation:

$$\forall i \in \mathcal{F}_{\text{up}}: \qquad \phi_{\text{cond}} + \phi_{\text{rad}} = \phi_{\text{flash}} - \phi_{\text{losses}}$$

$$2\lambda_i\delta\left(T_{\text{wall}} - T_i\right) + \pi\delta^2\left(\mathscr{I}_6^{e^i} - \mathscr{I}_6^{s^i}\right) = E_{\text{flash}}\delta^2\mathscr{D}(t) - h\delta^2\left(T_{\text{wall}} - T_{\text{amb}}\right) \quad (27)$$

Because of these last considerations, the temperatures at the lower/upper faces of the voxels on the lower/upper walls also have to be considered as unknowns in the problem.

3.6. Computation and numerical solving

X is the vector holding all the unknowns of the problem *i.e.*, seven times the number of voxels. This vector is the solution of a non-linear equation that can be written **F(X)** = **0** and that is solved numerically by Newton's iteration method for systems of 2nd or superior dimension according to:

$$\overline{\overline{\mathbf{F}'}}(\mathbf{X}_n)\left[\mathbf{X}_n - \mathbf{X}_{n+1}\right] = \mathbf{F}(\mathbf{X}_n) \quad \text{where}: \quad \overline{\overline{\mathbf{F}'}}(\mathbf{X}) = \left(\frac{\partial F_i(\mathbf{X})}{\partial X_j}\right)_{i,j} \quad (28)$$

Since $\mathbf{X}_n$ is known, and with $\mathbf{Y}_n = \mathbf{X}_n - \mathbf{X}_{n+1}$, the numerical problem amounts to solving a linear system for each iteration of Newton's method until the convergence to the solution giving **X**. There is an incredible amount of data to stock in memory (especially the matrix $\mathbf{F}'(\mathbf{X}_n)$). To numerically solve these linear systems while avoiding keeping any data in memory, the iterative algorithm of Fletcher-Reeves (modified by Polak-Ribiere) has been chosen. It is easy to set up and very efficient in convergence speed.

Below is the procedure of the numerical solving of our problem. For each time step, Newton's iteration method was applied until the convergence of the sought vector **X** (whose components were initialized at the ones resulting from the previous time step or at arbitrarily assumed values for the first time step). For each iteration, the Polak-Ribiere algorithm was applied until the convergence of the vector **Y** (arbitrarily initialized) from which $\mathbf{X}_{n+1}$ was deduced since $\mathbf{X}_n$ was known. Finally, the components of **X** giving the temperature in each voxel for each time step rendered it possible to obtain the temperature field within the sample for the experiment, and especially a thermogram of the opposite face to the Dirac stimulation in the case of the flash method.

4. Thermophysical and radiative properties

The code presented in the previous part is a direct model making it possible to find an unknown parameter by calculation. Here, this was the transient temperature in each voxel. The different local thermophysical and radiative properties that are particular to each constituent of the sample have to be input in the code.

The voxel size is given by the tomography resolution. The volume specific heat is used for each constituent at the experiment temperature and these values were either found in the literature or measured by techniques of calorimetry existing in the laboratory. The thermal diffusivity from the micrometric scale to one of a few dozen micrometers was measured by photoreflectance and infrared microscopy as a function of the temperature [23].

The optical index and the absorption coefficient of the constituent were issued from spectral emission measures thanks to the infrared spectroscopy technique [24]. The emission spectrum of the constituent made it possible to determine, by means of the dielectric function, the spectrum of the real part and the imaginary part of the complex refraction index: the spectrum of the optical index and the extinction index, respectively (from which the absorption coefficient could be deducted). Considering the mean radiative properties over the entire spectrum, the optical indexes and the absorption coefficients were calculated from the obtained spectrum with the following definitions:

$$n = \frac{\int_0^{+\infty} n_\lambda I_\lambda^0(T)\,\mathrm{d}\lambda}{\int_0^{+\infty} I_\lambda^0(T)\,\mathrm{d}\lambda} \quad \text{and} \quad \kappa = \frac{\int_0^{+\infty} \kappa_\lambda I_\lambda^0(T)\,\mathrm{d}\lambda}{\int_0^{+\infty} I_\lambda^0(T)\,\mathrm{d}\lambda} \quad \text{with: } \kappa_\lambda = 4\pi\frac{k_\lambda}{\lambda} \tag{29}$$

Finally, the reflection (reflectivities) and transmission (transmitivities) coefficients were determined at the interface between the constituents by the Fresnel relation under collimated light. It was integrated over all the half-space since an isotropic diffuse radiation is considered.

5. Code validation

In order to validate the code for which we have presented the theoretical conception, we have attempted to compare the results it provides with data obtained with another calculation code. The considered experiment consisted of a stationary guarded hot plate technique. The code simulated the unidirectional coupled conductive and radiative heat transfer between two infinite plates for which the emissivity was known. One was at a low temperature (cold), the other at a higher one (hot). To solve the problem, the other code used a discrete ordinates method and an adaptive step of space.

To be close to this configuration, we considered a monocolumn structure composed of voxels from the same material. However, our configuration was not strictly 1D. For the sake of comparison, the lateral walls of our column had to be considered as at infinity, implying that the form ratios with absorption had to be recalculated with quite different expressions than the ones defined. They were thus function of the form ratios. We could show that the form ratios with absorption between adjacent faces were equal to zero when the lateral walls were at infinity, as were those between the opposite faces at infinity. The form ratio with absorption between the opposite faces in the direction of the heat transfer (perpendicular to the plates) was not equal to zero. Its value tends to 1 when the optical thickness $\kappa\delta$ tends to 0.

The case was the following. The hot wall temperature was 2000 K, and that of the cold wall was 1000 K. Their emissivity was 1, the thickness of the sample was 2 mm, and its conductivity was 2.5 $\mathrm{mW.m^{-1}.K^{-1}}$. Figure 2 presents three temperature profiles for three values of the mean absorption coefficient (100, 1000 and 10000 $\mathrm{m^{-1}}$). It can be noticed that the profiles from the two different modeling runs fit each other exactly, which constitutes a first validation of our model.

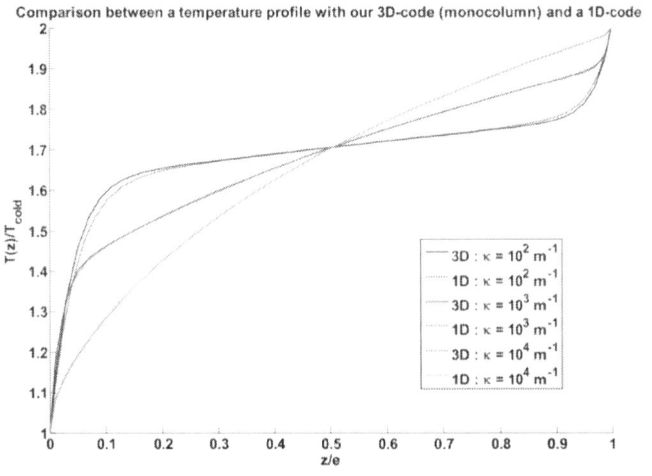

Figure 2. Inter-code comparison

We notice that the two profiles from the two different models are very close, which constitutes a first validation of our model. The low difference got for the low absorption coefficients is explained by the refined mesh near the walls taking into account in the 1D-model contrary to our model.

This study should be continued. The validations carried out during the progressive creation of the code (inter-codes or comparing with simplified versions for homogeneous media) have shown that this investigation is headed in the right direction. Future work involves validating the model with flash experiments on materials with different semi-transparency characteristics.

6. Conclusions

We have developed a calculation code in order to model the coupled conductive and radiative heat transfer within porous multiconstituent materials that are numerically represented by a numerical voxelized structure. This discretization corresponds to a very fine description of the morphology of the studied samples. The purpose of this work was to numerically simulate thermal experiments or the real behavior of the studied materials when subjected to any kind of thermal stress. The code was validated thanks to simplified conductive and radiative models solving the same problem in other ways. The next step will be to validate it by attempting to reproduce the real experimental behavior of materials of different characteristics.

References

[1] M. Niezgoda, D. Rochais, F. Enguehard, P. Echegut, B. Rousseau, Modeling of time-resolved coupled radiative and conductive heat transfer in multilayer semitransparent materials, *Appl. Phys. Lett.*, **99**, 224101 (2011).

[2] K. Boomsma, D. Poulikakos, On the effective thermal conductivity of a three dimensionally structured fluid saturated metal foams, *Int. J. Heat Mass Transfer*, **44** [4], 827–836 (2001).

[3] A. Bhattacharya, V.V. Calmidi, R.J. Mahajan, Thermophysical properties of high porosity metal foams, *Int. J. Heat Mass Transfer*, **45** [5], 1017–1031(2002).

[4] B.P. Singh, M. Kaviany, Modeling radiative heat transfer in packed beds, *Int. J. Heat Mass Transfer*, **35** [6], 1397–1405 (1992).

[5] J. Dombrovsky, H. Randrianalisoa, W. Lipinski, D. Baillis, Approximate analytical solution to normal emittance of semi-transparent layer of an absorbing, scattering, and refracting medium, *J. Quant. Spectrsc. Radiat. Transfer*, **112**, 1987–1994 (2011).

[6] S. André, A. Degiovanni, A new way of solving transient radiative–conductive heat transfer problems, *J. Heat Transfer*, **120**, 943–955 (1998).

[7] O. Hahn, F. Raether, M.C. Arduini-Schuster, J. Fricke, A. Degiovanni, Transient coupled conductive/radiative heat transfer in absorbing emitting and scattering media: application to laser-flash measurements on ceramic materials, *Int. J. Heat Mass Transfer*, **40** [3], 689–698 (1997).

[8] M. Lazard, S. André, D. Maillet, Diffusivity measurement of semi-transparent media: model of the coupled transient heat transfer and experiments on glass silica glass and zinc selenide, *Int. J. Heat Mass Transfer*, **47**, 477–487 (2004).

[9] R. Coquard, D. Rochais, D. Baillis, Experimental investigations of the coupled conductive and radiative heat transfer in metallic/ceramic foams, *Int. J. Heat Mass Transfer*, **52**, 4907–4918 (2009).

[10] A.M. Druma, M.K. Alam, C. Druma, Analysis of thermal conduction in carbon foams, *Int. J. Thermal Sci.*, **43** [7], 689–695 (2004).

[11] R. Coquard, M. Loretz, D. Baillis, Conductive heat transfer in metallic/ceramic open-cell foams, *Adv. Eng. Mater.*, **10** [4], (2008).

[12] R. Coquard, D. Baillis, Numerical investigation of conductive heat transfer in high-porosity foams, *Acta Mater.*, **57** [18], 5466–5479 (2009).

[13] M. Saadatfar, C.H. Arns, M.A. Knackstedt, T. Senden, Mechanical and transport properties of polymeric foams derived from 3D images, *Colloids Surf. A Physicochem. Eng. Aspects*, **263** [1–3], 284–289 (2004).

[14] M. Wang, N. Pan, Modeling and prediction of the effective thermal conductivity of random open-cell porous foams, *Int. J. Heat Mass Transfer*, **51** [5–6], 1325–1331 (2008).

[15] C.Y. Zhao, S.A. Tassou, T.J. Lu, Analytical considerations of thermal radiation in cellular metal foams with open cells, *Int. J. Heat Mass Transfer*, **51** [3–4], 929–940 (2008).

[16] J. Petrasch, P. Wyssb, A. Steinfeld, Tomography-based Monte Carlo determination of radiative properties of reticulate porous ceramics, *J. Quant. Spectrosc. Radiative Transfer*, **105** [2], 180–197 (2007).

[17] B. Zeghondy, E. Iacona, J. Taine, Determination of the anisotropic radiative properties of a porous material by radiative distribution function identification (RDFI), *Int. J. Heat Mass Transfer*, **49**, 2810–2819 (2006).

[18] M. Loretz, R. Coquard, D. Baillis, E. Maire, Metallic foams: radiative properties/comparison between different models, *J. Quant. Spectrosc. Radiative Transfer*, **109** [1], 16–27 (2008).

[19] B. Rousseau, D. De Sousa Meneses, P. Echegut, M. Di Michiel, J.-F. Thovert, Prediction of the thermal radiative properties of an x-ray m-tomographied porous silica glass, *Appl. Opt.*, **46** 4266–4276 (2007).

[20] R. Coquard, B. Rousseau, P. Echegut, D. Baillis, H. Gomart, E. Iacona, Investigations of the radiative properties of Al–NiP foams using tomographic images and stereoscopic micrographs, *Int. J. Heat Mass Transfer*, **55**, 1606–1619 (2012).

[21] A. Akolkar, J. Petrasch, Tomography based pore-level optimization of radiative transfer in porous media, *Int. J. Heat Mass Transfer*, **54** [23–24], 4775–4783 (2011).

[22] N. Vivet, S. Chupin, E. Estrade, T. Piquero, P.L Pommier, D. Rochais, E. Bruneton, 3D Microstructural characterization of a SOFC anode reconstructed by FIB tomography, *J. Power Sources*, **196** [18], 7541–7549 (2011).

[23] D. Rochais, G. Le Meur, G. Domingues, V. Basini, Microscopic thermal characterization of HTR particle layers, *Nuclear Engineering and Design* (2008).

[24] J.F. Brun, D. De Sousa Meneses, B. Rousseau, P. Echegut, Dispersion Relations and Phase Retrieval in Infrared Reflection Spectra Analysis, *Soc. Appl. Spect.*, **55** [6], 774–780 (2001).

Numerical prediction of the radiative behavior of metallic foams from the microscopic to macroscopic scale

B Rousseau[1], JY Rolland[1], P. Echegut[2], E. Brun[3] and J Vicente[4]

[1]LTN, UMR CNRS 6607, Rue Christian Pauc, BP 50609, 44306 Nantes cedex 3, France
[2]CEMHTI, UPR CNRS 3079, 45071, Orléans cedex 2, France
[3]ESRF, F38043, Grenoble, France
[4]IUSTI, UMR CNRS 6595, F13453, Marseille France
E-mail of the corresponding author: benoit.rousseau@univ-nantes.fr

Abstract. A parallel C++ numerical code was implemented to quickly and accurately calculate the radiative properties (reflectance and transmittance) of metallic foams that had scattering domains (pores, struts) with mean sizes larger than the wavelengths associated with thermal radiation. For application, the code required experimental knowledge of the material characteristics on several length scales ranging from the microscopic (approximately one micrometer) to macroscopic (approximately one centimeter). In particular, the 3D voxelized image of the investigated material was used, which was previously acquired using x-ray μ-tomography. The radiative properties were obtained by tracking a large number of rays within the numerical medium. On the local scale, rays interacted directly with the fluid/solid interface, which was obtained prior to the calculation by the application of a robust marching cubes algorithm to the entire set of voxels. The code was applied to aluminum foam samples, for which the experimental radiative properties were previously measured at room temperature with a customized instrument based on an infrared spectrometer. The code performance is discussed and the results enabled determination of the effects caused by the fluid/solid interface on the interfacial reflection.

1. Introduction

Understanding the thermal radiative properties of porous materials is of crucial interest when calculating heat transfers within industrial systems that operate at high temperatures. Porous materials are often used as part of the design for high-temperature systems (e.g., heat exchangers, solar thermal receivers [1] and glass melting furnaces [2-3]). However, porous materials have a complex texture that makes study of the radiative properties difficult [4]. The term "texture" describes the spatial arrangement and size distribution of the scattering objects (i.e., pores, cracks, struts and fibers) in the host solid matrix. Currently, an important challenge is to predict the thermal radiative properties of both the pertinent textural parameters (e.g., porosity, specific surface area and pore size distributions) and chemical composition. The achievement of this task will enable engineers to reliably design porous materials with the intended radiative properties.

Since the pioneering work of Gliksman and Torpey [5], several approaches have been developed to firmly connect textural parameters to the radiative properties. Recently, Loretz et *al.* reported

Published under licence by IOP Publishing Ltd

analytical and experimental methods used to determine radiative properties such as the extinction coefficient, scattering albedo and scattering phase function of metallic foams [6]. The authors suggested that the extinction coefficient can be simply calculated from the strut dimensions. However, for metallic foams, Zhao et *al.* used relations where the extinction coefficient was dependent on the ratio between the foam's porosity and the mean strut diameter [7]. The relations defined by Loretz et *al.* and by Zhao et *al.* can be used for struts that have nearly optically polished surfaces. However, results from the application of these formulas are less reliable as the struts become optically thin (i.e., the solid phase is semitransparent).

During the last five years, numerical approaches based on the use of 3D images obtained by x-ray μ-tomography provided new insights into the investigation of this topic [8-13]. In particular, in the most favorable cases, x-ray μ-tomography[14] provides easy access to the geometrical position of the fluid/solid interface at the limit of the spatial resolution of the tomographic experiment, as well as a numerical method to draw the interface. Most of the numerical methods that attempt to calculate the radiative properties of metallic foams are based on the stochastic propagation of a huge number of rays within the porous medium. It is noteworthy that if rays propagate within a volumetric domain higher than the representative elementary volume [9, 15-16], the calculated discrete-scale radiative data can be used to solve continuum-scale simulations of heat transfer within the medium [17-18]. As the texture of the porous material becomes more complex (i.e., the material has several populations of scattering objects with different length scales and chemical compositions), the numerical methods, which are only based on the geometrical optics approximations, begin to fail. Rousseau et *al.* [13] combined the geometrical optics approximation and effective medium approximation to model the radiative properties at T = 1000 K of an optically thick rough coating that contained micropores. For metallic foams, Loretz et *al.* [6] introduced a parameter of specularity to account for the scattering of light by the rough surfaces of struts. Coquard et al. [19] combined stereoscopic and tomographic imaging approaches to account for surface roughness.

These methods show that accurate modeling of the radiative properties is primarily dependent on an accurate description of the topology of each interface (e.g., fluid/solid) encountered. Furthermore, some studies indicate that trilinear interpolation methods [20] can be used to reconstruct a small piece of curvilinear surface during the propagation of a given ray within the foam at each interaction point [9, 12, 21]. Other investigations used the marching cube approach [22] to create a small flat interface section. For foams with a single solid phase, commercial codes [6, 11] were applied to obtain an entire interface composed of numerous triangles. Once the triangular grid was known, stochastic rays were launched into a given volume.

In this work, a homemade code, titled iMorph [23-24] (IUSTI, Marseille, France), was used to determine the entire fluid/solid interface of an aluminum foam (ERG® Al 20), which was scanned by x-ray μ-tomography prior to the calculations. The fluid/solid interface was determined by applying a robust marching cube algorithm. Then, according to a thorough textural and chemical analysis, an improved Monte Carlo Ray Tracing (MCRT) [25] scheme was applied to the resulting digital image to compute both the normal hemispherical spectral reflectance and transmittance of the foam. Finally, a comparison between the computed data and experimental data acquired by infrared spectroscopy allowed the light-scattering mechanisms that occurred at the fluid/solid interface to be discussed.

2. Experimental procedure and textural study

A highly porous aluminum foam with 20 pores per linear inch (ppi), which was supplied by the Erg company (Oakland, California, US), was investigated. Two small samples (S1 and S2) were machined from the first specimen. S1, which had dimensions of 57×57×55 mm, was used for the textural and chemical investigation. S2 was used to measure the radiative properties. S2 had a paved shape and had dimensions of 57×57×13 mm.

The 2D investigation of the foam morphology was performed with an Environmental Scanning Electron Microscope (ESEM-Philip XL 40 at CEMHTI, Orléans, France) with the detection mode set to be sensitive to Secondary Electrons (SE). This mode was adapted for determining the topographic contrast. Furthermore, the ESEM apparatus was equipped with an energy dispersive x-ray detector. This device was useful for performing the chemical analysis of the struts that constituted the foam. The results from the investigation were used to determine the complex refractive index of the solid phase. To obtain more quantitative information regarding the foam texture, x-ray μ-computed tomography experiments were performed at the European Synchrotron Radiation Facility on the ID19 beamline (ESRF, Grenoble, France). The absorption images were collected using a fast read out 2048 × 2048 CCD camera, FRELON (Fast Readout LOw Noise), developed at ESRF and described in detail in by Thi et *al.* [26] and by Coan et *al.* [27]. A customized gold-coated integrating sphere with a diameter of 150 mm was used to measure the normal hemispherical spectral reflectance at room temperature (CEMHTI, Orléans, France). The sphere was adapted onto a Fourier Transform InfraRed (FTIR) spectrometer (Bruker IFS 113v Ettlingen, Germany) using a customized holder.

In Figure 1, 2D views of the foam are shown, in which the spatial resolution is serially increased: 20 mm (a), 2 mm (b) and 20 μm (c). Inspection of the image indicates that the cell was composed of curved triangular struts of inconstant sizes. Moreover, careful observation of the cross-sections of the struts indicates that they were filled. Investigations at higher magnification show that the surface of the struts was rough (Figure 1c).

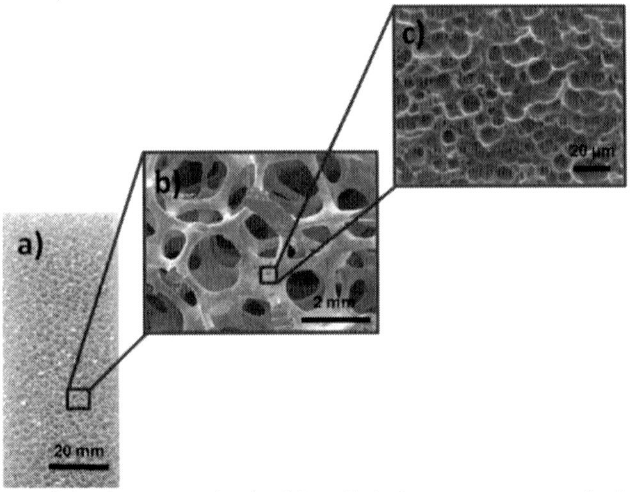

Figure 1: (a) A 2D view of the foam acquired with a digital camera at a resolution of 20 mm. (b) The same image with a higher magnification, which was obtained using an ESEM apparatus at a resolution of 2 mm. (c) The same image with a resolution of 20 μm.

Overlapping spherical pits with a mean diameter of 10 μm can be observed in Figure 1c. In the case of opaque media, this type of surface is well known to decrease the directional hemispherical reflectivity [28]. X-ray μ-tomography experiments were performed to acquire quantitative textural parameters (i.e., porosity, pore size distribution and volumetric surface) [6, 29-30]. After binning and filtering the raw data, the final numerical samples were shown as grey-scale images with a maximal size of 760×760×737 voxels and a digital resolution of 74.6 μm. A simple threshold procedure was performed with the iMorph software [24] to separate the solid phase from the fluid phase. The total porosity of the 3D image was determined by computing the ratio between the voxels belonging to the fluid phase to the total number of voxels. A calculated porosity of 92.0 % was determined by applying the iMorph software. It is worthy to note that this calculated porosity was similar to the value determined by the classical weighing method (91.5 %) [31]. The similar values indicate that the segmentation method was exact. To obtain the fluid/solid interface, a marching cubes algorithm was used [22, 32]. The numerical interface was constructed from a regular mesh composed of ordered triangles. A full 3D view of a reconstructed foam is shown in Figure 2.

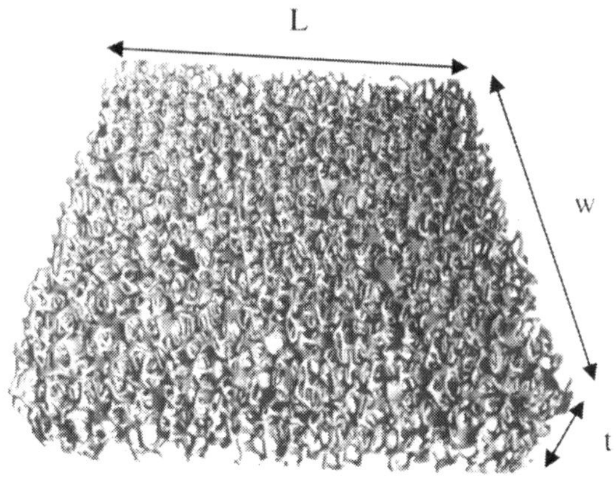

Figure 2: A 3D reconstructed view of the S2 aluminum foam sample used in the numerical experiment: length×width×thickness = 57×57×13 mm^3.

The implementation of the marching cubes algorithm in iMorph guaranteed that the mesh was continuous: neither holes between each triangle nor degenerated triangles were observed. Both of these artifacts can be very problematic as a ray tracing algorithm is used. The iMorph software also allowed each individualized pore to be segmented to retrieve the shape and size. An unbiased watershed transform was used to extract each cell, which resulted in the composition of the full 3D grey-scale image. The automated extraction of markers for the watershed was based on the maximal, included ball method [33]. The shape of the cell is generally assumed to be ellipsoidal, which was an assumption inspired by previous work on the same type of metallic foam [24, 31]. Upon completion of the segmentation, classical orientation and morphometric analyses (i.e., volume and mass moment of inertia) were used to obtain the shape and size distribution of a set of 1330 cells. The results indicated that the pores could be ellipsoids [31]. The 3 principal axes of the ellipsoids were denoted as $\boldsymbol{a} < \boldsymbol{b} < \boldsymbol{c}$, and the mean values were $\bar{\boldsymbol{a}}$ = 2150 μm, $\bar{\boldsymbol{b}}$ = 1430 μm, $\bar{\boldsymbol{c}}$ = 1120 μm. Moreover, a solid phase granulometry procedure (i.e., growth of the ball) was applied to the complete set of struts that composed the foam, which indicates that the struts had a typical lateral size of approximately 300 μm.

After the textural features were determined, a study of the chemical composition of the struts was performed. Energy-Dispersive X-ray (EDX) spectra indicated that the struts were mainly composed of aluminum. No other species (e.g., rare earth or transition metal ions) were observed. A small peak was observed at lower energy, which could correspond to oxygen; however, EDX is not a reliable technique for the characterization of species with low atomic numbers. In conclusion, only aluminum was considered in the following calculations for determining the local optical properties of the foam samples. Therefore, the complex refractive index of pure aluminum [34] was used.

3. Development of the numerical simulation

In this work, an Monte Carlo Ray Tracing (MCRT) program was implemented for computing the spectral dependence of the radiative properties of opened foams (reflectance and transmittance). The program was directly developed within the iMorph software previously used for the textural analysis and for the construction of the 3D image of the foam. The construction of the code was strictly based on the textural and chemical analysis of the studied sample. The principles of the MCRT code were extensively detailed in previous papers [4, 10, 25]. Briefly, the code launches a large number of parallel and unpolarized rays towards the center face of the numerical sample and tracks the rays until they exit the sample. Thus, the code was able to numerically reproduce experimental conditions required for measuring normal hemispherical reflectance, $\hat{R}(\sigma, T)$, and normal hemispherical transmittance, $\hat{T}_r(\sigma, T)$, with an integrating gold sphere. Normal normal transmittance can also be computed. For these variables, σ represents wave number and T represents temperature.

During the propagation of a ray on the local scale, only reflection occurred. If the ray had a certain energy (I_i) before reaching an interface, then the ray transports the reflected energy, $I_{i+1} = rI_i$, after the reflection event and releases $(1-r)I_i$ to the solid phase (i.e., a strut), where r corresponds to the local directional reflectivity given by the Fresnel law for a given incident angle, σ and T [35]. This scheme is possible because the struts are optically thick. A brief inspection of the value of the index of absorption of pure aluminum [34], k, indicates that for the entire spectral range of this study (i.e., 2-25 µm), the incident light would be completely absorbed for a mean thickness of 190 nm. Consequently, rays are not transmitted as they cross a strut. Reflection events can occur within the foam until the energy of a ray is higher than a cut off criterion, which a value equal to 0.01 was used in this paper. A discussion of the effects produced by the cut-off value was provided in a previous paper [36].

Finally, if rays can exit the sample through the back face, the energies of the rays contribute to the macroscopic normal hemispherical reflectance, $\hat{R}(\sigma,T)$. For the 5 other faces, energies supplied by the rays contribute to the normal hemispherical transmittance, $\hat{T}_r(\sigma,T)$.

Improvements to the MCRT code were performed. Recall that the order of the triangles in the mesh was determined using the iMorph software. At the voxel scale, special care focused on the optimization of the propagation time of a given ray. The voxels connected to the regular mesh are used to follow the propagation of rays through the sample. A step-by-step procedure is used to propagate rays into the media. This method assures that the algorithm uses a maximum of three triangles per voxel to determine if intersection occurs. If no intersection occurs in a voxel, then the next voxel is evaluated. The technique used in this article was faster and more efficient than the implementation of a more straightforward technique based on the direct calculation of the intersection with each voxel considered [8-9, 12, 21, 37]. Moreover, the step-by-step technique is much more efficient than a computation that uses every triangle for each intersection [6, 11]. For reference, the intersection calculation time for our method was on the order of $O(1)$, whereas direct computation with n triangles had a dependence of $n\log(n)$. Furthermore, in the algorithm used, the initial rays were placed into a "deque" memory structure. To be launched, each ray emerges from the end of the "deque". A ray is deleted from the "deque" upon intersection with the interface and a new ray, which results from the local interaction with the solid interface, is placed at the end of the deck. The process ends when the deck is empty. Therefore, the algorithm is very easy to parallelize. The initial 10^6 rays were distributed into different "deques" that were managed by separate threads.

Furthermore, a local optical scattering law that was based on the geometrical optics approximation was pragmatically introduced to account for the diffuse reflections induced by the rough surface of the struts (see Figure 1c). In this study, a simulation strategy that was previously employed for studies in the literature was used [6, 19, 38]. The strategy provided results largely comparable with those obtained from the approximated, as well as rigorous, resolution of Maxwell's equations at the interfaces [39]. If a ray intercepts a triangle while traveling, then the reflected energy is modified by a factor loss of $r^{n_{impact}}$, where n_{impact} is the mean number of local reflections generated by the surface roughness. The outgoing direction of the rays is randomly decided in the upward-hemisphere, in which the basal plane is defined by the intercepted triangle. An isotropic scattering behavior was assumed. In this work, n_{impact} values were obtained by studying the optical behavior of the virtual roughness of aluminum samples with an increasing slope (σ_{rms}/τ), where σ_{rms} is the root mean square of the height, and τ is the correlation length of the height. A previously known MCRT code that was developed to investigate the optical behavior of an opaque rough surface was applied [13], which allowed for the evolution of n_{impact} as a function of σ_{rms}/τ to be determined, as shown in Figure 3. The evolution was consistent with the earlier works of Bergstrom *et al.* [38, 40], where two scattering regimes could be distinguished. For $\sigma_{rms}/\tau < 0.25$, simple scattering occurs. However, for $\sigma_{rms}/\tau > 0.25$, multiple scattering regimes can occur with a local mean number of reflections, which can reach 5 local reflections if the slope is 2. This method enabled the spatial structure to be integrated in the calculations, which was not "viewed" by the tomographic experiment, at a resolution of 74.62 µm.

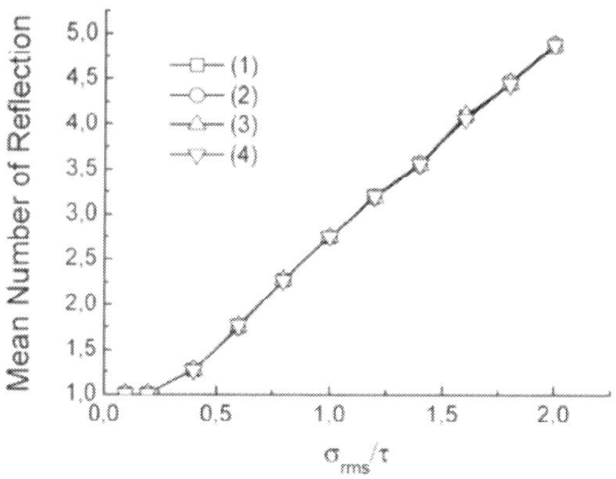

Figure 3: Evolution of n_{impact} as a function of the slope of the roughness for aluminum samples at T = 300 K. 10^5 rays were used for the MCRT simulation and the complex refractive index were extracted from Ref. [34]. Numerical simulations were performed for rays with various wave numbers: (1) 400 cm^{-1}, (2) 970 cm^{-1}, (3) 2580 cm^{-1} and (4) 4850 cm^{-1}.

4. Results and discussion

The results of the calculations of both the room-temperature normal hemispherical reflectance and transmittance of the numerical foam samples were determined for wave numbers in a spectral range of 400 cm^{-1} to 5000 cm^{-1} (i.e., 2-25 µm). As determined in section 2, the mean characteristic sizes of the pores and struts identified in the foam enabled the geometrical optics approximation to be largely suitable for calculating the light-matter interactions. The Mie parameter sizes ($X_{Mie,pore}$ and $X_{Mie,struts}$) were larger than 1 over the entire spectral range investigated. Furthermore, each strut was optically thick at T = 300 K. Consequently, the MCRT code (in C++) implemented in the iMorph software was applied to compute $\hat{R}(\sigma, T)$ and $\hat{T}(\sigma, T)$. For each experiment, 10^6 rays were launched at the bottom face of a given sample, which was a sensible compromise between statistical stability and computational cost. The statistical fluctuation of 10^6 rays was less than 0.5 % over several simulations. An incident beam diameter of 10 mm was used to reproduce the experimental conditions required for the use of the gold integrating sphere installed on the infrared spectrometer. More precisely, the aperture of the collection port for reflectance was virtually modified. Numerical calculations were performed with a commercial HP Z800 workstation equipped with an Intel(R)Xeon(R) Quad-Core processor (2.4 GHz, 12 MB cache, 1066 MHz memory). Using the QT framework that allows the iMorph software to run, parallel computations can be programmed. For a given wave number, 60 seconds was necessary to calculate the entire set of results.

Figure 4 shows the normal hemispherical spectral reflectance and transmittance as a function of the reconstructed foam thickness. When rays interact with the surface of the struts, only specular reflection events are considered, which indicates that the roughness of the struts does not cause peculiar scattering effects. For the normal spectral reflectance, the diameter of the aperture of the collector was set to 50 mm. As expected, the normal normal transmittance decreased exponentially, whereas the normal hemispherical reflectance increased and reached an asymptote at a thickness of 25 mm. The normal normal transmittance reached a value close to 0 for a foam thickness of 40 mm. The normal normal spectral transmittance decreased more rapidly than the normal hemispherical spectral transmittance. According to the procedure used for the calculations, normal normal transmittance was obtained by the ratio between the accumulated transmitted energies that arrived on the top face of the sample (i.e., the face opposite to the face that was impacted by the incident beam) to the initial energies deposited on the bottom face, whereas normal hemispherical transmittance included also the accumulated transmitted energies that reached the 4 side faces (i.e., the 4 faces other than the top face). As the sample thickness increased, the probability for rays to be transmitted to the lateral faces increased. The energies received on the 4 side faces were identical in value, which indicates that the distribution of energy was symmetric for the samples.

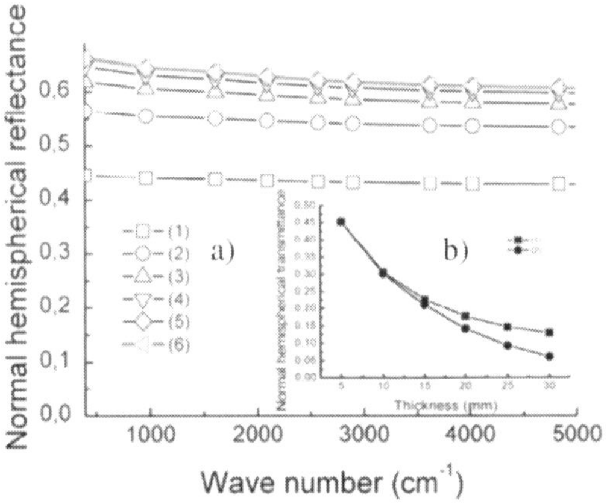

Figure 4: (a) Evolution of the normal hemispherical spectral transmittance for increasing thicknesses of the foam at T = 300 K: (1) 5 mm, (2) 10 mm, (3) 15 mm, (4) 20 mm, (5) 25 mm and (6) 30 mm. (b) A room temperature comparison of (1) the normal hemispherical transmittance of the foam at σ = 2900 cm^{-1} and (2) the normal normal transmittance at σ = 2900 cm^{-1}.

Figures 5 and 6 show the evolution of the spatial repartition of the back scattered energies for two aluminum foams with a thickness of 5 mm and 30 mm. The images indicate that as the thickness of foam increases, the back scattering spot becomes more extensive. For the sample with a thickness of 5 mm, 95 % of the reflected energy was localized in a circular area of 20 mm, whereas this value decreased to 81 % for the 30 mm-thick sample. A collector with an aperture diameter of 30 mm concentrated 99 % of the reflected energy from the 5 mm-thick foam, whereas the value increased to 93 % for the 30-mm thick foam. This result indicates that the thickness of the sample must be considered to correctly measure the normal hemispherical reflectance with an integrating sphere. For example, in studies with foams that have varied thicknesses and similar textural properties, use of an integrating sphere with the same aperture for the collection port of the reflected light can lead to incorrect experimental results.

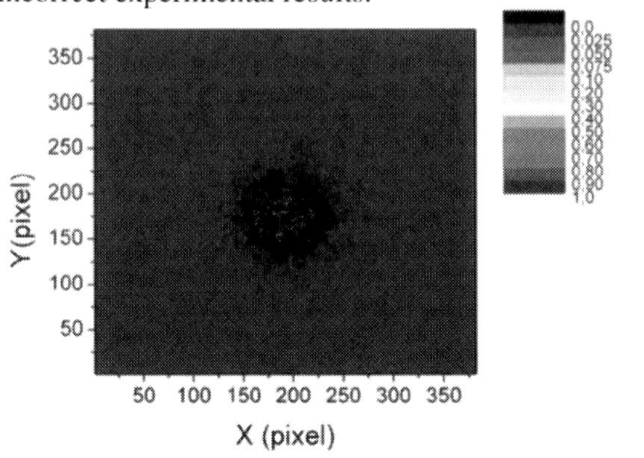

Figure 5: Spatial distribution of the diffuse reflected energy for a reconstructed aluminum foam with a thickness of 5 mm at T = 300 K.

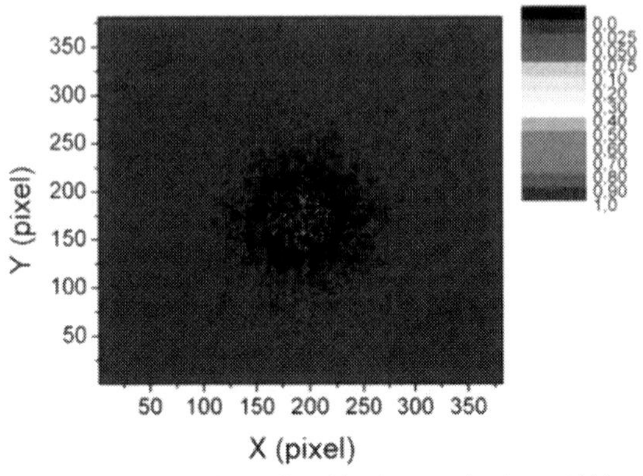

Figure 6: Spatial distribution of the diffuse reflected energy for a reconstructed aluminum foam with a thickness of 30 mm at T = 300 K.

To obtain a more quantitative appraisal of the effects on normal spherical reflectance caused by the surface roughness of the struts, numerical simulations were performed on struts with growing roughness slopes, σ_{rms}/τ. As explained in the section 3, calculations were used to integrate the surface optical scattering due to the solid struts. Increasing values of n_{impact} were used for different collector diameters, which the results are shown in Figure 7.

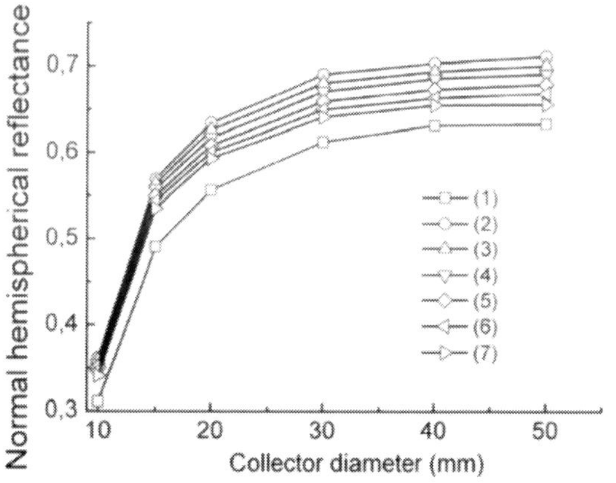

Figure 7: Effect of the surface roughness of the struts on the normal hemispherical reflectance of a reconstructed aluminum foam with dimensions of 57×57×13 mm (T = 300 K, σ = 2900 cm^{-1}): (1) the specular case and diffuse cases for (2) n_{impact} = 1, (3) n_{impact} = 1.2, (4) n_{impact} = 1.4 (5) n_{impact} = 1.6, (6) n_{impact} = 1.8 and (7) n_{impact} = 2.

According to the model, the surface roughness has a noticeable effect on the level of normal hemispherical reflectance. For collector diameters greater than 40 mm, the reflectance values seem to be approaching an upper limit.

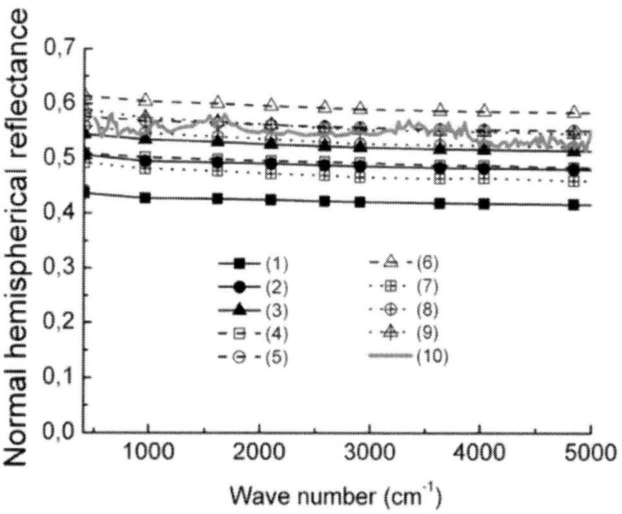

Figure 8: Comparison between the calculated normal hemispherical spectral reflectance for different collector sizes and surface scattering behaviors at a variety of wave numbers: (1) $\emptyset_{collector}$ = 15 mm, specular case (2) $\emptyset_{collector}$ = 20 mm, specular case (3) $\emptyset_{collector}$ = 25 mm, specular case (4) $\emptyset_{collector}$ = 15 mm, n_{impact} = 1.2, (5) $\emptyset_{collector}$ = 20 mm, n_{impact} = 1.2, (6) $\emptyset_{collector}$ = 25 mm, n_{impact} = 1.2, (7) $\emptyset_{collector}$ = 15 mm, n_{impact} = 1.4 (8) $\emptyset_{collector}$ = 20 mm, n_{impact} = 1.4 (9) $\emptyset_{collector}$ = 25 mm, n_{impact} = 1.4 and (10) experimental data ($\emptyset_{incident\ beam}$ = 10 mm).

The main augmentation occurs for n_{impact} = 1 (i.e., the scattering regime for which $\sigma_{rms}/\tau < 0.25$). For n_{impact} =1, 90 % of the reflected rays were concentrated in a spherical area with a diameter of 20 mm, whereas the value decreased to 85 % for the specular case. If a backward distribution of the diffuse reflected light at the impact point between a ray and a strut is assumed, then a global enhancement can be observed in the back scattering behavior of the foam. As n_{impact} increases, $\hat{R}(\sigma, T)$ starts to slowly decrease because the amount of energy that is locally absorbed by the matter is equal to $1 - r^{n_{impact}}$. For a fixed value of the collector size, the value of the roughness of the slope of the real Erg foam can be indirectly estimated by adjusting the experimental measurement and numerical calculation. Figure 8 depicts the comparison of experimental and calculated reflectances with the MCRT code at T = 300 K.

The best agreements were obtained for $\emptyset_{collector}$ = 20 mm and n_{impact} =1.2 and for $\emptyset_{collector}$ = 25 mm and n_{impact} =1.4, where n_{impact} =1.2 and 1.4 correspond to a slope of 0.35 and 0.45, respectively. The experimental collector had a diameter of 20 mm, which indicates that the mean slope roughness could be close to 1.2. This result indicates that the development of a contactless method to quantitatively probe the mean roughness parameters of the struts that constitute foams should be possible, which is easier than only using SEM images (Figure 1c). In cases where only simulated data can be provided, we would use the results obtained under the experimental acquisition conditions defined by the gold integrating sphere as a starting point.

5. Conclusion

The characterization of an aluminum foam was performed to provide experimental data for numerical software, which was intended to quickly compute the radiative properties. The numerical tool was integrated into a pre-existing software application, iMorph. The software was able to compute normal hemispherical reflectance and transmittance of materials for the following reasons: (i) at the microscopic scale, the interaction between light and fluid/solid interfaces can be treated pragmatically according to the geometrical optics approximation, and (ii) at the macroscopic scale, the dimensions of the reconstructed 3D images correspond to the dimensions of the sample used for the experimental characterization of the radiative behavior.

For a fixed probed volume, the code demonstrated that the effects caused by the experimentally observed micro-roughness on the surface of the struts using SEM (not observed in x-ray µ-tomography) can be important. According to the accuracy of the set-up used for measuring the diffuse reflectance and transmittance and to the dimensions of the probed sample, a contactless method may be possible to probe the roughness of struts.

6. References

[1] Steinfeld A and Fletcher E 1988 *Energy* **13** 301-11

[2] Pilon L, Zhao G and Viskanta R 2002 *Glass Science and Technology* **75**

[3] Pilon L, Guochang Z and Viskanta R 2002 *Glass Science and Technology* **75** 115-24

[4] Rousseau B, De Sousa Meneses D, Echegut P and Thovert J F 2011 *International Journal of Thermal Sciences* **50** 178-86

[5] Glicksman L R and Torpey M R 1988 A study of radiative heat transfer through foam insulation. Oak Ridge National Lab., TN (USA); Massachusetts Inst. of Tech., Cambridge (USA))

[6] Loretz M, Coquard R, Baillis D and Maire E 2008 *J. Quant. Spectrosc. Radiat. Transfer* **109** 16-27

[7] Zhao C, Lu T and Hodson H 2004 *International journal of heat and mass transfer* **47** 2927-39

[8] Zeghondy B, Iacona E and Taine J 2006 *Int. J. Heat Mass Transfer* **49** 2810-9

[9] Petrasch J, Wyss P and Steinfeld A 2007 *J. Quant. Spectrosc. Radiat. Transfer* **105** 180-97

[10] Rousseau B, Canizares A, De Sousa Meneses D, Matzen G, Echegut P, Di Michiel M and Thovert J 2007 *Colloids and Surfaces A: Physicochemical and Engineering Aspects* **300** 162-8

[11] Loretz M, Maire E and Baillis D 2008 *Adv. Eng. Mater.* **10** 352-60
[12] Haussener S, Lipinski W, Petrasch J, Wyss P and Steinfeld A 2009 *Journal of Heat Transfer* **131** 072701
[13] Rousseau B, Gomart H, Zanghi D, Bernard D and M. S 2010 *J. Mater. Res.* **25** 1890-7
[14] Baruchel J 2000 *X-RAY tomography in material science* Lavoisier)
[15] Lipinski W, Keene D, Haussener S and Petrasch J 2010 *Journal of Quantitative Spectroscopy and Radiative Transfer* **111** 2474-80
[16] Petrasch J, Haussener S and Lipinski W 2011 *Journal of Quantitative Spectroscopy and Radiative Transfer*
[17] Gusarov A 2008 *Physical Review B* **77** 144201
[18] Gusarov A 2010 *Physical Review B* **81** 064202
[19] Coquard R, Rousseau B, Echegut P, Baillis D, Gomart H and Iacona E 2012 *International journal of heat and mass transfer* **55** 1606-19
[20] Stelldinger P, Latecki L J and Siqueira M 2007 *IEEE transactions on pattern analysis and machine intelligence* 126-40
[21] Haussener S, Lipi ski W, Wyss P and Steinfeld A 2010 *Journal of Heat Transfer* **132** 061201
[22] Lorensen W E and Cline H E 1987 *ACM Siggraph Computer Graphics* **21** 163-9
[23] Brun E, Vicente J, Topin F and Occelli R 2009 iMorph : an open source software for 3D structural and geometrical analysis of porous materia. In: *Proc of Metfoam 2009,* (Bratislava. Slovakia
[24] Brun E, Vicente J, Topin F, Occelli R and Clifton M J 2009 *Advanced Engineering Materials* **11** 805-10
[25] Rousseau B, de Sousa Meneses D, Echegut P, Di Michiel M and Thovert J F 2007 *Applied optics* **46** 4266-76
[26] Thi H N and et al. 2003 *J. Phys. D: Appl. Phys.* **36** A83
[27] Coan P, Peterzol A, Fiedler S, Ponchut C, Labiche J C and Bravin A 2006 *Journal of Synchrotron Radiation* **13** 260-70
[28] Demont P, Huetz-Aubert M and N'Guyen H T 1982 *Int. J. Thermophys.* **3** 335-64
[29] Salvo L, Cloetens P, Maire E, Zabler S, Blandin J, Buffiere J, Ludwig W, Boller E, Bellet D and Josserond C 2003 *Nuclear instruments and methods in physics research section B: Beam interactions with materials and atoms* **200** 273-86
[30] Petrasch J, Wyss P, Stämpfli R and Steinfeld A 2008 *J. Am. Ceram. Soc.* **91** 2659-65
[31] Dillard T, N'guyen F, Maire E, Salvo L, Forest S, Bienvenu Y, Bartout J-D, Croset M, Dendievel R and Cloetens P 2005 *Philosophical Magazine* **85** 2147 - 75
[32] Delesse J-F, Le Sa B and Vignoles G 2001 An efficient data structure for random walk algorithms in faceted porous media. In: *Proceedings of the sixth ACM symposium on Solid modeling and applications,* (Ann Arbor, Michigan, United States: ACM) p 301
[33] Beucher S 2007 *Image and Vision Computing* **25** 405-15
[34] Raki A D 1995 *Appl. Opt.* **34** 4755-67
[35] Born M and Wolf E 1999 *Principles of Optics: Electromagnetic Theory of Propagation, Interference and Diffraction of Light* Cambridge University Press)
[36] Rousseau B, De Sousa Meneses D, Echegut P and Thovert J-F 2011 *International Journal of Thermal Sciences* **50** 178-86
[37] Bellet F, Chalopin E, Fichot F, Iacona E and Taine J 2009 *Int. J. Heat Mass Transfer* **52** 1544-51
[38] Bergstrom D, Powell J and Kaplan A 2008 *J. Appl. Phys.* **103** 103515--12
[39] Tang K and Buckius R 1998 *Int. J. Heat Mass Transfer* **41** 13
[40] Bergström D, Powell J and Kaplan A 2007 *Journal of Applied Physics* **101** 113504

Effects of nonlinear gradient index on radiative heat transfer in a one-dimensional medium by the DRESOR method

Z C Wang[1], Q Cheng[1], G H Wang[1], Z F Huang[1] and H C Zhou[1,2,3]

[1] State Key Laboratory of Coal Combustion, Huazhong University of Science and Technology, Wuhan 430074, Hubei, P. R. China
[2] State Key Laboratory of Control & Simulation of Power System & Generation Equipment, Department of Thermal Engineering, Tsinghua University, Beijing 10084, P. R. Chin

Email: hczh@mail.tsinghua.edu.cn (H C Zhou)

Abstract. The DRESOR (Distribution of Ratios of Energy Scattered by the medium Or Reflected by the boundary surface) method is applied for radiative heat transfer in a one-dimensional medium with a nonlinear gradient index and gray boundary surfaces. In this proposed method, the DRESOR values calculated by the Monte Carlo method express quantitatively the impact of scattering on radiative transfer and the radiative intensity with high directional resolution of high precision can be easily obtained. With given media characteristics and boundary conditions, the temperature and radiative flux distributions inside the medium are calculated under the condition of radiative equilibrium. It is shown, in the cases studied, that the DRESOR method has a good accuracy. The temperature distributions have a node with different kinds of sine changed gradient index distributions under the same boundary emissivity. The impact of the gradient index on the radiative heat transfer is considerable, and the same as that of the ratios of its amplitude and average index. Besides, the effects of optical thickness, boundary emissivity and scattering phase function on radiative transfer also should be paid adequate attention.

1. Introduction

As one basic energy and information transfer mode, radiative heat transfer has been greatly developed and widely used in high-tech fields, especially under vacuum and high temperature condition. Normally, the refractive index of the participating medium is always assumed to be constant. This dealing simplifies the calculation, while brings uncertainty of results and unavoidable errors. The fact is that the refractive index of the medium is affected by its components, temperature and some other factors. With the development of scientific research, the gradient index materials have important application background and broad prospects for development, which makes us try to obtain its transmission characteristics and mechanisms, and grasp quick and convenient calculating methods. However, as the rays propagate in curved paths, which are determined by the Fermat principle [1], the solution of radiative heat transfer in a medium with gradient index is much more difficult than that in a medium with uniform index.

[3] To whom any correspondence should be addressed.

A new process, Distribution of Ratios of Energy Scattered by the medium Or Reflected by the boundary surface method, known as the DRESOR method, was proposed by Zhou [2-5] to address the radiative transfer problems. However, the refractive index considered is assumed to be uniform in all of the above studies. Just recently, Wang [6] successfully expanded the DRESOR method to a one-dimensional linear gradient index medium with black boundaries. As the development of multilayer films material and media, cyclical changed gradient index gradually evoked people's attention and the sine changed gradient index is one of the most normally considered. However, things turn out to be manifestly difficult when considering complicated gradient index distributions and gray boundaries. The present work is therefore aimed at discussing the effects of nonlinear gradient index on radiative heat transfer using the DRESOR method in a one-dimensional medium with sine changed gradient index. The basic principle of the DRESOR method, the calculating of radiative intensity and the relevant quantities will be shown in Section 2. Radiative transfer with two sine changed gradient index is considered in three cases. The proper discrete grid number and direction number combination is determined first. Then the first case is taken as an example to examine the accuracy of the proposed method. Then the following two cases are taken to discuss the impact of the gradient index, the ratios of its amplitude and average index on the temperature distribution. Besides, the effects of optical thickness, boundary emissivity, scattering phase function and scattering albedo on radiative equilibrium are investigated, too. These will be described in detail in Section 3. Finally, some conclusions will be drawn in Section 4.

Nomenclature		N	discrete elements of medium	τ	optical thickness
		R_d^s	DRESOR value, $1/m^3$, or $1/m^2$	θ	polar angle
I	radiative intensity, $W/(m^2.sr)$	*Greek letters*		*Subscripts*	
$\hat{s}$	directional vector, (rad)	κ	absorption coefficient, 1/m	b	blackbody
q	radiative flux, $W/(m^2)$	δ_s	scattering coefficient, 1/m	w	surface or wall
T	temperature, K	Φ	scattering phase function	0	the left wall
n	refractive index	β	extinction coefficient, 1/m	N+1	the right wall
L	physical distance, m	σ	Stefan-Boltzmann constant	$i1$	the $i1$ grid
M	discrete numbers of angle	ρ/ε	surface reflectivity/emissivity		
		ω	scattering albedo		

2. Principles

2.1. Basic principle of the DRESOR method

The integral form of the radiative transfer equation in an absorbing, emitting-scattering medium with gradient index can be expressed as [7, 8]:

$$\frac{I(r,s)}{n^2}=\frac{I_{r_w}(r_w,s_w)}{n_{r_w}^2}\exp(-\int_{s_w}^{s}\beta ds^{'})+\int_{s_w}^{s}S(r^{'},s^{'})\exp(-\int_{s^{'}}^{s}\beta ds^{''})\beta ds^{'}, \tag{1}$$

Where $S(r^{'},s)$ is the source term, which is:

$$S(r',s)=(1-\omega)I_b(r')+\frac{\omega}{4\pi}\int_{4\pi}\frac{1}{n_{r'}^2}I(r',s_i)\Phi(s_i,s)d\Omega_i, \tag{2}$$

The corresponding boundary condition of the non-transparent surface with arbitrary characteristics is [7]:

$$I_w(r_w,s)=n_{r_w}^2\varepsilon(r_w)I_b(r_w)+\frac{1}{\pi}\int_{\hat{n}\cdot s'<0}\rho''(r_w,s',s)I(r_w,s')|n\cdot s'|d\Omega'. \tag{3}$$

According to the DRESOR method, after rewriting equations (2) and (3) then submitting them back into equation (1) we can get the following expression. In order to avoid duplication, details can be found in the previous work done by Wang et al [6].

$$\frac{I(r,s)}{n_r^2}=\left\{\int_W \frac{n_{r_w}^2}{n_{r_w}^2} R_d^s(r_w^{'},r_w,s)[\pi\varepsilon(r_w^{'})I_b(r_w^{'})]dA'+\int_V \frac{n_{r^{"}}^2}{n_{r_w}^2} R_d^s(r^{"},r_w,s)[4\pi\beta(1-\omega)I_b(r^{"})]dV"+[\pi\varepsilon(r_w)I_b(r_w)]\right\}$$

$$\cdot\frac{1}{n_{r_w}^2}(\frac{n_{r_w}^2}{\pi})\exp[-\int_0^s \beta ds"]+\int_0^s \frac{1}{4\pi\beta}\left\{4\pi\beta(1-\omega)I_b(r')+\int_W \frac{1}{n_{r'}^2} R_d^s(r_w^{'},r',s)[\pi\varepsilon(r_w^{'})n_{r_w}^2 I_b(r_w^{'})]dA'\right.$$

$$\left.+\int_V \frac{1}{n_{r'}^2} R_d^s(r",r',s)[4\pi\beta(1-\omega)n_{r"}^2 I_b(r")]dV"\right\}\exp[-\int_{s'}^s \beta ds"]\beta ds'. \tag{4}$$

Where $R_d^s(r_w^{'},r_w,s)$, $R_d^s(r^{"},r_w,s)$, $R_d^s(r_w^{'},r^{'},s)$ and $R_d^s(r^{"},r^{'},s)$ are all DRESOR values as defined in the DRESOR method [2, 3]. The DRESOR values within the system are essential for calculating the radiative intensity in the DRESOR method. The DRESOR value $R_d^s(r^{"},r^{'},s)$ denotes the ratio of the energy scattered by the unit volume around the point $r^{'}$ into a unit solid angle around the direction s to that emitted from a unit volume around the point $r^{"}$. Similarly, we could get the definitions of $R_d^s(r_w^{'},r_w,s)$, $R_d^s(r^{"},r_w,s)$ and $R_d^s(r_w^{'},r^{'},s)$. The detailed calculation method for the DRESOR values can be found in the literature [9, 10]. Calculating quantitatively the impact of scattering on radiative transfer, as combined within the DRESOR values, is one of the main features of the DRESOR method,.

2.2. Discretised calculation of intensity and relevant quantities

Taken the sine changed gradient index into consideration, the gradient index can be described as:

$$n(x)=n_1+(n_2-n_1)\sin(\pi x/L)\ \ (0\le x\le L)\,. \tag{5}$$

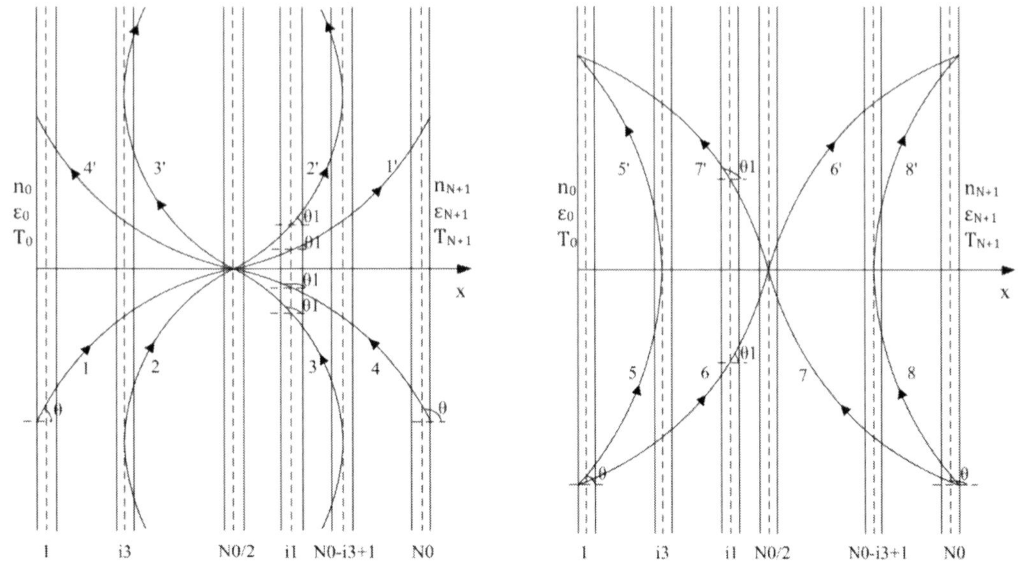

(a) sine model with $n_2>n_1$ (b) sine model with $n_2<n_1$

Figure 1. Physical model and ray tracing system.

As shown in figure 1, a one-dimensional, semi-transparent, gray, absorbing, emitting and scattering slab, with thickness x_L and gray boundaries, is considered. The system is discretized into *N* grids and *M* directions in each grid. The physical length of each grid is $\Delta x = x_L / N$. The midpoint of each grid is the calculation point, while the two boundaries are named as 0 and N+1 grid, respectively. The emissivity of the boundary walls are ε_0 and ε_{N+1}, while the temperatures of the boundary walls are kept at T_0 and T_{N+1}, respectively. The absorption and scattering coefficients of the medium, κ and σ_s, are kept constant and the extinction coefficient is $\beta = \kappa + \sigma_s$. After discretization, the refractive index of the medium in an arbitrary grid i, $n(i)$, is constant, and varies among different grids.

The intensity at grid $i1$ in direction θ_{i1}, $I(i1,\theta_{i1})$, can be got by using the discretizing equation (4). The integration is taken along curved path lines for different directions θ_{i1}, as shown in figure 1.

For the path lines 22' and 33' in figure 1, there will be full-reflection at grid $i3$ and, because of symmetry, another full-reflection will appear at grid N-$i3$+1. The lines will propagate between these two grids till the energy they bring is totally absorbed and scattered by the medium. So the discretized formula of intensity can be described as follows:

$$\frac{I(i1,\theta_{i1})}{n_{i1}^2} = \frac{1}{4\pi\beta}\sum_{i2=i3}^{N-i3+1}\left\{4\pi\beta(1-\omega)I_b(i2) + \frac{1}{n_{i2}^2}\pi\varepsilon n_0^2 I_b(0)R_d^s(0,i2,\theta_{i2}) + \frac{1}{n_{i2}^2}\pi\varepsilon n_{N+1}^2 I_b(N+1)R_d^s(N+1,i2,\theta_{i2})\right.$$
$$\left. + \Delta x\sum_{i4=1}^{N}\frac{1}{n_{i2}^2}4\pi\beta(1-\omega)n_{i4}^2 I_b(i4)R_d^s(i4,i2,\theta_{i2})\right\}[|\exp(-\tau_{i2-1\to i1}) - \exp(-\tau_{i2\to i1})|], \tag{6}$$

However, as for the other path lines in figure 1, no matter whether there will be full-reflection, the path lines will end at the boundary surfaces. So the corresponding formula of intensity can be shown as:

$$\frac{I(i_1,\theta_{i_1})}{n_{i_1}^2} = \frac{1}{n_w^2}\left(\frac{n_w^2}{\pi}\right)\left\{\pi\varepsilon I_b(w) + \frac{1}{n_w^2}\pi\varepsilon n_{N+1}^2 I_b(N+1)R_d^s(N+1,w,\theta_w) + \Delta x\sum_{i_3=1}^{N}\frac{1}{n_w^2}4\pi\beta(1-\omega)n_{i_3}^2 I_b(i_3)R_d^s(i_3,w,\theta_w)\right.$$
$$\left. + \frac{1}{n_w^2}\pi\varepsilon n_0^2 I_b(0)R_d^s(0,w,\theta_w)\right\}\exp(-\tau_{w\to i_1}) + \frac{1}{4\pi\beta}\sum_{i_2}^{\text{along the path}}\left\{\Delta x\sum_{i_4=1}^{N}\frac{1}{n_{i_2}^2}4\pi\beta(1-\omega)n_{i_4}^2 I_b(i_4)R_d^s(i_4,i_2,\theta_{i_2})\right.$$
$$\left. + 4\pi\beta(1-\omega)I_b(i_2) + \frac{1}{n_{i_2}^2}\pi\varepsilon n_{N+1}^2 I_b(N+1)R_d^s(N+1,i_2,\theta_{i2}) + \frac{1}{n_{i_2}^2}\pi\varepsilon n_0^2 I_b(0)R_d^s(0,i_2,\theta_{i2})\right\}$$
$$\times[|\exp(-\tau_{i_2\to i_1}) - \exp(-\tau_{i_2-1\to i_1})|]. \tag{7}$$

It is worth noting that along the integration path, n_w refers to the boundary of the system and the end of the integration path, and varies with different integration lines. Varying with θ_{i1}, the integration lines such as lines 11', 44', 55', 66', 77' and 88', shown in figure 1, demonstrate different paths according to the distribution of refractive index inside the medium, and end at different surfaces. The optical thickness can be calculated along the paths. The integration paths are curved between the adjacent grids i and j according to Descartes-Snell's law. This is the other main features of the DRESOR method, obtaining the radiative intensity with high directional resolution of high precision easily, which plays a vital role in radiative image processing in some industrial applications.

After the intensity of each grid in every direction is got, it is of no difficulty to calculate the radiative flux $q(i)$ of each grid using the following formula [11]:

$$q(i) = \sum_{\theta_i=0}^{\pi} 2\pi I(i,\theta_i)\sin(\theta_i)\cos(\theta_i)\Delta\theta. \tag{8}$$

When the radiation attains equilibrium inside the medium, we can get the temperature distribution of the medium [11]:

$$T(i) = \left[\frac{1}{4n_i^2\sigma}\sum_{\theta_i=0}^{\pi} 2\pi I(i,\theta_i)\sin(\theta_i)\Delta\theta\right]^{0.25}. \tag{9}$$

3. Results and discussions

3.1. Objective studied, cases and conditions of calculation

Consider the radiative transfer process inside the one-dimensional medium with two boundaries, as shown in figure 1. The boundaries are diffuse gray walls with different, given temperatures. The scattering albedo, the optical thickness and the nonlinear gradient index distribution inside the medium are set before calculation. The temperature and radiative flux distribution inside the medium are to be analyzed quantitatively.

In these cases, the temperatures of boundaries are $T(0)=1000\text{K}$ and $T(N+1)=1500\text{K}$. All the calculations are performed using an Intel Pentium E5200 @ 2.5GHz computer.

3.2. Case 1: the purely emitting and absorbing medium

First, the influence on the accuracy of the results and the computing time are examined to determine the proper discrete grid number N and direction number M combination. The scattering albedo is $\omega=0.0$ and the gradient index of the medium is $n(x)=1.8-0.6\sin(x/L)$. The boundaries are all black and the optical thickness is $\tau=1.0$. As shown in figure 2 and table 1, it is obvious to see that the results of the DRESOR method are stable under all these conditions. When $N\geq 100$ and $M\geq 18$, the influence of N and M is quite weak. While the computing time of $M=180$ doesn't increase too much compared with that of $M=18$ under the same $N=100$; however, the computing time of $N=1000$ is unacceptable compared with that of $N=100$ with the same $M=180$. So $N=100$ and $M=180$ are used in the calculations below.

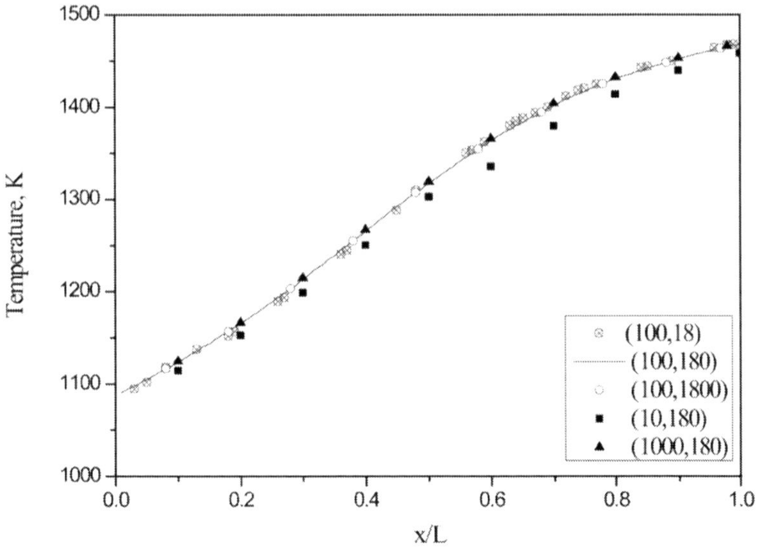

Figure 2. The results under different values of N and M combination.

Table 1. The computing time under different values of N and M combination.

(N, M)	Computing time	(N, M)	Computing time
(10, 180)	12.56185 s	(100, 18)	700.3893 s
(100, 180)	907.04685 s	(100, 1800)	3103.81255 s
(1000, 180)	273275.311 s		

In this case, two different gradient index distributions and four different kinds of boundary emissivity are considered.

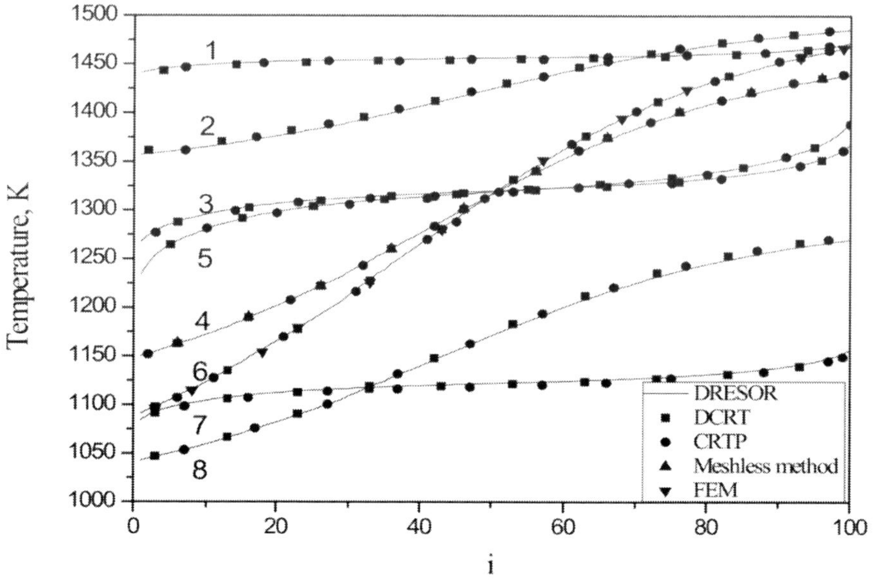

Figure 3. Temperature field under radiative equilibrium in the purely emitting and absorbing medium.
ε_0=0.2, ε_{N+1}=1.0; curve 1: n(x)=1.2+0.6sin(πx/L), curve 2: n(x)=1.8-0.6sin(πx/L);
ε_0=0.7, ε_{N+1}=0.7; curve 3: n(x)=1.2+0.6sin(πx/L), curve 4: n(x)=1.8-0.6sin(πx/L);
ε_0=1.0, ε_{N+1}=1.0; curve 5: n(x)=1.2+0.6sin(πx/L), curve 6: n(x)=1.8-0.6sin(πx/L);
ε_0=1.0, ε_{N+1}=0.2; curve 7: n(x)=1.2+0.6sin(πx/L), curve 8: n(x)=1.8-0.6sin(πx/L).

The results of temperature distribution are shown in figure 3. These are compared with those got by Huang [12] using the discrete curved ray-tracing method (DCRT), Tan [13] with combined curved ray-tracing and pseudo-source adding method (CRTP), Liu [14] using the meshless method and Liu [15] through a finite element method (FEM). The results predicted by the DRESOR method show excellent agreement with these methods under different gradient indexes and boundary emissivity. The relative errors are below 0.35%.

Unlike that of linear gradient index distribution [6], the temperature distributions have a node of these sine changed gradient index distributions under the same conditions of boundary emissivity. The reason is that the sine changed gradient index causes the special ray propagation path: complete symmetry within two sides of the center, which makes the node affected just by the characters of two surfaces rather than the gradient index. Just as shown in figure 3, if $\varepsilon_0 = \varepsilon_{N+1}$, the node will be the midpoint of the medium and its radiative equilibrium temperature stays constant. However, this node has a shifting trend towards the bigger boundary emissivity side if $\varepsilon_0 \neq \varepsilon_{N+1}$.

The effect of optical thickness on radiative transfer under nonlinear gradient index is shown in figure 4. As we can see, with the increase of optical thickness, the temperature difference between its two sides increases while the temperature difference between a sine changed gradient index and a constant gradient index first increases and then decreases. In order to incarnate the sensitivity of the optical thickness, $\tau = 1.0$ is used in the following cases.

3.3. Case 2: the absorbing, emitting and isotropic scattering medium

In this case, the boundary emissivity is assumed to be $\varepsilon_0 = \varepsilon_{N+1} = 0.7$, the scattering albedo is $\omega = 0.3$. The effect of the gradient index distributions is investigated.

As shown in figure 5, A/P denotes the ratios of amplitude and average index of the gradient index. As we can see, doing isolated research on the impact of the amplitude or the average index is incorrect and the results seem chaotic. So the effect caused by the amplitude coupled the average index is investigated. First, once the A/P is decided, such as when A/P equals 0.5 in figure 5, no matter what the amplitude and the average index are, the temperature distributions are completely overlapped with each other. Then, with the decrease of the A/P, the propagate path of the radiation shown in figure 1(a) increases, which makes the impact of the boundaries on its nearby medium greater. In consequence, the temperature difference between the two sides increases and so does the temperature gradient.

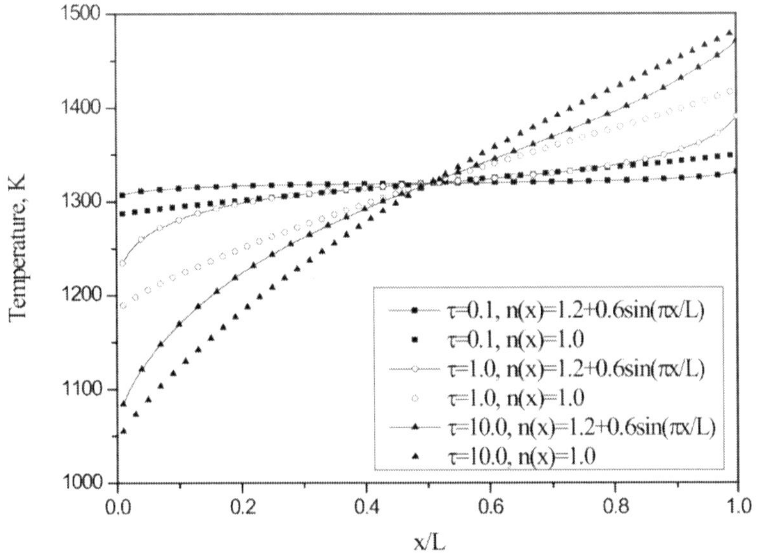

Figure 4. Temperature field under radiative equilibrium with different kinds of optical thickness.

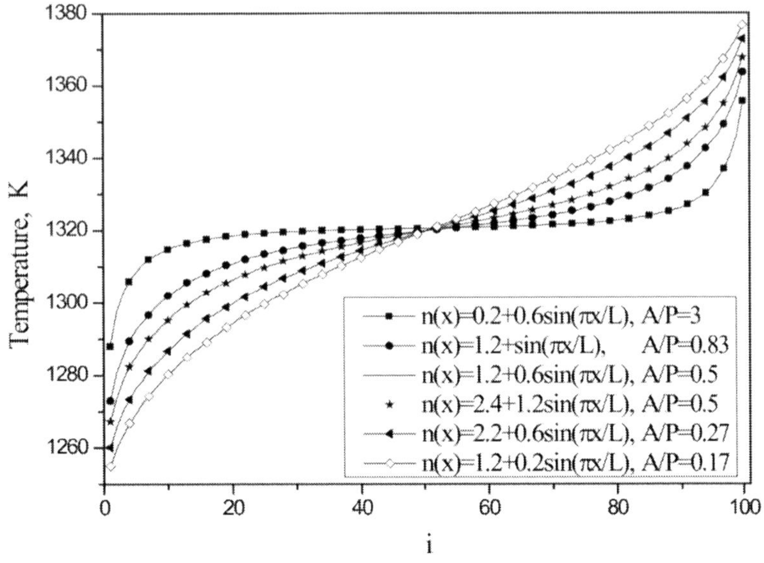

Figure 5. Temperature field under radiative equilibrium in the isotropic scattering medium.

3.4. Case 3: the absorbing, emitting and anisotropic scattering medium

In this case, the gradient index distribution is $n(x) = 1.8 - 0.6\sin(x/L)$. Two linear scattering phase functions; $\Phi = 1 + b\mu\mu', (b = 1, -1)$, are investigated.

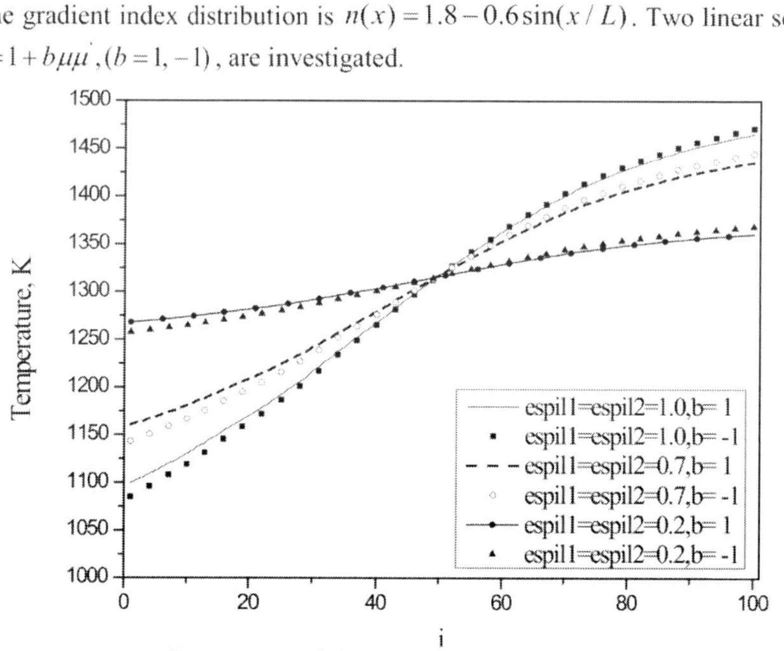

Figure 6. Temperature field in anisotropic scattering medium.

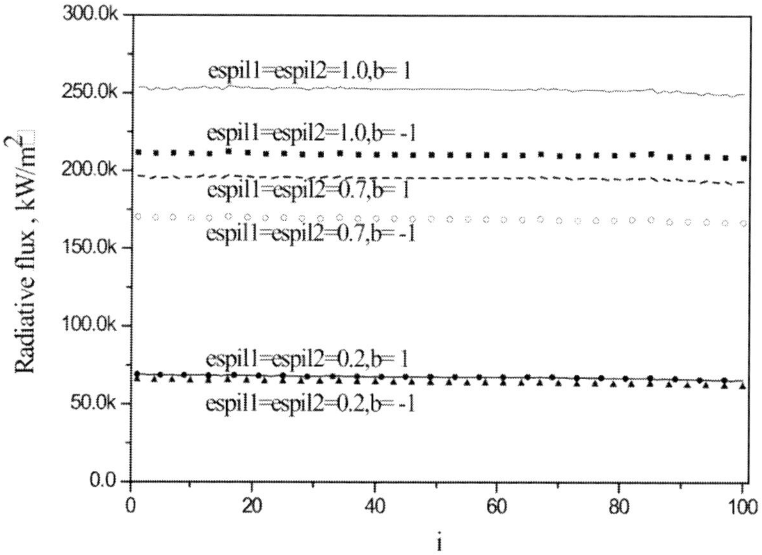

Figure 7. The corresponding radiative flux distribution of figure 6.

Figures 6 and 7 show the effect of boundary emissivity on the radiative equilibrium. Here, the scattering albedo is $\omega = 0.7$. Three combinations of boundary emissivity are calculated. As we can see, with a decrease in the boundary emissivity, the energy emitted from the two sides reduces, which in turn makes the energy absorbed and scattered by the medium reduce. So the radiative equilibrium flux then decreases, as shown in figure 7. In other words, the heat effect on the medium of the two boundaries is decreasing gradually. Of course, this makes the temperature difference of its two sides lower and the temperature curve flatter.

At the same time, it is also worth noting that; even though the effect of anisotropic scattering phase function on temperature distribution, as shown in figure 6, is weaker than that caused by the boundary emissivity, it cannot be ignored. Compared with forward scattering $b=1$, the backward scattering $b=-1$ enhances the radiative intensity in the negative direction, which, according to formula. (8), reduces the radiative flux under radiative equilibrium. However, the backward scattering $b=-1$ increases the transmitting length to some degree, which means that the heat effect on the medium of the two boundaries is greater than that of $b=1$. That is why under the same conditions, the flux of the backward scattering $b=-1$ is less than that of forward scattering $b=1$, while the temperature difference of the backward scattering $b=-1$ is higher than that of forward scattering $b=1$.

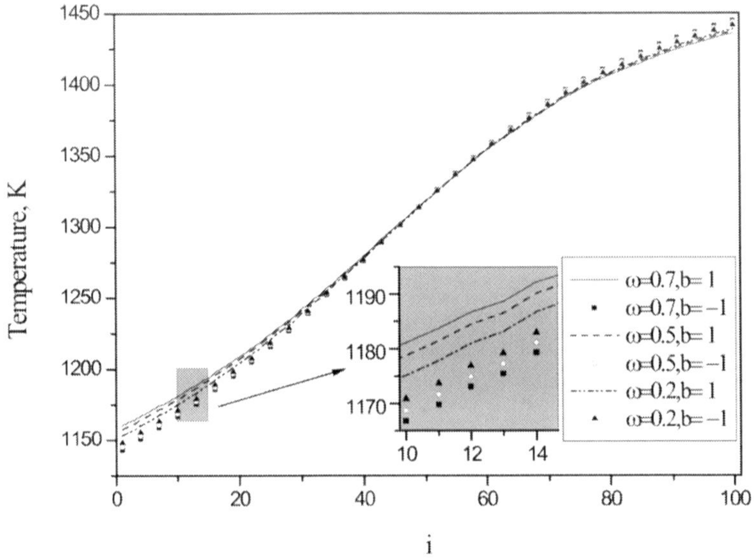

Figure 8. Temperature field under radiative equilibrium in anisotropic scattering medium with $\varepsilon_0=\varepsilon_{N+1}=0.7$.

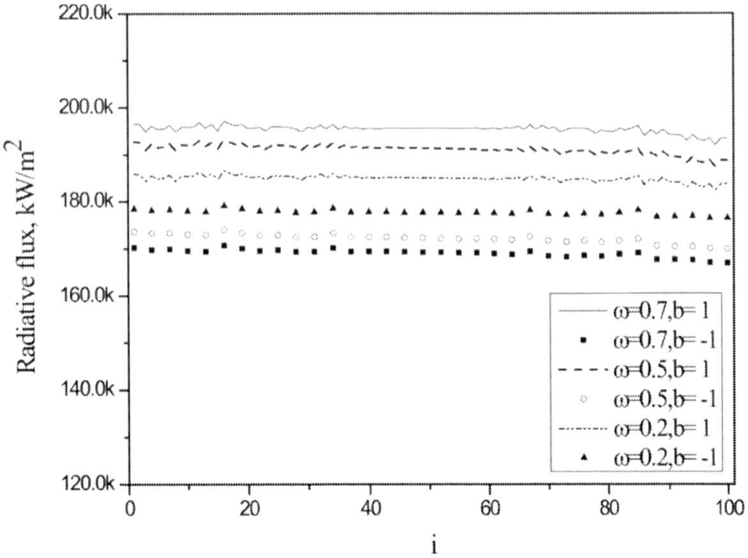

Figure 9. The corresponding radiative flux distribution of figure 8.

Figures 8 and 9 show the effect of scattering albedo ω on radiative equilibrium. Here, the boundary emissivities are $\varepsilon_1=\varepsilon_{N+1}=0.7$. Three different scattering albedos; $\omega=0.7$, $\omega=0.5$ and $\omega=0.2$, are calculated. When the scattering albedo decreases, the absorbing capacity of the medium increases correspondingly. With forward scattering $b=1$, when the scattering albedo ω changes from 0.7, 0.5 to 0.2, the radiative intensity decreases, so does the radiative flux. However, the temperature difference between the two sides increases because of the increase in the absorbing capacity of the medium. However, with backward scattering $b=-1$, the decrease in the scattering albedo ω reduces the radiative intensity in then negative direction, which makes the radiative flux increase. At the same time, the enhancement effect of backward scattering as we discussed above is likely to decline, which makes the temperature difference between the two sides decrease. Besides, when the scattering albedo ω reduces, so does the effect of anisotropic scattering. As shown in figure 8, the difference between $b=1$ and $b=-1$ becomes smaller and smaller.

4. Conclusions

In this paper, the DRESOR method has been applied to radiative transfer in nonlinear gradient index media, with a DRESOR value introduced. The temperature and radiative flux distributions inside the medium were calculated in three cases. The impacts of the gradient index distributions, the variation of the amplitude and the average index, the optical thickness, the scattering albedo, the boundaries emissivity and the scattering phase function were all investigated. The main conclusions follow below.

- This method could deal well with the radiative transfer problem in a one-dimensional semi-transparent slab with nonlinear gradient index and complex boundary characteristics.
- The temperature distributions have a node with different sine changed gradient index distributions under the same boundary emissivity.
- The impact of the gradient index on the radiative heat transfer is the same as that of the ratios of its amplitude and average index. The boundary emissivity has a significant impact on radiative heat transfer, while the effect of optical thickness, the scattering phase function is much weaker. However, they cannot be ignored and should be paid adequate attention.

References

[1] Born M and Wolf E 1997 *Principle of optics 6nd* (Cambridge: Cambridge University Press)
[2] Zhou H C 2005 *Visual detection principle and technology of the furnace flame* (Beijing: Science Press)
[3] Zhou H C, Chen D L and Cheng Q 2004 *J. Quant. Spectrosc. Radiat. Transfer* **83** 459-81
[4] Zhou H C, Cheng Q, Huang Z F and He C 2007 *J. Quant. Spectrosc. Radiat. Transfer* **104** 99-115
[5] Cheng Q and Zhou H C 2007 *J. Heat Tran-Transactions of the Asme* **129** 634-45
[6] Wang Z C, Cheng Q, Wang G H and Zhou H C 2011 *J. Quant. Spectrosc. Radiat. Transfer* **112** 2835-45
[7] Miharas D and Mihals B W 1984 *Foundations of radiation hydrodynamics* (New York: Oxford University Press)
[8] Fumeron S and Asllanaj F 2009 *J. Quant. Spectrosc. Radiat. Transfer* **110** 1005-12
[9] Huang Z F, Cheng Q, He C and Zhou H C 2008 *Heat Transfer-Asian Research* **37** 138-52
[10] Cheng Q, Luo Z X and Zhou H C 2010 *Int. J. Nonlin Sci Numer* **11** 927-945
[11] Modest M F 2003 *Radiative heat transfer 2nd* (San Diego: Academic Press)
[12] Huang Y 2002 *A study on the thermal radiative transfer in gredient index semitransparent medium Ph.D thesis* (Harbin: Harbin Institute of Technology)
[13] Tan H P, Huang Y and Xia X L 2003 *Int. J. Heat Mass Transfer* **46** 2005-14
[14] Liu L H 2006 *Int. J. Heat Mass Tran* **49** 219-29
[15] Liu L H 2005 *Int. J Heat Mass Tran* **48** 2260-65

Acknowledgement

This study was supported by the National Natural Science Foundation of China (Nos. 50906027, 51025622, 51021065) and the Fundamental Research Funds for the Central Universities (No. 2011TS143).

Some issues related to polarized radiative transfer in a multilayer medium with a changing index of refraction

R D M Garcia

Instituto de Estudos Avançados, Trevo Cel. Av. José Alberto Albano do Amarante n° 1, São José dos Campos, SP 12228-001, Brazil

E-mail: rdgarcia@ieav.cta.br

Abstract. A couple of issues raised in a recent work on Fresnel boundary and interface conditions for polarized radiative transfer are discussed. The first issue concerns a normalization that has to be performed on the transmission matrix relating the transmitted and incident Stokes vectors at a smooth interface between different materials. It was found that the normalization step is not performed in a correct way in many existing works and so the effects of the two most frequent discrepancies encountered in these works are evaluated for a simple model of an atmosphere-water system. The second issue has to do with a choice of the sign of the parallel component of the amplitude coefficient of reflection R_l when constructing the reflection matrix that relates the reflected and incident Stokes vectors at a smooth interface. A computational approach based on the limit to the case with constant index of refraction is used to determine the appropriate sign of R_l for polarized radiative transfer calculations in multilayers with different indices of refraction.

1. Introduction

In a recent work [1], a couple of issues were raised concerning polarized radiative transfer in a multilayer medium with a layer-dependent index of refraction.

The first of these issues has to do with a normalization of the transmission matrix that relates the transmitted and incident Stokes vectors at a smooth interface between two material layers with different indices of refraction. Normalization of the transmission matrix is required to ensure conservation of energy. Surprisingly, the author noted [1] that in many published works on this topic such a normalization is not done in a correct way. The most frequent discrepancies detected in these works were either a complete absence of a normalizing factor or the lack of a multiplicative term of the form $(n_b/n_a)^2$, where n_b denotes the index of refraction of the medium to where radiation is being transmitted and n_a that of the medium of incidence. The effects of these discrepancies in the calculated reflectance and transmittance of a single water layer with a vacuum on top and a black bottom were estimated using an analytical discrete-ordinates (ADO) solution [2] that does not take polarization effects into account.

The second issue is more subtle than the first. The reflection matrix that relates the reflected and incident Stokes vectors at a smooth interface is derived from the well-known Fresnel formulas for the amplitude coefficients of reflection [3, 4]. As discussed many years ago [5], the theory is unable to determine the signs that should be used with the amplitude coefficients of reflection R_l

and R_r. Physically, a change of sign applied to one of these coefficients is equivalent to a change of π in the phase of the corresponding component (parallel or perpendicular) of the reflected electric field with respect to that of the incident electric field. A recommendation on the signs to be used with the R_l and R_r coefficients of reflection was produced [6] in an ellipsometry symposium held in Nebraska in 1968. An undesirable feature of that sign choice—which we call "conventional"—is that the R_l and R_r coefficients are not consistent with each other when normal incidence is considered [6]. Even so, the conventional sign choice is the one that appears most frequently in optics and electromagnetism books [6].

If a sign change is applied to the usual form [6] of the R_l coefficient, it becomes consistent with the R_r coefficient when the incidence is normal to the interface. Moreover, the physical notion that a change of π should be expected in the phase of the reflected electric field with respect to that of the incident electric field when the index of refraction of the medium of incidence is smaller than the index of refraction of the reflecting medium [7] is verified. We refer to such a choice of sign as "consistent".

Except for one work [8], all of the works in radiative transfer that the author is aware of use the conventional sign choice. The net effect of a sign change in R_l is a sign change in the (3,3), (3,4), (4,3), and (4,4) elements of the reflection matrix. While these sign changes are of almost no consequence in simple situations like the cases of no scattering and/or normal incidence (only sign changes in the calculated Stokes parameters U and V are expected), the same is not true when there is scattering and the beam incidence is not in the normal direction. In this case, all four Stokes parameters (I, Q, U, and V) are coupled and so completely different numerical results are expected when a sign change is applied to the R_l coefficient.

In this work, we study a simple model of an atmosphere-water system with a two-fold purpose: (i) to evaluate the influence of the most frequently encountered discrepancies in the transmission matrix when polarization effects are taken into account, and (ii) to compare the calculated Stokes parameters for the two sign choices of the reflection coefficient R_l mentioned above, aiming at determining the appropriate sign choice for polarized radiative transfer calculations.

2. Mathematical description of a multilayer medium subject to Fresnel conditions

Considering a medium composed of layers $k = 1, 2, \ldots, K$, we write the equation of transfer for layer k as [1]

$$\mu\frac{\partial}{\partial\tau}\mathbf{I}_k(\tau,\mu,\phi) + \mathbf{I}_k(\tau,\mu,\phi) = \frac{\varpi_k}{4\pi}\int_0^{2\pi}\int_{-1}^{1}\mathbf{P}_k(\mu,\mu',\phi-\phi')\mathbf{I}_k(\tau,\mu',\phi')\mathrm{d}\mu'\mathrm{d}\phi', \tag{1}$$

where $\mathbf{I}_k(\tau,\mu,\phi)$ is the Stokes vector with the four Stokes parameters $I_k(\tau,\mu,\phi)$, $Q_k(\tau,\mu,\phi)$, $U_k(\tau,\mu,\phi)$, and $V_k(\tau,\mu,\phi)$ as components [9], $\tau \in (a_{k-1}, a_k)$ is the optical variable that measures the position in layer k, and $\mu \in [-1,1]$ and $\phi \in [0,2\pi]$ are, respectively, the cosine of the polar angle (measured with respect to the *positive* τ-axis) and the azimuthal angle that define the direction of propagation of the radiation. In our notation, a_0 gives the location of the surface of the first layer, $a_1, a_2, \ldots, a_{K-1}$ give the locations of the interfaces between the layers, and a_K the location of the surface of the last layer. To each layer, we associate an index of refraction n_k, an albedo for single scattering ϖ_k, and a phase matrix $\mathbf{P}_k(\mu,\mu',\phi-\phi')$ that we allow to be expressed in the general form [10]

$$\mathbf{P}_k(\mu,\mu',\phi-\phi') = \frac{1}{2}\sum_{m=0}^{L_k}(2-\delta_{0,m})[\mathbf{C}_k^m(\mu,\mu')\cos m(\phi-\phi') + \mathbf{S}_k^m(\mu,\mu')\sin m(\phi-\phi')], \tag{2}$$

where L_k is a truncation parameter that defines the scattering order,

$$\mathbf{C}_k^m(\mu,\mu') = \mathbf{A}_k^m(\mu,\mu') + \mathbf{D}\mathbf{A}_k^m(\mu,\mu')\mathbf{D}, \tag{3a}$$

and

$$\mathbf{S}_k^m(\mu,\mu') = \mathbf{A}_k^m(\mu,\mu')\mathbf{D} - \mathbf{D}\mathbf{A}_k^m(\mu,\mu'). \tag{3b}$$

In these equations, $\mathbf{D} = \text{diag}\{1, 1, -1, -1\}$ and

$$\mathbf{A}_k^m(\mu,\mu') = \sum_{l=m}^{L_k} \mathbf{P}_l^m(\mu)\mathbf{B}_{k,l}\mathbf{P}_l^m(\mu'), \tag{4}$$

where

$$\mathbf{B}_{k,l} = \begin{pmatrix} \beta_{k,l} & \gamma_{k,l} & 0 & 0 \\ \gamma_{k,l} & \alpha_{k,l} & 0 & 0 \\ 0 & 0 & \zeta_{k,l} & -\epsilon_{k,l} \\ 0 & 0 & \epsilon_{k,l} & \delta_{k,l} \end{pmatrix} \tag{5}$$

and

$$\mathbf{P}_l^m(\mu) = \begin{pmatrix} P_l^m(\mu) & 0 & 0 & 0 \\ 0 & R_l^m(\mu) & -T_l^m(\mu) & 0 \\ 0 & -T_l^m(\mu) & R_l^m(\mu) & 0 \\ 0 & 0 & 0 & P_l^m(\mu) \end{pmatrix}. \tag{6}$$

The constants $\{\alpha_{k,l}, \beta_{k,l}, \gamma_{k,l}, \delta_{k,l}, \epsilon_{k,l}, \zeta_{k,l}\}$ are known in the literature as *Greek constants*. In addition, $P_l^m(\mu)$ is a normalized version of the associated Legendre function of the first kind, and the normalized functions $R_l^m(\mu)$ and $T_l^m(\mu)$ are expressed in terms of the generalized spherical functions $P_{m,-2}^l(\mu)$ and $P_{m,2}^l(\mu)$ discussed by Gel'fand and Šapiro [11]. We note that the matrix $\mathbf{P}_l^m(\mu)$ and its elements have been defined and denoted in slightly different ways in the literature; in this work we have followed the notation and the definitions given in a paper by Siewert [12].

The boundary and interface conditions subject to which we must solve Eq. (1) for layers $k = 1, 2, \ldots, K$ have been derived from first principles [1]. Assuming that a uniform beam of radiation characterized by the Stokes vector

$$\mathbf{F}_0 = \begin{pmatrix} F_I \\ F_Q \\ F_U \\ F_V \end{pmatrix} \tag{7}$$

is traveling, along a direction defined by (μ_0, ϕ_0), towards the surface $\tau = a_0$ in an external $(\tau < a_0)$, non-participating medium with index of refraction n_0, we can write the boundary condition for the first layer as [1]

$$\mathbf{I}_1(a_0,\mu,\phi) = \mathbf{X}(n_{1,0},\mu)\mathbf{I}_1(a_0,-\mu,\phi) + \mathbf{Y}(n_{1,0},\mu)\mathbf{F}_0\,\delta[f(n_{1,0},\mu) - \mu_0]\delta(\phi - \phi_0), \tag{8}$$

for $\mu \in (0, 1]$ and $\phi \in [0, 2\pi]$. In this expression, we make use of some general definitions: the ratio between indices of refraction $n_{k,k'} = n_k/n_{k'}$, the function

$$f(n,\mu) = [1 - n^2(1-\mu^2)]^{1/2}, \tag{9}$$

the 4×4 reflection matrix

$$\mathbf{X}(n,\mu) = \begin{cases} \mathbf{G}(n,\mu), & n \le 1, \\ \mathbf{G}(n,\mu)H[\mu - \mu_c(n)] + \mathbf{\Gamma}(n,\mu)\{1 - H[\mu - \mu_c(n)]\}, & n \ge 1, \end{cases} \tag{10a}$$

and the 4×4 transmission matrix

$$\mathbf{Y}(n,\mu) = \begin{cases} \mathbf{D}(n,\mu), & n \le 1, \\ \mathbf{D}(n,\mu)H[\mu - \mu_c(n)], & n \ge 1. \end{cases} \tag{10b}$$

In Eqs. (10), we use the Heaviside function

$$H(x) = \begin{cases} 1, & x \geq 0, \\ 0, & x < 0. \end{cases} \tag{11}$$

the quantity $\mu_c(n) = (1-1/n^2)^{1/2}$ is the cosine of the critical angle [3], and the non-zero elements of the 4×4 basic matrices $\mathbf{G}(n,\mu)$, $\mathbf{\Gamma}(n,\mu)$, and $\mathbf{D}(n,\mu)$ are given by [1]

$$G_{11}(n,\mu) = G_{22}(n,\mu) = \frac{1}{2}\left\{\left[\frac{nf(n,\mu)-\mu}{nf(n,\mu)+\mu}\right]^2 + \left[\frac{n\mu - f(n,\mu)}{n\mu + f(n,\mu)}\right]^2\right\}, \tag{12a}$$

$$G_{12}(n,\mu) = G_{21}(n,\mu) = \frac{1}{2}\left\{\left[\frac{nf(n,\mu)-\mu}{nf(n,\mu)+\mu}\right]^2 - \left[\frac{n\mu - f(n,\mu)}{n\mu + f(n,\mu)}\right]^2\right\}, \tag{12b}$$

$$G_{33}(n,\mu) = G_{44}(n,\mu) = \left[\frac{nf(n,\mu)-\mu}{nf(n,\mu)+\mu}\right]\left[\frac{n\mu - f(n,\mu)}{n\mu + f(n,\mu)}\right], \tag{12c}$$

$$\Gamma_{11}(n,\mu) = \Gamma_{22}(n,\mu) = 1, \tag{13a}$$

$$\Gamma_{33}(n,\mu) = \Gamma_{44}(n,\mu) = 1 - \frac{2(1-\mu^2)^2}{1-(1+1/n^2)\mu^2}, \tag{13b}$$

$$\Gamma_{34}(n,\mu) = -\Gamma_{43}(n,\mu) = \frac{2\mu(1-\mu^2)[\mu_c^2(n)-\mu^2]^{1/2}}{1-(1+1/n^2)\mu^2}, \tag{13c}$$

$$D_{11}(n,\mu) = D_{22}(n,\mu) = 2n^3\mu f(n,\mu)\left\{\frac{1}{[nf(n,\mu)+\mu]^2} + \frac{1}{[n\mu + f(n,\mu)]^2}\right\}, \tag{14a}$$

$$D_{12}(n,\mu) = D_{21}(n,\mu) = 2n^3\mu f(n,\mu)\left\{\frac{1}{[nf(n,\mu)+\mu]^2} - \frac{1}{[n\mu + f(n,\mu)]^2}\right\}, \tag{14b}$$

and

$$D_{33}(n,\mu) = D_{44}(n,\mu) = \frac{4n^3\mu f(n,\mu)}{[nf(n,\mu)+\mu][n\mu + f(n,\mu)]}. \tag{14c}$$

At the interfaces between the layers, we have [1]

$$\mathbf{I}_k(a_k,-\mu,\phi) = \mathbf{X}(n_{k,k+1},\mu)\mathbf{I}_k(a_k,\mu,\phi) + \mathbf{Y}(n_{k,k+1},\mu)\mathbf{I}_{k+1}[a_k,-f(n_{k,k+1},\mu),\phi] \tag{15a}$$

and

$$\mathbf{I}_{k+1}(a_k,\mu,\phi) = \mathbf{X}(n_{k+1,k},\mu)\mathbf{I}_{k+1}(a_k,-\mu,\phi) + \mathbf{Y}(n_{k+1,k},\mu)\mathbf{I}_k[a_k,f(n_{k+1,k},\mu),\phi], \tag{15b}$$

for $\mu \in (0,1]$, $\phi \in [0,2\pi]$, and $k = 1,2,...,K-1$. With regard to the boundary condition for the surface located at $\tau = a_K$, we assume that there is no radiation coming from an external ($\tau > a_K$), non-participating medium with index of refraction n_{K+1}, and so we write [1]

$$\mathbf{I}_K(a_K,-\mu,\phi) = \mathbf{X}(n_{K,K+1},\mu)\mathbf{I}_K(a_K,\mu,\phi), \tag{16}$$

for $\mu \in (0,1]$ and $\phi \in [0,2\pi]$.

In order to simulate the case of an atmosphere-water system, we take $K = 2$ in our formulation. Since both the atmosphere ($k = 1$) and the water ($k = 2$) layers are taken to be described by Rayleigh scattering, we use $\varpi_k = 1.0$, $L_k = 2$ and the Greek constants of table 1 for both $k = 1$ and $k = 2$. The beam incident on the top of the atmosphere is unpolarized, and so we take $F_I = 1$ and $F_Q = F_U = F_V = 0$ in Eq. (7). To complete the specification of the atmosphere-water system that we wish to study, we use the data of Sommersten et al. [8]: the atmosphere has an optical depth of 0.15 and an index of refraction $n_1 = 1.0$ while the water layer has an optical depth of 1.0 and an index of refraction $n_2 = 1.338$. Above the atmosphere, a vacuum (index of refraction $n_0 = 1.0$) is assumed; at the bottom of the water layer, a black surface is simulated by assuming the presence of an external, non-participating medium with an index of refraction identical to that of the water body ($n_3 = 1.338$). The cosine of the polar angle of incidence is $\mu_0 = 1/2$.

Table 1. The Greek constants for Rayleigh scattering.

l	α_l	β_l	γ_l	δ_l	ϵ_l	ζ_l
0	0	1	0	0	0	0
1	0	0	0	3/2	0	0
2	3	1/2	$-\sqrt{6}/2$	0	0	0

The calculations were performed with an extension of a scalar ADO method [13]. The extension of that work [13] for the case with polarization is still in documentation phase, so the complete description of the method will appear in a future publication.

3. The issue concerning the transmission factor

As discussed in detail in our recent work [1], when radiation crosses an interface between two different media, conservation of energy requires that a normalizing scalar (that we call transmission factor) be applied to the transmission matrix obtained from the Fresnel formulas for the amplitude coefficients of transmission. When both indices of refraction are real, the transmission factor is given by [1]

$$f_T = \begin{cases} (n_b/n_a)^3(\cos\vartheta_b/\cos\vartheta_a), & n_a/n_b \le 1 \text{ or } n_a/n_b > 1 \text{ with } \vartheta_a < \vartheta_c, \\ 0, & n_a/n_b > 1 \text{ with } \vartheta_a \ge \vartheta_c, \end{cases} \tag{17}$$

where n_a and n_b are the indices of refraction of the incidence and transmission media, respectively. In addition, ϑ_a and ϑ_b are, respectively, the angles of incidence and refraction measured with respect to the interface normal, and $\vartheta_c = \arcsin(n_b/n_a)$ is the critical angle [3].

Figures 1 through 8 compare the results of our calculations for three situations: (i) correct transmission factor f_T as given by Eq. (17), (ii) top line of Eq. (17) divided by n^2, where $n = n_b/n_a$, and (iii) f_T missing or, equivalently, top line of Eq. (17) replaced with 1. Items (ii) and (iii) replicate the most frequent discrepancies detected in the transmission matrix by our study [1]. The plots shown in these figures correspond to the scattered Stokes parameters just below the water surface, which is one of the places where the differences are most conspicuous. Figures 1 and 2 are for the azimuthal angle $\phi = \phi_0$, figures 3 through 6 for the azimuthal angle $\phi = \phi_0 + \pi/2$, and figures 7 and 8 for the azimuthal angle $\phi = \phi_0 + \pi$. The U and V parameters are not shown for $\phi = \phi_0$ and $\phi = \phi_0 + \pi$ because they are identically zero for these angles.

The results shown in figures 1 through 8 show that the magnitudes of all four Stokes parameters obtained when f_T is ignored or replaced with f_T/n^2 are much smaller than the correct ones on the water side of the interface. This is in agreement with our previous conclusion [1] that the transmission of radiation to the water layer is underestimated when the transmission factor f_T is not used (or used incorrectly with a missing n^2 term). In particular, it should be

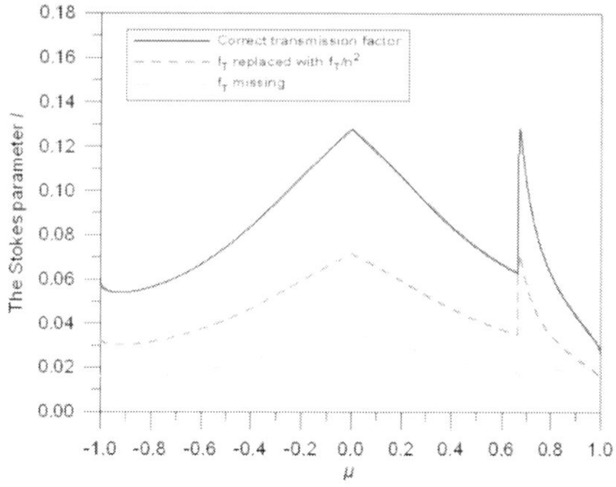

Fig. 1. The Stokes parameter $I_2(a_1, \mu, \phi_0)$.

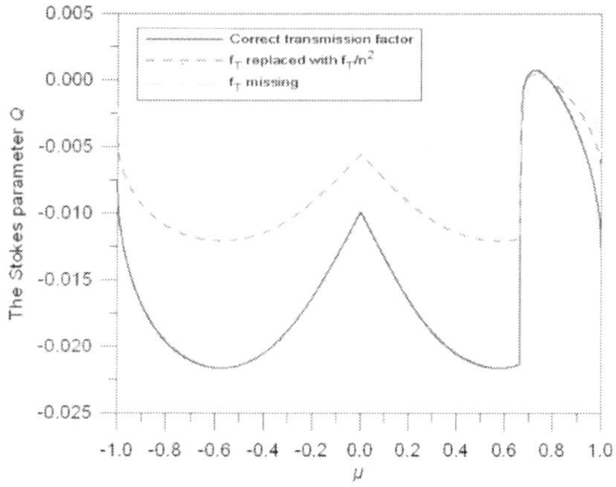

Fig. 2. The Stokes parameter $Q_2(a_1, \mu, \phi_0)$.

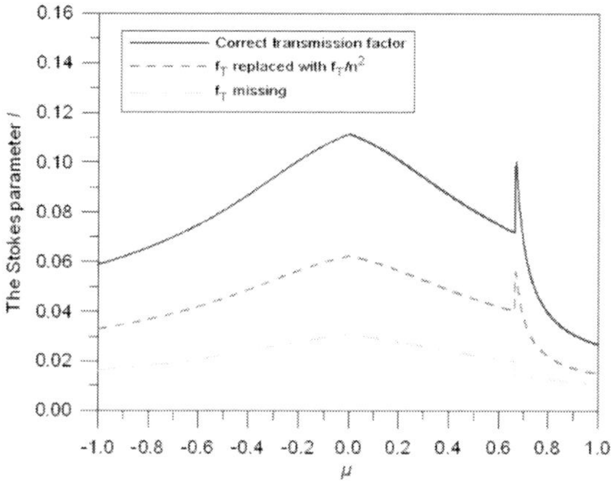

Fig. 3. The Stokes parameter $I_2(a_1, \mu, \phi_0 + \pi/2)$.

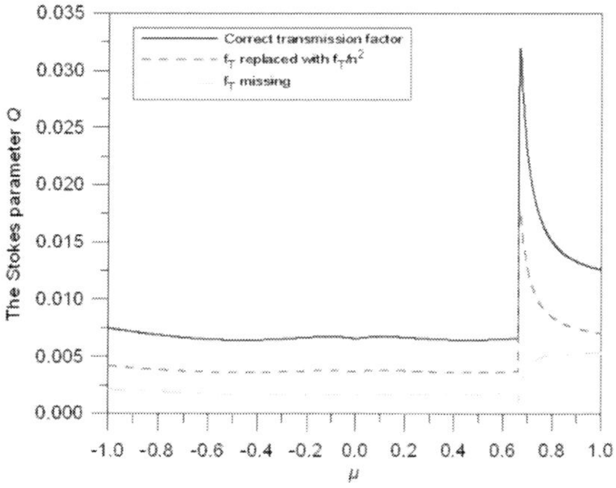

Fig. 4. The Stokes parameter $Q_2(a_1, \mu, \phi_0 + \pi/2)$.

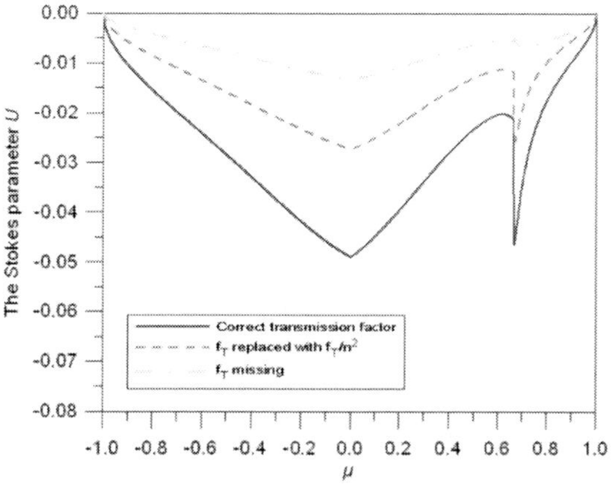

Fig. 5. The Stokes parameter $U_2(a_1, \mu, \phi_0 + \pi/2)$.

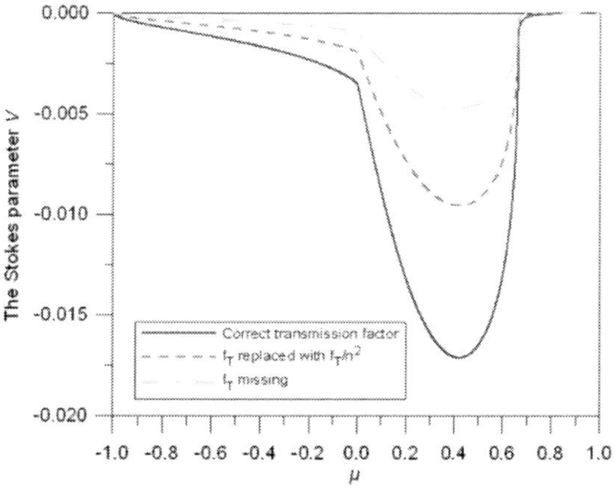

Fig. 6. The Stokes parameter $V_2(a_1, \mu, \phi_0 + \pi/2)$.

noted that when f_T is neglected the sharp peaks that appear near the cosine of the critical angle, $\mu_c(1.338) \approx 0.664$, are almost completely lost.

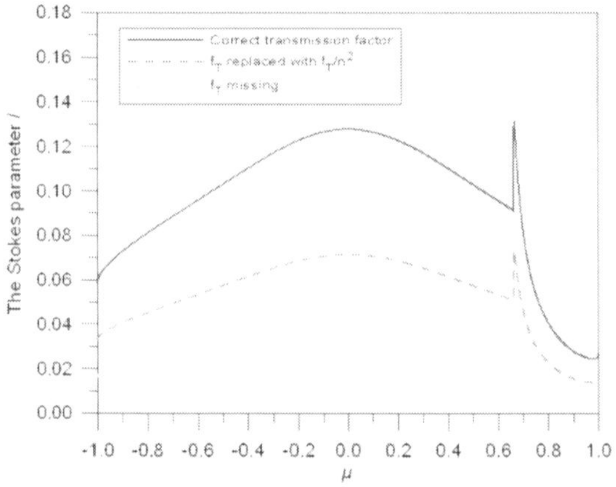

Fig. 7. The Stokes parameter $I_2(a_1, \mu, \phi_0 + \pi)$.

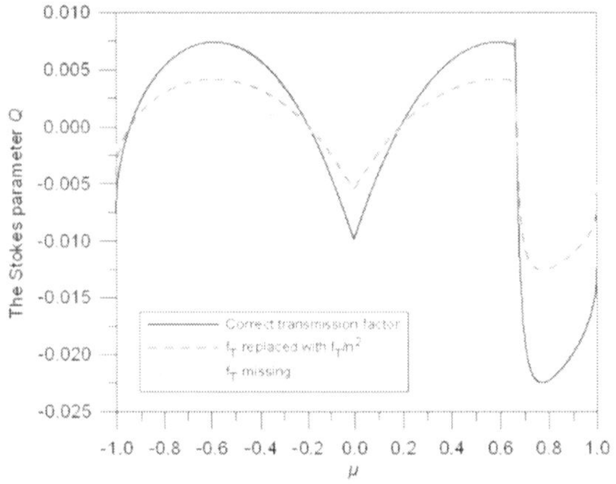

Fig. 8. The Stokes parameter $Q_2(a_1, \mu, \phi_0 + \pi)$.

Finally, we note that there is apparent agreement between our results and the results presented also in graphical form by Sommersten et al. [8] for test 1 of their work. It should be pointed out that these authors define μ as the negative of our μ. This causes a sign difference in the U and V parameters and explains why our curves look like mirror images of their curves.

4. The issue concerning the sign of one of the reflection coefficients

The amplitude coefficients of reflection by an interface are usually written as

$$R_l = \frac{\tan(\vartheta_a - \vartheta_b)}{\tan(\vartheta_a + \vartheta_b)} \tag{18a}$$

and

$$R_r = -\frac{\sin(\vartheta_a - \vartheta_b)}{\sin(\vartheta_a + \vartheta_b)}, \tag{18b}$$

where the subscripts l and r indicate the coefficients that correspond, respectively, to the parallel and perpendicular components of the electric field. In addition, the angle ϑ_a defines the initial direction of propagation in medium a with index of refraction n_a and the angle ϑ_b the direction of propagation in medium b with index of refraction n_b, after transmission. Among 25 references compiled by Muller [6], 19 use these coefficients as given by Eqs. (18) while six use Eq. (18a) with a change in sign. The choice of signs recommended by Muller [6] follows the one used by most authors and is referred to in this paper as "conventional". The main argument in favor of this sign choice is that it preserves the coordinate system used before and after reflection [6].

Equations (18) can be written in the alternative forms

$$R_l = \frac{n_b \cos\vartheta_a - n_a \cos\vartheta_b}{n_b \cos\vartheta_a + n_a \cos\vartheta_b} \tag{19a}$$

and

$$R_r = \frac{n_a \cos\vartheta_a - n_b \cos\vartheta_b}{n_a \cos\vartheta_a + n_b \cos\vartheta_b}. \tag{19b}$$

Using Eqs. (19) and Snell's law

$$n_a \sin\vartheta_a = n_b \sin\vartheta_b, \tag{20}$$

we can easily conclude that, for normal incidence ($\vartheta_a = 0$), R_l reduces to $(n_b - n_a)/(n_b + n_a)$ while R_r reduces to $(n_a - n_b)/(n_b + n_a)$. This difference in sign between the parallel and perpendicular reflection coefficients is unjustifiable on physical grounds, since the parallel and perpendicular components of the electrical field become indistinguishable when the incidence is normal.

A sign change in either R_l or R_r means simply a change of π in the phase of the corresponding component of the reflected electric field with respect to that of the incident electric field. Because a change of π in the phase of the reflected electric field with respect to that of the incident electric field is expected, at least for normal incidence, when radiation travels from an optically rarer medium to an optically denser medium [7], the use of Eq. (18a) [and Eq. (19a)] with a sign change can be considered as the most consistent choice from a physical point of view. This is the reason why we call it the "consistent" sign choice. However, one problem with this sign choice is that it does not preserve the coordinate system upon reflection [6]. Among eleven works on radiative transfer with polarization where the choice of the signs of R_l and R_r is specified [1], only one [8] uses the consistent sign choice.

A sign change in R_l causes only a small difference in our formulation: for the conventional sign choice the $\mathbf{G}(n, \mu)$ and $\mathbf{\Gamma}(n, \mu)$ matrices are used as defined in section 2, whereas for the consistent sign choice the signs of the (3,3), (3,4), (4,3), and (4,4) elements of these matrices must be changed.

In a preliminary version [14] of our previous work [1], we have adopted the conventional sign choice. However, motivated by a comment of a referee during the review process of [1], we came to consider the possibility that the consistent sign choice could be the most adequate. This idea has also permeated the abstract submitted to CTRPM-IV. In the end, a numerical study performed with a code based on an extension of the ADO method discussed in [13] to the case with polarization, led us to conclude that the conventional sign choice is indeed the appropriate one for our formulation. We discuss that numerical study next.

The numerical study was based on the same test case defined at the end of section 2 but with a difference: n_2, the index of refraction of layer 2 (originally water), was varied towards 1.0. For consistency, n_3 was taken to be equal to n_2 for all of the six choices of n_2 so that the condition of a black surface at $\tau = a_2$ was preserved throughout the study. We note that when $n_2 = 1.0$, the studied two-layer system reduces to a single homogenous layer without Fresnel boundary/interface conditions, a well-studied case for which the issue of the appropriate sign choice for the reflection coefficient R_l does not exist.

Our numerical results for the scattered U and V parameters in layer 2, at a position just below the interface between layers 1 and 2, are given in tables 2 and 3 for the conventional choice of the R_l sign. For our purposes here, it was sufficient to consider values of μ in the vicinity of zero and one value of the azimuthal angle for which the parameters U and V do not vanish. Tables 4 and 5 provide similar results for the choice of the R_l sign that we have called consistent.

We can see from tables 2 and 3 that the results obtained with the conventional sign choice of R_l are continuous as $\mu \to 0$ from above and from below, and that the results for $n_2 = 1.0$ are approached smoothly. On the other hand, as can be seen from tables 4 and 5, the consistent sign choice yields results for U and V that exhibit a sign flip as $\mu = 0$ is crossed. Moreover, it is apparent from table 4 that the U results for $n_2 = 1.0$ are not approached smoothly for all values of μ shown in the table when the consistent sign choice of R_l is made. We note that the results reported for $n_2 = 1.0$ in tables 2 through 5 were confirmed using an independent computational implementation of the ADO method for radiative transfer with polarization [15]. The study was repeated for more challenging cases, including cases defined by larger numbers of layers and/or phase matrices with more expansion terms than the Rayleigh phase matrix. Similar results were obtained.

In conclusion, we expect the reader will agree with us that there is enough evidence in support of our claim that the conventional sign choice for the R_l coefficient [6] is the appropriate one for our formulation of polarized radiative transfer with Fresnel boundary and interface conditions.

Table 2. The behavior of $U_2(a_1, \mu, \phi_0 + \pi/2)$ as $n_2 \to 1.0$ for the conventional sign choice.

μ	$n_2 = 1.1$	$n_2 = 1.01$	$n_2 = 1.001$	$n_2 = 1.0001$	$n_2 = 1.00001$	$n_2 = 1.0$
−0.1	−5.4010(−2)	−5.4582(−2)	−5.4143(−2)	−5.4079(−2)	−5.4072(−2)	−5.4072(−2)
−0.05	−5.6692(−2)	−5.7862(−2)	−5.7451(−2)	−5.7386(−2)	−5.7379(−2)	−5.7378(−2)
−0.01	−5.8748(−2)	−6.0505(−2)	−6.0126(−2)	−6.0061(−2)	−6.0054(−2)	−6.0053(−2)
−0.001	−5.9183(−2)	−6.1084(−2)	−6.0714(−2)	−6.0652(−2)	−6.0647(−2)	−6.0646(−2)
−0.0	−5.9230(−2)	−6.1147(−2)	−6.0779(−2)	−6.0717(−2)	−6.0712(−2)	−6.0712(−2)
0.0	−5.9230(−2)	−6.1147(−2)	−6.0779(−2)	−6.0717(−2)	−6.0712(−2)	−6.0712(−2)
0.001	−5.9182(−2)	−6.1084(−2)	−6.0714(−2)	−6.0652(−2)	−6.0647(−2)	−6.0777(−2)
0.01	−5.8737(−2)	−6.0504(−2)	−6.0126(−2)	−6.0061(−2)	−6.1287(−2)	−6.1358(−2)
0.05	−5.6606(−2)	−5.7857(−2)	−6.1433(−2)	−6.0430(−2)	−6.0134(−2)	−6.0101(−2)
0.1	−5.3766(−2)	−5.4571(−2)	−5.1931(−2)	−4.9825(−2)	−4.9630(−2)	−4.9609(−2)

Table 3. The behavior of $V_2(a_1, \mu, \phi_0 + \pi/2)$ as $n_2 \to 1.0$ for the conventional sign choice.

μ	$n_2 = 1.1$	$n_2 = 1.01$	$n_2 = 1.001$	$n_2 = 1.0001$	$n_2 = 1.00001$	$n_2 = 1.0$
−0.1	−7.8510(−4)	−2.0199(−5)	−3.0494(−7)	−3.5996(−09)	−3.8114(−11)	0.0
−0.05	−9.0750(−4)	−2.7002(−5)	−4.8436(−7)	−6.4323(−09)	−7.1833(−11)	0.0
−0.01	−1.0462(−3)	−3.8121(−5)	−9.7414(−7)	−1.9178(−08)	−2.8185(−10)	0.0
−0.001	−1.0873(−3)	−4.2667(−5)	−1.3183(−6)	−3.7796(−08)	−9.4900(−10)	0.0
−0.0	−1.0923(−3)	−4.3301(−5)	−1.3812(−6)	−4.3336(−08)	−1.3632(−09)	0.0
0.0	−1.0923(−3)	−4.3301(−5)	−1.3812(−6)	−4.3336(−08)	−1.3632(−09)	0.0
0.001	−1.1366(−3)	−5.9816(−5)	−6.7433(−6)	−1.7489(−06)	−5.2965(−07)	0.0
0.01	−1.5355(−3)	−2.0757(−4)	−5.3354(−5)	−1.2031(−05)	−6.6611(−12)	0.0
0.05	−3.2563(−3)	−7.8781(−4)	−9.9907(−8)	−3.7190(−10)	−4.0113(−12)	0.0
0.1	−5.1878(−3)	−1.1063(−3)	−2.4599(−8)	−2.7124(−10)	−2.9271(−12)	0.0

Table 4. The behavior of $U_2(a_1, \mu, \phi_0 + \pi/2)$ as $n_2 \to 1.0$ for the consistent sign choice.

μ	$n_2 = 1.1$	$n_2 = 1.01$	$n_2 = 1.001$	$n_2 = 1.0001$	$n_2 = 1.00001$	$n_2 = 1.0$
−0.1	−4.3329(−2)	−5.1631(−2)	−5.3681(−2)	−5.4024(−2)	−5.4066(−2)	−5.4072(−2)
−0.05	−4.4265(−2)	−5.3964(−2)	−5.6735(−2)	−5.7289(−2)	−5.7368(−2)	−5.7378(−2)
−0.01	−4.3973(−2)	−5.4908(−2)	−5.8699(−2)	−5.9782(−2)	−6.0012(−2)	−6.0053(−2)
−0.001	−4.3423(−2)	−5.4573(−2)	−5.8690(−2)	−6.0085(−2)	−6.0508(−2)	−6.0646(−2)
−0.0	−4.3279(−2)	−5.4434(−2)	−5.8584(−2)	−6.0025(−2)	−6.0494(−2)	−6.0712(−2)
0.0	4.3279(−2)	5.4434(−2)	5.8584(−2)	6.0025(−2)	6.0494(−2)	−6.0712(−2)
0.001	4.3423(−2)	5.4573(−2)	5.8690(−2)	6.0085(−2)	6.0508(−2)	−6.0777(−2)
0.01	4.3978(−2)	5.4908(−2)	5.8699(−2)	5.9782(−2)	−6.0907(−2)	−6.1358(−2)
0.05	4.4253(−2)	5.3960(−2)	−4.4713(−2)	−6.0358(−2)	−6.0131(−2)	−6.0101(−2)
0.1	4.3230(−2)	5.1621(−2)	−5.1484(−2)	−4.9809(−2)	−4.9629(−2)	−4.9609(−2)

Table 5. The behavior of $V_2(a_1, \mu, \phi_0 + \pi/2)$ as $n_2 \to 1.0$ for the consistent sign choice.

μ	$n_2 = 1.1$	$n_2 = 1.01$	$n_2 = 1.001$	$n_2 = 1.0001$	$n_2 = 1.00001$	$n_2 = 1.0$
−0.1	5.4662(−4)	1.8015(−5)	2.9514(−7)	3.5647(−09)	3.8000(−11)	0.0
−0.05	6.2694(−4)	2.4036(−5)	4.6865(−7)	6.3697(−09)	7.1618(−11)	0.0
−0.01	7.1105(−4)	3.3731(−5)	9.4116(−7)	1.8988(−08)	2.8100(−10)	0.0
−0.001	7.3098(−4)	3.7525(−5)	1.2703(−6)	3.7389(−08)	9.4596(−10)	0.0
−0.0	7.3222(−4)	3.8001(−5)	1.3290(−6)	4.2834(−08)	1.3583(−09)	0.0
0.0	−7.3222(−4)	−3.8001(−5)	−1.3290(−6)	−4.2834(−08)	−1.3583(−09)	0.0
0.001	−6.9480(−4)	−2.2205(−5)	3.9738(−6)	1.6577(−06)	5.2655(−07)	0.0
0.01	−3.4472(−4)	1.2004(−4)	5.0196(−5)	1.1937(−05)	4.8954(−12)	0.0
0.05	1.2082(−3)	6.8552(−4)	−3.8939(−8)	3.6267(−10)	3.9986(−12)	0.0
0.1	2.9893(−3)	1.0094(−3)	2.1978(−8)	2.6837(−10)	2.9182(−12)	0.0

5. Concluding remarks

Two standing issues relevant to polarized radiative transfer in a multilayer medium with a layer-dependent index of refraction were analyzed. In particular, the effects of two commonly encountered discrepancies in forms of the transmission matrix used by different authors, namely the neglect of a normalization factor required for conservation of energy across an interface and the omission of a factor given by the square of the ratio between the refractive index of transmission and that of incidence, were evaluated for a typical calm and clear atmosphere-water system. In addition, we have used a computational approach to determine which sign choice for the reflection coefficient R_l should be used in our formulation. We have reached the conclusion that the conventional sign choice recommended by Muller [6] is the appropriate one. Finally, since the Fresnel reflection and transmission matrices also play a role in the case of a rough interface (see, for example, [16]), we believe that the issues discussed in this work are equally important in that context.

Acknowledgments

This work was supported by CNPq.

References

[1] Garcia R D M 2012 *J. Quant. Spectrosc. Radiat. Transfer* **113** 306
[2] Garcia R D M, Siewert C E and Yacout A M 2008 *J. Quant. Spectrosc. Radiat. Transfer* **109** 2151
[3] Born M and Wolf E 1964 *Principles of Optics* (New York: Pergamon)
[4] Modest M F 1993 *Radiative Heat Transfer* (New York: McGraw-Hill)
[5] Friedmann G and Sandhu H S 1965 *Am. J. Phys.* **33** 135
[6] Muller R H 1969 *Surface Sci.* **16** 14
[7] Bennett J M 1964 *J. Opt. Soc. Am.* **54** 612
[8] Sommersten E R, Lotsberg J K, Stamnes K and Stamnes J J 2010 *J. Quant. Spectrosc. Radiat. Transfer* **111** 616
[9] Chandrasekhar S 1950 *Radiative Transfer* (London: Oxford University Press)
[10] Siewert C E 1982 *Astron. Astrophys.* **109** 195
[11] Gel'fand I M and Šapiro Z Ya 1956 *Amer. Math. Soc. Transl.* **2** 207
[12] Siewert C E 2000 *J. Quant. Spectrosc. Radiat. Transfer* **64** 227
[13] Garcia R D M and Siewert C E 2010 *J. Quant. Spectrosc. Radiat. Transfer* **111** 128
[14] Garcia R D M 2011 *Proc. Int. Conf. on Mathematics and Computational Methods Applied to Nuclear Science and Engineering* 8–12 May 2011 Rio de Janeiro (Rio de Janeiro: LAS/ANS) CD-ROM
[15] Garcia R D M and Siewert C E 2011 *J. Quant. Spectrosc. Radiat. Transfer* **112** 2801
[16] Ota Y, Higurashi A, Nakajima T and Yokota T 2010 *J. Quant. Spectrosc. Radiat. Transfer* **111** 878

Radiative heat transfer in honeycomb structures-New simple analytical and numerical approaches

D. Baillis[1], R. Coquard[2], and J. Randrianalisoa[3]

[1] Université de Lyon, CNRS, INSA-Lyon, LAMCOS, UMR5259, F-69621, Villeurbanne, France
Phone : + 33 (0) 4 72 43 82 72 - Email : dominique.baillis@insa-lyon.fr
[2] Société « Etude Conseils Calcul en Mécanique des Structures » (EC²MS) ; 66, boulevard Niels Bohr, 69603 Villeurbanne CEDEX, France.
Phone: : +33 (0)4 37 48 84 09 ; Email: remi.coquard@ec2-ms.fr
[3] Université de Reims Champagne Ardennes, GRESPI/CATHERM, F-51687, Reims, France

Abstract Porous Honeycomb Structures present the interest of combining, at the same time, high thermal insulating properties, low density and sufficient mechanical resistance. However, their thermal properties remain relatively unexplored. The aim of this study is the modelling of the combined heat transfer and especially radiative heat transfer through this type of anisotropic porous material.
The equivalent radiative properties of the material are determined using ray-tracing procedures inside the honeycomb porous structure. From computational ray-tracing results, simple new analytical relations have been deduced. These useful analytical relations permit to determine radiative properties such as extinction, absorption and scattering coefficients and phase function functions of cell dimensions and optical properties of cell walls. The radiative properties of honeycomb material strongly depend on the direction of propagation.
From the radiative properties computed, we have estimated the radiative heat flux passing through slabs of honeycomb core materials submitted to a 1-D temperature difference between a hot and a cold plate. We have compared numerical results obtained from Discrete Ordinate Method with analytical results obtained from Rosseland-Deissler approximation. This approximation is usually used in the case of isotropic materials. We have extended it to anisotropic honeycomb materials. Indeed a mean over incident directions of Rosseland extinction coefficient is proposed. Results tend to show that Rosseland-Deissler extended approximation can be used as a first approximation. Deviation on radiative conductivity obtained from Rosseland-Deissler approximation and from the Discrete Ordinated Method are lower than 6.7% for all the cases studied.
Keywords: *Honeycomb structures, conduction-radiation coupled heat transfer, radiative properties, Ray-Tracing, equivalent thermal conductivity, analytical model*

NOMENCLATURE

a : Cellular dimension in m
b : Cellular dimension in m
D_{cell} : Cell diameter in m
g_λ : Asymmetry parameter
$I_\lambda(z,\theta)$: Spectral radiation intensity in W/(m²Sr)
I^0_λ : Spectral radiation intensity emitted by black body in W/(m²Sr)
k_{air} : thermal conductivity of air in W/(mK)

k_e : Effective conductivity in W/(mK)
k_c : Equivalent conductivity in W/(mK)
k_r : "Radiative" conductivity in W/(mK)
k_s : thermal conductivity of sheets in W/(mK)
l : mean free paths in m
L : Thickness of the slab in m
$p_{s,\lambda}$: Spectral specularity parameter
q_r : Radiative heat flux in W/m²
R : Radius of the cells in m
R_λ : Spectral reflectivity of the walls
T_h and T_c : Temperatures of the hot and cold boundaries in K
T_m : Mean temperature in K
Tr_λ Spectral transmittivity of the walls
$x = \pi D_{cell} / \lambda$: Size parameter
$W(\theta)$: Angular distribution of the energy reflected during Ray-Tracing
z : Coordinate along the 1-D temperature gradient in m

Greek Symbols
α_λ : Spectral absorptivity of the walls
ΔT : Temperature difference
ε : Porosity
ε_h, ε_c : Emissivity of the hot and cold plates
λ : Radiation wavelength in m
β_λ, κ_λ, σ_λ, : Spectral extinction, absorption and scattering coefficients in m^{-1}
β^*_λ, σ^*_λ, κ^*_λ: Weighted Spectral extinction, scattering and absorption coefficients in m^{-1}
$\mu = cos\theta$: direction cosine of the radiation direction
θ : Propagation angle of the radiation in rad
$\Phi_\lambda(\vec{\Delta}' \rightarrow \vec{\Delta})$: Spectral scattering phase function
$\vec{\Delta}$: Radiation direction
τ_p : Diffuse transmittance
ω_λ : Scattering albedo

Subscripts
abs, sca, ext : absorption, scattering, extinction
λ : at wavelength λ

1 . Introduction

The geometry of porous honeycomb structures (Fig. 1.a) presents the advantage to minimize the amount of used material to reach minimal weight and minimal material cost. Porous honeycomb structures present the interest of combining, at the same time, high thermal insulating properties, low density and sufficient mechanical resistance. Honeycomb materials in aluminum, fiberglass, cellulosic and advanced composite materials are widely used in the aerospace industry, they have been featured in aircraft and rockets since the the 1950s. They can also be found in many other fields, from packaging materials in the form of paper-based honeycomb cardboard, to sporting goods like skis and snowboards. They are also used for the thermal insulation of solar collectors, storage tanks or pipelines of hot fluids.

Four different groups of researchers have conducted experimental or theoretical investigations on heat transfer through insulating materials with honeycomb structure: Morris et al. [1], Hollands et al.[2,3], Suehrcke et al. [4] and Arulanantham and Kaushika [5,6]. All of these studies have shown that heat transfer, in this kind of materials, is due to heat conduction but also to the propagation of thermal radiation while the convective heat transfer is generally negligible. They used ray-tracing model or zone method to evaluate theoretically the radiative transfer through the honeycomb structure. The local radiation balance in the medium requires integrating the incoming radiation from the cell faces and the surrounding faces. These studies are not well-adapted to a parametric study. These numerical tools are not based on Homogeneous Phase Approach. They do not require to determine radiative properties of homogeneous media. As a result radiative properties remain relatively unexplored in the literature. In order to predict the thermal performances of honeycomb structures, we have adopted another modelling approach, the so-called Homogeneous Phase Approach (HPA). This approach implies that the thermal behaviour of the composite material can be matched faithfully by an equivalent homogeneous semi-transparent conductive medium. It has been successfully used for the modelling of other type of highly porous materials such as open or closed cell foams [7,8] or fibrous insulations [9]. The principle is to assimilate the porous structure to a homogeneous semi-transparent material with equivalent conductive and radiative properties. The heat transfer is then governed by the Energy and Radiative Transfer Equations expressing the coupling between conductive and radiative heat transfer. This system of equations can be solved numerically using a combination of the volume method and the Discrete Ordinates Method (DOM).

The major difficulty is to determine the properties of the equivalent homogeneous medium that permits to match satisfactorily the thermal behaviour of this anisotropic porous material. A similar homogenization problem for anisotropic system of parallel ducts has been studied by Tancrez et al. [10]. In the present study, the equivalent radiative properties of the honeycomb structures have been evaluated using ray-tracing (RT) procedures inspired from the methodologies proposed by prior researchers. A special emphasis is put to deduce new simple analytical relations permitting to determine equivalent radiative properties functions of cell dimensions and cell walls optical properties. About the conductive properties, we used the classical parallel thermal resistance model to compute the effective thermal conductivity k_c. It is particularly well-suited for this type of materials. One of the advantages of the present approach using analytical relations to determine radiative and conductive properties is that it is well-adapted to a parametric study and to the optimization of these architectured materials.

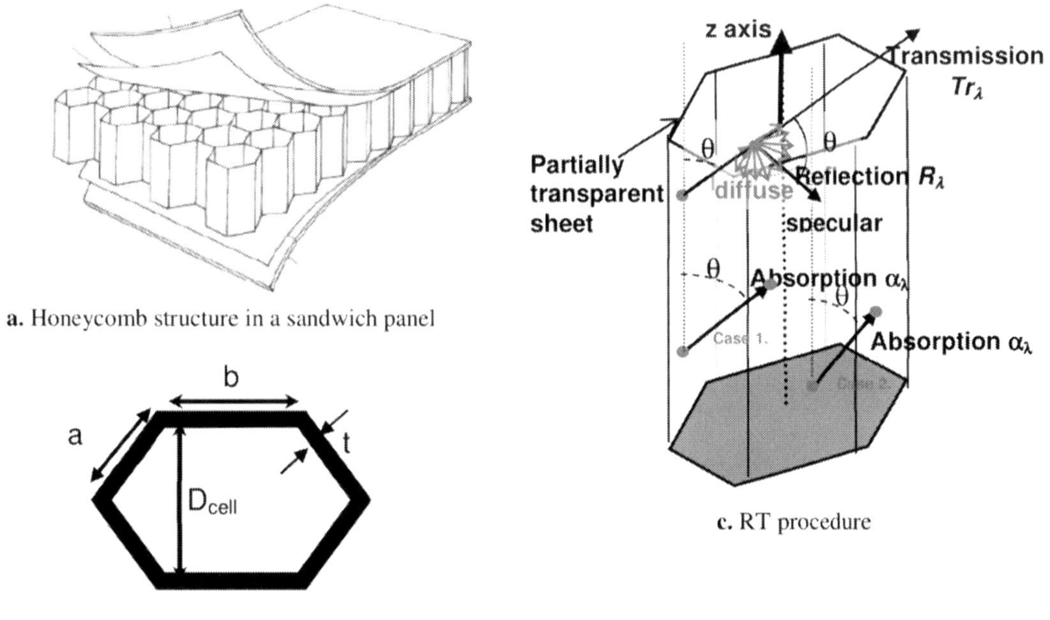

a. Honeycomb structure in a sandwich panel

b. Cross section of a honeycomb cell

c. RT procedure

Figure 1: Illustration of the structure of the honeycomb and of the RT procedures developed to compute the radiative properties

2 . Radiative transfer in semi-transparent media/ Homogeneous Radiative Transfer Equation

In the present study, we propose to model the heat transfer in the porous core materials by assimilating them to homogeneous semi-transparent conductive media. Indeed, this approximation has already proved its suitability to depict the radiation propagation in other types of porous materials. We can cite as example fibrous media [11,12] or cellular materials such as polymer [6,7], carbon [13] or metal foams [8]. The prediction of the equivalent radiative properties of the core material is the key issue of the present model. These properties must reproduce the radiation-matter interaction inside the honeycomb structure.

Porous media are often treated as continuous semi-transparent media that refract, absorb, emit and scatter radiation [14]. This radiative behavior may also vary with the radiation wavelength λ considered. The propagation of radiation in these materials is entirely characterized by the spectral radiative properties : the monochromatic extinction coefficient, β_λ (m^{-1}), the monochromatic scattering coefficient, σ_λ (m^{-1}), the monochromatic absorption coefficient, κ_λ (m^{-1}), the scattering phase function, $\Phi_\lambda(\vec{\Delta}'\rightarrow\vec{\Delta})$, describing the angular distribution of the energy scattered by the medium characterizes the probability, for a radiation incident in direction $\vec{\Delta}'$, to be scattered in an elementary solid angle around the direction $\vec{\Delta}$.

It is necessary to solve the Homogeneous Radiative Transfer Equation (HRTE) governing the spatial and angular distribution of monochromatic radiation intensity $I_\lambda(z,\mu)$ in the medium. Up to now, most of the composite materials studied using this approach (Expanded Polystyrene foams, fibrous materials), were generally considered as isotropic materials from a radiative point of view so that the radiative properties were independent of the propagation direction. In the case of honeycomb materials, the porous structure is anisotropic and the radiative properties are strongly dependent on the angle θ between the z-axis and the radiation direction. Moreover, the honeycomb structure presents an azimuthal symmetry. Therefore, for a 1-D heat transfer in the z-direction, the HRTE can be expressed by:

$$\mu\frac{\partial I_\lambda(z,\theta)}{\partial z} = -\beta_\lambda(\theta)I_\lambda(z,\theta)+\kappa_\lambda I_\lambda^0(T)+\frac{1}{2}\int_{\Omega'=4\pi}\sigma_\lambda(\theta)\Phi_\lambda(\mu'\to\mu)I_\lambda(z,\theta').d\Omega' \quad (1)$$

with the boundary conditions

$$\begin{aligned} I_\lambda(0,\theta_{\mu>0}) &= \varepsilon_{h,\lambda}I_\lambda^0(T_h)+2.(1-\varepsilon_{h,\lambda}).\int_{\mu'<0} I(0,\theta').\mu'd\mu' \\ I_\lambda(L,\theta_{\mu<0}) &= \varepsilon_{c,\lambda}I_\lambda^0(T_c)+2.(1-\varepsilon_{c,\lambda}).\int_{\mu'>0} I(L,\theta').\mu'd\mu' \end{aligned} \quad (2)$$

Several other characteristics derived from β_λ, ω_λ and $\Phi_\lambda(\vec{\Delta}'\to\vec{\Delta})$ are usually considered:

- the asymmetry parameter $g_\lambda(\vec{\Delta}) = \frac{1}{4\pi}\int_{\Omega=4\pi}\Phi_\lambda(\vec{\Delta}'\to\vec{\Delta}).\vec{\Delta}'\times\vec{\Delta}.d\Omega$ (3)
  which describes the way the energy is scattered: $0<g<1$ means that the radiation is predominantly scattered in forward directions whereas $-1<g<0$ means that the radiation is predominantly scattered in backward directions
- the weighted scattering and extinction coefficients

$$\sigma_\lambda^*(\vec{\Delta})=\sigma_\lambda(\vec{\Delta})(1-g_\lambda(\vec{\Delta}))\ ;\ \beta_\lambda^*(\vec{\Delta})=\kappa_\lambda(\vec{\Delta})+\sigma_\lambda(\vec{\Delta})(1-g_\lambda(\vec{\Delta}))=\kappa_\lambda(\vec{\Delta})+\sigma_\lambda^*(\vec{\Delta}) \quad (4)$$

Contrary to σ_λ, σ_λ^* not only quantifies the probability of an intercepted ray to be scattered but encompasses the whole scattering phenomenon by taking into account the way the radiative energy is redistributed via the asymmetry factor g_λ.

3 . Modeling of the radiative of Porous Insulating Materials with honeycomb structures

The common feature of natural or man-made honeycomb structures is an array of hollow cells formed between thin vertical walls. The cells are often columnar and hexagonal in shape. The insulating materials that are investigated in this work are composed of a man-made honeycomb core sandwiched between two composite skins. The walls forming the hexagonal structure are partially transparent. The honeycomb cross section is considered to be hexagonal with the dimensions a, b and D_{cell} indicated on Fig. 1.b.

3.1 General Hypotheses.

Several hypotheses concerning the structure and characteristics of the core material are made to simplify the solution:

- The Geometric Optics Approximation (GOA) is used for the honeycomb structure, i.e., the radiation can be treated as pencils of rays propagating according to straight lines. The GOA is valid for the structures considered since the size parameter $x=\pi D_{cell}/\lambda>>1$. In the present study, the radiation wavelengths are in the I.R ($\lambda\approx 10\ \mu m$) whereas the size of the cells is several mm. Thus, the GOA is fulfilled: $x = \frac{\pi . D_{cell}}{\lambda} >> 1$. Therefore, the propagation of the radiation in the materials can be modelled using the geometric optics laws of reflection and refraction. Moreover, the scattering of radiation due to diffraction is neglected since the directions of the rays diffracted by large scatterers ($x >> 1$) are very close to the incident directions. Finally, the interparticle distance is much greater than the wavelength and the particles are assumed randomly dispersed so that the interference effects are negligible.

- The sheets forming the honeycomb structures are assumed partially transparent and reflecting with a transmittivity Tr_λ and a reflectivity R_λ which are assumed independent of the angle of incidence. The absorptivity α_λ is given by $\alpha_\lambda = 1\text{-}Tr_\lambda\text{-}R_\lambda$. Moreover, we take into account an intermediate type of reflection between perfectly specular or totally diffuse using a single parameter: the specularity parameter $p_{s,\lambda}$. For $p_{s,\lambda} = 1$, the reflection is assumed perfectly specular whereas it is totally diffuse for $p_{s,\lambda} = 0$. For $0 < p_{s,\lambda} < 1$, the incident rays have a probability $p_{s,\lambda}$ to be reflected specularly and $1\text{-}p_{s,\lambda}$ to be reflected diffusely.

3.2 Theoretical basis.

Most of the previous studies about the radiative properties of porous materials have treated the problem by dividing the porous structure into particles whose shape and size permit to reproduce the internal structure of the porous material. Then, under the assumption of independent scattering, the equivalent radiative properties of the porous medium could be calculated by simply adding the radiative characteristics of each particle present in an elementary volume [14,15].
However, although they are commonly used, these methods suffer from inaccuracies since the dependent scattering between the particles is neglected. Moreover, they are poorly adapted to materials with anisotropic radiative behaviour. That is the reason why we propose to use another approach for predicting the radiation-matter interaction in the core material. The method is based on RT procedures inside the exact structure and has been firstly proposed in recent publications [16,17]. It has been recently modified by Randrianalisoa and Baillis [18] in order to reduce the computing time and to take into account the possible directional dependence of the properties[19]. This method is only applicable when the interaction of radiation with matter can be treated by the GOA.
The proposed method consists in computing the absorption, scattering and extinction mean free paths $l_{abs}(\theta)$, $l_{sca}(\theta)$ and $l_{ext}(\theta)$ and the distribution of scattering directions of the radiation bundles $W(\theta,\theta')$. The length of honeycomb cell is assumed very large compared to the cell diameter. These mean free paths correspond to the mean distance travelled by the rays between two consecutive extinctions, absorptions or scatterings and are related to the absorption, scattering and extinction coefficients by:

$$\kappa(\theta) = \frac{1}{l_{abs}(\theta)}; \quad \sigma(\theta) = \frac{1}{l_{sca}(\theta)}; \quad \beta(\theta) = \sigma(\theta) + \kappa(\theta) = \frac{1}{l_{ext}(\theta)} = \frac{1}{l_{abs}(\theta)} + \frac{1}{l_{sca}(\theta)} \tag{5}$$

While the scattering phase function can be computed from $W(\theta)$:

$$\Phi(\theta,\theta') = \frac{W(\theta,\theta')}{\frac{1}{4\pi}\int_{\Omega'=4\pi} W(\theta,\theta').d\Omega'} \tag{6}$$

$l_{abs}(\theta)$, $l_{sca}(\theta)$ and $l_{ext}(\theta)$ and $W(\theta,\theta')$ are computed through RT procedures inside the 3-D structure.
The principle of these procedures is to trace the histories of a great number of radiation bundles, denoted by N_{RAY}, propagating through the real medium under investigation until they are absorbed or scattered by the porous structure. During the travel of bundles, their histories are stored. In the case of honeycomb structures, the mean free paths and $W(\theta,\theta')$ depend on the direction of propagation considered and therefore, they had to be evaluated for every direction in the range $\theta=0,\pi/2$ (where θ is the angle between the starting direction and the vertical z-axis; see Fig. 1.c).

In the RT procedure (illustrated on Fig. 1.c), the initial location of the ray can be chosen as if it corresponds to the location of a ray after an extinction event (i.e. scattering or absorption). For honeycomb Transparent Insulating Materials (TIM), given that the fluid phase is considered transparent (air), the rays are generated from the solid-fluid interface only. Practically, we randomly choose a point of the solid surface from which the ray will be emitted. The starting direction of the ray corresponds to the direction in which the radiative properties have to be computed. As explained before, all the starting directions in the range $\theta=0,\pi/2$ have to be investigated. In practice, the radiative properties are evaluated for every degree of angle : $\theta_0=0°$, $\theta_1=1°$, ... $\theta_{90}=90°$.

Alternative RT procedure consists in choosing the initial location of the ray as if it corresponds to the location of an emission ray of the two isothermal hot and cold boundary plates (perpendicular to the lateral walls of honeycomb).

The path of the ray is tracked until it is intercepted by the solid phase (absorbed or reflected). The distance l travelled by the ray before interception is stored by incrementing the counter L_{ext} ($L_{ext} = L_{ext} + l$) and the type of extinction (absorption or reflection) is memorized by incrementing the counter N_{abs} ($N_{abs} = N_{abs} + 1$, if absorption) or N_{SCA} ($N_{SCA} = N_{SCA} + 1$, if scattering). If the ray is reflected, the reflecting angle θ' between the incident and scattering directions is computed and the parameter $W(\theta,\theta')$ is incremented by unity. At the end of the RT procedure, we have:

$$l_{ext} = \frac{L_{ext}}{N_{ray}} ; \; l_{sca} = \frac{l_{ext}}{N_{sca}} N_{RAY} ; \; l_{abs} = \frac{l_{ext}}{N_{abs}} N_{RAY} \tag{7}$$

The destiny of a ray impacting the solid phase is determined using randomly generated numbers. More details of the method can be found in the previous papers [18,19].

It is worth noting that in our computation, all the rays hitting the cell walls are not necessarily considered as scattering. Some rays intercepted by the cell walls are transmitted through the cell faces without any deviation and continue their path in the medium as if they had not impacted the solid phase. This is due to the fact that the cell walls are considered to be partially transparent. As a consequence, the extinction mean free path and thus the extinction coefficient may vary with the wavelength considered due to the variations of the transmittance of the sheets.

3.3 *Evolution of the radiative properties of Honeycomb structures.*

The radiative properties of honeycomb material strongly depend on the angle θ between the direction of propagation and the vertical z-axis (see Fig. 1.c). It can be noted that radiative properties have been calculated functions of azimuthal angle, φ. Results have shown that β and κ are nearly independent of φ whereas the asymmetry factor g is totally independent of φ. As a result azimuthal symmetry is assumed. We have computed the extinction, scattering and absorption coefficients as well as the scattering phase function for every angle θ_i with i=0,90. The computations were conducted for different cell sizes (geometrical parameters a, b and D_{cell}) and different optical properties of the walls (Tr_λ, R_λ, α_λ and $p_{s,\lambda}$). Moreover, for these investigations, we have also considered that the optical properties are grey in order to simplify the problem. The results are obtained in terms of extinction coefficient β, scattering albedo $\omega=\sigma/\beta$ and scattering phase function $\Phi(\theta,\theta')$.

3.3.1 *Extinction coefficient*

The numerical results, for a given direction, only depend on the size of the cells and on the transmittivity of the walls. The two possible choices of initial location are investigated. From computational results, simple analytical relations have been deduced. Computational results calculated for transmittivity values comprised between 0 and 1 reveal that the following relation is verified.

$$\beta_R = \frac{(1-Tr)}{a}\sqrt{(1-\mu^2)}C \tag{8}$$

The constant C depends on the choice of the initial location. When initial location of the rays is chosen on one of the six lateral walls of the hexagonal cell of honeycomb (corresponding to the location of a ray after an extinction event, case 1 on Fig. 1.c), the constant C is C= 0.87

When initial location of the ray is chosen as if it corresponds to the location of an emission ray of the two isothermal hot and cold boundary plates (perpendicular to the lateral walls of honeycomb, case 2 of Fig. 1.c): C=1.2207

In the next part we only consider the case 1 in order to shorten the paper. But similar study could be performed for the case 2.

3.3.2 Scattering albedo

The scattering albedo ω provides information concerning the relative importance of absorption and scattering in the extinction phenomena. We have already shown in a previous study (Coquard and Baillis, 2004) that when the reflectivity R is assumed independent of the incident angle, ω is independent of the shape of the structure or the type of reflection.

$$\omega = \frac{R}{R+\alpha} \tag{9}$$

3.3.3 Scattering Phase Function

We illustrate the shapes of the scattering phase functions computed for walls with totally diffuse (p_s=0) on Fig. 2 and with perfectly specular reflection (p_s=1) on Fig. 3.

One can notice that the scattering phase function obtained assuming diffusely reflecting walls are isotropic for all the propagation directions, θ (Fig.2). The fact that the scattering phase functions are identical for all propagation angles θ is not surprising since, for each reflection event, the direction of scattering is chosen independently of the incident direction on the wall.

When specular reflection is assumed, the scattered energy is redirected in a scattering direction with an angle θ'=θ since the walls encountered are parallel to the z-axis (Fig. 3). Thus, from an energetic point of view, all is going on as if the rays were not reflected by the walls and the asymmetry parameter g is equal to 1 for all propagation directions θ.

As a result the phase function can be expressed with the simple relation by using the Dirac function *δ*:

$$\Phi(\theta,\theta') = p_S.\,\delta(\theta-\theta') + (1-p_S) \tag{10}$$

4 . Evolution of the equivalent thermal conductivity of honeycomb Porous Insulating Materials

From the radiative properties computed using the method described in the previous section, we have estimated the radiative heat flux passing through slabs of honeycomb core materials submitted to a 1-D temperature difference between a hot and a cold plate. We compare numerical results obtained from DOM with results obtained from Rosseland-Deissler analytical approximation.

4.1 Numerical solution of the coupled heat transfer

The HRTE can be solved numerically by using a combination of the control volume method and the Discrete Ordinates Method (DOM) [14,20,21]. It is one of the most frequently used method among others (spherical harmonics method, zone method of Hottel, RT methods...) and gives accurate results. The principle is to divide the slab thickness of core material in elementary control volumes. The temperature of the medium at the center of each volume, i.e. temperature distribution in the medium, has to be calculated. The equivalent and "radiative" conductivities are computed by solving the Energy Equation and HRTE. For the computation of k_r, radiative equilibrium has to be considered. In both cases, we have to solve the following system of equations:

$$\begin{cases} \dfrac{\partial q_r^z}{\partial z} = -\dfrac{\partial q_c^z}{\partial z} = k_c.\dfrac{\partial^2 T}{\partial z^2} \text{ with } k_c = k_{air}.\varepsilon + k_s.(1-\varepsilon) \text{ for } k_{equ} \quad ; \quad \dfrac{\partial q_r^z}{\partial z} = 0 \quad \text{for } k_r \\ \mu\dfrac{\partial I_\lambda(z,\vec{\Delta})}{\partial z} = -\beta_\lambda I_\lambda(z,\vec{\Delta}) + \kappa_\lambda I_\lambda^0(T) + \dfrac{\sigma_\lambda}{4\pi}\displaystyle\int_{\Omega'=4\pi}\Phi_\lambda(\vec{\Delta}'\rightarrow\vec{\Delta}) I_\lambda(z,\vec{\Delta}')d\Omega' \end{cases} \tag{11}$$

$$\text{with } q_r(z,\vec{\Delta}) = \int_0^\infty\int_{\Omega'=4\pi} I_\lambda(z,\vec{\Delta}').\vec{\Delta}.d\Omega'.d\lambda \quad \Leftrightarrow \quad q_r^z(z) = 2\pi\int_0^\infty\int_{-1}^{1} I_\lambda(z,\vec{\Delta}').\mu'.d\mu'.d\lambda \tag{12}$$

and k_r is simply given by : $$k_r = \frac{q_r^z}{\Delta T/L} \tag{13}$$

The DOM has already been explained in numerous previous publications as (Siegel and Howell, 1992, Dombrovsky and Baillis, 2010). The accuracy of the results is entirely dependent on the angular discretization. Therefore, in the present study, we choose a very fine discretization dividing the 2π radians in 180 directions of 1° and whose weighting factors are proportional to the solid angle they encompass.

Figure 2. Evolution of the scattering phase functions $\Phi(\theta,\theta')$ computed for different direction of propagation θ and assuming a totally diffuse reflection by the walls (p_s=0)

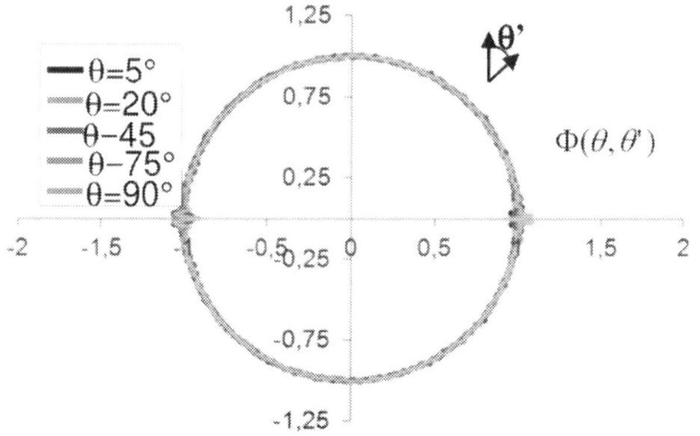

4.2 *Rosseland-Deissler analytical Approximation*

The Rosseland-Deissler approximation is the following (Deissler 1964):

$$k_r = \frac{4\sigma_{SB}\overline{T}^3 L}{\left(\frac{1}{\varepsilon_c}+\frac{1}{\varepsilon_h}-1\right)+\frac{3}{4}\beta_R L} \qquad (14)$$

where $T_m = (T_c + T_h)/2 \qquad \beta_R = \langle\kappa\rangle + \langle\sigma(1-g)\rangle = \langle\beta\rangle\times(1-\omega g)$,

$\langle\beta\rangle$ is the mean calculation over θ angle of $\beta(\theta)$.

Extinction coefficients have been proved to verify analytical relation Eq. (8). As a result the calculation of $\langle\beta\rangle$ depends on the mean calculation $\left\langle\sqrt{1-\mu^2}\right\rangle$. But the question arises how to calculate this mean value. In order to answer this question, the mean ,$\left\langle\sqrt{1-\mu^2}\right\rangle$, is obtained by identification by minimizing the relative deviation between results obtained from reference DOM and Rosseland-Deissler calculation for different specular parameter p_s=0.5 or 1, different albedo values comprised between 0 and 1 (0, 0.1, 0.2, 0.3, 0.4, 0.5, 0.6, 0.7, 0.8, 0.9)and different thickness of honeycomb slab comprised between 0.001 and 1 m (0.001, 0.008,0.027, 0.064, 0.125, 0.216, 0.343, 0.512, 0.729 and 1 m).

The identified value verifies $\left\langle\sqrt{1-\mu^2}\right\rangle = 0.4855$ *and thus* $\beta_R = \frac{(1-Tr)}{a} 0.4855.C.$

The mean relative deviation between DOM and Rosseland-Deissler approximation using this identified value is 6.7%. Figure 4 shows as example some cases of the evolution of the radiative equivalent conductivities obtained with the thickness of the slab for honeycomb structures obtained for DOM and Rosseland with this identified value. Note that it tends to show that Rosseland mean extinction can be approximatively obtained by the mean of the extinction mean free path weighted by $\mu d\mu$

$$\frac{1}{\beta_R} \approx \frac{\int_0^1 \frac{1}{\beta}\mu d\mu}{\int_0^1 \mu d\mu} = 0.5 \qquad (15)$$

5 . Conclusions

The radiative heat transfer in Porous Insulating Materials with honeycomb structures has been modelled by assuming the core material to an equivalent semi-transparent medium with spectral radiative properties depending on its porous structure and composition. This approach is rather new in the field of the modelling of heat transfer in honeycomb structures. The equivalent radiative properties of the core material have been modelled using a methodology based on the GOA which was recently proposed by Coquard and Baillis (Coquard and Baillis, 2004; Coquard and Baillis, 2005) and Randrianalisoa and Baillis (Randrianalisoa and Baillis, 2010). It is particularly well-adapted to the honeycomb structures studied given that the geometrical conditions ($x = \frac{\pi . D_{cell}}{\lambda} >> 1$) are largely fulfilled and that the RT procedures used in the modelling permit to take into account the particular porous structure very accurately. The particularity of honeycomb cellular material concerns the high anisotropy of the radiative properties.

From computational RT results, new simple useful analytical relations have been deduced to determine radiative properties such as extinction, absorption and scattering coefficients and phase function.

From the radiative properties computed, we have estimated the total heat flux passing through slabs of honeycomb core materials submitted to a 1-D temperature difference between a hot and a cold plate. We have compared numerical results obtained from DOM with analytical results obtained from Rosseland-Deissler approximation. Rosseland mean extinction has been proved to verify a simple analytical relation obtained from the inverse of the mean of the extinction mean free path. The mean is calculated over the polar angle.

Results tend to show that Rosseland-Deissler can be used as a first approximation. Deviations on radiative conductivity obtained between the Rosseland-Deissler approximation and the DOM results are lower than 6.7%.

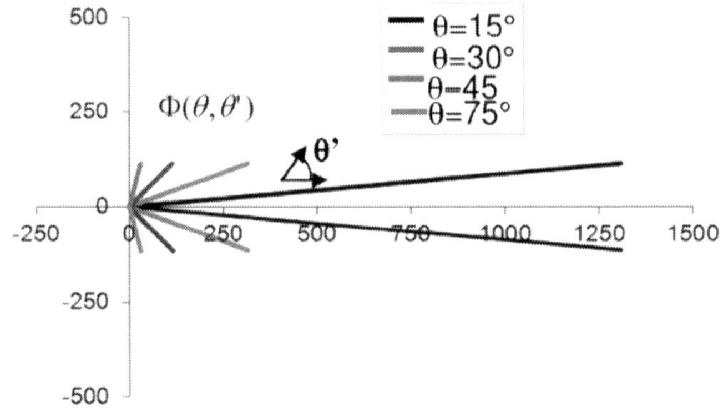

Figure.3 Evolution of the scattering phase function $\Phi(\theta', \theta)$ computed for different direction of propagation θ' and assuming a specular reflection by the walls (p_s=1)

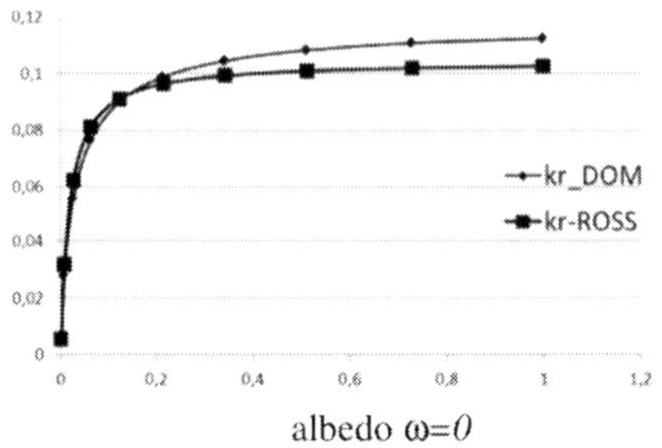

albedo ω=*0*

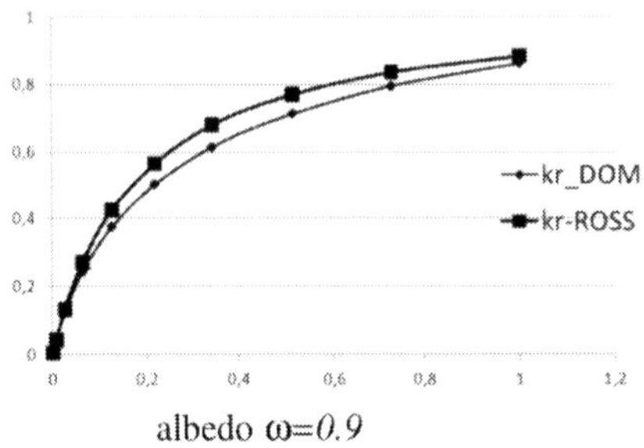

albedo ω=*0.9*

Figure 4 Evolution of the radiative equivalent conductivities with the thickness of the slab for honeycomb structures with D_{cell}=10.0 mm, Tr=0, $\boldsymbol{p_s}$**=1**, $\boldsymbol{\varepsilon_c=\varepsilon_h}$=1 T_m=293K and ΔT=10K

Acknowledgement

The initial research leading to these results has received funding from the European Community's Seventh Framework Programme FP7/2007-2013 under grant agreement n°213371 (www.maaximus.eu). The authors are grateful to Bruno Estebe, Matthieu Thomas from Air Bus and Alain Combescure from INSA Lyon to initiate the works.

References

[1] Morris P. A., James R. K., McMurrin J. C., Felland J. R., Edwards D. K., Buchberg H., Aug. 9-11, 1976, *Radiative transfer through thin-walled glass honeycomb, Am. Soc. of Mech. Eng. (ASME) and Am. Inst.of Ch. Eng. (AICE), Heat Transfer Conference*, (St. Louis, Mo.)

[2] Hollands, K.G.T., Raithby, G.D., Russell, F.B., Wilkinson,R.G., 1984, « Coupled radiative and conductive heat transfer across honeycomb panels and through single cells », *International Journal of Heat and Mass Transfer*, **27**, 2119–2131

[3] Hollands, K.G.Y., Iynkaran, K., 1993, "Analytical model for the thermal conductance of compound honeycomb transparent insulation, with experimental validation", *Solar Energy* **51** (3), 223–227.

[4] Suehrcke H. , Daldehog D., Harris J.A., Lowe R.W., « Heat transfer across corrugated sheets and honeycomb transparent insulation », *Solar Energy*, 76 (2004) 351–358

[5] Arulanantham M. and Kaushika N. D.,1996, "Coupled Radiative and Conductive Thermal Transfers across Transparent Honeycomb Insulation Materials", *Applied Thermal Engineering*, **16** (3), 209-217

[6] Coquard R., Baillis D., 2009, "Radiative properties of expanded polystyrene foams", *Journal of Heat Transfer ASME Journal*, **131** (1), 012702

[7] Coquard R., Maire E. and Baillis D., 2010, "Numerical investigation of the radiative properties of polymer closed cell foams from tomographic images", *Journal of Thermophysics and Heat Transfer*, **24** (3), 647-657

[8] Loretz M., Coquard R., Baillis D. and Maire E., 2008, "Metallic foams: Radiative properties/comparison between different models", *Journal of Quantitative Spectroscopy and Radiative Transfer*, **109** (1), 16-27

[9] Coquard R., D. Baillis, 2006, "Radiative properties of dense fibrous medium containing fibers in the geometric optics limit" Author(s), *Journal of heat transfer*, **128** (10) 1022-1030

[10] Tancrez M., Hilka M., Taine J., 2001, "Comparison between local and homogenised models of radiative transfer in a set of parallel ducts considered as a porous medium", *Int. J. Heat Mass Transfer* **45** (1), 173–180

[11] Lee S.-C. and Cunnington G.R., 1998, *Theoretical models for radiative transfer in fibrous media, Annual Review in Heat Transfer.*, Ed. C.L. Tien, (New York , Begell House) , pp.159–218

[12] Tagne, K.H., Baillis D., 2010, "Reduced models for radiative heat transfer analysis through anisotropic fibrous medium", *ASME Journal of Heat Transfer*, **132** (7), 072403.1- 072403.8 (8 pages)

[13] Baillis D., Raynaud M. and Sacadura J. F., 2000, "Determination of spectral radiative properties of Open Cell Foam. Model validation", Journal of Thermophysics and Heat Transfer AIAA, **14** (2), 137-143

[14] Dombrovsky L. A., and Baillis, D., 2010, *Thermal Radiation in Disperse Systems: An Engineering Approach*, (Redding, CT: Begell House)

[15] Brewster M.Q., 1992, *Thermal Radiative Transfer and Properties*, (New York: Wiley–Interscience)

[16] Coquard R., Baillis D., 2004, "Radiative Characteristics of Opaque Spherical Particles Beds: A New Method of Prediction", *Journal of Thermophysics and Heat Transfer*, **18** (2) 178-186

[17] Coquard R., Baillis D., 2005, "Radiative Characteristics of beds of spheres containing an absorbing and scattering medium", *Journal of Thermophysics and Heat Transfer*, **19** (2), 226-234

[18] Randrianalisoa, J. and Baillis, D., 2010, "Radiative properties of densely packed spheres in semitransparent media: A new geometric optics approach", *J. Quant. Spectrosc. Rad. Transfer*, **111** (10) 1372-1388.

[19] Coquard R., Rousseau R., Echegut P., Baillis D., Gomart H. and Iacona E., 2012, "Investigations of the radiative properties of Al-NiP foams using tomographic images and stereoscopic micrographs", *International Journal of Heat and Mass Transfer IJHMT*, **55** (5-6), 1606-1619

[20] Siegel R. and Howell J.R., 1992, *Thermal radiation heat Transfer, 3rd ed.*, (Washington DC Hemisphere Publishing Corp.)

[21] Modest M.F., 2003, *Radiative Heat Transfer, Second edition*, (New York, Academic Press)

Journal of Physics: Conference Series **369** (2012) 012007 doi:10.1088/1742-6596/369/1/012007

Comparison of Monte Carlo simulations with exact Maxwell solutions for polarized light scattering by multiple absorbing spheres

Ansgar Hohmann[1], **Florian Voit, Jan Schäfer and Alwin Kienle**
Institut für Lasertechnologien in der Medizin und Meßtechnik (ILM)
Helmholtzstraße 12, 89081 Ulm, Germany

E-mail: ansgar.hohmann@ilm.uni-ulm.de

Abstract. The goal of this study was to investigate the differences between radiative transfer theory and Maxwell theory for simulation of light propagation in a turbid medium. Polarization effects as well as absorbing scatterers with complex index of refraction are taken into account. The simulation volume contained different numbers of scattering and absorbing spheres (radius: 1 μm) and was irradiated from one side with a plane electromagnetic wave ($\lambda = 600$ nm). The absorption was varied as well as the concentration of the scatterers. The resulting 16 Müller matrix elements were compared for the Monte Carlo method as well as for the Maxwell method for all scattering angles. An increasing absorption of the spheres resulted in larger differences especially for the intensity results (M_{11} Müller matrix element) between the two solution methods, while increasing scatterer concentrations led generally to larger differences for all Müller matrix elements. That means that the results of radiative transfer theory have to be treated with care for high scatterer concentrations and large absorption. By using the presented method, differences between the two theories can be investigated for arbitrary particle size parameters (spheres) and optical properties of the scatterers.

1. Introduction

The investigation of polarized light propagation in turbid media is an important topic in different research areas as for example in biomedical optics [1, 2]. Maxwell theory can be used for an exact calculation of polarized light propagation in scattering media. However, the problem of this method are extremely high computational needs to investigate relevant simulation volumes, which makes it inappropriate for many applications.

Radiative transfer theory is an approximation of Maxwell theory, allowing to study light propagation also in more extended simulation volumes of scattering media. Polarization dependence as well as absorbing scatterers with complex refractive index can be considered by this approach. However, due to the approximative manner of radiative transfer theory neglecting the wave character of light, there are possible sources of errors which have not been fully investigated until now [3].

[1] phone (+)49 (0)731 1429 777

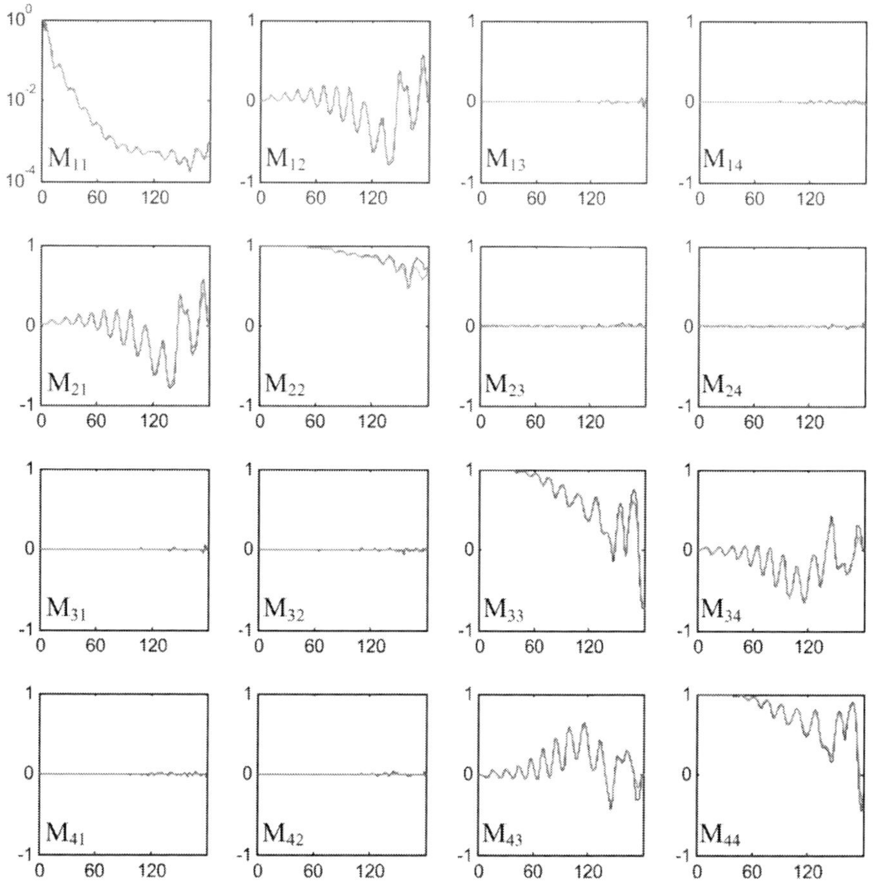

Figure 1. Angularly resolved Müller matrix elements for scattering angles θ (0° ≤ θ ≤ 180°), f_V = 5.03 Vol.-% (12 spheres), n_{Sphere} = 1.59 + 0.01i, Monte Carlo method (red) and Maxwell method (blue).

The goal of this study was to investigate the differences between radiative transfer theory and Maxwell theory considering polarization as well as absorbing scatterers. The complete Müller matrix was calculated for several configurations.

2. Material and Methods

A cubic simulation volume (10 µm x 10 µm x 10 µm) containing different numbers (N = 12, 24, 48 and 96) of scattering spheres (radius: r = 1 µm) was modelled corresponding to different scatterer concentrations (f_V = 5.03 Vol.-%, 10.05 Vol.-%, 20.11 Vol.-% and 40.21 Vol.-%). The refractive index of the infinite outer medium was n_m = 1.33 while the refractive index of the scatterers was varied (n_{Sphere} = 1.59+0.00i; 1.59+0.01i and 1.59+1.00i). The simulation volume was irradiated by a plane electromagnetic wave with vacuum wavelength λ = 600 nm.

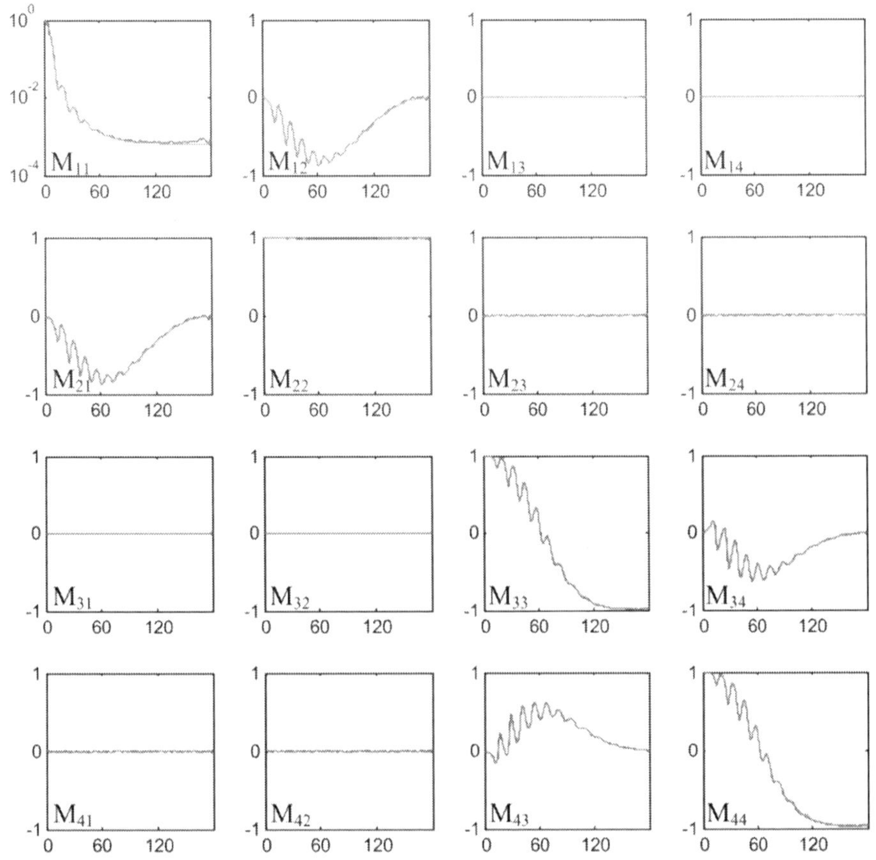

Figure 2. Angularly resolved Müller matrix elements for scattering angles θ (0° ≤ θ ≤ 180°), f_V = 5.03 Vol.-% (12 spheres), n_{Sphere} = 1.59 + 1.00i, Monte Carlo method (red) and Maxwell method (blue).

The angular dependent scattering properties (all 16 Müller matrix elements) of the cuboidal simulation volume were calculated by two methods based on the radiative transfer theory and the Maxwell theory, respectively.

The first method consists of a combination of the radiative transfer theory and Mie theory and was solved by a Monte Carlo (MC) program, which was developed by us for this purpose [3-12]. The single sphere Mie solution [13] was used as phase function for the Monte Carlo program. The scattering coefficient μ_s and absorption coefficient μ_a are needed as input for the Monte Carlo program. They were calculated by the Mie formalism as well and depend on the volume fraction of the spheres and their complex refractive index.

An assumed volume fraction of the spheres dispersed in the medium of 5.03 % (12 spheres) with n_{Sphere} = 1.59+0.01i (n_{Sphere} = 1.59+1.00i) led to $\mu_s = 96.18\ mm^{-1}$ and $\mu_a = 12.39\ mm^{-1}$ ($\mu_s = 47.63\ mm^{-1}$ and $\mu_a = 38.62\ mm^{-1}$). Table 1 shows the calculated absorption and scattering coefficients for all cases.

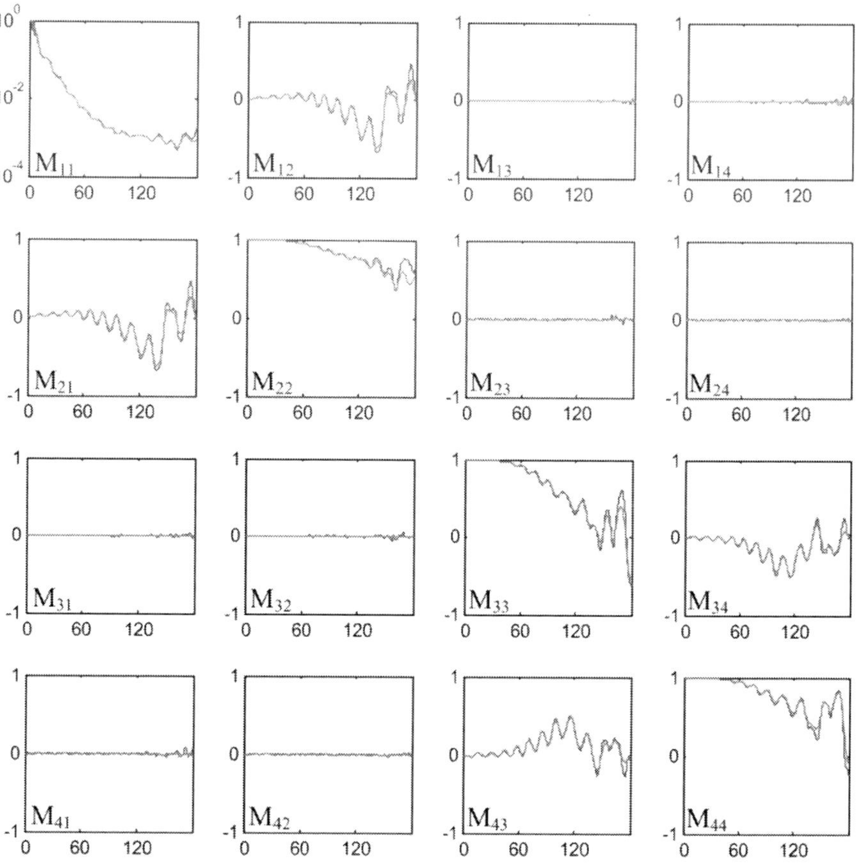

Figure 3. Angularly resolved Müller matrix elements for scattering angles θ (0° ≤ θ ≤ 180°), f_V = 10.05 Vol.-% (24 spheres), n_{Sphere} = 1.59 + 0.01i, Monte Carlo method (red) and Maxwell method (blue).

Table 1. Absorption and scattering coefficients used as input for the Monte Carlo program.

Number of spheres	Volume concentration	n_{Sphere}	μ_a [mm^{-1}]	μ_s [mm^{-1}]
12	5.03 %	1.59 + 0.01i	12.39	96.18
12	5.03 %	1.59 + 1.00i	38.62	47.63
24	10.05 %	1.59 + 0.01i	24.77	192.36
24	10.05 %	1.59 + 1.00i	77.24	95.26
48	20.11 %	1.59 + 0.01i	49.54	384.72
48	20.11 %	1.59 + 1.00i	154.47	190.52
96	40.21 %	1.59 + 0.01i	99.09	769.43
96	40.21 %	1.59 + 1.00i	308.94	381.04

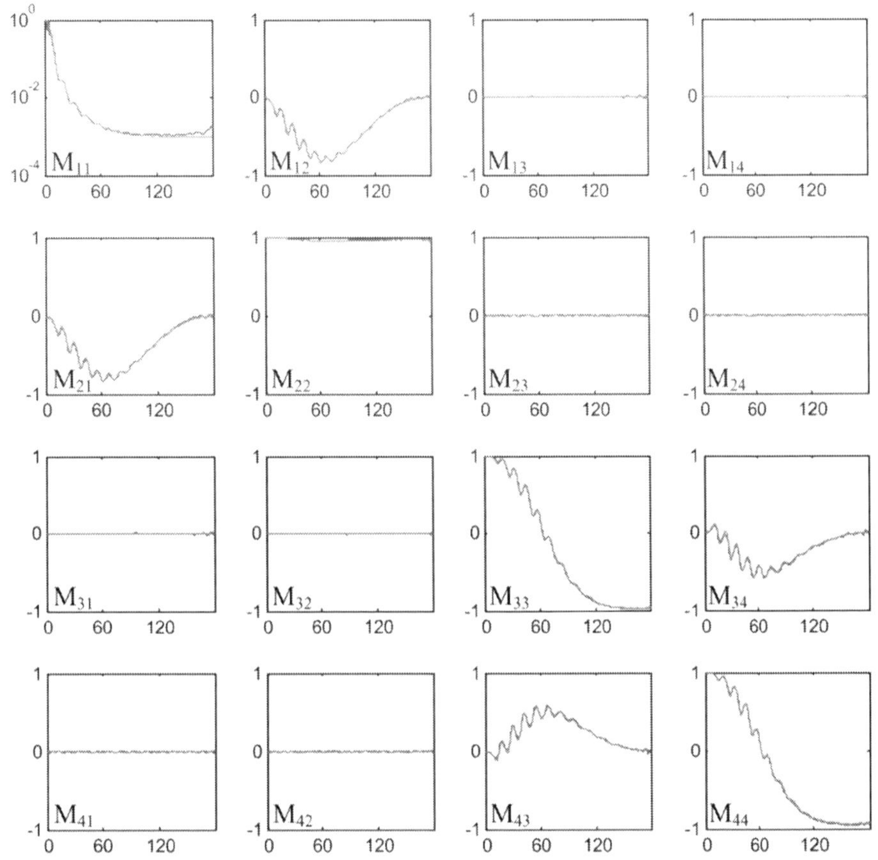

Figure 4. Angularly resolved Müller matrix elements for scattering angles θ (0° ≤ θ ≤ 180°), f_V = 10.05 Vol.-% (24 spheres), n_{Sphere} = 1.59 + 1.00i, Monte Carlo method (red) and Maxwell method (blue).

Using these input data, the light scattering problem was solved by the Monte Carlo method and registration of the scattering angle of each photon resulted in the angularly resolved Müller matrix after averaging over the azimuthal φangle.

The second method is provided by a solution of the Maxwell theory (Generalized Multisphere Mie) [14]. Different volume concentrations were modelled by spatial distributions of different numbers of spheres (N = 12, 24, 48 or 96, as in the first method) in a cubic simulation volume where no overlapping of the spheres was allowed. This model was used to calculate the angularly resolved Müller matrix using the same parameters as above (simulation volume: 10 µm x 10 µm x 10 µm, r = 1 µm, n_m = 1.33, n_{Sphere} = 1.59+0.00i, 1.59+0.01i and 1.59+1.00i, λ = 600 nm). The results were averaged over 10 randomly generated spatial distributions of the spheres in the simulation volume leading to a reduction of interference speckles.

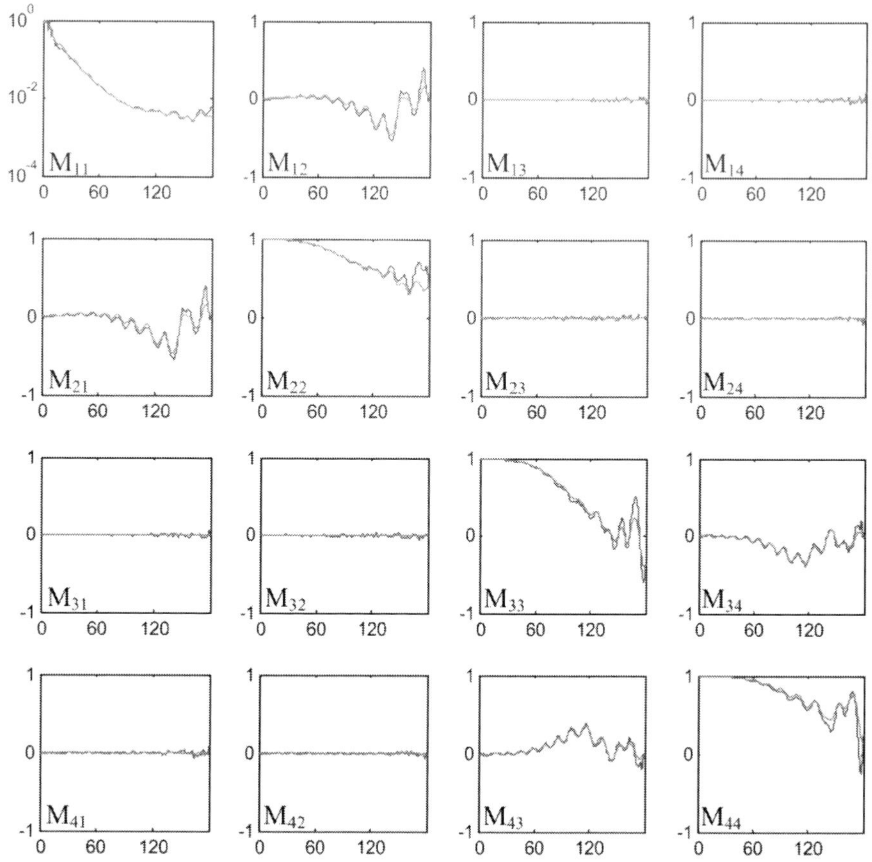

Figure 5. Angularly resolved Müller matrix elements for scattering angles θ (0° ≤ θ ≤ 180°), f_V = 20.11 Vol.-% (48 spheres), n_{Sphere} = 1.59 + 0.01i, Monte Carlo method (red) and Maxwell method (blue).

For both methods, the normalized form of the Müller matrix was used for representation of the calculated scattering characteristics. Therefore, all elements except M_{11} itself were normalized to M_{11} for each scattering angle. The M_{11} element was normalized to forward direction values of the Monte Carlo solution in both cases. This enabled a direct comparison of the angular variations as well as the absolute values of the scattering characteristics between the two approaches.

3. Results and discussion

Radiative transfer theory (Monte Carlo solution) as well as Maxwell theory have been applied to calculate all Müller matrix elements for comparison.

Figures 1-8 show the angular dependent 16 Müller matrix elements versus the scattering angle θ in the range from 0 to 180 degrees for different scatterer concentrations and for different imaginary parts of the scatterer refractive index. The results of the Monte Carlo simulations as well as of the Maxwell solutions are shown in each figure.

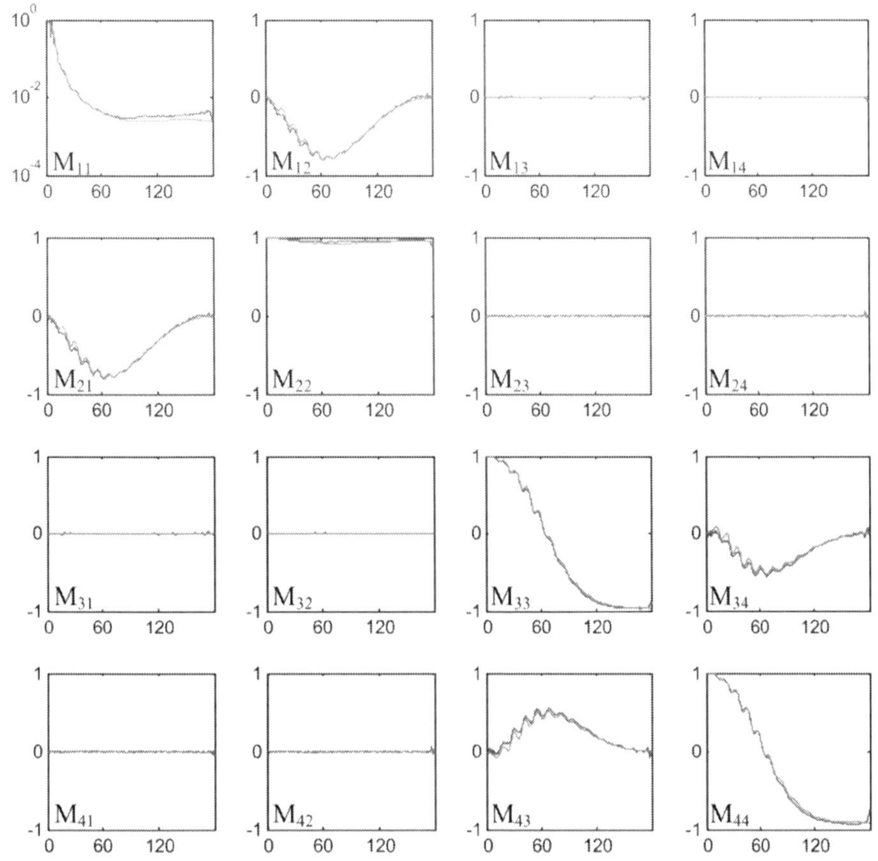

Figure 6. Angularly resolved Müller matrix elements for scattering angles θ (0° ≤ θ ≤ 180°), f_V = 20.11 Vol.-% (48 spheres), n_{Sphere} = 1.59 + 1.00i, Monte Carlo method (red) and Maxwell method (blue).

The results for n_{Sphere} = 1.59 + 0.00i, which are not given here, show a similar behaviour as the results for n_{Sphere} = 1.59 + 0.01i (figures 1, 3, 5 and 7) for all concentrations and all Müller matrix elements, whereas the quantitative values of the intensity (M_{11} element) for n_{Sphere} = 1.59 + 0.00i exceed the values for n_{Sphere} = 1.59 + 0.01i.

The overall shape of the Müller matrix remains roughly similar for all concentrations for n_{Sphere} = 1.59 + 0.01i (figures 1, 3, 5 and 7) as well as for n_{Sphere} = 1.59 + 1.00i (figures 2, 4, 6 and 8). However, large differences in Müller matrix structure can be observed between the two different refractive indices. In case of a small imaginary part of the refractive index, a strong dependence on the scattering angle can be observed while in the backscattering region the Müller matrix approximates the properties of a half-wave plate in case of a high imaginary part of the refractive index. This means, that the matrix approaches a diagonal matrix with M_{22} = 1, M_{33} = -1 and M_{44} = -1 for these angles.

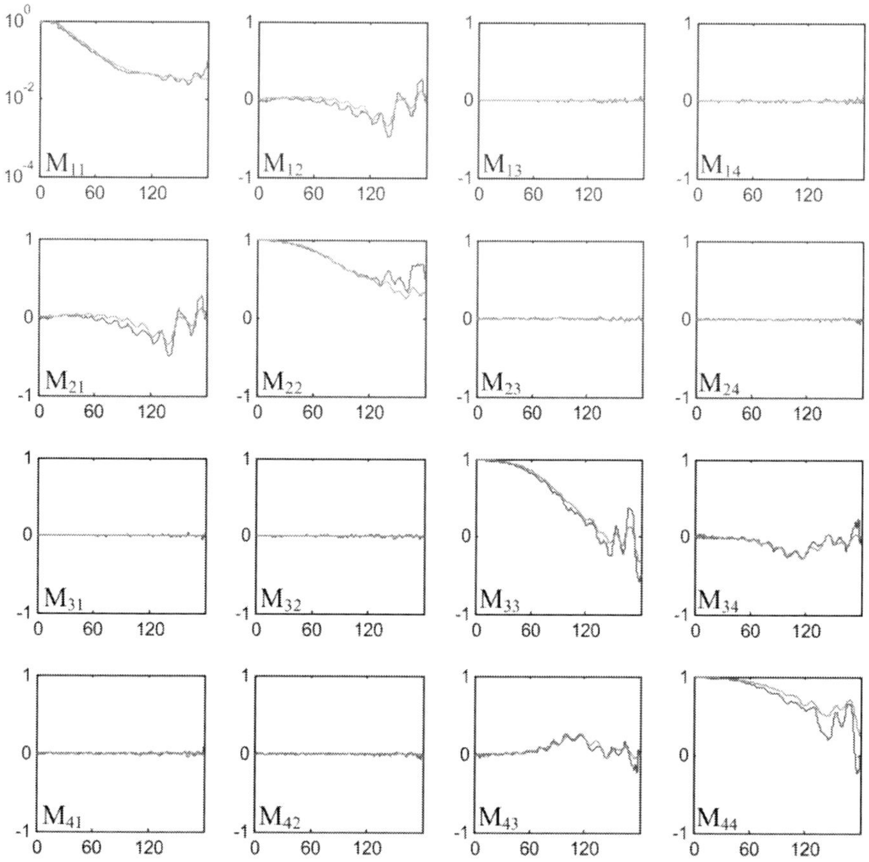

Figure 7. Angularly resolved Müller matrix elements for scattering angles θ (0° ≤ θ ≤ 180°), f_V = 40.21 Vol.-% (96 spheres), n_{Sphere} = 1.59 + 0.01i, Monte Carlo method (red) and Maxwell method (blue).

In case of small concentrations more oscillations over the scattering angle θ can be observed for both refractive indices. These oscillations result from the single Mie solution and decrease with increasing scatterer concentrations. This effect is visible for both solution methods and is mainly attributed to multiple scattering.

The differences between radiative transfer theory and Maxwell theory which are visible especially for the M_{11} element in the low angle range (θ ≤ 10°) are due to forward interferences and result from the limited simulation volume. This effect was already described for unpolarized [15] and polarized light [3] for non-absorbing scatterers and was now observed for absorbing scatterers (non-zero imaginary part of the index of refraction) here. The peak near 180 degrees (visible in the M_{11} element) results from coherent backscattering [16]. Dependent scattering effects cannot be described by the classical radiative transfer theory and are thus a further source of differences to the exact Maxwell theory as was already described in literature [17, 18]. These differences are mostly visible for high scatterer concentrations.

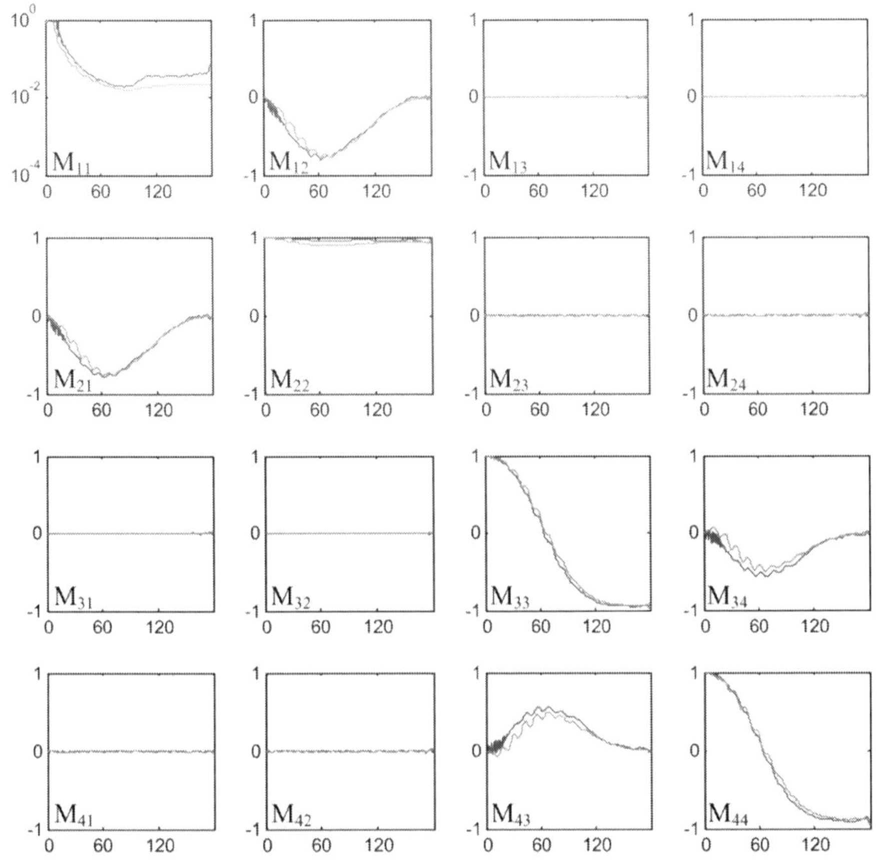

Figure 8. Angularly resolved Müller matrix elements for scattering angles θ (0° ≤ θ ≤ 180°), f_V = 40.21 Vol.-% (96 spheres), n_{Sphere} = 1.59 + 1.00i, Monte Carlo method (red) and Maxwell method (blue).

An interesting behaviour of the M_{11} element can be seen for angles θ >≈ 90° in case of n_{Sphere} = 1.59 + 1.00i which becomes more pronounced if high scatterer concentrations are taken into account: the Maxwell solution of the M_{11} element strongly deviates from the Monte Carlo solution in this angular range. This effect is not visible for a small imaginary part of the refractive index of the spheres.

A good agreement was observed between the two theories for all Müller matrix elements, especially for small concentrations. An increasing absorption of the scattering spheres showed larger differences in intensity results between the two methods. Also, increasing scatterer concentrations within the medium cause larger errors, due to dependent scattering. That means, that the results of radiative transfer theory have to be treated with care for higher scatterer concentrations. It was also found, that the differences between the two theories depend not only on the concentration and index of refraction, but also on the polarization (individual Müller matrix elements).

4. Conclusions

When polarized light propagation has to be simulated using radiative transfer theory, one has to pay attention to possible sources of errors if high scatterer concentrations or strongly absorbing particles are to be considered. A variation of the scatterers imaginary part of the refractive index between 0.01i and 1.00i showed a very different behaviour of the Müller matrix elements.

By the presented method, differences between the two theories can be investigated for arbitrary particle size parameters (spheres) and optical properties of the scatterers.

5. References

[1] Ghosh N and Vitkin I 2011 Tissue polarimetry: Concepts, challenges, applications, and outlook *J. Biomed. Opt.* **16** 110801
[2] Tuchin V, Wang L and Zimnyakov D 2006 *Optical polarization in biomedical applications Vol. 1* (Berlin: Springer)
[3] Hohmann A, Voit F, Schäfer J and Kienle A 2011 *Proceedings of SPIE 8090* (Bellingham, Wash.: SPIE) 80900H-1
[4] Bartel S and Hielscher A 2000 Monte Carlo simulations of the diffuse backscattering Mueller matrix for highly scattering media *Applied Optics* **39** 1580
[5] Rakovic M, Kattawar G, Mehrübeoglu M, Cameron B, Wang L, Rastegar S and Cote G 1999 Light backscattering polarization patterns from turbid media: Theory and experiment *Applied Optics* **38** 3399
[6] Côté D and Vitkin A 2005 Robust concentration determination of optically active molecules in turbid media with validated three-dimensional polarization sensitive Monte Carlo calculations *Optics Express* **13** 148
[7] Guo X, Wood M and Vitkin A 2008 A Monte Carlo study of penetration depth and sampling volume of polarized light in turbid media *Optics Communications* **281** 380
[8] Kattawar G and Adams C 1989 Stokes vector calculations of the submarine light field in an atmosphere-ocean with scattering according to a Rayleigh phase matrix: Effect of interface refractive index on radiance and polarization *Limnology and Oceanography* **34** 1453
[9] Maksimova I, Romanov S and Izotova V 2002 The effect of multiple scattering in disperse media on polarization characteristics of scattered light *Optics and Spectroscopy* **92** 915
[10] Ramella-Roman J, Prahl S and Jacques S 2005 Three Monte Carlo programs of polarized light transport into scattering media: Part I *Optics Express* **13** 4420
[11] Wang X and Wang L 2002 Propagation of polarized light in birefringent turbid media: A Monte Carlo study *Journal of Biomedical Optics* **7** 279
[12] Akarçay HG and Rička J 2011 *Proceedings of SPIE 8088* (Bellingham, Wash.: SPIE) 80880K-1
[13] Bohren C and Huffman D 1998 *Absorption and scattering of light by small particles* (New York: Wiley-Interscience)
[14] Xu Y and Gustafson B 2003 *An analytical solution to electromagnetic multisphere-scattering. Technical Report*, Department of Astronomy, University of Florida, available at: http://www.scattport.org/files/xu/codes.htm
[15] Voit F, Schäfer J and Kienle A 2009 Light scattering by multiple spheres: Comparison between Maxwell theory and radiative-transfer-theory calculations *Optics Letters* **34** 2593
[16] Mishchenko M, Travis L and Lacis A 2006 *Multiple scattering of light by particles: Radiative transfer and coherent backscattering* (Cambridge: Cambridge Univ. Press)
[17] Brewster MQ and Tien CL 1982 Radiative Transfer in Packed Fluidized Beds: Dependent Versus Independent Scattering *Journal of Heat Transfer* **104** 573
[18] Kumar S and Tien CL 1990 Dependent Absorption and Extinction of Radiation by Small Particles *Journal of Heat Transfer* **112** 178

Application of the Spectral Line-based Weighted-Sum-of-Gray-Gases model (SLWSGG) to the calculation of radiative heat transfer in steel reheating furnaces firing on low heating value gases

P D Nguyen[1], A Danda[1], M Embouazza[2], M Gazdallah[3], P Evrard[3] and V Feldheim[3]

[1] ArcelorMittal Global Research & Development, Maizières Process, Voie Romaine, BP 30320, F-57283 Maizières-lès-Metz Cedex, France. Phone: +33 (0)3 87 70 40 36, Fax: +33 (0)3 87 70 41 04

[2] ArcelorMittal Lorraine, Coke Plant of Serémange - 57191 Florange, France. Phone: +33 (0)3 82 51 47 62, Fax: +33 (0)3 82 51 44 40

[3] Université de Mons, Service de Thermique et Combustion, Rue de l'Espagne 56, 7000 Mons, Belgium. Phone: +32 (0)6 537 4461

Corresponding author's e-mail address: phucdanh.nguyen@arcelormittal.com

Abstract. The Spectral Line-based Weighted-Sum-of-Gray-Gases (SLWSGG) model is applied to calculate the gaseous radiative properties of the aero- or oxy-combustion products of low heating value gases issued from steel making process such as Blast Furnace Gas (BFG) as well as of high heating value gases such as Coke Oven Gas (COG) and conventional Natural Gas (NG). The comparison of total emissivities shows that the 3-gray-gases SLWSGG model is in very good agreement with the Hottel and Sarofim's database. The 3-gray-gases SLWSGG model is then integrated into AnsysFluent® Discrete Ordinates method under User Defined Function and CFD simulations are performed using these combined models. The simulations are done, with full combustion-radiation coupling, for steel reheating furnaces firing on three types of gases: BFG, COG and NG. The results are compared with the simulations realized with the 1-gray-gas WSGG model available in AnsysFluent®. The comparison shows that the 1-gray-gas WSGG model highly overestimates the steel discharging temperature as compared to the 3-gray-gases SLWSGG model. Significant temperature differences are observed between the two radiative models, i.e. 116°C, 55°C and 67°C for the BFG, COG and NG cases, respectively. It can be concluded that the 3-gray-gases SLWSGG model should be used to calculate the radiation heat transfer in large industrial furnaces with more accuracy not only for low heating value gases such as BFG but also for high heating value gases such as COG and NG.

1. Introduction

Using auto-produced low heating value gases issued from steel making process in steel reheating process has a very positive impact on the environment and the energy saving. Integrated steel plants CO_2 footprint and NOx pollutants will be reduced and natural gas (NG) will be substituted by industrial off-gases such as Blast Furnace Gas (BFG), a reusable low heating value gas. Designing combustion systems for efficient utilization of BFG in steel reheating furnaces is still a challenging task [1]. The Computational Fluid Dynamics (CFD) plays an important role in the design phase. As in large industrial furnaces, radiative heat transfer stands for nearly 90% of the total heat transfer into steel products to be reheated, the remaining 10% is attributed to the convection. Therefore, calculating the radiative heat transfer with accuracy is essential in succeeding in the design of such combustion systems.

In the literature, much more effort was focused on method developments and applications for NG combustion. Amongst different methods available, the non-gray global model family has an important interest for practical combustion system calculations for two reasons. It is more accurate when compared to the simple gray gas model and it is more computationally efficient when compared to the Statistical Narrow Band Correlated K (SNB-CK) model [2]. The non-gray global methods consist of the Spectral Line-based Weighted-Sum-of-Gray-Gases (SLWSGG) model [3], the Absorption Distribution Function (ADF) model [4], and the more recently developed Statistical Narrow Band Full Spectrum Correlated K (SNB-FSCK) [5]. The SLWSGG model has a special interest for practical coupling with commercial CFD combustion solvers as it is an extension of the WSGG model which is largely and successfully used in Ansys Fluent® [6]. However, this model is limited to NG combustion application and no gaseous radiative properties model is available to address low heating value gases combustion. In the present study and along this line, the SLWSGG model is applied to calculate the gaseous radiative properties of the aero- or oxy-combustion products of low heating value gases issued from steel making process such as Blast Furnace Gas (BFG) as well as of high heating value gases such as Coke Oven Gas (COG) and conventional Natural Gas (NG).
In a first attempt, the quality of the 3-gray-gases SLWSGG model will be evaluated by comparing the total emissivity with the Hottel and Sarofim's database [7] as well as with the results using Soufiani and Djavdan [8] and Smith et al. [9] methods. To keep the computational efforts reasonable the SLWSGG model takes into account only 3-gray-gases. It is then integrated into Ansys Fluent® Discrete Ordinates (DO) method as User Defined Function.to perform Computational Fluid Dynamics (CFD) simulations. A full combustion-radiation coupling in steel reheating furnaces firing on three types of gases: BFG, COG and NG was simulated. To reduce the computational cost, only half of the furnaces is considered in the simulations. The results will be compared with the simulations realized with the 1-gray gas WSGG model available in AnsysFluent®.

The section below describes briefly the methodology of SLWSGG model. In the next section, the results will be presented. Finally, some concluding remarks will be drawn out in the concluding section.

2. Methodology

The concept of a Weighted-Sum-of-Gray-Gases (WSGG) model was first presented by Hottel and Sarofim [7]. In this concept the non-gray gas is replaced by a number of gray gases, for which the heat transfer rates are calculated independently. The total heat flux is then calculated by adding the heat fluxes of each gray gas. For theoretical background, the reader is referred to the comprehensive books on radiative heat transfer [5,10,11]. For latest updates on the development of the Spectral Line-based Weighted-Sum-of-Gray-Gases (SLWSGG) model, the references [3,12-17] are recommended. A brief description is recalled in the following. In the SLWSGG concept, the Radiative Transfer Equation (RTE) to be solved is:

$$\frac{dI_n}{ds} = k_n p_a \left(a_n I_b - I_n \right) \tag{1}$$

where I_n (W/m) is the radiation intensity along the direction s of the kth gas, I_b (W/m) the blackbody radiation intensity, k_n (m^{-1}.atm^{-1}) the absorption coefficient of the kth gas, p_a(atm) the partial pressure of the absorbing species, a_n the weighting factor of the kth gray gas with temperature dependence and T_w (K) is wall temperature. Equation (1) is solved for each gray gas $n = 1, 2, ...N$, with the boundary condition given by $s = 0 : I_n = a_n I_b(T_w)$, using any standard solution method, but Discrete Ordinates Method (DOM) is commonly used. The total intensity field (or the total heat fluxes) will be determined by adding the intensities (heat fluxes) of all grey gases concerned: $I(s) = \sum_{n=1}^{N} I_n(s)$.

In the method of SLWSGG, the total emissivity of a gaseous mixture is written as:

$$\varepsilon(T, l, p_a) = \sum_{n=1}^{N} a_n \left(1 - e^{-k_n p_a l}\right) \tag{2}$$

where T (K) is the temperature, l (m) the optical thickness. The emissivity is the contribution of all gray gases. Each one is characterised by an absorption coefficient and a weighting factor commonly described by temperature dependent polynomials of J degree written in the form:

$$a_n = \sum_{j=1}^{J} \alpha_{n,j} T^{j-1} \tag{3}$$

$$k_n = \sum_{j=1}^{J} \beta_{n,j} T^{j-1} \tag{4}$$

where $\alpha_{n,j}$ is the polynomial coefficient for the weighting factors and $\beta_{n,j}$ is the polynomial coefficient for the absorption coefficient. For a transparent region of the spectrum (clear gas), the absorption coefficient is set to zero in order to account for windows in the spectrum between spectral regions of high absorptions.

In the present work, we have developed a dedicated code and used SLWSGG model to calculate the coefficients $\alpha_{n,j}$ and $\beta_{n,j}$ of the aero- or oxy-combustion products of low heating value gases issued from steel making process such as Blast Furnace Gas (BFG) as well as of high heating value gases such as Coke Oven Gas (COG) and conventional Natural Gas (NG). The main steps for determination of the polynomial coefficients in the code are: 1) for a given molar fraction of H_2O, CO_2, CO, a range of optical length and a fixed temperature, the SLWSGG model is used to calculate the total emissivity, 2) a least square method is used to calculate the absorption and weight coefficients for a fixed gray gas number to meet emissivity calculated above, 3) steps 1 and 2 are repeated for a set of N temperatures points ranging from 400K to 2400 K, and 4) the profiles of absorption and weight coefficients versus temperature obtained are used to calculate the polynomial coefficients. The parameters in the SLWSGG model are obtained from correlation of the line-by-line spectra of H_2O, CO_2 and CO. It should be additionally noted that our code can use Hottel and Sarofim's database [7] or total emissivities evaluated by SLWSGG model to calculate the gray gas weights and absorption coefficients in the form of polynomials laws. To keep the computational efforts reasonable, only 3-gray-gases were considered.

3. Results and discussion

3.1. Emissivity of the 3-gray-gases SLWSGG model

The accuracy of the 3-gray-gases SLWSGG model was tested by comparing the total emissivity with the Hottel and Sarofim's database [7], used as a benchmark, as well as with the results using Soufiani and Djavdan [8] and Smith et al. [9] methods. In the comparison shown in figures 1 and 2 for a given burnt gases mixture (H_2O and CO_2) issued from the aero-combustion of the different gases such as BFG and COG (similar result obtained for NG case but not shown here). The following gaseous compositions were used in the study (in volume percentage); BFG: 3%H_2, 20%CO, 22%CO_2 and

55%N_2, COG: 62%H_2, 25%CH_4, 6%CO, 2%CO_2, 2%C_2H_4, 1%C_2H_6 and 2%N_2, and NG: 91%CH_4, 6%C_2H_6 and 3%N_2.

In the Soufiani and Djavdan model [8], p_a is the partial pressure of H_2O, the absorption coefficient values are expressed in atm^{-1}.m^{-1}, and only one configuration $p_w / p_c = 2$ is available (where w = H_2O, c = CO_2). In the Smith et al. model [9], p_a is the sum of the partial pressures of H_2O and CO_2, i.e., $p_a = p_{H_2O} + p_{CO_2}$, and the absorption coefficient values are expressed in atm^{-1}.m^{-1}. Two configurations are available for this model; the configuration $p_w / p_c = 2$ is available for NG and COG, and the configuration $p_w \to 0\ atm$ for BFG.

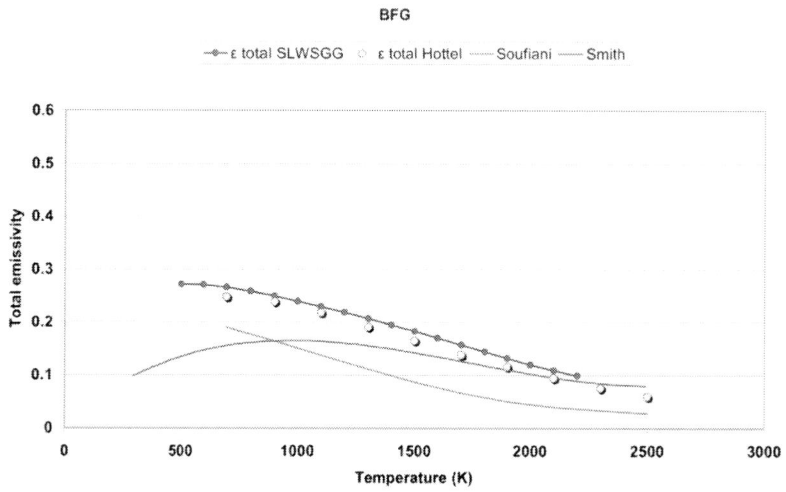

Figure 1. Total emissivity of the products from the aero-combustion of BFG.

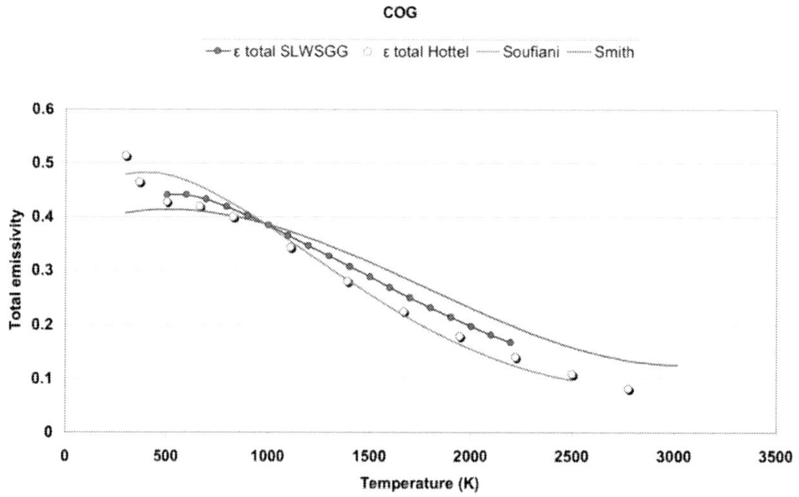

Figure 2. Total emissivity of the products from the aero-combustion of COG.

In the case of BFG (figure 1), the 3-gray-gases SLWSGG model is in very good agreement with the Hottel and Sarofim's database [7], while the total emissivity obtained with Soufiani and Djavdan [8] and Smith et al. [9] methods shows an important departure from the Hottel and Sarofim's database [7]. This is due to the fact that the 3-gray-gases SLWSGG model is specifically adapted for low heating value gas combustion such as BFG, while the models used in Soufiani and Djavdan [8] and Smith et al. [9] were developed for high heating value gases. Then in the case of high heating value gas COG (figure 2), no much difference between the methods is observed in the total emissivity. However, it should be noted that the SLWSGG model always offers an advantage over the WSGG model to take into account many gray gases to calculate radiation with more accuracy.

3.2. CFD computations

The 3-gray-gases SLWSGG model was integrated into Ansys Fluent® Discrete Ordinates (DO) method as a User Defined Function and Computational Fluid Dynamics (CFD) simulations were performed for a full combustion-radiation coupling in a steel reheating furnaces firing on three different types of gases: BFG, COG and NG. The same burners characteristics are used for NG and COG cases while in the case of BFG, specific dimensional characteristics burners are used. To reduce the computational cost, only half of furnace is considered in the simulations. The overall characteristics of the half of furnace are: 37m in length, 12.8m in width and 1.79m in height. The steel slabs to be processed are of 10.5m in length and 0.22m in thickness and move through the furnaces with a velocity of 0.003241m/s which corresponds to a throughput of 212ton/h. The slabs are charged onto the furnace at 27°C in all simulations. Figure 3 shows an overview of a half of a slab reheating furnace using BFG burners.

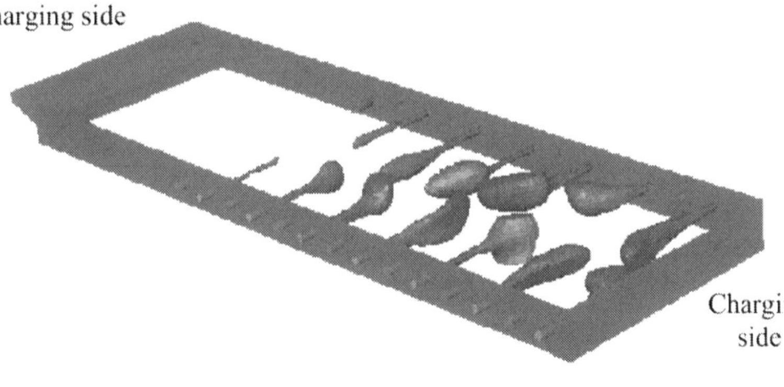

Figure 3. Overview of a half of a steel slab reheating furnace using BFG burners.

The simulations are realized for 3 gases: NG, COG and BFG. For each gas, two radiative models are used for comparison: the 3-gray-gases SLWSGG model and the 1-gray-gas WSGG model. The power is kept the same for all simulations and is of 33MW for the present study. In the case of NG and COG, the combustion air is preheated to 890°C and the gas is injected at 25°C. But in the case of lean gas BFG, both the air and gas are preheated up to high temperatures of 860°C and 730°C, respectively.

Figure 4 shows the results of the simulations obtained with the 3-gray-gases SLWSGG model for the three gases considered, i.e., NG, COG, and BFG. It is clearly noted that the highest maximum burnt gases temperature (1800°C) is observed in the COG case (Figure 4.b) due to a very high proportion of H_2 (62% in volume) in the COG composition. The lowest maximum burnt gases temperature (1537°C) is observed in the BFG case (Figure 4.c) even though the BFG is preheated to a high temperature of 730°C before injection. This low burnt gases temperature is due to the low calorific value of BFG; the gas contains about 20%CO and 3%H_2 (in volume) that are considered as reactives in its composition and the major components (22%CO_2 and 55%N_2) are inert gases. The maximum burnt gases temperature value of the NG case lies between the values obtained in the COG and the BFG cases as expected, and takes a value of 1773°C (Figure 4.a). It is generally observed, in three cases, that the temperature of the burnt gases issued from the burners located near the charging side is lower than that of the burnt gases at the discharging side. This is due to the fact that the steel slabs are loaded into the furnaces at low temperature (27°C) from the charging side, therefore an important heat flux transfers from the burnt gases to the steel slabs, and this important heat transfer decreases the burnt gases temperature at the burners located near the charging door.

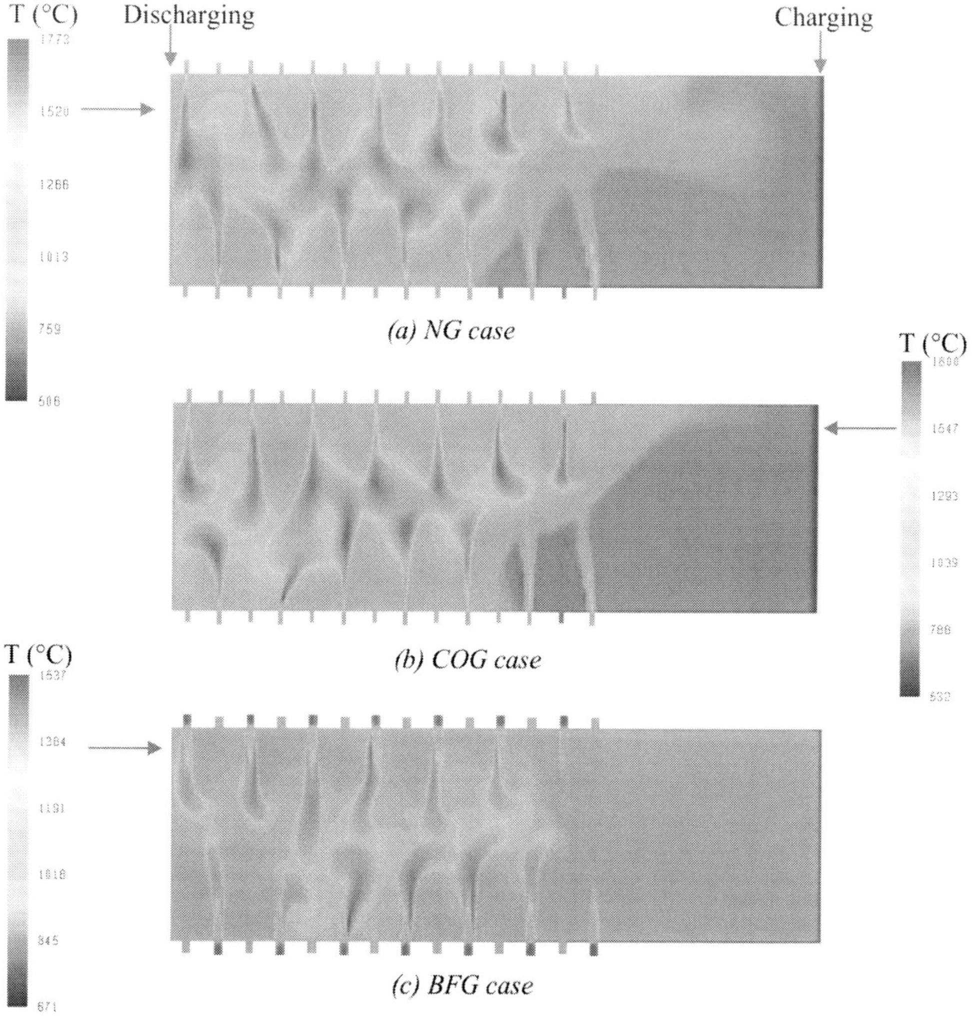

Figure 4. Burnt gases temperature in the furnaces, obtained with the 3-gray-gases SLWSGG model, for NG (a), COG (b) and BFG (c) cases.

Figures 5, 6 and 7 show the temperature contour on the upper skin of slaps from the charging to the discharging door, obtained with the 1-gray-gas WSGG and 3-gray-gases SLWSGG models, for three cases considered. It is important to recall that for comparison between the two radiation models, the power is kept constant (33MW) and the slabs are charged onto the furnace at 27°C for all simulations. It is systematically noted, from figures 5, 6 and 7 for each case of gas, that the temperature of the slab at the exit of the furnace is always lower with the 3-gray-gases SLWSGG model as compared to the 1-gray-gas WSGG model. Important mean discharged temperature differences of 67°C and 55°C are noticed for the NG and COG cases, respectively. For the BFG case, the discrepancy between the 1-gray-gas WSGG and the 3-gray-gases SLWSGG is more important and attained a value of 116°C. This difference is confirmed in figure 8 that shows the core temperature profile of the discharged slab, obtained with the two radiation models, for the BFG case. It is obviously noted, from figure 8, that the 1-gray-gas WSGG model over-predicts of more than 100°C as compared to the 3-gray-gases SLWSGG model along the length of slab.

(a) 1-gray-gas WSGG model

(b) 3-gray-gases SLWSGG model

Figure 5. Temperature contour on the upper skin of slaps from the charge to the discharge, obtained with the two radiation models, for NG case.

(a) 1-gray-gas WSGG model

(b) 3-gray-gases SLWSGG model

Figure 6. Temperature contour on the upper skin of slaps from the charge to the discharge, obtained with the two radiation models, for COG case.

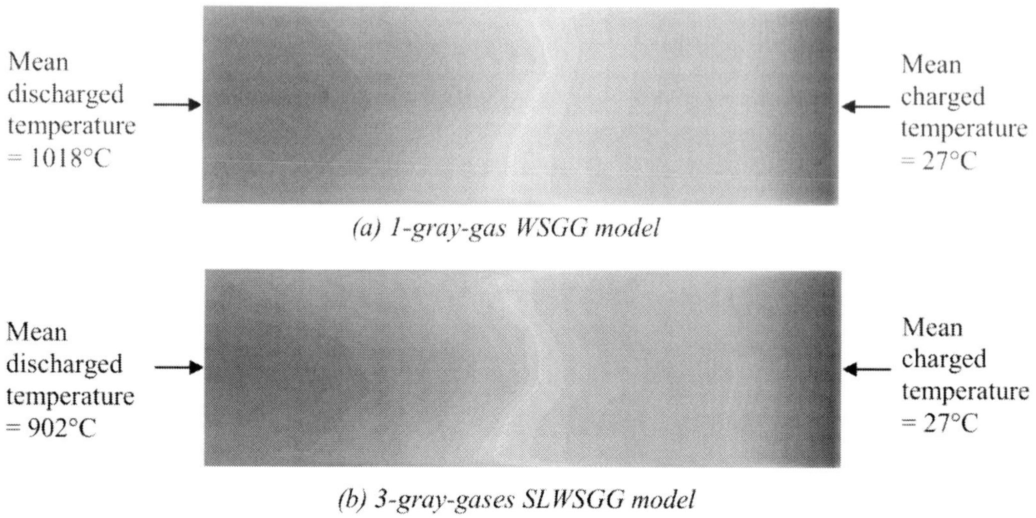

Figure 7. Temperature contour on the upper skin of slaps from the charge to the discharge, obtained with the two radiation models, for BFG case.

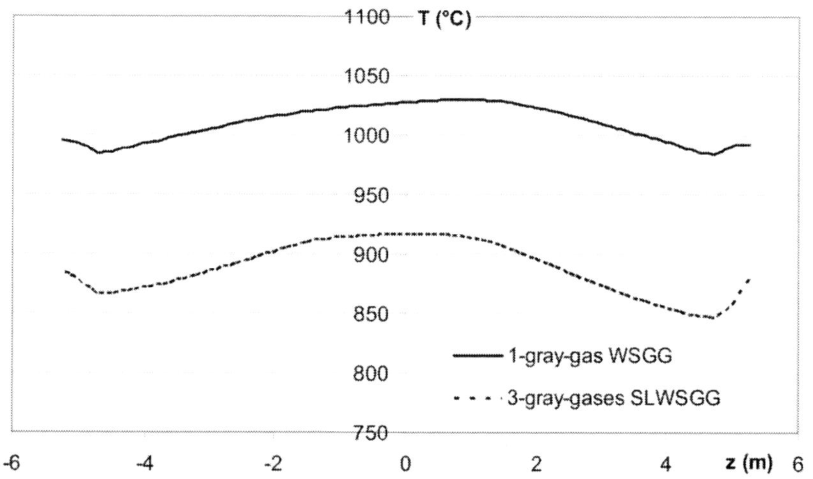

Figure 8. Discharged slab core temperature profile along the length of slab, obtained with the two radiation models, for BFG case.

In large-scale steel reheating furnaces, radiation dominates convection and stands for nearly 90% of the total heat transfer from hot gases to products to be processed. In this line, Table 1 shows the radiative heat transfer (in %) from burnt gases to steel slabs, obtained with the two radiative models, for three cases of gas. If we consider the 1-gray-gas WSGG model, it is noted that the radiation heat transfers are 92%, 91.8% and 90.2%, respectively for NG, COG and BFG cases (the remaining is of convection) and less than 2% of difference is noted between the rich gases (NG and COG) and the lean gas BFG. If we consider the 3-gray-gases SLWSGG model, we observe that the radiation heat transfers are 88.5%, 89.3% and 84.6%, respectively for NG, COG and BFG cases, and nearly 5% of difference is noted between the rich gases (NG and COG) and the lean gas BFG. This 5% in radiation difference reflects the authentic result obtained with the 3-gray-gases SLWSGG because the total emissivity of burnt gases issued from BFG combustion is largely smaller than that from rich gases such as NG and COG (comparing figure 1 with figure 2). That explains the lower heating quality of BFG as compared to the rich gases NG and COG.

Now, if we compare the two radiation model for each gas, it appears from Table 1 that the 1-gray-gas WSGG model overestimates the radiation heat transfer in comparison with the 3-gray-gases SLWSGG model. Differences of 3.5%, 2.5% and 5.6% are noted for the NG, COG and BFG cases, respectively.

Table 1. Radiation heat transfer proportion from burnt gases to steel slabs.

Model	NG	COG	BFG
1-gray-gas WSGG	92.0 %	91.8 %	90.2 %
3-gray-gases SLWSGG	88.5 %	89.3 %	84.6 %
Difference between two models	3.5 %	2.5 %	5.6 %

4. Conclusion

The Spectral Line-based Weighted-Sum-of-Gray-Gases (SLWSGG) model was applied to calculate the gaseous radiative properties of the aero- or oxy-combustion products of low heating value gases issued from steel making process such as Blast Furnace Gas (BFG) as well as of high heating value gases such as Coke Oven Gas (COG) and conventional Natural Gas (NG). The quality of the 3-gray-gases SLWSGG model was evaluated by comparing the total emissivity with the Hottel and Sarofim's database [7] as well as with the results using Soufiani and Djavdan [8] and Smith et al. [9] methods. The comparison shows, in the case of BFG, that the 3-gray-gases SLWSGG model is in very good agreement with the Hottel and Sarofim's database [7] while the total emissivity obtained with Soufiani and Djavdan [8] and Smith et al. [9] methods shows an important departure from the Hottel and Sarofim's database. This is due to the fact that the 3-gray-gases SLWSGG model was specifically adapted for low heating value gas BFG combustion, while the models used in Soufiani and Djavdan [8] and Smith et al. [9] were developed for high heating value gases. In the case of high heating value gases of COG and NG, no much difference between the methods is observed in the total emissivity. However, it should be noted that the SLWSGG model always offers an advantage over the WSGG model to take into account many gray gases to calculate radiation with more accuracy.

The simulations performed for steel slab reheating furnaces burning on BFG, COG and NG indicate that the 1-gray-gas WSGG model highly overestimates the steel discharging temperature in comparison with the 3-gray-gases SLWSGG model. Important mean discharged temperature differences are noted between the two radiation models, i.e. 116°C, 55°C and 67°C for the BFG, COG and NG cases, respectively. It can be concluded that the 3-gray-gases SLWSGG model should be used to calculate the radiation heat transfer in large industrial furnaces with more accuracy not only for low heating value gases such as BFG but also for high heating value gases such as COG and NG.

5. References

[1] Niska J, Rensgard A and Ekman T, Oxyfuel combustion of low calorific blast furnace gas for steel reheating furnaces, *Finnish-Swedish Flame Days 2009*, IFRF, Finland.

[2] Lacis A A and Oinas V., A description of the correlated-k distribution method for modelling nongray gaseous absorption, thermal emission, and multiple scattering in vertically inhomogeneous atmospheres. *Journal of Geophysical Research*, 96:9027-9063 (1991).

[3] Denison M K and Webb B W, A Spectral Line based Weighted-Sum-of-Gray-Gases Model for Arbitrary RTE Solvers. *Journal of Heat Transfer*, Vol. 115, 1993.

[4] Pierrot L, Rivière P, Soufiani A. and Taine J, A Fictitious-Gas-Based Absorption Distribution Function Global Model for Radiative Transfer in Hot Gases, *Journal of Quantitative Spectroscopy & Radiative Transfer*, 62:609-624 (1999).

[5] Modest M F, Radiative heat transfer, 2nd ed., New York: McGraw-Hill; 2003. ISBN: 0-12-503163-7.

[6] ANSYS FLUENT user's guide v13.0.

[7] Hottel H C and Sarofim A F, Radiative Transfer, McGraw-Hill, New York, 1967.

[8] Soufiani A and Djavdan E, A Comparison between Weighted Sum of Gray Gases and Statistical Narrow Band Radiation Models for Combustion Applications. *Combustion and Flame*, 97:240-250 (1994).

[9] Smith T F, Shen Z F and Friedman J N, Evaluation of Coefficients for the Weighted Sum of Gray Gases Model. *ASME*, vol. 104, pp. 602-608 (1982).

[10] Viskanta R, Radiative transfer in combustion systems: fundamentals and applications. New York: Begell House Inc.; 2005. ISBN: 1-56700-211-0.

[11] Siegel R and Howell J R, Thermal radiation heat transfer, 4th ed., Washington: Taylor & Francis; 2002. ISBN: 1-56032-839-8.

[12] Denison M K and Webb B W, The Spectral Line based Weighted-Sum-of-Gray-Gases Model for H2O/CO2 Mixtures. *ASME*, Vol 117, 1995.

[13] Denison M K and Webb B W, The Spectral Line based Weighted-Sum-of-Gray-Gases Model in Nonisothermal and Nonhomogeneous Media. *Journal of Heat Transfer*, Vol. 117, 1995.

[14] Solovjov V P and Webb B W, SLW modeling of radiative transfer in multicomponent gas mixtures. Journal of Quantitative Spectroscopy & Radiative Transfer, 65:655-672 (2000).

[15] Solovjov V P and Webb B W, An Efficient Method for Modeling Radiative Transfer in Multicomponent Gas Mixtures With Soot. *Journal of Heat Transfer*, Vol. 123, 2001.

[16] Solovjov V P, Lemonnier D and Webb B W, The SLW-1 model for efficient prediction of radiative transfer in high temperature gases. *Journal of Quantitative Spectroscopy & Radiative Transfer*, 2010, doi:10.1016/j.jqsrt.2010.08.009, in press.

[17] Solovjov V P and Webb B W, Global Spectral Methods in Gas Radiation: The Exact Limit of the SLW Model and Its Relationship to the ADF and FSK Methods. *Journal of Heat Transfer*, Vol. 133, 2011.

Laminar Smoke Point Based Subgrid Soot Radiation Modeling Applied to LES of Buoyant Turbulent Diffusion Flames

Prateep Chatterjee[1], John L de Ris[1], Yi Wang[1], Niveditha Krishnamoorthy[2] and Sergey B Dorofeev[1]

[1]Research Division, FM Global, 1151 Boston-Providence Turnpike, Norwood, MA 02062, USA
[2]Presently at CD-adapco, 1 Technology Drive, Irvine, CA 92618, USA

E-mail: `prateep.chatterjee@fmglobal.com`

Abstract. Large eddy simulations (LES) of gaseous buoyant turbulent flames have been conducted with the application of a flamelet based soot-radiation model. The subgrid model applies a turbulent eddy description of soot formation, oxidation and radiation and is based on the laminar smoke point concept. Two parameters, a local turbulent strain rate and prior enthalpy loss/gain fraction influence the soot formation and radiation. Radiation heat transfer is simulated by solving the finite volume discretized form of the radiative transfer equation (RTE) with the subgrid soot-radiation model implemented. The radiant heating of surfaces in close proximity of the flames is computed and predicted heat fluxes and surface temperatures are compared against experimental data. Fire growth in a rack storage arrangement is simulated with the application of a pyrolysis model. Computed heat release rate (HRR) is compared against experimental data.

1. Introduction

Thermal radiation, primarily due to soot, is responsible for fire spread in industrial scale rack storage scenarios. Flue spaces in rack storage setups are typically in the range of 0.15-0.6 m. Radiation in these spaces tends to be in the optically thin regime for moderately sooting pyrolysates from solid fuel surfaces. Accurate spatial and temporal prediction of radiant emission from the turbulent flame sheet therefore becomes the primary factor in predicting fire growth. This article presents results from CFD studies of (1) radiant heat transfer from a 60 kW propane fire to inert parallel panels [1], and (2) fire growth in an industrial-scale rack storage setup [2].

Studies by Markstein [3] have established that radiation from turbulent buoyant diffusion flames correlates with the fuel's laminar smoke point height. Soot formation, oxidation and radiation all scale with the laminar smoke point. Using this concept, a model for soot radiation was developed by Lautenberger et al [4]. CFD simulations with this model showed good agreement of computed results with laminar diffusion flame experimental data. In turbulent flames the Kolmogorov micro-time scale governs the processes of soot formation, oxidation and radiation. This Kolmogorov mixing occurring in small reaction volumes must be described.

A subgrid soot-radiation model based on the laminar smoke point model [5] is applied to compute radiant emission from buoyant turbulent diffusion flames. The subgrid soot-radiation model includes the effect of turbulence on the soot generation and oxidation rates. The model

Published under licence by IOP Publishing Ltd

requires the application of a characteristic turbulent micro-scale strain rate, a_0, and prior radiant loss/gain fraction, H_p, as parameters that determine soot release and radiation from the fire. The model has been coupled with an LES solver, FireFOAM [6] and tested against experimental data on radiation from optically thin buoyant turbulent diffusion flames from a range of fuel and fire sizes [5]. In the present study, the model provides an isotropic emission term from the turbulent flame sheet that is introduced in the radiative transfer equation (RTE) and consequently, radiant heat transfer from a propane fire to inert parallel panels and rack storage fire growth are simulated.

2. Numerical Model

In LES of diffusion flames the flame sheet is not resolved by the computational mesh (Fig. 1(a)). It is assumed that, in the grid cell where the flame sheet is present, the entrained air and fuel mix to a molecular scale where reaction takes place. This mixing is the result of dissipating eddies/vortex-structures (Fig. 1(b)). This assumes the presence of sheet-like laminar flamelets strained by the local turbulent eddies (Fig. 1(c)).

2.1. Flamelet Model

The flamelet model [5] comprises of a 1-D transient packet of fluid being strained by a constant strain rate normal to its axis. The model applies the laminar smoke point concept [4] to describe soot formation, oxidation and radiation. Soot is formed and oxidized in the immediate flame with the residual soot being released. The local mixture composition and prior ene5rgy loss by radiation are tracked by solving the conservation of mixture fraction and total enthalpy equations. The "radiatively perturbed laminar flamelet" concept [7] is used in which the temperature is computed as a function of mixture fraction and total enthalpy. The emitted radiation is assumed to be a function of fuel type (in particular, its laminar smoke point height ℓ_s), a prior local heat loss fraction (radiant and convective, H_p) and an imposed strain rate at the smallest dissipation scale (a_0). The characteristic strain rate at the dissipation scale is estimated as the inverse of the Kolmogorov micro-time, $a_0 = C_\eta/\tau_\kappa$, where C_η is a constant. A single transient flamelet is assumed to be of initial thickness η (the Kolmogorov length scale). This initial thickness is shown in Fig. 1(c).

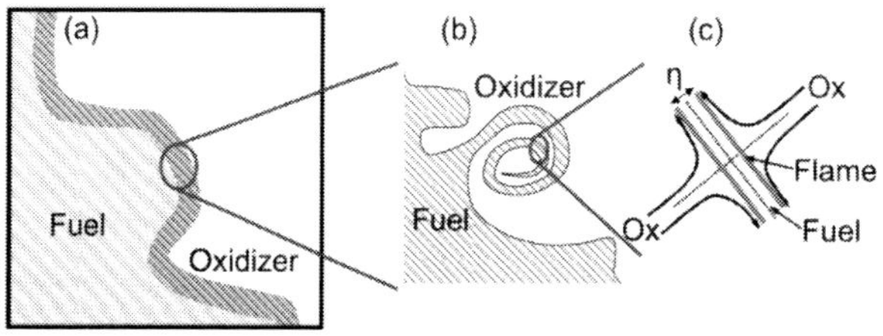

Figure 1. Subgrid model for soot-radiation: (a) turbulent diffusion flame sheet in a LES grid cell, (b) local straining of flame sheet by vortices, and (c) one-dimensional transient flamelet representation.

2.1.1. Flamelet Equations The original three-dimensional conservation equations follow from the approach of Lautenberger et al [4]. The soot conservation equation is

$$\frac{\partial \rho_g \phi_s}{\partial t} + \nabla \cdot (\rho_g \phi_s \mathbf{u}) = \nabla \cdot \left(0.55 \frac{Y_s \mu}{T} \nabla T\right) + \dot{\omega}_s'''. \tag{1}$$

Here, we have redefined the soot scalar variable as $\phi_s = Y_s/(1 - Y_s)$, where Y_s is the soot mass fraction. The source term in the soot conservation equation, $\dot{\omega}_s'''$, is computed as a sum of the soot formation and oxidation rates: $\dot{\omega}_s''' = \dot{\omega}_{sf}''' + \dot{\omega}_{so}'''$. Lautenberger et al [4] used experimental measurements to guide the selection of general shapes of the soot formation and oxidation functions. The soot formation rate, $\dot{\omega}_{sf}'''$, was assumed to be a parabolic function of mixture fraction and temperature and is evaluated as a product of $\dot{f}_{sf}'''(Z)$, and $g_{sf}(T)$: $\dot{\omega}_{sf}''' = \dot{f}_{sf}'''(Z)\, g_{sf}(T)$.

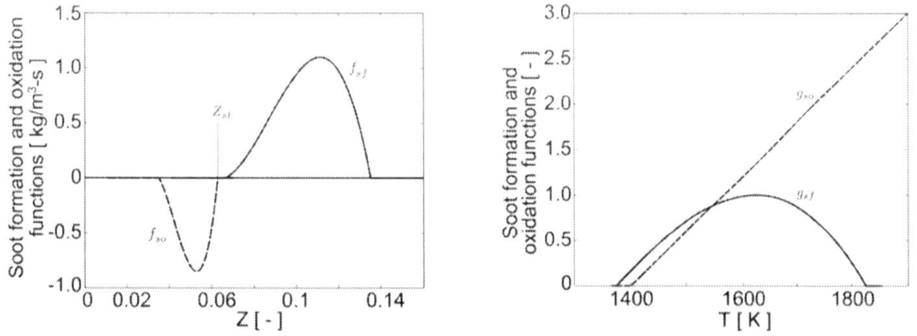

(a) (b)

Figure 2. Soot formation and oxidation functions: (a) mixture fraction dependent soot formation function, $\dot{f}_{sf}'''(Z)$ $[kg/m^3\text{-}s]$, and oxidation function, $\dot{f}_{so}'''(Z)$ $[kg/m^3\text{-}s]$, and (b) temperature dependent soot formation function, $g_{sf}(T)$ $[-]$, and oxidation function, $g_{so}(T)$ $[-]$.

Peak soot formation rates occur at mixture fraction values between 0.1 and 0.15 and over a temperature range of 1500-1600 K. The peak soot formation rate is assumed to be inversely proportional to the laminar smoke point height, $\dot{\omega}_{sf,P}''' \propto 1/\ell_s$. Soot is oxidized on the oxidizer side of the flame, controlled by the diffusion of oxygen, as long as the temperature is above 1400 K. The soot oxidation rate, $\dot{\omega}_{so}'''$, is also evaluated as a product of a function of mixture fraction, $\dot{f}_{so}'''(Z)$, and temperature, $g_{so}(T)$: $\dot{\omega}_{so}''' = \dot{f}_{so}'''(Z)\, g_{so}(T)$. The mixture fraction dependent soot formation and oxidation functions, $\dot{f}_{sf}'''(Z)$ and $\dot{f}_{so}'''(Z)$, respectively are shown in Fig. 2(a). The temperature dependent functions, $g_{sf}(T)$ and $g_{so}(T)$ are shown in Fig. 2(b).

The total enthalpy equation is solved with a radiant emission sink term included,

$$\frac{\partial \rho_g h}{\partial t} + \nabla \cdot (\rho_g h \mathbf{u}) = \nabla \cdot (\rho_g \mathcal{D} \nabla h) + S_{sf} + S_{so} - \dot{q}_r''', \tag{2}$$

where, $h = h_T + \phi_s h_{T,s}$. Here, h_T is the gas-phase total enthalpy (sensible + chemical) and $\phi_s h_{T,s}$ is the soot enthalpy [4]. $S_{sf} = \dot{\omega}_{sf}'''(h_{T,F} - h_{T,s})$ is the change in sensible enthalpy due to soot formation, and $S_{so} = \frac{\dot{\omega}_{so}'''}{M_F}\left[M_F h_{T,s} + \left(x + \frac{y}{4}\right) M_{O_2} h_{T,O_2} - x M_{CO_2} h_{T,CO_2} - \frac{y}{2} M_{H_2O} h_{T,H_2O}\right]$ is the release of energy per unit volume due to soot oxidation for an arbitrary hydrocarbon fuel, $C_x H_y$. The conservation of mixture fraction equation is

$$\frac{\partial \rho_g Z}{\partial t} + \nabla \cdot (\rho_g Z \mathbf{u}) = \nabla \cdot (\rho_g \mathcal{D} \nabla Z) + \dot{\omega}_Z''', \tag{3}$$

where, the source term, $\dot{\omega}_Z'''$ includes the effect of soot formation or oxidation on the mixture fraction (see [4] for details). The Howarth-Dorodnitzyn transformation [8] is applied to eliminate the effects of variable density by replacing the coordinates y, t by the dimensionless coordinates ξ, τ, where ξ is a non-dimensional space coordinate, $\xi = \left(\frac{a_0}{\rho_\infty^2 \mathcal{D}_\infty}\right)^{1/2} \int_0^y \rho_g(y', t)\, dy'$ and, τ is a non-dimensional time, $\tau = \int_0^t a(t')\, dt'$. Here, ρ_g is the gas-phase density, $\rho_\infty = \rho_g(\infty, t)$. With the approximation $\rho_g^2 \mathcal{D} = \rho_\infty^2 \mathcal{D}_\infty$ applied, we transform the conservation equations to their non-dimensional forms,

$$\begin{aligned} \frac{\partial \phi_s}{\partial \tau} - \xi \frac{\partial \phi_s}{\partial \xi} &= S_{th} + \frac{\dot{\omega}_s'''}{\rho_g a_0}, \\ \frac{\partial h}{\partial \tau} - \xi \frac{\partial h}{\partial \xi} &= \frac{\partial^2 h}{\partial \xi^2} + \frac{S_{sf} + S_{so} - \dot{q}_r'''}{\rho_g a_0}, \\ \frac{\partial Z}{\partial \tau} - \xi \frac{\partial Z}{\partial \xi} &= \frac{\partial^2 Z}{\partial \xi^2} + \frac{\dot{\omega}_Z'''}{\rho_g a_0}. \end{aligned} \tag{4}$$

Here, S_{th} is the soot thermophoresis term which is computed as $S_{th} = 0.55\, Sc \frac{\partial}{\partial \xi}\left[\frac{\phi_s}{(1+\phi_s)T}\frac{\partial T}{\partial \xi}\right]$.

2.2. Flamelet Radiant Emission

Thermal radiation is assumed to be optically thin and to occur primarily due to soot forming and radiating. The soot absorption coefficient is computed following Tien et al. [9], $\kappa_s \approx 1226 f_v T\, [m^{-1}]$, where f_v is the soot volume fraction and T is the temperature. The radiation source term is computed as a function of the soot volume fraction and temperature, $\dot{q}_r''' = 4\kappa_s \sigma T^4$. The non-dimensional initial thickness of the 1-D slab/flamelet (see Fig. 1(c)) is estimated as a function of $\eta \sim (\nu^3/\varepsilon)^{1/4}$ and the characteristic strain rate a_0 as

$$\Gamma = \left(\frac{a_0}{\rho_\infty^2 \mathcal{D}_\infty}\right)^{\frac{1}{2}} \int_0^\eta \rho dy \sim \left(\frac{\rho_F}{\rho_\infty}\right) Sc^{1/2}, \tag{5}$$

where, Sc is the Schmidt number, ρ_F is the density of fuel, ρ_∞ is the ambient density, ε is the dissipation rate and ν is the kinematic viscosity. The non-dimensional initial width for a propane flamelet is $\Gamma = 1.2774$.

Figure 3 shows the influence of the turbulent micro-time strain rate, a_0 and prior enthalpy loss fraction, H_p on the flamelet temperature and soot mass fraction. In Fig. 3(a) due to the strain rate being low ($a_0 = 5\, s^{-1}$) and no prior loss ($H_p = 0.0$), large amount of soot forms, oxidizes and radiates, lowering the flamelet temperature to below 1500 K and the flamelet releases unoxidized soot. In Fig. 3(b) for $a_0 = 5\, s^{-1}$ and $H_p = 0.3$, the flamelet temperature distribution at $\tau = 0$ shows lower temperatures due to the 30% lowering of sensible enthalpy uniformly across the flamelet. This reduction in initial temperatures results in lower amounts of soot generation and consequently lower radiation from the flamelet. No soot is released from the flamelet for this condition. Increasing a_0 to $30\, s^{-1}$ causes lower radiant emission from the flamelet and no soot release. This is true both for no prior enthalpy loss (Fig. 3(c)) and with $H_p = 0.3$ (Fig. 3(d)). The computed radiant energy fraction, Ψ_r and soot released per unit combustion energy, Ψ_s are dependent on a_0 and H_p. In general, lower values of a_0 and H_p result in higher radiant emission and greater soot release from the flamelet. A lookup-table is generated for Ψ_r as a function of a_0 and H_p. In the LES of turbulent flames, the local radiant emission from the turbulent flame sheet in a grid cell is computed as $\dot{q}_r''' = \Psi_r \dot{Q}'''$, where $\dot{Q}'''$ is the turbulent volumetric heat release rate. The soot source term is computed as $\dot{\omega}_{Y_s}''' = \Psi_s \dot{Q}'''$.

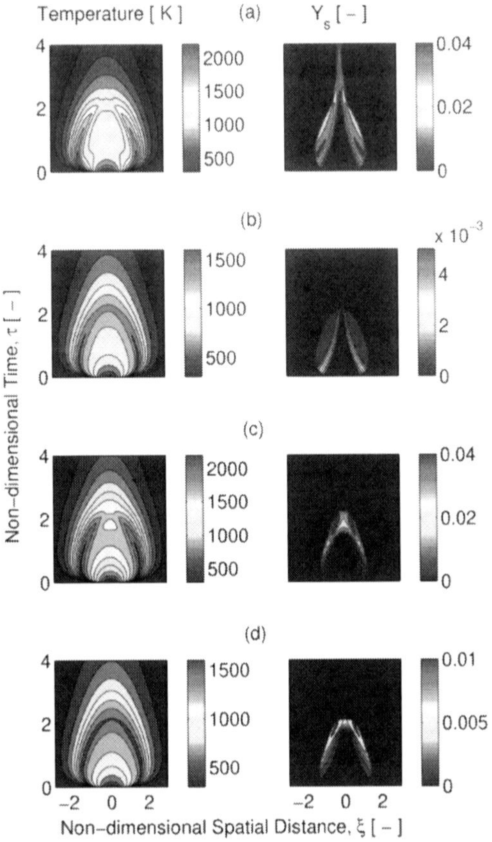

Figure 3. Contour plots of flamelet temperature, T, and soot mass fraction, Y_s: (a) $a_0 = 5\,s^{-1}$, $H_p = 0.0$, (b) $a_0 = 5\,s^{-1}$, $H_p = 0.3$, (c) $a_0 = 30\,s^{-1}$, $H_p = 0.0$, (d) $a_0 = 30\,s^{-1}$, $H_p = 0.3$.

2.3. LES Solver

The current set of simulations has been carried out with a large eddy simulation (LES) code, FireFOAM [6], which is a pressure based segregated solver, solving the Favre filtered fully compressible Navier-Stokes equations. Details of FireFOAM equations can be found in [10]. The turbulent sub-grid scale stress is modeled by the eddy viscosity concept using the one-equation model [11], which solves a sub-grid scale (SGS) kinetic energy equation. The dissipation rate is estimated by the model as $\varepsilon = c_s \Delta^{-1} \widetilde{k}^{3/2}$. The turbulent micro-time strain rate required by the flamelet model is computed as the inverse of the Kolmogorov time scale as $a_0 = C_\eta \left(\varepsilon/\nu\right)^{1/2}$. The turbulent combustion model adopts an infinitely-fast chemistry assumption. This mixed-is-burnt model removes the requirement of chemical kinetics, and reasonably represents the physics in well-ventilated fires. The fuel consumption rate is estimated from the eddy dissipation model as $\dot{\omega}'''_{Y_F} \sim \frac{\rho}{\tau_{mix}} min\left(\widetilde{Y_F}, \frac{\widetilde{Y_O}}{s}\right)$, where τ_{mix} is a turbulent macro-mixing time. The sensible enthalpy equation is solved with a volumetric heat release rate source term $\dot{Q}''' = \Delta H_c \dot{\omega}'''_{Y_F}$ and a radiation sink term $\dot{q}'''_r$. The prior heat (enthalpy) loss is modeled following Brookes and Moss [12] as $H_p = \left(\widetilde{h}_T - \widetilde{Z} h^0_{F,\infty}\right) / \left(\widetilde{h}^0 - \widetilde{Z} h^0_{F,\infty}\right)$ where $\widetilde{h}^0$ is the chemical enthalpy and $\widetilde{h}_T$ is the total enthalpy (sensible + chemical). A soot transport equation is solved with the source term $\dot{\omega}'''_{Y_s}$ included. The source term provides the released unoxidized soot from the flame sheet. No further oxidation of the transported soot takes place.

A gray media assumption is made for computing radiation heat transfer. This assumption is justified since soot is primarily responsible for radiation in diffusion flames of propane. It is also assumed that radiation in the parallel panel configuration is in the optically thin limit. Radiant emission due to soot from the flame sheet is the only source of radiation. Products of combustion and the soot released from the flame-sheet do not participate in radiation heat transfer. However, the panels, ground, sand burner and the domain boundaries emit and absorb radiation. The flame-sheet volumetric radiant emission, $\dot{q}_r'''$, is included as an isotropic emission term in the RTE [10],

$$\frac{d\overline{I}}{ds} = \overline{\kappa I_b} - \overline{\kappa I} = \frac{\dot{q}_r'''}{4\pi}. \tag{6}$$

In the above equation, we assume that only the flame emits radiation isotropically, i.e. $\overline{\kappa I_b} \approx \frac{\dot{q}_r'''}{4\pi}$. In the simulations, the emitted radiation from the flame sheet is incident on the solid surfaces. It is reflected, absorbed and re-radiated from these surfaces, however this emitted radiation is not absorbed by the flame or the combustion products, i.e. $\overline{\kappa I} = 0$. The net radiative heat flux on solid-fuel surfaces is obtained by calculating the difference between the incident radiative flux and the emitted and re-radiated heat flux components. For opaque materials like corrugated-cardboard (rack storage) and Marinite-I (inert parallel panel), there is no transmitted component of incident radiation. The surface emissivity, which is a function of the char-fraction, is provided by the pyrolysis model. The surface temperature for the emission component is also computed by the pyrolysis model. The net radiative heat flux along with the convective flux is used as input to the pyrolysis model. A finite-volume method [13] is applied for the solution of the RTE using the existing computational mesh.

3. Results and Discussion

Results are presented from two studies: radiant heat transfer from a propane fire onto inert parallel panels and fire spread over corrugated commodities in a rack storage arrangement. The standard parallel panel configuration used by FM Global [1] for characterizing industrial clean room materials [14] is considered. The rack storage fire growth simulations follow fire growth studies conducted earlier for corrugated parallel panels [15]. In addition to vertical fire growth encountered in the parallel panel setup, in rack storage scenarios both vertical and lateral fire spread occurs. Flue spaces in rack storage arrangements vary between 0.15-0.6 m and complex geometrical arrangements of boxes, obstructions and large heat release rates (HRR) make simulation of fire growth in rack storage scenarios even more challenging.

3.1. Flame Heat Transfer to Inert Parallel Panels

Flame heat transfer, primarily radiative, from a 60 kW propane diffusion flame to adjacent walls in a parallel panel configuration [14] is simulated. In this configuration, the flame is anchored on a 0.3 m × 0.6 m sand burner placed between two inert vertical panels. The panels are made of a fire-resistant material, Marinite-I. Figure 4 shows a section of the computational domain. The sand burner is elevated by 0.3 m from the ground and the panels are each 2.4 m tall and 0.025 m in thickness with the back boundary insulated. Heat flux (using Schmidt-Boelter gauges) and surface temperature (using thermocouples) are recorded on the panels at vertical centerline and offset locations (as shown in Fig. 4). Six locations on one of the panels are shown in the figure – two centerline locations at 0.3 m and 0.9 m from the bottom and four offset locations at 0.6 m and 1.2 m from the bottom (±0.15 m offset from the centerline). The computational domain is 1.2 m × 1.2 m × 3 m high. The finest mesh resolution used comprised of hexahedral cells of volume $(0.009375\,m)^3$. Details of the mesh sensitivity study conducted can be found in [10]. Angular resolution for the RTE solver was selected based on a sensitivity study also described in [10]. The total number of angles used in the present study is 48 (12 azimuthal and 4 polar).

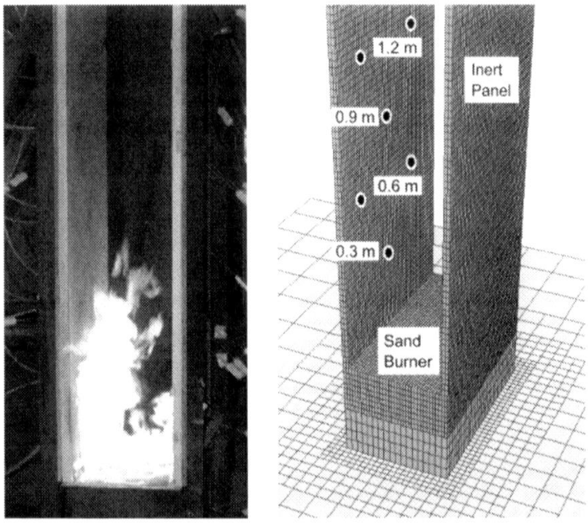

(a) (b)

Figure 4. Parallel panel setup: (a) $60\,kW$ propane fire between inert parallel panels, and (b) computational mesh with measurement locations.

The sand-burner anchored propane fire emits radiation isotropically and over time emitted radiant power (approximately $18\,kW$) remains invariant. The $18\,kW$ radiant emission corresponds to an overall radiant fraction, $\chi_r = 0.3$. Out of the $18\,kW$ flame emission, approximately $6\,kW$ is incident on each of the two panels (see Fig. 5). Initially, the overall incident radiant power on one panel is $7\,kW$, but this incident power increases to approximately $9\,kW$ at $50\,s$, as can be observed in Fig. 5. The additional $2\,kW$ is due to the re-radiation from the adjacent panel, which re-radiates approximately $4\,kW$ at $50\,s$ (50% of this re-radiated component is incident on the neighboring panel). The re-radiation from the panels increases with time as their surface temperature increases.

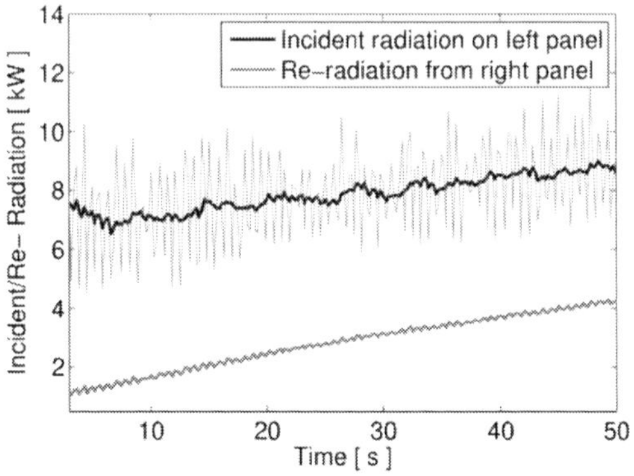

Figure 5. Incident radiation on left panel and re-radiation from the right panel.

The computed incident heat fluxes on the Marinite-I panels are now compared against experimentally observed values for point measurements. As the panels heat up, the surface temperature increases. The predicted heat flux and surface temperature comparisons against experimental data are shown in Fig. 6. The predicted heat flux for the locations on the left panel show good agreement with the experimental data. Surface temperatures are predicted with reasonably good accuracy, except for the 0.9 m location on the centerline. Further investigation is needed to determine the difference in surface temperature at the 0.9 m location. The prediction error at this height also is large due to the low temperatures encountered.

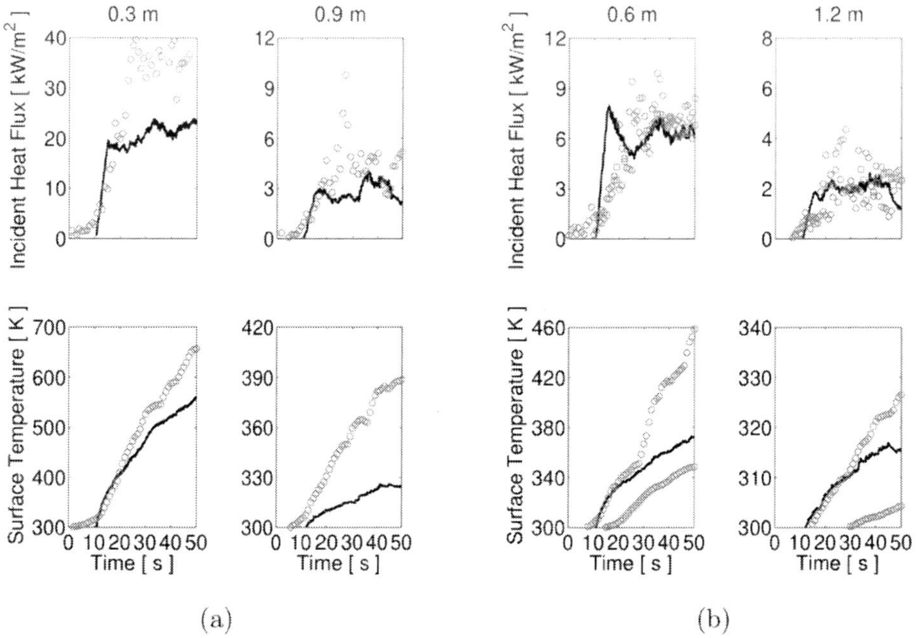

Figure 6. Parallel panel incident heat flux and surface temperature: (a) centerline locations, and (b) offset locations. —— computation, ○ experiment.

3.2. Fire Growth in a Rack Storage Setup

Fully coupled CFD simulations (pyrolysis, soot and radiation, gas-phase combustion) of fire spread were conducted in a 2 × 4 × 3 rack storage configuration [2]. Pyrolysis of charring materials was simulated with the solution of a 1-D, finite-rate model coupled to the gas-phase solver through boundary conditions for mixture fraction, temperature and velocities [15]. The simulations used solid fuel material properties determined from bench scale fire propagation apparatus (FPA) [16] tests and with the application of optimization schemes [17]. Several instrumented fire tests were conducted under a 20 MW fire products collector (FPC) with the purpose of providing validation data. Each of the 24 units in the $2 \times 4 \times 3$ configuration consisted of three nested boxes made of double-wall corrugated cardboard enclosing a sheet metal liner. Chemical heat release rates were measured for comparison against computed results.

Figure 7 shows four snapshots showing fire growth in the rack storage arrangement. The fire initially grows vertically between four boxes and lateral spread begins at a later time (HRR is above 3 MW). Initial vertical and lateral spread are due to ignition of corrugated cardboard primarily by incident radiation. Incident heat fluxes of the order of 30 kW/m^2 to greater than 200 kW/m^2 were observed in the computations. Measured heat fluxes in the experiment also showed heat fluxes in excess of 100 kW/m^2 on the central vertical faces of the corrugated boxes.

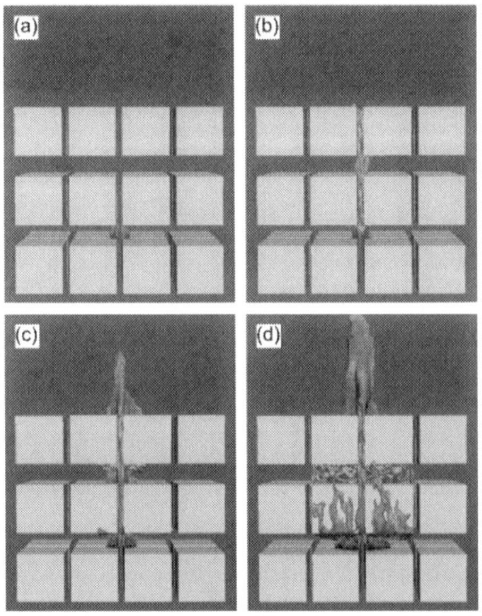

Figure 7. Rack storage fire growth depicted by the stoichiometric mixture fraction contour at different times corresponding to heat release rates of (a) $0.5\,MW$, (b) $1\,MW$, (c) $3\,MW$, and (d) $9\,MW$.

The computed HRR is compared against experimentally measured HRR in Fig. 8. Comparison is made after a HRR of $500\,kW$ is reached as the initial ignition process and fire growth do not show a similar behavior in the experiments. This is because for practical testing scenarios, igniter setup and ambient conditions can vary and affect initial ignition of the corrugated cardboard. However, ignition and initial fire growth are known to not influence large scale fire growth. The experimental curves and the computed curve have been time-shifted so that they coincide when $500\,kW$ HRR is reached. Computed HRR shows good agreement with experimental data. The uncertainty/variability of the experimental measurements is illustrated by the scatter of data for repeated tests.

4. Conclusions and Future Work

Results from a CFD study simulating radiation heat transfer from a buoyant turbulent flame to inert parallel panels have been presented. A LES code, FireFOAM, was used in the study with a laminar smoke point based subgrid soot-radiation model implemented and with the solution of a finite volume based RTE solver. A conjugate heat transfer model was applied to compute panel heat-up. Computed radiant fluxes are compared against experimental data and predicted panel surface temperatures are also compared. Overall, the simulation has shown good agreement with experimental data. Fire growth in a rack storage arrangement was also simulated with the application of a pyrolysis model for corrugated cardboard. Computed HRR from the simulation shows good agreement with experimental measurements.

This study assumed optically thin radiation. The model needs to be extended for larger fires where radiation is in the optically thick limit. Also, even though angular discretization tests were performed, the dependence of radiation transfer on the computational mesh has to be further investigated. Ray effects were observed in the parallel panel simulations and their causes will be investigated. Gas-phase radiation was neglected to speed-up the computation of these large-scale simulations. Future work will improve the radiation model with the inclusion of wide-band models for gas-phase radiation developed using statistical narrow-band models.

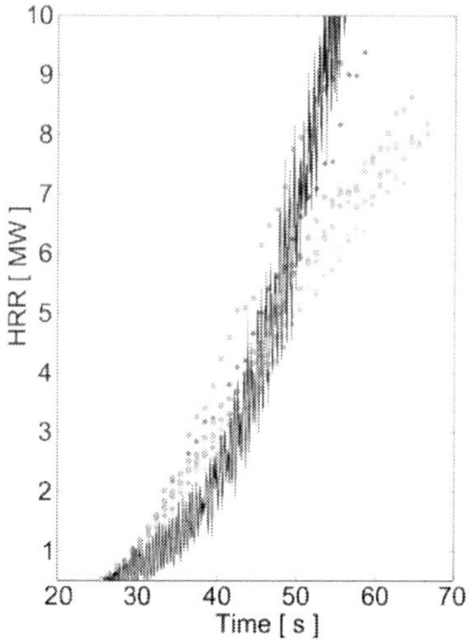

Figure 8. Rack storage fire growth heat release rate compared against experimental data. —— computation, ○ experiments. Experimental data beyond peak HRR are not shown since water-spray based suppression was applied in the experiment.

Acknowledgements
Thanks are due to Drs. Karl Meredith and Yibing Xin for providing the rack storage heat release rate data and to Dr. Marcos Chaos for providing the inert parallel panel data. The work presented in this paper was funded by FM Global and performed within the framework of the FM Global Strategic Research Program on Fire Modeling.

References
[1] Tewarson, A and Khan, M M 1989 *22nd Symposium (International) on Combustion* pp 1231–1240
[2] Xin, Y and Tamanini, F 2009 *Fire Safety Science* **9** 527–538
[3] Markstein, G H 1984 *20th Symposium (International) on Combustion* pp 1055–1061
[4] Lautenberger, C W, de Ris, J L, Dembsey, N A, Barnett, J R and Baum, H R 2005 *Fire Safety Journal* **40** 141–176
[5] Chatterjee, P, de Ris, J L, Wang, Y and Dorofeev, S B 2011 *Proceedings of the Combustion Institute* **33** 2665–2671
[6] Available from http://www.fmglobal.com/modeling
[7] Chatterjee, P, de Ris, J L and Lautenberger, C W 2008 *Proceedings of the 5th International Seminar on Fire and Explosion Hazards* (Edinburgh, UK: University of Edinburgh) pp 404–414
[8] Carrier, G F, Fendell, F E and Marble, F E 1975 *SIAM Journal of Applied Mathematics* **28** 463–500
[9] Tien, C L, Lee, K Y and Stretton, A J 2002 *Radiation Heat Transfer* (Quincy, MA: National Fire Protection Association) SFPE Handbook of Fire Protection Engineering 3rd ed
[10] Chatterjee, P, Krishnamoorthy, N, Wang, Y, de Ris, J L and Dorofeev, S B 2010 *Proceedings of the 12th Fire Science and Engineering Conference (INTERFLAM)* vol 1 pp 777–788
[11] Fureby, C, Tabor, G, Weller, H G and Gosman, A D 1997 *Physics of Fluids* **9** 1416–1429
[12] Brookes, S J and Moss, J B 1999 *Combustion and Flame* **116** 486–503
[13] Raithby, G D and Chui, E H 1990 *Journal of Heat Transfer* **112** 415–423
[14] ANSI/FM Approvals 4910 2004 *American National Standard for Cleanroom Materials Flammability Test Protocol*
[15] Krishnamoorthy, N, Chaos, M, Khan, M M, Chatterjee, P, Wang, Y and Dorofeev, S B 2010 *Proceedings of the 12th Fire Science and Engineering Conference (INTERFLAM)* vol 1 pp 709–720
[16] ASTM E 2058-09 2009 *Standard Test Methods for Measurement of Synthetic Polymer Material Flammability Using a Fire Propagation Apparatus (FPA)*
[17] Chaos, M, Khan, M M, Krishnamoorthy, N, de Ris, J L and Dorofeev, S B 2011 *Proceedings of the Combustion Institute* **33** 2599–2606

A theoretical analysis of the influence of turbulence on radiative emission in turbulent diffusion flames of methane

P J Coelho

Mechanical Engineering Department, Instituto Superior Técnico/IDMEC, Technical University of Lisbon, Av. Rovisco Pais, 1049-001 Lisboa, Portugal

E-mail: pedro.coelho@ist.utl.pt

Abstract. A theoretical analysis is reported to quantify the increase of radiative emission due to turbulence for methane diffusion flames burning in air. The instantaneous thermochemical state of the reactive mixture is described by a flamelet model and a detailed chemical mechanism. Mean values of the absorption coefficient, blackbody radiation intensity and radiative emission are evaluated for different turbulence levels by assuming the pdf shape of mixture fraction. The results show that turbulent fluctuations generally contribute to reduce the Planck mean absorption coefficient of the medium, in contrast with the blackbody radiation intensity, which is significantly increased by turbulence. If the turbulence level is relatively small, the influence of turbulence on the absorption coefficient is marginal. Otherwise, fluctuations of the absorption coefficient of the medium should be taken into account. The mean radiative emission is underestimated if turbulent fluctuations are fully ignored and overestimated if only temperature fluctuations are considered, while neglecting fluctuations of the absorption coefficient of the medium, the error being generally higher in the latter case. The effects of turbulence on radiative emission are stronger in the fuel-lean region and close to stoichiometric conditions than in the fuel-rich region.

1. Introduction

Thermal radiation plays an important role in many turbulent reactive flows, particularly in industrial combustion equipment, rocket nozzles and fires. The interaction between turbulence and radiation (TRI) may significantly increase radiative emission, and influence the radiative heat source and the temperature field [1-3]. The importance of TRI has been demonstrated, and the effect of various approximations to account for TRI in the numerical solution of the radiative transfer equation has been investigated using either prescribed pdf [4-6] or pdf transport methods [7-9] to simulate turbulent flames. Despite of this, most current numerical simulations of turbulent reactive flows neglect TRI or account only for temperature fluctuations in the calculation of the mean blackbody radiation intensity. It is relatively simple to account for temperature fluctuations only, but the accuracy of this simplification may be poor. Therefore, it is important to investigate the influence of TRI in radiative emission, and how accurate is to account only for temperature fluctuations. This is the subject of the present article, which reports a theoretical analysis for turbulent diffusion flames of methane.

A simple theoretical analysis reported in [10] quantifies the influence of turbulence on the radiative energy emitted by a hot medium. It shows that, if fluctuations of the emissivity are neglected, temperature fluctuations of ±20% increase emission by 24%, while temperature fluctuations of ±30% increase emission by more than 50%. A similar analysis is presented in [11], assuming a beta pdf shape for the temperature. It confirms the findings of Cox [10], and reveals that the effect of TRI on radiative emission is relatively insensitive to the pdf shape, varying by less than 10% for temperature fluctuations of ±20%.

A recent theoretical analysis of the importance of TRI on radiative emission for a turbulent diffusion flame of methane burning in air was reported in [3]. Chemistry was described by a single one-step, infinitely fast, irreversible global reaction. Fluctuations of the temperature and species molar fractions were taken into account. The purpose of the present work is to extend the previous analysis [3] to estimate the importance of TRI on the radiative emission for turbulent diffusion flames of methane burning in air. The flame structure is described by a flamelet model using a detailed chemical mechanism. The influence of the turbulence level is investigated. Previous works were either focused on a particular flame/combustion system or based on relatively crude theoretical analysis. The main contribution of the present paper is that it reports results concerning the importance of TRI in turbulent diffusion flames that are broadly independent of the reactive system (burner geometry, Reynolds number, stoichiometry), albeit restricted to the flamelet regime.

2. Theory

The radiative transfer equation (RTE) may be written as follows for an emitting, absorbing and non-scattering medium [12]:

$$\frac{d\, I_\nu}{d\, s} = -\kappa_\nu\, I_\nu + \kappa_\nu\, I_{b\nu} \tag{1}$$

In this equation I_ν is the spectral radiation intensity, $I_{b\nu}$ the spectral blackbody radiation intensity (Planck function), s the direction of propagation, κ_ν the spectral absorption coefficient of the medium, and the subscripts ν and b denote wavenumber and blackbody, respectively. Integration of this equation over the full spectrum yields

$$\frac{d\, I}{d\, s} = -\kappa_G\, I + \kappa_P\, I_b \tag{2}$$

where the incident mean absorption coefficient, κ_G, and the Planck mean absorption coefficient, κ_P, are defined as

$$\kappa_G = \int_0^\infty \kappa_\nu\, I_\nu\, d\nu \Big/ \int_0^\infty I_\nu\, d\nu \tag{3}$$

$$\kappa_P = \int_0^\infty \kappa_\nu\, I_{b\nu}\, d\nu \Big/ \int_0^\infty I_{b\nu}\, d\nu \tag{4}$$

The time-averaged form of equation (2) is written as

$$\frac{d\, \bar{I}}{d\, s} = -\overline{\kappa_G\, I} + \overline{\kappa_P\, I_b} \tag{5}$$

where the overbar denotes a time-averaged value.

The TRI is generally taken into account using the optically thin fluctuation approximation [13], which neglects the correlation between the radiation intensity and the absorption coefficient of the medium, i.e., $\overline{\kappa_G\, I} \approx \bar{\kappa}_G\, \bar{I}$. Therefore, the main contribution to TRI comes from the radiative emission, i.e., from the second term on the right side of equation (5). In order to investigate the importance of TRI in this term, the following quantities are defined:

$$R_k = \kappa_P\left(\bar{T}, \bar{x}_i\right) / \bar{\kappa}_P \tag{6}$$

$$R_{I_b} = I_b(\bar{T})/\bar{I}_b \tag{7}$$

$$R_{kI_b} = \bar{\kappa}_P\ \bar{I}_b/\overline{\kappa_P\ I_b} \tag{8}$$

$$R_{emiss} = \kappa_P(\bar{T},\bar{x}_i)\,I_b(\bar{T})/\overline{\kappa_P\ I_b} = R_k\ R_{I_b}\ R_{kI_b} \tag{9}$$

In these equations, $\bar{T}$ stands for the mean temperature and $\bar{x}_i$ for the mean molar fraction of species i. Equations (6), (7) and (8) quantify the absorption coefficient self-correlation, the temperature self-correlation and the absorption coefficient–blackbody radiation intensity correlation. Equation (9) yields the overall influence of TRI on the radiative emission term of the RTE. All the ratios defined in equations (6) to (9) are equal to unity in the absence of turbulent fluctuations. The TRI becomes increasingly more important as they depart from unity.

It is worth to stress that the impact of TRI on the above correlations does not directly depend on the optical thickness of the medium. In fact, radiative emission only depends on local properties, and this allows the present analysis, which is independent of the turbulent reactive flow details, to be carried out. Several previous studies have investigated the influence of the optical thickness of the medium on TRI, which may affect the temperature and incident radiation fields, and the heat fluxes to the boundaries. However, the radiative emission does not depend on the optical thickness of the medium, being only directly influenced by the local mean and rms of mixture fraction, and local fraction of radiative heat loss.

Turbulent diffusion flames of methane burning in air are considered in the remainder of this work. The instantaneous thermochemical state of the reactive mixture is described by a flamelet model using the GRI 3.0 mechanism [14] and determined from the numerical solution of the flamelet equations [15]. These equations yield the temperature and the chemical composition of the reactive mixture as a function of mixture fraction and scalar dissipation rate, in the absence of turbulent fluctuations. Flame Master code [16] was used to solve the flamelet equations for scalar dissipation rates, χ, ranging from near equilibrium to near flame quenching conditions.

The radiative source term was not included in the flamelet equations, since it yields unrealistic results in the case of steady flamelets, as discussed in [17]. Temperature flamelet profiles for non-adiabatic conditions were generated from the temperature flamelet adiabatic profiles by assuming that the local fraction of radiative heat loss, X_R, which is prescribed, is independent of mixture fraction, z [18]. The fraction of radiative heat loss is defined as

$$X_R = (h_{ad} - h)\Big/\left(h_{ad} - \sum_i y_i\, h_i^o\right) \tag{10}$$

where h_{ad} and h are the enthalpy of the mixture for adiabatic and non-adiabatic conditions, respectively, h^o is the standard enthalpy of formation, y the mass fraction of a species and subscript i identifies the species under consideration. The enthalpy of the mixture for adiabatic conditions is a linear function of mixture fraction, being equal to the enthalpy of the air for $z = 0$ and to the enthalpy of the fuel for $z = 1$. Fixing the fraction of radiative heat loss, the enthalpy of the mixture for non-adiabatic conditions is determined from equation (10), and the non-adiabatic temperature is calculated from the enthalpy and the chemical composition using the caloric equation of state for values of mixture fraction ranging from 0 to 1. The mass fractions of the species are assumed to be independent of the fraction of radiative heat loss [19]. Further details on the generation of non-adiabatic temperature flamelet profiles are given in [5].

The Planck mean absorption coefficient of the absorbing species (CO_2, H_2O, CO and CH_4) is determined as a function of temperature by curve fits [20] to the narrow band model implemented in the RADCAL computer code [21]. The Planck mean absorption coefficient of any species in the reactive mixture depends on the temperature, molar fraction, and pressure. Both the temperature and the molar fraction of the species are related to the mixture fraction and the scalar dissipation rate using the flamelet data.

The pdf shape of mixture fraction is assumed to be a clipped Gaussian. The pdf is defined from the mean value and the rms of mixture fraction. The mean value of any variable in equations (6)-(9) is evaluated from integration over the mixture fraction range (0 to 1) of the instantaneous value of that variable, expressed as a function of mixture fraction, weighted by the pdf of mixture fraction, i.e.,

$$\bar{\phi} = \int_0^1 \phi(z)\, pdf(z)dz \tag{11}$$

where ϕ stands for the quantity under consideration. As an example, in the case of the radiative emission, $\phi(z) = \kappa_P(T(z,\chi,X_R), x_i(z,\chi))\, I_b(T(z,\chi,X_R))$. Fluctuations of the scalar dissipation rate and local fraction of radiative heat loss are not considered.

3. Results and Discussion

Standard calculations have been performed for diffusion flames of methane burning in air, at atmospheric pressure, for scalar dissipation rate $\chi = 1\ s^{-1}$ and fraction of radiative heat loss $X_R = 10\%$. A previous investigation has shown that the influence of the turbulent scalar fluctuations on the radiative emission is approximately independent of the scalar dissipation rate and fraction of radiative heat loss [22].

Figure 1(a) shows the temperature and the H_2O and CO_2 molar fraction profiles as a function of mixture fraction, as determined by solving the flamelet equations for adiabatic conditions, and then modifying the temperature profile to account for the prescribed fraction of radiative heat loss. The vertical dashed line indicates the stoichiometric mixture fraction (z_{st} = 0.055), while the solid vertical line marks a change in the scale. The maximum temperature occurs for a slightly rich mixture, z ≈ 0.065.

The Planck mean absorption coefficient of the mixture is displayed in figure 1(b). Three curves are shown, depending on whether radiation from CO and CH_4 is considered or not. The radiation from CO has a minor impact, but that from CH_4 plays an important role in the fuel-rich region of the flame. The Planck mean absorption coefficient of the species under consideration decreases with the increase of temperature (except for CO and CO_2 and temperatures lower than about 600 K), and it is directly proportional to the molar fraction of the species. It turns out that the influence of temperature is dominant for $0.015 < z < 0.43$, and therefore κ_P decreases with the increase of temperature in this range. For very lean mixtures, namely for $z < 0.015$, the influence of the molar fraction of the absorbing species on κ_P is greater than that of temperature, and so κ_P decreases as z tends to 0. Similarly, the decrease of the molar fraction of combustion products for very rich mixtures, namely for $z > 0.43$, explains the decrease of κ_P as z tends to unity if radiation from CH_4 is ignored. However, if radiation from CH_4 is taken into account, κ_P continues to increase as z approaches unity, since the high molar fraction of CH_4 compensates for the decrease of H_2O and CO_2 molar fractions.

The turbulence level in a diffusion jet flame, quantified here by $\text{rms}(z)/\bar{z}$, typically increases monotonically in radial direction from the centreline to the flame edge. Values of $\text{rms}(z)/\bar{z}$ up to 0.9 have been considered in the present work. In the fuel-rich region, fluctuations up to $\text{rms}(z)/\bar{z} \approx 0.3$ are commonly found, but higher fluctuations, even stronger than $\text{rms}(z)/\bar{z} = 0.9$, may occur at the flame edge due to intermittency (see, e.g., the experimental data for turbulent diffusion flames available in [23]). It can be mathematically demonstrated that $\text{rms}(z) \le [\bar{z}(1-\bar{z})]^{1/2}$. Accordingly, there is an upper limit of $\bar{z}$ for a given turbulence level, i.e., only values of $\bar{z}$ lower than that limit may occur for a prescribed turbulence level. That limit decreases with the increase of the turbulence level, e.g., the limit is equal to 0.917 for $\text{rms}(z)/\bar{z} = 0.3$ and 0.552 for $\text{rms}(z)/\bar{z} = 0.9$.

Figure 2 shows the mean temperature as a function of mean mixture fraction and turbulence level. The maximum mean temperature occurs for values of mean mixture fraction in the range 0.06-0.07, which are slightly above the stoichiometric mixture fraction for methane. As the turbulence level increases, the mean temperature decreases, except for fuel-rich flames and values of mean mixture fraction greater than a certain threshold, which depends on the turbulence level. The reason for this behaviour lies on the shape of the temperature–mixture fraction profile (convex for lean and moderately rich mixtures, and concave elsewhere, as shown in figure 1a) and on the Gaussian pdf of mixture fraction.

The total mean radiative emission, given by the second term on the right side of equation (5), is shown in figure 3 as a function of $\bar{z}$ and $\mathrm{rms}(z)/\bar{z}$. The mean radiative emission decreases with the increase of turbulence intensity for stoichiometric and for moderately fuel-lean or fuel-rich mixtures. In the case of very lean (typically $\bar{z} < 0.01$) or very rich mixtures, the mean radiative emission becomes much smaller than the maximum mean radiative emission. In those cases, even though the influence of turbulent fluctuations may be large, the contribution to total flame emission is quite small, and so it is not worth to consider TRI.

Figure 4(a) shows the Planck mean absorption coefficient evaluated at mean temperature and mean species molar fractions, $\kappa_P(\bar{T},\bar{x}_i)$, as a function of $\bar{z}$ and $\mathrm{rms}(z)/\bar{z}$. In the range where the influence of T on κ_P is dominant, i.e., for moderately fuel-lean or fuel-rich mixtures, turbulence yields a decrease of the mean temperature, and therefore an increase of $\kappa_P(\bar{T},\bar{x}_i)$. However, for very rich mixtures, the influence of turbulence on $\kappa_P(\bar{T},\bar{x}_i)$ is the opposite due to the role of CH_4 for such mixtures. Figure 4(b) shows the mean value of the Planck mean absorption coefficient, $\bar{\kappa}_P$, as a function of the mean mixture fraction and turbulence level. The profile of $\kappa_P(T,x_i)$ displayed in figure 1(b) exhibits a local maximum at $z \approx 0.017$ and a local minimum at $z \approx 0.068$. Turbulence smoothes out these extremes, in such a way that $\bar{\kappa}_P$ decreases in the vicinity of $z \approx 0.017$ and increases in the vicinity of $z \approx 0.068$. In the case of highly fuel-rich mixtures, turbulence causes a decrease of $\bar{\kappa}_P(T,x_i)$.

Figure 5 shows the values of R_k, R_{I_b}, R_{kI_b} and R_{emiss} as a function of mixture fraction for different levels of turbulent fluctuations. Figure 5(a) shows that turbulent fluctuations generally contribute to reduce the mean value of the Planck mean absorption coefficient of the medium, yielding $\kappa_P(\bar{T},\bar{x}_i) > \bar{\kappa}_P$, i.e., $R_k > 1$. This is a direct consequence of the dependence of $\kappa_P(T,x_i)$ and $\bar{\kappa}_P$ on $\bar{z}$, which was discussed above. If the turbulence level is relatively small, such that $\mathrm{rms}(z)/\bar{z} \le 0.3$, then R_k is rather close to unity, and the influence of turbulence on the absorption coefficient is marginal. Otherwise, turbulent fluctuations should be taken into account in the calculation of the absorption coefficient of the medium. This influence is more important on the fuel-lean side of the flame, where R_k exhibits a local maximum, and becomes progressively smaller as $\bar{z}$ increases.

The blackbody radiation intensity significantly increases due to turbulence yielding $I_b(\bar{T}) < \bar{I}_b$, i.e., $R_{I_b} < 1$, as shown in figure 5(b), in contrast with the Planck mean absorption coefficient. Therefore, the influence of turbulence on the mean radiative emission is generally lower than that predicted by accounting for temperature fluctuations, while neglecting the fluctuations of the absorption coefficient of the medium. The influence of turbulence on the blackbody radiation intensity is stronger than that on the Planck mean absorption coefficient, and is more important on the fuel-lean side of the flame, where R_{I_b} has a local minimum. Even though R_{I_b} is also small for highly fuel-rich mixtures, decreasing with the increase of mixture fraction, high turbulence levels do not generally occur for such mixtures.

The correlation $\overline{\kappa_P' I_b'}$ may be either positive or negative (see figure 6), yielding $R_{kI_b} < 1$ or > 1, respectively (see figure 5c). This figure shows that $\overline{\kappa_P' I_b'}$ is very small for $\mathrm{rms}(z)/\bar{z} = 0.1$, and negative for $\mathrm{rms}(z)/\bar{z} \le 0.5$, except for very lean mixtures, yielding $R_{kI_b} > 1$ for fluctuations of this magnitude. The negative absorption coefficient–blackbody radiation intensity correlation is explained

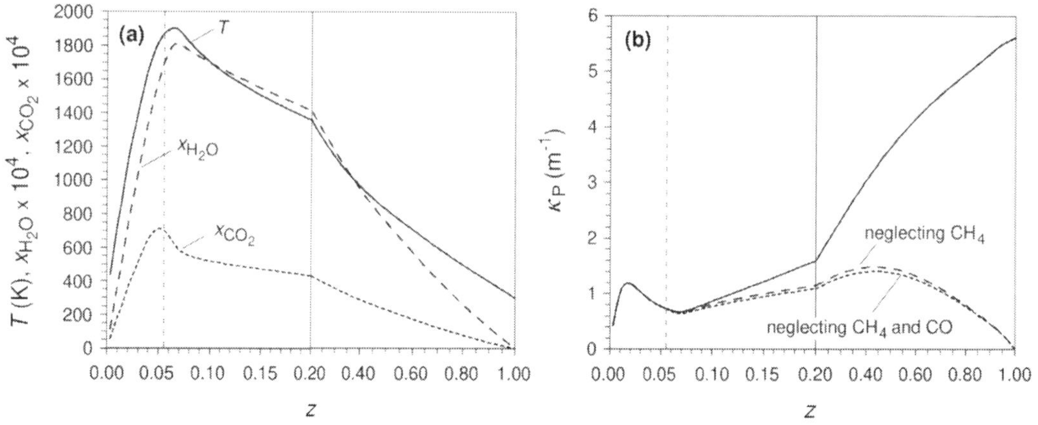

Figure 1. Instantaneous temperature, H_2O and CO_2 molar fractions (a) and Planck mean absorption coefficient (b) as a function of mixture fraction.

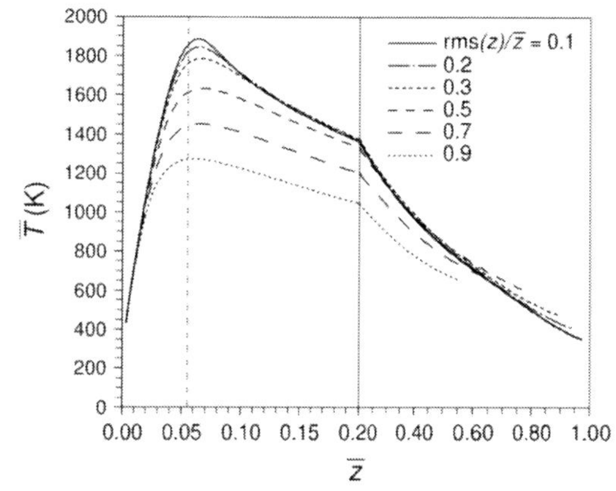

Figure 2. Mean temperature as a function of mean and rms of mixture fraction.

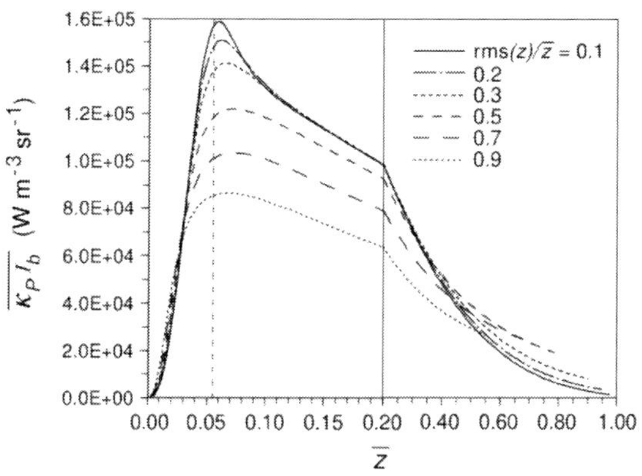

Figure 3. Mean radiative emission as a function of mean and rms of mixture fraction.

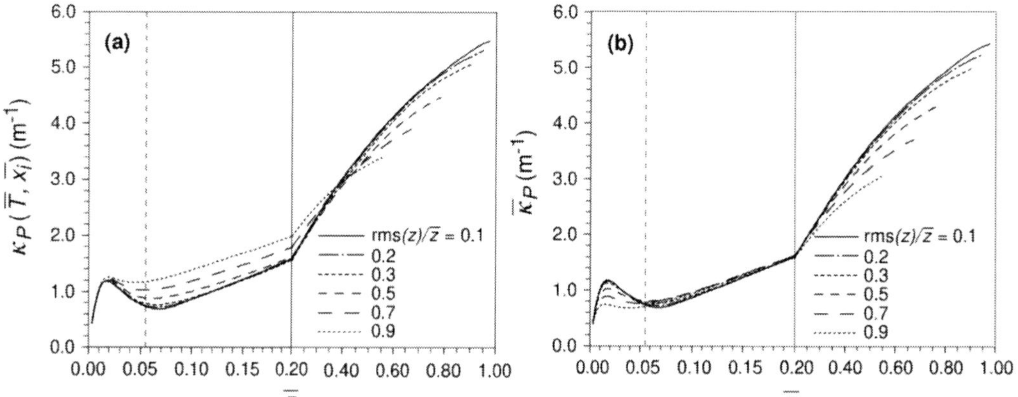

Figure 4. Planck mean absorption coefficient evaluated at mean temperature and mean species molar fractions (a) and mean value of the Planck mean absorption coefficient (b) as a function of mean and rms of mixture fraction.

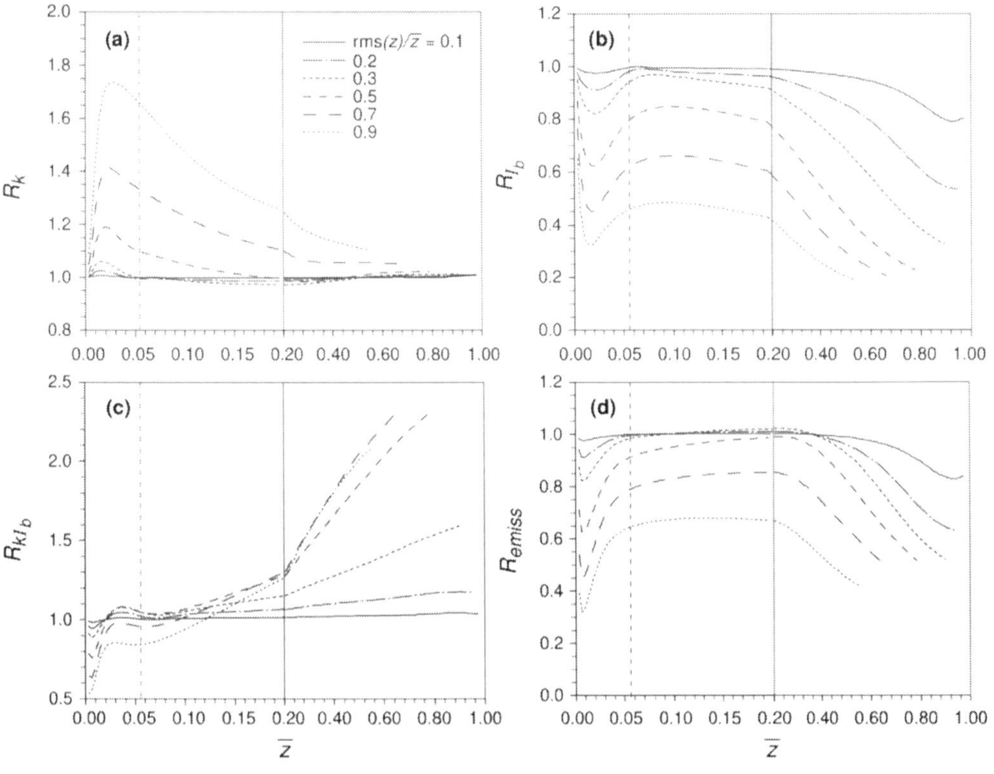

Figure 5. Influence of turbulent scalar fluctuations on the Planck mean absorption coefficient (a), blackbody radiation intensity (b), absorption coefficient–blackbody radiation intensity correlation (c) and radiative emission (d) as a function of mean and rms of mixture fraction.

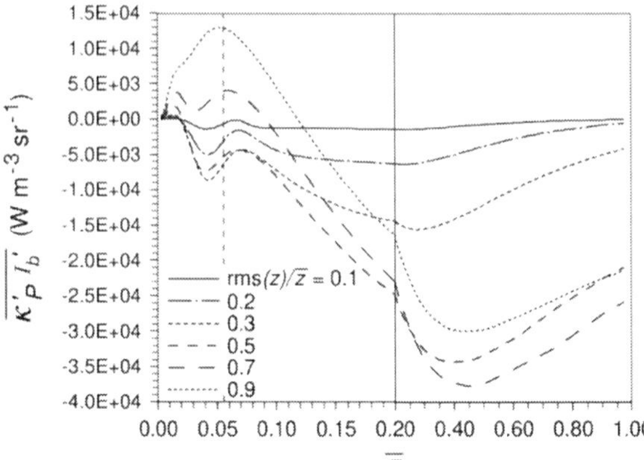

Figure 6. Influence of turbulent scalar fluctuations on the absorption coefficient–blackbody radiation intensity correlation as a function of mean and rms of mixture fraction.

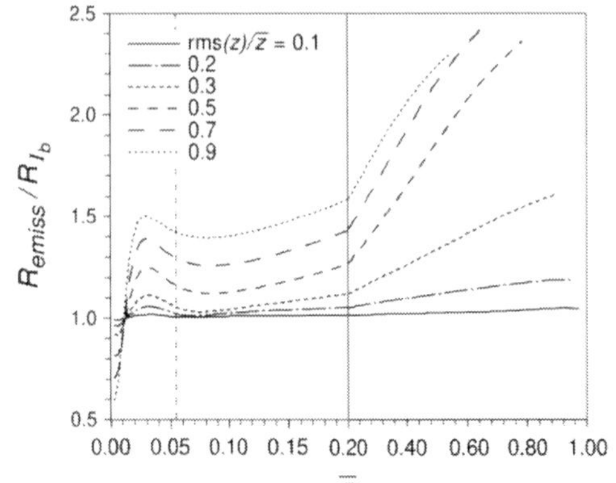

Figure 7. Ratio of the mean radiative emission evaluated by considering only temperature fluctuations to the mean radiative emission calculated by fully taking TRI into account.

by the decrease of κ_P with the increase of temperature for moderately lean or fuel-rich mixtures. However, for large fluctuations ($\text{rms}(z)/\bar{z} \geq 0.7$), the correlation $\overline{\kappa'_P I'_b}$ becomes positive for fuel-lean or moderately fuel-rich mixtures, due to the significant probability of occurrence of instantaneous values of mixture fraction close to zero, for which κ_P and T are positively correlated (see figure 1).

This probability is increasingly smaller as the mean mixture fraction increases, so that $\overline{\kappa_P' I_b'}$ eventually becomes negative. The correlation $\overline{\kappa_P' I_b'}$ is small for low turbulence levels, i.e, $R_{\kappa I_b}$ is close to unity for $\mathrm{rms}(z)/\bar{z} \le 0.3$, as also observed for the absorption coefficient self-correlation. In the case of higher turbulence levels, it is less important than the absorption coefficient self-correlation for fuel-lean or moderately fuel-rich mixtures, and more important for very rich mixtures, as the values of $R_{\kappa I_b}$ and R_κ reveal. However, it is always less important than the temperature self-correlation.

Figure 5(d) reveals that R_{emiss}, which represents the ratio of radiative emission calculated by ignoring TRI to radiative emission computed by fully accounting for TRI, is generally lower than one, i.e., turbulence yields an increase of radiative emission, in agreement with previous theoretical analysis and experimental evidence. This increase may exceed 100% (R_{emiss} < 0.5) for high turbulence levels, which are more likely to occur at the fuel-lean side of a diffusion flame. Although R_{emiss} is also small for very fuel-rich mixtures, these have a small contribution to radiative emission in real flames, because the mean radiative emission and the turbulence level are smaller for these mixtures.

Sometimes, in order to save computing time, TRI is partially taken into account by considering temperature fluctuations, but ignoring fluctuations of κ_P. To investigate the accuracy of this approximation, figure 7 shows the ratio of the mean radiative emission evaluated by considering only temperature fluctuations to the mean radiative emission calculated by fully taking TRI into account: $R_{emiss}/R_{I_b} = \kappa_P(\bar{T},\bar{x}_i)\bar{I}_b / \overline{\kappa_P I_b}$. Large errors arise when fluctuations of κ_P are neglected and the turbulence level is high. Moreover, if turbulent fluctuations are neglected altogether, the radiative emission is underestimated, i.e., R_{emiss} < 1. In contrast, if only the influence of turbulent fluctuations on temperature is considered, the radiative emission is overestimated, i.e., $R_{emiss}/R_{I_b} > 1$.

Figure 8 shows the ratio of the absolute error of the mean radiative emission in the case of fully neglecting TRI to the corresponding error in the case of accounting only for temperature fluctuations, i.e., $\left|\kappa_P(\bar{T},\bar{x}_i)I_b(\bar{T}) - \overline{\kappa_P I_b}\right| / \left|\kappa_P(\bar{T},\bar{x}_i)\bar{I}_b - \overline{\kappa_P I_b}\right| = R_{I_b}\left|R_{emiss}-1\right| / \left|R_{emiss}-R_{I_b}\right|$. This ratio is higher than unity for z < 0.025, i.e., for very lean mixtures, as well as for low turbulence levels and very high fuel-rich mixtures. Elsewhere, the ratio is lower than unity, which means that the error in the calculation of the mean radiative emission is smaller when TRI is fully neglected than when accounting only for temperature fluctuations.

The results displayed in figure 8 may seem surprising, since previous simulations of turbulent diffusion flames have shown that predictions of the fraction of radiative heat loss and radiative heat fluxes are more accurate when temperature fluctuations are taken into account than when TRI is fully neglected [8, 24, 25]. A similar trend is reported in [24] for the total radiative emission. The reason for the contradiction between the present results and those in [24] is unclear as far as the total radiative emission is concerned. However, the present results are compatible with previous ones regarding the fraction of radiative heat loss and radiative heat fluxes. In fact, let us first suppose that TRI is fully ignored. Then, the radiative emission is underestimated, since R_{emiss} < 1, while the radiative absorption is overestimated, since R_κ > 1. Accordingly, the radiative heat loss will be underestimated. Now, suppose that only temperature fluctuations are taken into account. Then, both the radiative emission and the radiative absorption are overestimated, since $\kappa_P(\bar{T},\bar{x}_i)\bar{I}_b = \overline{\kappa_P I_b}\, R_{emiss}/R_{I_b} > \overline{\kappa_P I_b}$, and R_κ > 1. Therefore, the error in the solution of the time-averaged RTE may be lower, because both the emission and the absorption are overestimated, and errors in the evaluation of the mean emission and mean absorption compensate to some extent. In contrast, the errors add up if TRI is fully ignored.

Figure 9 shows the absolute error of the mean radiative emission when TRI is ignored, normalized by the maximum radiative emission without turbulent fluctuations, which occurs for a slightly fuel-rich mixture (z = 0.059): $\left|\kappa_P(\bar{T},\bar{x}_i)I_b - \overline{\kappa_P I_b}\right| / (\kappa_P I_b)_{\max} = \left|1 - R_{emiss}\right| \overline{\kappa_P I_b} / (\kappa_P I_b)_{\max}$. The larger this normalized error, the greater the importance of TRI. Although R_{emiss} is lowest for $\bar{z}$ < 0.01, radiative emission is too small for such values of $\bar{z}$, and therefore TRI does not have any relevant

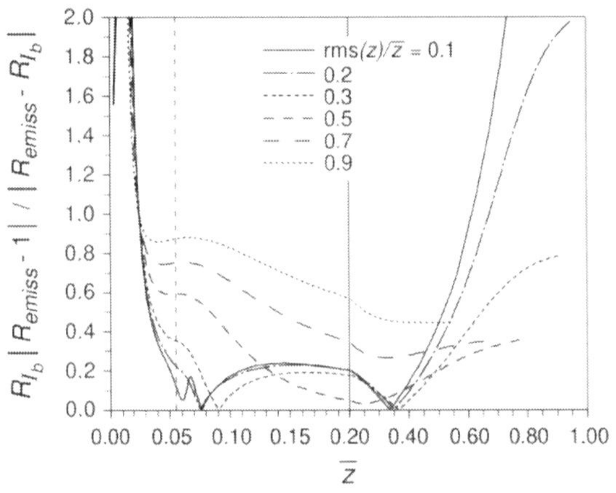

Figure 8. Ratio of the absolute error of $\overline{\kappa_P I_b}$ in the case of fully neglecting TRI to the corresponding error in the case of accounting only for temperature fluctuations.

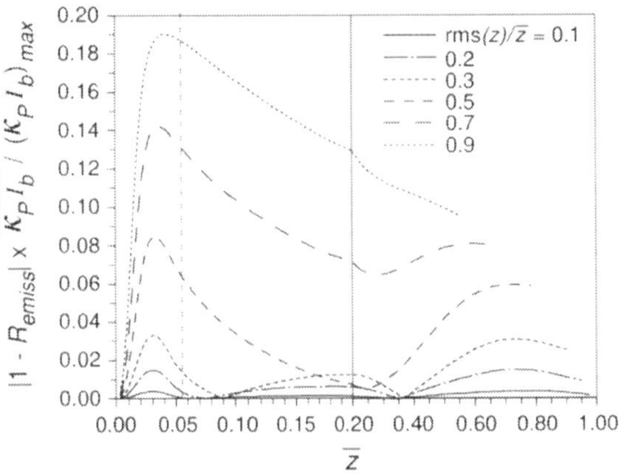

Figure 9. Influence of turbulent fluctuations on the normalized absolute error of the mean radiative emission when TRI is ignored.

influence in radiative transfer calculations for these very lean mixtures. TRI is increasingly more important as the radiative emission increases and R_{emiss} departs from unity. In the fuel-lean region, radiative emission increases and R_{emiss} approaches unity as $\bar{z}$ increases. Figure 9 shows that TRI is most important for $\bar{z}$ in the range 0.03–0.04, and becomes progressively less significant as $\bar{z}$ further increases. Although TRI tends to increase for very rich mixtures, turbulence levels are typically small for those mixtures, and therefore TRI is not relevant in the very rich region of real flames.

A few words of caution are appropriate before concluding. First, the present analysis is restricted to turbulent diffusion flames in the flamelet regime. It is not applicable to flames where extinction and re-ignition phenomena are significant. Second, it does not account for radiation from soot. This is acceptable for methane flames, which produce very little soot. Third, it is assumed that the pdf of mixture fraction is a clipped Gaussian. Although this assumption is often used in the literature, the pdf shape may slightly influence the results. However, it is unlikely to alter qualitative results and conclusions. Finally, the analysis only addresses radiative emission. It does not consider radiative absorption, and therefore does not fully characterize TRI effects in radiative transfer calculations for turbulent reactive flows. It does show, however, where TRI effects might be important.

4. Conclusions

A theoretical analysis of the influence of turbulence on radiative emission in turbulent diffusion methane flames was presented. The influence of turbulence on radiative emission is only dependent on local quantities. Therefore, it is independent of the optical thickness of the medium. The flamelet equations were solved for a range of scalar dissipation rates, and mean values of fluctuating quantities were evaluated by prescribing the pdf shape of mixture fraction. The relevance of TRI was investigated as a function of mean and variance of mixture fraction. The following conclusions are useful to identify the flame regions where turbulence influences radiative emission, and how strong that influence is, regardless of details of the combustion system and operating conditions.

The results show that turbulent fluctuations generally contribute to reduce the Planck mean absorption coefficient of the medium, in contrast with the blackbody radiation intensity, which is significantly increased by turbulence. If the turbulence level is relatively small, the influence of turbulence on the absorption coefficient is marginal. Otherwise, fluctuations of the absorption coefficient of the medium, arising from fluctuations of temperature and molar fractions of the species, should be taken into account.

The influence of turbulence on the mean radiative emission is generally lower than that predicted by accounting only for temperature fluctuations. Moreover, if turbulent fluctuations are neglected altogether, the mean radiative emission is underestimated, while if only the influence of turbulent fluctuations on temperature is considered, the mean radiative emission is overestimated. The error in the calculation of the mean radiative emission is smaller when TRI is fully neglected than when accounting only for temperature fluctuations, in contrast with results reported in the literature for the radiative heat loss and heat fluxes in turbulent flame simulations, which show improved predictions when the influence of turbulence on temperature is considered.

The influence of turbulence on radiative emission is more relevant in the fuel-lean region and close to stoichiometric conditions than in the fuel-rich region. In the former case, although the mean temperature is lower than the maximum one, the turbulence level is high due to flame intermittency, and the effects of turbulence on radiation are more important than in the latter case.

Acknowledgements
Financial support of FCT-Fundação para a Ciência e a Tecnologia within the framework of project PDTC/EME-MFE/102405/2008 is acknowledged.

References

[1] Faeth J G, Gore M P, Chuech S G and Jeng S M 1989 Radiation from turbulent diffusion flames *Annual Reviews of Numerical Fluid Mechanics and Heat Transfer* (vol 2), ed C L Tien and T C Chawla (New York: Hemisphere) pp 1–38.
[2] Coelho P J 2007 *Prog. Energy Combust. Sci.* **33** 311–83.
[3] Coelho P J 2012 *J. Heat Transfer* **134** 031001.
[4] Adams B R and Smith P J 1995 *Combust. Sci. Technol.* **109** 121–40.
[5] Coelho P J, Teerling T and Roekaerts D 2003 *Combust. Flame* **133** 75–91.
[6] Habibi A, Merci B and Roekaerts D 2007 *Combust. Flame* **151** 303–20.
[7] Mazumder S and Modest M F 1999 *Int. J. Heat Mass Transfer* **42** 971–91.
[8] Li G and Modest M F 2003 *J. Heat Transfer* **125** 831–8.
[9] Tessé L, Dupoirieux F and Taine J 2004 *Int. J. Heat Mass Transfer* **47** 555–72.
[10] Cox G 1977 *Combust. Sci. Technol.* **17** 75–8.
[11] Burns S P 1999 Radiation Interaction Modelling in Hydrocarbon Pool Fire Simulations *Report No. SAND99-3190*, Sandia National Laboratories.
[12] Modest MF 2003 *Radiative Heat Transfer* (New York: Academic Press).
[13] Kabashnikov V P and Kmit G I 1979 *J. Applied Spectroscopy* **31** 963–7.
[14] Smith G P *et al.*, GRI-Mech 3.0, available from http://www.me.berkeley.edu/gri_mech/.
[15] Peters N 1984 *Prog. Energy Combust. Sci.* **10** 319–39.
[16] Pitsch H 1998 FlameMaster v3.1, A C++ Computer program for 0D combustion and 1D laminar flame calculations, available from http://www.stanford.edu/group/pitsch/.
[17] Pitsch H, Chen M and Peters N 1998 *Proc. Combust. Inst.* **17** 1057-64.
[18] Young K J and Moss J B 1995 *Combust. Sci. Technol.* **105** 33-53.
[19] Gore J P, Ip U-S and Sivathanu Y R 1992 *J. Heat Transfer* **114** 487-493.
[20] Barlow R S, Karpetis A N, Frank J H and Chen J-C 2001 *Combust. Flame* **127** 2102-18.
[21] Grosshandler W H 1993 RADCAL: A narrow-band model for radiation calculations in a combustion environment, *NIST Technical Note 1402*.
[22] Coelho P J 2011 A theoretical analysis of the influence of turbulence on radiative emission in turbulent reactive flows *Proc. European Combustion Meeting*, Cardiff, U.K., 28 June – 2 July.
[23] International Workshop on Measurement and Computation of Turbulent Nonpremixed Flames, available from http://www.sandia.gov/TNF/simplejet.html.
[24] Coelho P J 2004 *Combust Flame* **136** 481–92.
[25] Wang A and Modest M F 2008 *J. Quant. Spectrosc. Radiat. Transfer* **109** 269–79.

Journal of Physics: Conference Series 369 (2012) 012011 doi:10.1088/1742-6596/369/1/012011

Influence of radiative property models on soot production in laminar coflow ethylene diffusion flames

Rodrigo Demarco[1,2,3], **Jean-Louis Consalvi**[1,3] **and Andrés Fuentes**[4]

[1] Université de Provence, IUSTI/ UMR CNRS 6595, 5 rue E. Fermi, 13453 Marseille Cedex 13, France
[2] Institut de Radioprotection et de Sûreté Nucléaire (IRSN), BP 3, 13115 St Paul-Lez-Durance, France
[3] ETIC Lab., IRSN-CNRS-UAM (I,II), 5 rue Enrico Fermi, 13453 Marseille Cedex 13, France
[4] Departamento de Industrias, Universidad Técnica Federico Santa María, Av. España 1680, Valparaíso, Chile

E-mail: rodrigo.demarco@polytech.univ-mrs.fr

Abstract. Two axisymmetric laminar coflow non-smoking and smoking ethylene diffusion flames are studied numerically in order to assess the influence of different radiative property models on the soot formation and oxidation processes. Simulations are carried out by considering the Steady Laminar Flamelet (SLF) concept and a modified two-equation acetylene-based model to describe the soot nucleation, surface growth and oxidation processes. Several radiative property models are considered: the simple Optically Thin Approximation (OTA), the Weighted-Sum of Grey-Gases (WSGG), the Grey-Wide-Band model (GWB), the Statistical Narrow Band Correlated-k model (SNBCK) and the Full Spectrum Correlated-k model (FSCK). Comparisons between calculations carried out with the SNBCK model and experimental data show a reasonable agreement. Model results show that the choice of the radiative property models influence largely the soot prediction, especially in the upper part of the flame where oxidation occurs. Simulations show that the reabsorption of spectral gas and soot is an important feature and thus the commonly used OTA or grey models introduce large discrepancies. The GWB model leads to improved solutions, but it should be avoided if accurate predictions are desired. The FSCK provides equivalent results as compared to the SNBCK model with a substantial gain in CPU time.

1. Introduction

An important topic of research is to predict the sooting characteristics of a specific fuel or understanding how the soot is produced and consumed in order to effectively control it. Great efforts have been done in understanding the mechanisms of soot inception, growth and oxidation. Detailed models involving several hundreds of elementary reactions and comprehensive models for both the soot-particle dynamics and the soot kinetics were successfully applied to compute ethylene coflow diffusion flames [1]. Simplified semi-empirical models [2-6], describing the main features of soot formation processes, were also developed. One characteristic that is well recognized is the important coupling between soot kinetics and thermal radiation. For laminar coflow diffusion flames the radiative heat transfer is of vital importance to predict the soot kinetics, since it significantly affects the temperature and therefore the complex reactions associated with the soot processes. This coupling is generally treated in a simple manner using the Optically Thin Approximation (OTA) [7–9], despite some studies which concluded that a detailed modelling of thermal radiation is necessary for accurately predict the flame structure, species concentrations and formation of soot [10–13]. In one of these studies, Liu *et al.* [11] showed that the use of the OTA induces significant temperature underpredictions in the upper part of coflow laminar ethylene diffusion flames, affecting directly the prediction of soot volume fractions. Despite these efforts, an extensive comparison of several radiative property models on soot predictions has not been reported.

The aim of the present study is to assess the influence of different radiative property models over the soot formation/oxidation processes in axisymmetric laminar coflow ethylene flames. To accomplish this task, a Non-Smoking (NS) and a Smoking (S) flame, previously investigated experimentally by Santoro *et al.* [14] and Megaridis and Dobbins [15, 16], are simulated by considering different radiative property models. The selection of the radiative models follows a bibliographic survey, choosing those that are most representative and commonly used. The Steady Laminar Flamelet (SLF) model is used as combustion model and a modified version of the two-equation soot model proposed by Leung *et al.* [2] is employed to predict soot nucleation, surface growth and oxidation.

2. Model formulation and numerical method

The numerical simulation of the laminar flames includes the solution of the overall continuity equation, the Navier-Stokes equations in a low Mach number formulation and transport equations for the mixture fraction (Z), the soot mass fraction (Y_s), the soot number density per unit mass of mixture (N_s) and the total enthalpy (h).

Chemical interactions are modelled by using the Steady Laminar Flamelet concept. This model allows obtaining state relationships for the scalar properties of the medium as function of the mixture fraction, the scalar dissipation rate ($\chi = 2D_{th}(\overrightarrow{\nabla Z} \cdot \overrightarrow{\nabla Z})$) and an enthalpy defect parameter ($X_R = (h - h_{ad})/(h_u - h_{ad})$). In the last expressions D_{th} is the thermal diffusivity and h_{ad} and h_u are the adiabatic and unburned enthalpies, respectively. The flamelet library was generated by using the GRI-Mech 3.0 chemical reaction mechanisms [17]. The enthalpy defect parameterization is introduced by using the methodology described by Carbonell *et al.* [18].

2.1. Soot model

The soot kinetics is modelled using a semi-empirical acetylene-based two-equation model proposed by Leung *et al.* [2] and modified later by Liu *et al.* [11]. This model accounts for the contributions of soot nucleation, surface growth and oxidation. Thermophoretic velocities of soot are expressed as $V_{T,i} = -0.55\,\mu/(\rho T)\,\partial T/\partial x_i$, where μ, ρ and T represent the viscosity, the density and the temperature of the gas mixture, respectively. $x_i = r, z$ represents the radial and axial directions.

Acetylene (C_2H_2) is assumed to be the main gaseous species involved in the nucleation (r_1) and surface growth (r_2) processes. These reactions and their reaction rates are assumed as follows:

$$C_2H_2 \rightarrow 2C(S) + H_2 \qquad \text{with} \qquad r_1 = 2 \times 1.7\exp(-7548/T)[C_2H_2] \tag{1}$$

$$C_2H_2 + nC(S) \rightarrow (n+2)C(S) + H_2 \qquad \text{with} \qquad r_2 = 2 \times 5.0\exp(-6038/T)A_s[C_2H_2] \tag{2}$$

In these reactions $C(S)$ represents carbon in its solid form. $A_S = \pi d_S^2(\rho N_S)$ and $[C_2H_2]$ represent the total soot surface area and the molar concentration of acetylene. $d_S = (6Y_S/\pi N_S \rho_S)^{1/3}$ is the diameter of the (assumed) spherical soot particles and $\rho_S = 2000$ kg/m^3 is the soot density.

The soot oxidation process is attributed to OH, O_2 and O reactions. Soot oxidation by O_2 is given by the NSC model, while soot oxidation by OH and O is based on the Fenimore and Jones model. These reactions and their reaction rates are assumed as follows:

$$0.5O_2 + C(S) \rightarrow CO \qquad \text{with} \qquad r_3 = 120\left[\frac{k_a X_{O_2}\chi_1}{1 + k_z X_{O_2}} + k_b X_{O_2}(1 - \chi_1)\right]A_S f_{O_2} \tag{3}$$

$$OH + C(S) \rightarrow CO + H \qquad \text{with} \qquad r_4 = 1270\varphi_{OH}T^{-1/2}X_{OH}A_S f_{OH} \tag{4}$$

$$O + C(S) \rightarrow CO \qquad \text{with} \qquad r_5 = 55.4\varphi_O T^{-1/2}X_O A_S \tag{5}$$

In the latter expressions X_i represents the mole fraction of the i-th species, whereas the collision efficiency factors, φ_{OH} and φ_O, are fixed to 0.05 and 0.2 respectively. χ_1 is expressed as $\chi_1 = (1 + k_t/k_b x_{O_2})^{-1}$, with k_a, k_b, k_z and k_t defined from [19]. The rate constants used in r_4 and r_5 are obtained from [20] and [21] respectively. The commonly used NSC model and the OH soot oxidation rates are modified in this study by introducing two temperature dependent correction factors proposed by Liu *et al.* [11].

$$f_{O_2} = (1 + \exp[-(T - 1650)/80])^{-1} \tag{6}$$

$$f_{OH} = \begin{cases} 0.1824(1 + \exp[-(T - 1600)/85])^{-1}, & T \leq 1600\text{ K} \\ (1 + \exp[-(T - 1675)/50])^{-1}, & 1600\text{ K} < T < 1675\text{ K} \\ (1 + \exp[-(T - 1650)/80])^{-1}, & T \geq 1675\text{ K} \end{cases} \tag{7}$$

Finally, the source terms of the transport equations for the soot number density (S_N) and soot mass fraction (S_m) are defined as

$$S_N = \frac{N_A}{C_{min}} r_1 \qquad \text{and} \qquad S_m = (r_1 + r_2)W_s - r_3 - r_4 - r_5\,, \tag{8}$$

where N_A is the Avogadro's number (6.022×10^{23} part/mol), C_{min} is the number of carbon atoms in the incipient soot particle (700) and W_s is the soot molecular weight (12.011 g mol^{-1}). In the source term of the soot number density (see equation (8)), the agglomeration process is neglected based on the experimental data of Megaridis and Dobbins [15, 16], which established that soot aggregates consist of more or less identical primary particles and obtained an almost constant primary soot particle density in the soot growth region. Therefore the primary soot particle coalescence can be neglected [10].

2.2. *Radiation models*

Several radiative property models are asses in the present study. A brief description of every model is presented below. More details can be found in [22].

In this study the soot particles are assumed to attain the Rayleigh scattering limit. In this regime the scattering is negligible and the spectral soot absorption coefficient, at wavelength λ, is given by $\kappa_{s,\lambda} = 5.5\ f_s/\lambda$. The soot volume fraction f_S is defined as $f_S = \rho Y_S/\rho_S$. Depending on the radiative model, a Planck-averaged absorption coefficient for the soot can be required. This coefficient is given by the following expression [23]

$$\kappa_{s,\text{Planck}} = 5.5\ (3.83/14388 \times 10^{-6}) f_s T \tag{9}$$

In this study only CO_2 and H_2O are considered as the participating gases, neglecting the other combustion species. A previous study showed that neglecting the contribution of CO has only a minor influence on the divergence of the radiative flux [24].

2.2.1. Optically Thin Approximation model (OTA). This model is widely used as a first approximation to estimate the divergence of the radiative flux. It neglects the gas and soot reabsorption, assuming that the medium is optically thin. In the present study the Planck-mean absorption coefficients for the combustion gases are evaluated from the correlations given in [25], whereas the Planck-mean absorption coefficient for the soot is computed from equation (9).

2.2.2. Planck model. This grey model solves the Radiative Transfer Equation (RTE) with the Planck-mean absorption coefficients computed as described by the OTA.

2.2.3. Statistical Narrow Band Correlated-k model (SNBCK). This model is based on the assumption that the spectrum can be divided into narrow bands (NB), in which the blackbody intensity and soot properties can be considered as constant. The erratic spectral absorption coefficient is then reordered on each NB to form a monotonically increasing function, called the k-distribution, which can be easily integrated. A 7 points Gauss-Legendre quadrature scheme is used for this integration. NB k-distributions are generated from the 43 NB database of Soufiani and Taine [26] and mixed using the scheme of Modest and Riazzi [27]. To speed up the solution a mixed NB database of k-distributions is pre-generated for 15 values of $R = X_{CO_2}/X_{H_2O}{}^0$, between 0.01 and 2.2, and 23 temperatures in the range between 300 and 2500 K. This database was generated by considering a fixed H_2O concentration of $X_{H_2O}{}^0 = 0.1$. This procedure implies that the weak dependence of spectral line broadening on the species mole fraction is neglected [28], allowing to express the absorption coefficient of the mixture as:

$$\kappa_{mix}(R,T,X_{H_2O}) = \frac{X_{H_2O}}{X_{H_2O}{}^0}\kappa_{mix}{}^0(R,T) \tag{10}$$

where $\kappa_{mix}{}^0(R_i,T_j) = \kappa_{H_2O}(X_{H_2O}{}^0,T_j) + \kappa_{co_2}(R_i * X_{H_2O}{}^0,T_j)\big|_{i=1,15;j=1,23}$ are the tabulated values. For mixtures of arbitrary species mole fraction ratios and temperature, the absorption coefficient is computed from the database by carrying out linear interpolations on R and spline interpolations on T. Introducing this database has a minor influence as compared to direct SNBCK calculations. Soot absorption coefficients for each NB are estimated as $\kappa_{s,\eta_i} = 5.5f_s(r,z)\eta_i$, where η_i represents the i-th NB. The SNBCK model is also used to assess the validity of the grey assumption for the soot. In this case the grey soot absorption coefficient is estimated by equation (9). This model is denoted as SNBCKgs hereafter.

2.2.4. The Weighted-Sum of Grey Gases (WSGG). This model is based on expressing the total gas emissivity as the weighted-sum emissivity of four grey gases. In this study a non-grey formulation of this model is considered, working directly with the absorption coefficients and emission weighted factors provided by the model. Then the RTE is solved directly for the four "grey gases". Emission weighted factors and absorption coefficients for each grey gas are calculated from the polynomial relationships of Smith *et al.* [29]. The coefficients involved in the polynomials are tabulated for particular ratios of partial pressure p_{H_2O}/p_{CO_2}. In the present study, parameters tabulated for $p_{H_2O}/p_{CO_2} = 1$ are used, corresponding to the stoichiometric ratio for the complete combustion of ethylene in air. The soot absorption coefficient is considered as grey (equation (9)) in this model.

2.2.5. Grey-Wide-Band model (GWB). This model considers that the spectrum can be divided into 6 bands and in each one a Planck-mean absorption coefficient is considered. This model is implemented in the CFD code Fire Dynamic Simulator (FDS) [30]. As in FDS, the absorption coefficient of each wide band is computed from the RADCAL database. It should be noted that Planck coefficients computed from RADCAL and from the database of Soufiani and Taine are in close agreement [31, 32]. As a consequence, the use of these different databases to compute radiative properties should not be a great source of discrepancies. In addition, the part of the spectrum considered in the RADCAL database has been restricted to be consistent with that of Soufiani and Taine. In FDS a grey formulation is also implemented. This radiative model was used for the calibration of new soot models based on the concept of the smoke point [5, 6]. In this case the absorption coefficient is computed as $\kappa = \min(\kappa_e, \kappa_p)$, where κ_e is an effective absorption coefficient and κ_p the Planck-mean absorption coefficient. κ_e is determined to match the total radiative intensity leaving a uniform medium of path length l bounded by black walls at 0 K. The path length is taken as $l = 2r_f = 8.13(D_\infty^2 l_f/g)^{1/4}$, where r_f, D_∞, l_f and g are the characteristic radius of the flame, the diffusion coefficient, the flame height and the acceleration of gravity, respectively. This model is denoted as GWBg hereafter.

2.2.6. Full Spectrum Correlated-k model (FSCK). This model is an extension of the *SNBCK* model, in which the reordering is carried out into the entire spectrum at once. This allows an important decrease in the number of resolutions of the RTE and then in the computational resources. The Full Spectrum (FS) k-distributions are assembled from the same 43 mixed-NB database used for the SNBCK model. The assembly scheme chosen is the one presented by Modest and Riazzi [27]. Two different procedures to generate the FS k-distributions are implemented and compared in this study: in the first, non-grey soot is added to the pre-databased mixed-NB k-distributions and then the k-distributions are assembled (FSCK MNB). In the second, mixed-FS k-distributions are pre-databased (assembled from the mixed-NB database) and then a Planck-mean soot absorption coefficient is directly added (FSCK MFS). The latter procedure is much more efficient on a computational point of view as the assembly procedure is not carried out during the sequence of calculations, but this formulation requires considering soot particles as grey. For both formulations the integration of the FS k-distributions is carried out considering a 7 points Gauss-Legendre quadrature scheme. One characteristic of this model is the need to define a reference state $(T, X_{CO_2}, X_{H_2O}, f_s)_{ref}$ to treat non-homogeneous non-isothermal conditions. A simple average over the flame volume is considered for the gas species and soot volume fraction, while an emission-weighted temperature is used as the reference temperature [22, 24].

2.3. Numerical method
The governing equations are discretized and solved in cylindrical coordinates using a control volume method on a staggered grid. For the convective terms the ULTRASHARP approach is applied, while a second-order central difference scheme is used for the diffusion terms. The pressure-velocity coupling is solved using the Iterative PISO algorithm. The tridiagonal matrix algorithm is used to solve the transport equations. The overall computational domain is $4.5cm\ (r) \times 13.12cm\ (z)$. A non-uniform mesh with $89(r) \times 400(z)$ cells is used, with a mesh size of $0.1875mm \times 0.375mm$ in the flaming zone. Further refinement of the computational mesh has negligible effects on the results. The radiative transfer equation is solved by the finite volume method using the special mapping developed by Chui *et al.* for axisymmetric configurations [33]. Computations are carried out using the first-order UPWIND spatial discretization scheme and an angular mesh with 12x16 control angles. Convergence is achieved when the maximum relative error for all the computed variables is lower than 10^{-4}.

3. Results and discussion
The different radiative property models have been used to simulate the Smoking (S) and Non-Smoking (NS) ethylene coflow flames studied experimentally by Santoro *et al.* [14], Megaridis and Dobbins [15, 16], Kennedy *et al.* [3] and Smyth [34]. Experiments used a burner consisting in two concentric tubes of 11.1mm and 101.6mm inner diameter respectively. Pure ethylene flows through the inner tube, while air passes through the annular region between the two tubes. The mean velocities for the fuel and air streams are of $3.98 cm.s^{-1}$ and $8.9 cm.s^{-1}$ for the NS flame and $5.05 cm.s^{-1}$ and $13.3 cm.s^{-1}$ for the S flame. A parabolic profile is imposed for the fuel injection zone while a flat profile is considered for the oxidant injection zone. A small solid wall of 0.55 mm separates both injection zones. Fuel and oxidizer are injected at ambient temperature, neglecting the effects of preheating. Ambient temperature and pressure are fixed to 300 K and 1atm.

3.1. Comparison with available data
A first comparison is carried out in order to assess the main characteristics of the simulations and have an overall idea of the accuracy obtained. Computations carried out with the SNBCK model are compared to the experimental data of Santoro *et al.* [14] and Megaridis and Dobbins [15, 16]. Figures 1a and 1b present the radial profiles of temperature and soot volume fraction at three heights of the NS flame. Overall predictions present an acceptable agreement with the measurements, especially for the temperature. The locations of the temperature peaks as well as their magnitudes are well predicted at z = 7 mm and 20 mm. The preheating of the fuel caused by conduction with the pipe flow is disregarded

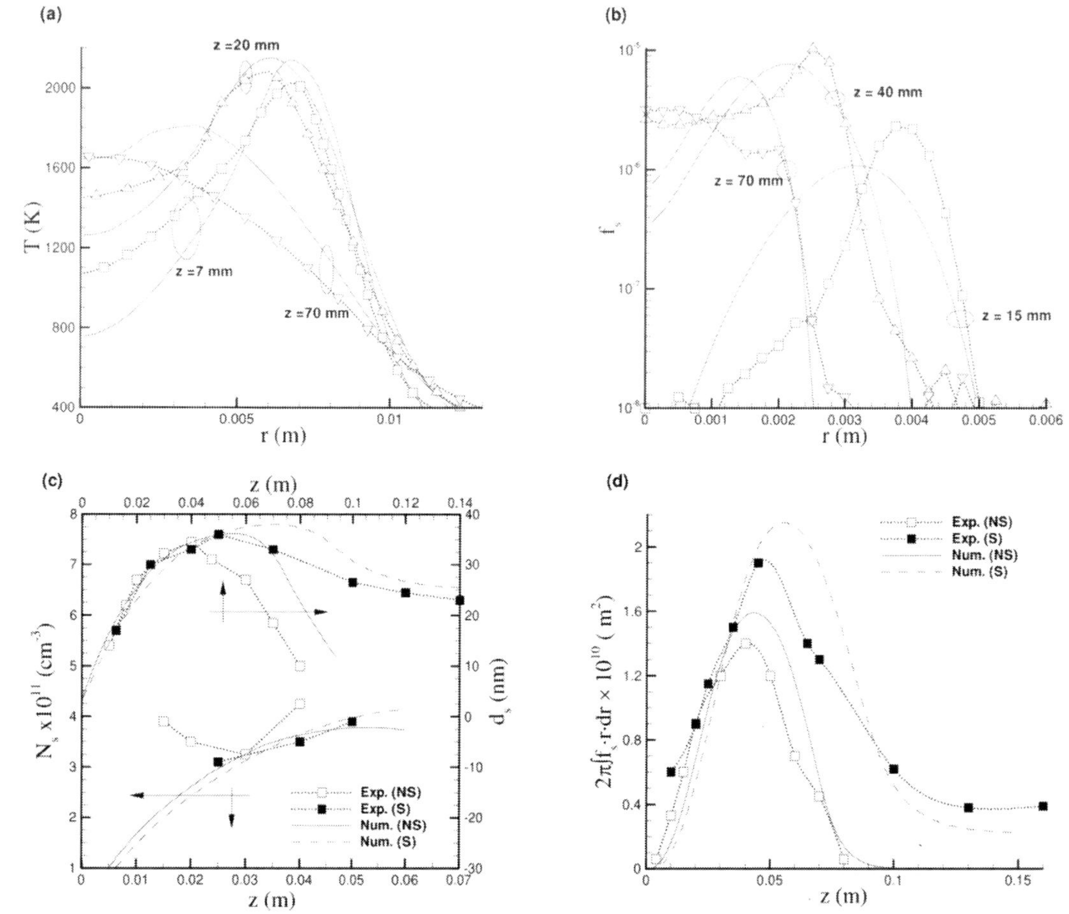

Figure 1. Comparisons between experimental data and predictions obtained with the SNBCK model. Radial profiles of (a) temperature and (b) soot volume fraction at different heights for the NS flame, experimental values from Santoro *et al.* [14]; (c) soot primary particle number density and soot primary particle diameter along the path line exhibiting the maximum soot volume fraction for the NS and S flames, experimental values from Megaridis and Dobbins [15, 16]; and (d) integrated soot volume fraction along the flame height, experimental values from Santoro *et al.* [14].

in this study, which may explain the underprediction observed for the centreline values (see figure 1a). Figure 1b shows that the predicted profiles of soot volume fraction at these two heights present wider peaks, with values slightly lower than the experimental values. The results at z = 70 mm show that the numerical simulations still predict the peaks of soot volume fraction and temperature in an annular region whereas the experiments exhibit maximum values along the centreline (see figures 1a and 1b).

Figure 1c presents the soot primary particle number density and the soot primary particle diameter along the path line exhibiting maximum soot volume fraction for both flames. The experimental data show that the values of soot primary particle number density in both flames are quite similar in most of the soot growth region. Numerical results are in reasonable agreement with the experimental data for both flames. The soot primary particle diameter also presents the same behaviour as observed experimentally. For both flames an excellent agreement between simulations and experiments is achieved within the first 4cm. Above this height the diameter is somewhat overpredicted, showing that the model underestimate the oxidation process. Figure 1d presents the integrated soot volume fraction along the flame height for both flames. Results show that two distinct regions are observed: in the first region, soot nucleation and surface growth processes produce an important increase in the soot volume fraction with height. This is followed by an oxidative region where the overall soot volume fraction decreases until the soot is completely burned, as in the NS flame, or partially burned, as in the S flame. A comparison between the experiments and the numerical results shows that the integrated soot volume fraction is slightly underestimated in the lower part of both flames. At the peaks this tendency is inverted and a more important overestimation is observed. The predicted peak locations are shifted by a few millimetres away from the burner as compared with the experiments.

3.2. Influence of the radiative models

The accuracy of the different radiative property models is assessed by using the SNBCK model as reference for the S flame. It is important to mention that only Line-by-Line (LBL) calculations provide 'exact' solutions of the radiative heat transfer. However, the use of the LBL method in multidimensional combustion problems requires prohibitive computational resources and is beyond the scope of this paper. The SNBCK assumes a statistical description of each line of the spectrum and for non-isothermal non-homogeneous medium it is assumed that the absorption coefficient is correlated. Previous numerical studies have demonstrated that the SNBCK model, in the absence of LBL solutions, is accurate enough to be used as the reference solution. It has been shown also that the 43 NB database provides solutions in close agreement with the 367 NB database (both from Soufiani and Taine) [22].

Local relative errors between the current model and the reference are estimated as

$$E_{r,i}(\phi) = \left(\phi_{model,i} - \phi_{ref,i}\right)/\phi_{ref,i} \tag{11}$$

where i represents the current mesh point and ϕ is the variable considered. In order to avoid interpreting discrepancies encountered in regions where ϕ is close to zero, points where $\phi_{ref,i} < 0.02 * \max_{i=1,N} \phi_{ref,i}$ are excluded from the error analysis. Mean and maximum errors are defined as $E_{mean} = 1/N\left(\sum_{i=1}^{N}|E_{r,i}|\right)$ and $E_{max} = \max_{i=1,N}|E_{r,i}|$, respectively.

Figures 2a, 2b and 2c present the temperature, the divergence of the radiative flux (radiation source term for the energy equation) and the soot volume fraction along the path of maximum soot volume fraction for the SNBCK model. In these figures the temperature difference and the local relative error are also presented for all the radiative property models studied. Figure 2a shows that along the path line exhibiting the maximum soot volume fraction the temperature increase rapidly in the first 3cm and then reaches an approximately constant value in the range of 1500-1600 K; at around z = 9.5 cm the temperature starts to decrease linearly until it attains a temperature of 1300 K at z = 15 cm. The zone of constant temperature is characterized by very large radiative losses (see figure 2b); the large soot volume fraction observed in this region is the main responsible of this intense radiation (see figure 2c). Figure 2b shows that the radiative source term decreases after this zone due to a reduction in both temperature and soot volume fraction (see figures 2a and 2c). Figure 2d shows the integrated soot volume fraction along the flame height for the SNBCK model and the relative errors obtained with the different radiative model considered. Between z=6 cm to z=12 cm the dominant mechanism is the soot oxidation. The results show that the oxidation process is quite sensitive to the radiative property model, more sensitive than the nucleation and the surface growth of the soot. This is directly related to the temperature attained for every model, conditioning the oxidation rates in each simulation. This effect produces important relative errors between the models, especially in the oxidation region (see Figure 2). Therefore the choice of a radiative property model largely affects the amount of soot released by the flame.

Table 1 summarizes the mean and maximum relative error, or temperature difference, obtained for all the radiative models presented in figure 2. The first important result is that the OTA leads to very poor predictions, with a maximum underprediction of the temperature of about 139 K. Maximum errors are higher than 100% for all the variables considered and the amount of soot released from the flame is largely overestimated. In addition the OTA also affects significantly the soot production in the soot growth zone, where the soot volume fraction is around 15% lower as compared to the SNBCK

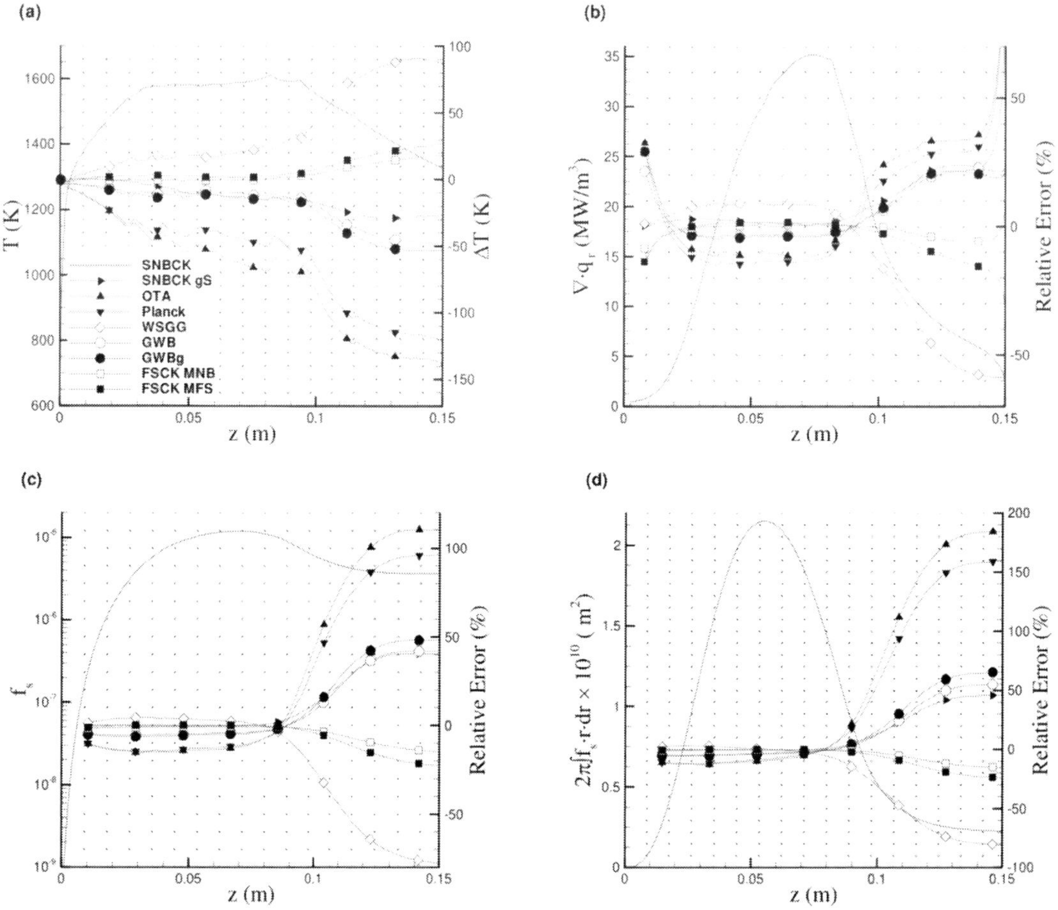

Figure 2. Comparisons between radiative property models for the S flame. The solution predicted by SNBCK model acts as reference. (a) Temperature and temperature differences $\Delta T = T_{model} - T_{ref}$, (b) divergence of the radiative flux and (c) soot volume fraction along the path line exhibiting the maximum soot volume fraction; and (d) integrated soot volume fraction along the flame height. Relative error is calculated with equation (11). The reference curve is represented on the left axis whereas the relative errors for the other methods are represented on the right axis.

(see figure 1b). These comparisons show clearly that reabsorption by gas and soot should be considered. If reabsorption is taken into account by considering the simple grey model based on Planck-mean absorption coefficient (Planck Model), table 1 shows that temperature differences and relative errors are similar to those obtained with the OTA. The grey model implemented in FDS (GWBg) improves the predictions but discrepancies are still high. The influence of the grey approximation for soot can be isolated by considering non-grey gases, modelled by the SNBCK, and grey soot modelled by a Planck-mean absorption coefficient. Table 1 shows that this approximation (SNBCKgs) should also be avoided, demonstrating that the spectral nature of soot must be taken into account. Although temperature is relatively well predicted with a maximum difference of 29 K as compared to the reference, it produces substantial mean errors of about 8.5, 12.5 and 14.3% for the radiative source term, the soot volume fraction and the integrated soot volume fraction respectively.

GWB and WSGG take into account the spectral dependence of combustion gases and soot in a simplified manner. The GWB model presents similar discrepancies as the GWBg discussed previously with a maximum temperature difference of about 50 K. The parameters of the WSGG model, *i.e.* a_j

Table 1. Error estimations, or temperature difference, for the temperature, the divergence of the radiative flux and the soot volume fraction along the path of maximum soot volume fraction; and the integrated soot volume fraction along the flame height calculated with different radiative property models for the S flame. The SNBCK model is used as reference. Relative errors are estimated with equation (11). Last column presents the CPU time ratio between the reference and the other models. $CPU_{ratio} = t_{ref}/t_{model}$

Model	T (K)		$\nabla \cdot \dot{q}_r''$ (%)		f_s (%)		Integ (%)		*CPU*
	ΔT_{mean}	ΔT_{max}	$E_{r,mean}$	$E_{r,max}$	$E_{r,mean}$	$E_{r,max}$	$E_{r,mean}$	$E_{r,max}$	*ratio*
SNBCKgs	14.1	29.0	8.5	23.9	12.5	40.8	14.3	46.1	1.0
OTA	72.1	139.0	19.7	135.6	42.4	111.1	65.3	184.8	153.4
Planck	59.2	122.0	18.7	130.9	36.7	96.6	55.9	159.4	111.2
WSGG	37.8	91.0	20.2	59.2	24.7	77.7	27.6	81.2	52.9
GWB	19.4	50.0	9.6	79.3	14.5	42.1	17.9	55.4	33.4
GWBg	23.0	54.0	9.3	29.1	16.9	48.4	21.5	65.6	63.9
FSCK MNB	4.7	18.0	2.2	16.0	4.1	14.8	4.3	15.4	4.0
FSCK MFS	7.9	27.0	5.1	29.3	5.8	22.7	6.7	24.1	20.5

and κ_j, are obtained for fixed ratios of H_2O and CO_2 partial pressures. In the present study, a ratio of $P_{H_2O}/P_{CO_2} = 1$ is considered, resulting from the complete combustion of ethylene in air. It should be noted that this condition is never encountered in the computational domain as a consequence of incomplete combustion and differential diffusion. This may explain the large discrepancies observed for this model.

The FSCK model using the mixed-NB database and spectral soot (FSCK MNB) predicts temperatures that are quite close to the reference, with only mean and maximum differences of around 5 and 18 K, respectively. This model also present the lowest relative errors among all the radiative models tested, with mean and maximum errors lower than 4.4 and 16%. These discrepancies may be explained by the assumption of a correlated absorption coefficient, which is introduced for both NB and FS k-distributions in the case of a non-homogeneous non-isothermal medium. This assumption is less restrictive on a NB basis than on a FS basis. By considering the mixed-FS database and grey soot (FSCK MFS) the solution are obviously less accurate. Nevertheless, the errors are globally lower than the SNBCKgs and closer to the FSCK MNB.

Table 1 also summarizes the CPU time ratio between the current model and the SNBCK ($CPU_{ratio} = t_{ref}/t_{model}$). It is important to mention that each calculation differs only by the radiative model considered. The FSCK MNB model, which present the best results compared to the SNBCK, provides a solution with a CPU time reduced by a factor of 4, making it attractive for this kind of calculation.

4. Conclusions

A numerical study of two axisymmetric laminar coflow smoking and non-smoking ethylene diffusion flames was carried out in order to assess the influence of different radiative property models on soot formation and oxidation processes. The SLF concept was used as combustion model and a modified two-equation acetylene-based model was used to describe the soot nucleation, surface growth and oxidation processes. The main conclusions of this study can be summarized as follows:

1. Comparisons between the experimental data and the numerical solutions obtained with the SNBCK model show a fairly good agreement.
2. Predicted soot volume fractions are very sensitive to the radiative property model used, especially in the zone where the soot is oxidized. In particular the amount of soot released by the smoking flame is highly dependent of the radiative model. As a consequence numerical study concerning soot formation/oxidation processes should consider an accurate radiation model.

3. Gas and soot reabsorptions are crucial features for accurately estimating the radiative losses, and then temperature and soot volume fractions. Consequently the OTA leads to inaccurate predictions.
4. The spectral nature of gas and soot is a crucial feature to obtain reliable predictions. Therefore the use of grey models and GWB models should be avoided.
5. The FSCK provides similar results as compared to the SNBCK with a substantial gain in CPU time. Therefore it is a good alternative to be implemented in order to study soot formation processes.

References

[1] Dworkin SB, Zhang Q, Thomson MJ, Slavinskaya NA and Riedel U 2011 *Combust. Flame* **158** 1682–95
[2] Leung KM, Lindstedt RP and Jones WP 1991 *Combust. Flame* **87** 289–305
[3] Kennedy IM, Yam C, Rapp DC and Santoro RJ 1996 *Combust. Flame* **107** 368–82
[4] Di Domenico M, Gerlinger P and Aigner M 2010 *Combust. Flame* **156** 246-258
[5] Lautenberger C, de Ris J, Dembsey N, Barnett J and Baum H 2005 *Fire Saf. J.* **40** 141–76
[6] Beji T, Zhang JP, Yao W and Delichatsios M 2011 *Combust. Flame* **185** 281–90
[7] Balthasar M, Heyl A, Mauss F, Schmitt F and Bockhorn H 1996 *26th Int. Symp. Combustion* (Pittsburg, PA: The Combustion Institute) 2369–77
[8] Bai XS, Balthasar M, Mauss F and Fuchs L 1998 *27th Int. Symp. Combustion* (Pittsburg, PA: The Combustion Institute) 1623–30
[9] Smooke M, McEnally C, Pfefferle L, Hall R and Colket M 1999 *Combust. Flame* **117** 117–39
[10] Liu F, Guo H, Smallwood G and Gülder Ö 2002 *J. Quant. Spectrosc. Radiat. Transf.* **73** 409–21
[11] Liu F, Guo H, Smallwood G and Gülder Ö 2003 *Combust. Theory Modelling* **7** 301–15
[12] Guo H, Liu F, Smallwood G and Gûlder Ö 2006 *Combust. Flame* **145** 324–38
[13] Zhang Q, Guo H, Liu F, Smallwood G and Thomson M 2009 *Proc. Comb. Inst.* **32** 761–68
[14] Santoro RJ, Yeh TT, Horvath JJ and Semerjian HG 1987 *Combust. Sci. and Tech.* **53** 89–115
[15] Megaridis CM and Dobbins RA 1988 *Proc. Comb. Inst.* **22** 353–62
[16] Megaridis CM and Dobbins RA 1989 *Combust. Sci. and Tech.* **66** 1–16
[17] Smith G *et al.* http://www.me.berkeley.edu/gri_mech/
[18] Carbonell D, Oliva A and Perez-Segarra CD 2009 *Combust. Flame* **156** 621–32
[19] Nagle J and Strickland-Constable RA 1962 *Proc. 5th Conf. on Carbon* (London: Pergamon) 154–64
[20] Fenimore CP and Jones GW 1967 *J. Phys. Chem.* **71** 593–7
[21] Guo H, Liu F, Smallwood GJ and Gülder ÖL 2002 *Combust. Theory Modelling* **6** 173–87
[22] Demarco R, Consalvi JL, Fuentes A and Melis S 2011 *Int. J. Thermal Sci.* **50** 1672–84
[23] Modest M 2003 *Radiative Heat Transfer*, Academic Press
[24] Demarco R, Consalvi JL, Fuentes A and Melis S 2011 *Proc. 7th Mediterranean Combustion Symp.* (Sardinia, Italy)
[25] Zhang H and Modest M 2002 *J. Quant. Spectrosc. Radiat. Transf.* **73** 649–53
[26] Soufiani A and Taine J 1997 *Int. J. Heat Mass Transf.* **40** 987–91
[27] Modest M and Riazzi R 2005 *J. Quant. Spectrosc. Radiat. Transf.* **90** 169–89
[28] Wang L, Modest MF, Haworth DC and Turns SR 2005 *Combust. Theory Model.* **9** 673–91
[29] Smith T, Shen Z and Friedman J 1982 *J. Heat Transfer* **104** 602–8
[30] Grosshandler W *A narrow band model for radiation calculations in a combustion environment* NIST Technical Note TN 1402
[31] Zheng Y, Barlow RS and Gore JP 2003 *J. Heat Transf.* **125** 678–86
[32] Coelho PJ 2004 *Combust. Flame* **136** 481–92
[33] Chui EH, Raithby GD and Hughes PMJ 1992 *J. Thermophys. Heat Transf.* **6** 605–11
[34] Smyth KC http://www.fire.nist.gov/fire/flamedata/

Modelling thermal radiation from one-meter diameter methane pool fires

J L Consalvi[1] **and R Demarco**[1,2]

[1] Université de Provence, IUSTI/ UMR CNRS 6595, 5 rue E. Fermi, 13453 Marseille Cedex 13, France
[2] Institut de Radioprotection et de Sûreté Nucléaire (IRSN), BP 3, 13115 St Paul-Lez-Durance, France

E-mail: jean-louis.consalvi@polytech.univ-mrs.fr

Abstract. The first objective of this article is to implement a comprehensive radiation model in order to predict the radiant fractions and radiative fluxes on remote surfaces in large-scale methane pool fires. The second aim is to quantify the importance of Turbulence-Radiation Interactions (TRIs) in such buoyant flames. The fire-induced flow is modelled by using a buoyancy-modified k-ε model and the Steady Laminar Flamelet (SLF) model coupled with a presumed probability density function (pdf) approach. Spectral radiation is modelled by using the Full-Spectrum Correlated-k (FSCK) method. TRIs are taken into account by considering the Optically-Thin Fluctuation Approximation (OTFA). The emission term and the mean absorption coefficient are closed by using a presumed pdf of the mixture fraction, scalar dissipation rate and enthalpy defect. Two 1m-diameter fires with Heat Release Rates (HRR) of 49 kW and 162 kW were simulated. Predicted radiant fractions and radiative heat fluxes are found in reasonable agreement with experimental data. The importance of TRIs is evidenced, computed radiant fractions and radiative heat fluxes being considerably higher than those obtained from calculations based on mean properties. Finally, model results show that the complete absorption coefficient-Planck function correlation should be considered in order to properly take into account the influence of TRIs on the emission term, whereas the absorption coefficient self-correlation in the absorption term reduces significantly the radiant fractions.

1. Introduction

The pool fire scenario represents a simplified version of fires that occur in practice. The numerical simulation of large-scale pool fires appears thus to be an important issue for fire safety purposes. Tiezen *et al.* [1] reported measurements of velocity in a methane pool fire with a diameter of 1 m. Xin *et al.* [2] and Cheung and Yeoh [3] provided large eddy simulations of these experiments by using simplified models to treat radiative heat transfer: Xin *et al.* [2] either used a constant radiant fraction method or a radiation model based on the finite volume method (FVM) coupled with a Gray-Wide Band (GWB) model whereas Cheung and Yeoh considered the discrete ordinates method with the grey assumption [3]. Nevertheless, the implementation of efficient methods to predict radiative heat transfer in large scale pool fires is a crucial issue since thermal radiation plays a major role in the heat transferred by such flames to the surroundings.

An accurate prediction of radiative heat transfer in turbulent flames is a difficult task since it requires a precise solution of the radiative transfer equation (RTE), a proper modelling of the spectral dependence of radiating species (radiatively participating gases and soot particles) and a proper

Published under licence by IOP Publishing Ltd

evaluation of the TRIs. The exact line by line method [4-6] as well as narrow band (NB) models, such as the statistical NB model [7] and the NB k-distribution method [8, 9], were implemented in CFD models to deal with combustion situations. However, they are probably too time-consuming for practical applications. The Full-Spectrum k-distribution (FSK) method [10], which extends the concept of the k-distribution over the entire spectrum, was found also to provide a reliable description of gas and soot radiation in combustion problems and is a good candidate to be implemented in CFD models [11].

Taking TRIs into account requires the modelling of two terms. The first term, known as 'absorption TRI', represents the nonlinear coupling between incident radiation and the local absorption coefficient. It is the consequence of property fluctuations across the domain. Its modelling is complex since it requires having a detailed knowledge of the instantaneous fields of temperature and species. Assuming that the local intensity is only slightly correlated with the local absorption coefficient this term can be ignored. This approximation, commonly referred to as OTFA, was found to be valid for a wide range of conditions [4-6, 12, 13], with the exception of very large-scale sooting jet flames where TRIs were found to reduce absorption by about 10% [14]. The second term, referred to as 'emission TRI', is determined on the basis of local properties only. It can be evaluated by using presumed pdf methods [15-19] or joint-probability transport models [4-6, 20-22].

The aim of this study is threefold: 1) to implement a comprehensive radiative model to predict both radiant fractions and radiative heat fluxes from methane pool fires produced by a burner with a diameter of 1 m for which the gas phase dominates radiation, 2) to quantify the importance of TRIs in the prediction of radiative heat fluxes in such flames, and 3) to determine the relative importance of different TRI-related terms. To the best authors knowledge, a comprehensive evaluation of the role of TRIs in large scale pool fires has not been reported in the past. The FSCK coupled with the mixing scheme of Modest and Riazzi [23] is used as the radiative property model. The OTFA is considered and a presumed-pdf approach coupled with the SLF model is applied to model TRI-related correlations.

The article begins with the presentation of the numerical model. Next, predicted radiant fractions and radiative heat fluxes on remote targets are compared with available experimental data. The effects of TRIs on predicted radiant fractions and heat fluxes are then discussed. Finally the conclusions drawn from the present study are summarized.

2. Numerical model

2.1. Flow field model

The numerical simulation of the turbulent flame flow fields includes the solution of the overall continuity equation, the Favre-Averaged Navier–Stokes equations in a low Mach number formulation, and transport equations for the turbulent kinetic energy, for the dissipation rate of turbulent energy, for the mixture fraction, $\tilde{\xi}$, for the mixture fraction variance, for the soot mass fraction, for the soot number density per unit mass of mixture, and for the total enthalpy. The conservation equations are solved in axisymmetric cylindrical coordinates using a finite volume method. For the convective terms, the ULTRASHARP approach is applied [24]. A second-order central difference scheme is used for diffusion terms. The pressure-velocity coupling is solved using the Iterative PISO algorithm [25]. Steady-state solutions are reached by time marching.

The production of turbulence due to buoyancy is modelled by a second-order expression [26, 27], whereas the effects of the anisotropy of turbulence due to buoyancy on the turbulent Reynolds stresses are introduced by using an algebraic stress model [28].

Turbulence-chemistry interactions are modelled by using the SLF model [29] coupled with a presumed-pdf of the mixture fraction, scalar dissipation rate (χ), and enthalpy defect parameter (X_R) [30]. The flamelet library was generated by considering a detailed chemical mechanism consisting of 70 species and 463 reactions [31]. Statistical moments for scalar quantities can then be computed from the flamelet library and a joint pdf of the mixture fraction, scalar dissipation rate and enthalpy defect parameter. The statistical independence of ξ, χ, and X_R is assumed and the joint pdf is thus modelled as the product of a Beta function for the mixture fraction and two Dirac functions for both the scalar dissipation rate [32] and the enthalpy defect parameter [33].

The acetylene-based two-equation soot model proposed by Leung *et al.* is used [34]. The surface growth term is assumed to depend linearly on soot particle surface area. The kinetic constants in the nucleation and growth rates are $k_n = 10^4\exp(-21000/T)\ s^{-1}$ and $k_{sg} = 0.75 \times 10^3 exp(-12100/T)\ m^{-2}s^{-1}$. The oxidation of soot is attributed to OH and O_2. Soot oxidation by O_2 is given by the NSC model [35] while soot oxidation by OH is based on the Fenimore and Jones model [36]. The rate of soot oxidation by OH is given by $k_{OH} = 1.27 \times 10^{-3}/\sqrt{T}\ \ m^{-2}s^{-1}$ and the collision factor is assumed to be 0.08. Finally, the number of carbon atoms in the incipient soot particle is taken as being equal to 60 [37] and the agglomeration rate is taken as being equal to 3 [37].

2.2. Radiation Model

2.2.1. Radiative property model. In the present simulations only CO_2 and H_2O are considered as gaseous radiatively participating species. The spectral soot absorption coefficient was determined using the small particle limit of Rayleigh scattering for a cloud of non-uniform particle size, leading to $\kappa_{\eta,S} = Cf_S\eta$. C is taken equal to 4.9 which is consistent with the value of the soot complex index of refraction, $m = 1.57 - 0.56i$, commonly used for reporting soot volume fraction measurements in flames. f_S is the soot volume fraction and η is the wavenumber.

The FSK method is used as radiative property model [10]. In this method the absorption coefficient is reordered over the entire spectrum by defining a Full-Spectrum (FS) Planck-function-weighted k-distribution, $f\left(T,\underline{\phi},k\right) = \frac{1}{I_b(T)}\int_0^\infty I_{b\eta}(T)\,\delta\left[k - \kappa_\eta\left(\underline{\phi}\right)\right]d\eta$, and a FS Planck-function-weighted cumulative k-distribution, $g\left(T,\underline{\phi},k\right) = \int_0^k f\left(T,\underline{\phi},k\right)dk$. In the latter definitions, $\kappa, k\ and\ I_b$ represent the absorption coefficient, the absorption coefficient variable and the Planck function. Spectral quantities are indexed by η. δ is the Dirac function. $\underline{\phi}$ is an array of state variables of the medium properties (T,p,x_i,f_s) where T, p and x_i are the temperature, the pressure, and the molar fraction of the species i, respectively. Due to the smooth nature of the cumulative function, the integration of the reordered wavenumber can be easily performed with a simple quadrature scheme. In the case of non-homogeneous and/or non-isothermal media the assumption that the absorption coefficient is scaled or correlated is applied. In addition, a reference state of the medium properties must be defined to carry out the derivation. Quantities relative to the reference state are indexed by 0 in the following. In this study the FSCK formulation is considered, leading to a reordered RTE in the smoothly-varying g_0-space:

$$\frac{dI_g}{ds} = k^*\left(T_0,\underline{\phi},g_0\right)\left[a\left(T,T_0,g_0\right)I_b(T) - I_g\right] \tag{1}$$

where $a\left(T,T_0,g_0\right) = dg\left(T,\underline{\phi}_0,k\right)/dg_0\left(T_0,\underline{\phi}_0,k\right)$. I_g represents the radiative intensity at a given cumulative k-distribution value. The total intensity, I, is then estimated as:

$$I = \int_0^1 I_g\, dg_0 \approx \sum_{j=1}^{N_G} I_{gj} w_j \tag{2}$$

N_G, g_j and w_j are the number of quadrature points, the quadrature points and the corresponding weights respectively. Decoupled radiative heat transfer computations showed that N_G can be set equal to 5 without altering the quality of the results [38]. The reference state is defined here by the molar fractions of the gas species and by the soot volume fraction averaged over the flame volume. In agreement with a previous work the reference temperature is computed as a time-averaged emission-weighted temperature [38].

The FS *k*-distributions are assembled from the 43 NB database of Soufiani and Taine [39] by using the Modest and Riazzi mixing scheme [23]. It should be noticed that the use of the 43 NB database instead of the 367 NB database allows large computational savings without altering the quality of the solutions [38]. The procedure can be described as follow: firstly on each NB the single *k*-distributions for the gas mixture are generated by using the following expression:

$$g_{mix}(k_{mix}) = \int_0^1 g_{CO_2}(k_{mix} - k_{H_2O})\,\mathrm{d}g_{H_2O} \tag{3}$$

Secondly the absorption coefficient of soot is added to the NB cumulative *k*-distributions. Finally the single FS *k*-distributions are assembled from the previously computed single NB *k*-distributions by using the following relationship:

$$g(T,\underline{\phi},k) = \sum_{j\in[\text{all NBs}]} \frac{I_{bj}(T)}{I_b(T)} g_j(\underline{\phi_0},k) \text{ and } I_{bj}(T) \approx I_{b\eta}(T)\Delta\eta_j \tag{4}$$

Mixing at the NB level is demanding computationally. In order to further reduce the CPU time, mixture NB (MNB) k-distributions are pre-databased for different values of $R = x_{CO_2}/x_{H_2O}$, and interpolations are carried out for mixtures of arbitrary species mole fraction ratios [40].

2.2.2. Turbulence Radiation Interactions. Equation (1) is time-averaged to account for TRIs:

$$\frac{\mathrm{d}\overline{I_g}}{\mathrm{d}s} + \underbrace{\overline{k^*(T_0,\underline{\phi},g_0)I_g}}_{\text{Absorption TRI}} = \underbrace{\overline{k^*(T_0,\underline{\phi},g_0)a(T,T_0,g_0)I_b(T)}}_{\text{Emission TRI}} \tag{5}$$

The absorption term can be expressed as $\overline{k^* I_g} = \overline{k^*}\overline{I_g} + \overline{k^{*\prime} I_g^{\prime}}$. In the present study the term $\overline{k^{*\prime} I_g^{\prime}}$ is neglected based on the OTFA. As discussed in the introduction, previous studies demonstrated that this approximation is adequate for weakly-sooting flames as considered in the present study. Mean values of the absorption coefficient and of the emission term are computed as:

$$\overline{k^*} = \bar{\rho}\int_0^1\int_0^1\int_0^\infty\int_0^1 \frac{k^{fl}(\xi,\chi,X_R,f_S)}{\rho^{fl}(\xi,\chi,X_R)}\tilde{P}(\xi,\chi,X_R,f_S)\,\mathrm{d}\xi\,\mathrm{d}\chi\,\mathrm{d}X_R\,\mathrm{d}f_S \tag{6}$$

$$\overline{k^* a I_b} = \bar{\rho}\int_0^1\int_0^1\int_0^\infty\int_0^1 \frac{k^{fl}(\xi,\chi,X_R,f_S)a^{fl}(\xi,\chi,X_R)I_b^{fl}(\xi,\chi,X_R)}{\rho^{fl}(\xi,\chi,X_R)}\tilde{P}(\xi,\chi,X_R,f_S)\,\mathrm{d}\xi\,\mathrm{d}\chi\,\mathrm{d}X_R\,\mathrm{d}f_S \tag{7}$$

where the pdf is expressed as $\tilde{P}(\xi,\chi,X_R,f_S) = \tilde{P}(\xi)\delta(\chi-\tilde{\chi})\delta(X_R - X_{R,mean})\delta(f_s - \tilde{f_s})$. $\tilde{\chi} = C_{v2}\frac{\epsilon}{k}\widetilde{\xi''^2}$ and $X_{R,mean}$ are the local mean values of the scalar dissipation rate and of the heat loss fraction respectively. C_{v2} is taken as being equal to 2 and $X_{R,mean}$ is computed as proposed by Brookes and Moss [33]. The present formulation neglects the influence of the fluctuations of soot volume fraction. This approximation may alter the evaluation of emission TRI since the soot absorption coefficient (soot volume fraction)-Planck function (temperature) correlation is ignored.

Nevertheless, it seems reasonable for the pool fires considered in the present study where the radiative heat transfer is dominated by the gas-phase contribution. This model will be referred to hereafter as Mod 1.

2.2.3. RTE Solver. The radiative transfer equation is solved by the Finite Volume Method (FVM) using the special mapping developed by Chui *et al.* for axisymmetric configurations [41]. Computations were carried out using the first-order STEP scheme for spatial discretization and a uniform angular mesh with 36×48 control angles.

3. Results and Discussions

3.1. Experimental and computational details

The two 1m-diameter methane pool fires with HRRs ($\dot{Q}$) of 49 kW and 162 kW, experimentally investigated by Hostikka *et al.* [42], were simulated. The corresponding dimensionless fire powers, given by $\dot{Q}_D^* = \dot{Q}/\rho_\infty C_p T_\infty \sqrt{gD}D^2$, are of 0.044 and 0.14 respectively. In the previous expression D and g represent the pool diameter and the gravitational acceleration. The subscript ∞ refers to ambient conditions. In both cases the radiant fraction, χ_R, was reported. The radial and vertical radiative heat flux profiles outside the flame region were measured. The vertical distributions were measured at 1m and 0.8m from the center of the burner for the 49 kW and the 162 kW pool fires, respectively. The radial distributions were measured along a radius on the plane aligned with the burner surface. Unfortunately no measurement concerning the flame structure was reported in the study of Hostikka *et al.* [42].

The overall computational domain is 3 m (R) × 6 m (L). A preliminary grid refinement study indicated that a non-uniform mesh with 85 × 260 cells and a mesh size of the order of 1 cm × 1 cm in the flaming zone is sufficient to achieve grid-independent solutions.

Figure 1 shows the fields of mean temperature. In the case of the 49 kW pool fire (figure 1a), the combustion is found to occur close to the burner. This follows the experimental observations of Zukoski [43] who pointed out that, when $\dot{Q}_D^* < 0.1$, the combustion region is divided into many flamelets that lean toward the axis of the source and that have flame lengths much smaller than the diameter of the source. Predicted flame heights are determined here as the location where the mean mixture fraction reaches its stoichiometric value (0.055 for methane in air). The flame height normalized by the burner diameter is of about 0.05, which is consistent with the data compiled by Zukoski [43] for small range of $\dot{Q}_D^*$.

As the normalized HRR is increased to 0.14 (figure 1b), the mean temperature field suggests that the classical description of a pool fire with a single continuous flame region, an intermittent flame region and the inert plume is recovered. This behaviour is also in qualitative agreement with experimental observations which mention that the short flamelets are expected to merge and form a single flame for $\dot{Q}_D^* \approx 0.1$ [43]. The normalized flame length is 0.61, which compares rather well with the value of 0.8 computed from both the correlations of Heskestad [44] and Quintiere and Grove [45]. The correlations were applied by setting the radiant fraction, χ_R, to 0.16 as measured by Hostikka *et al.* [42]. The numerical predictions of flame heights agrees also well with the data compiled by Zukoski [43], which show that measured normalized flame heights range from about 0.6 to 1 for $\dot{Q}_D^* \approx 0.14$.

3.2. Comparison with experimental available data

Hostikka *et al.* [42] reported radiant fractions of 0.08 and 0.16 for the 49 kW and the 162 kW pool fires with an uncertainty on these values of 11%. Measured radiant fractions range thus from 0.071 to 0.089 and from 0.142 to 0.178 for the 49 kW and the 162 kW pool fires respectively. The corresponding predicted radiant fractions are 0.095 and 0.206, which compares favourably well with the upper bounds of the experimental data. In addition these results show that the model reproduces the increase in radiant fraction with $\dot{Q}_D^*$ correctly.

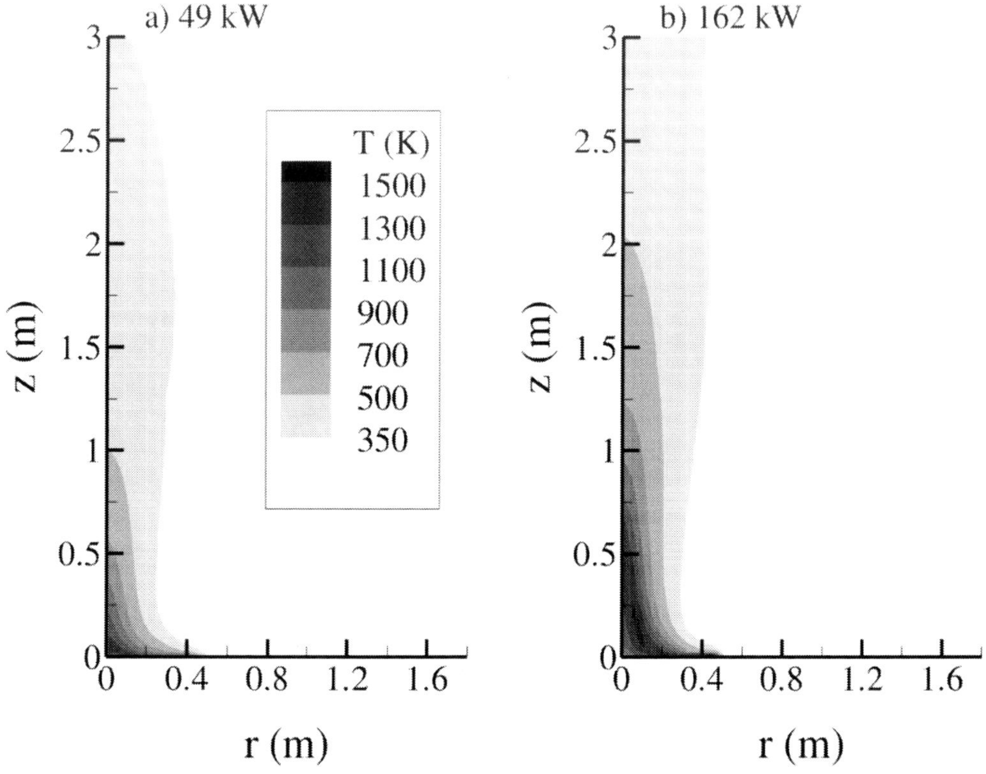

Figure 1. Mean temperature for: (a) the 49 kW pool fire and (b) the 162 kW pool fire.

Figure 2 shows the vertical and radial distributions of radiative flux. Predictions are in reasonable agreement with the experimental data. Figures $2a_1$ and $2b_1$ show that in the vertical direction the radiative heat fluxes reach a peak and then decrease rapidly. Calculated vertical profiles are found to match closely the experiments. Figures $2a_2$ and $2b_2$ show that in the radial direction the radiative heat flux decreases quickly as the radial distance increases. For the 162 kW pool fire the simulation overpredicts the experiment near the burner whereas a more satisfactory agreement is obtained for $r > 0.75m$ (see figures $2b_2$). The largest discrepancies are observed for the 49 kW pool fire (see figure $2a_2$), where the heat fluxes are very low (not significant for fire safety purposes) and therefore very sensitive to small errors in the flow field predictions.

3.3. Influence of TRI-related-terms on radiative outputs

This section discusses how the predicted radiant fractions and radiative heat fluxes on surrounding surfaces are affected by considering different TRI closures for the absorption and emission terms. The role of TRIs on radiative outputs can be better understood by isolating their effects on the radiation calculations alone. As a consequence, decoupled radiation calculations are carried out by using the flow field computed with the most complete model (Mod 1).

The different scenarios as well as the corresponding predicted radiant fractions are summarized in table 1. The comparison between the calculations carried out with (Mod 1) and without TRIs (Mod 2) shows clearly the strong influence of TRIs on the flame radiative losses. The radiant fractions are underpredicted by about 130% and 178% for the 49 kW and the 162 kW pool fires when TRIs are ignored. In addition, these results suggest that the enhancement in radiative losses due to TRIs increases with $\dot{Q}^*$. The comparison of the radiant fractions computed with Mod 2, Mod 3, Mod 4, and Mod 5 shows the effects of the different levels of closure on the emission term. In all these simulations the absorption coefficient in the absorption term is modelled from mean property values. In Mod 3 only the influence of TRIs on mean Planck function is considered. In Mod 4 the fluctuations of the absorption coefficient and of the Planck function are taken into account separately, but the correlation between the two is disregarded. Finally the most complete formulation, Mod 5, considers the effects of the correlation between the gas absorption coefficient and the Planck function. Results in table 1 reveal that accounting for the temperature self-correlation increases significantly the radiant fractions (Mod 2 vs. Mod 3) but its contribution alone is not sufficient to model correctly the emission term (Mod 3 vs. Mod 5). Mod 4 improves the predictions as compared to Mod 3 (Mod 4 vs. Mod 5). However the comparison between Mod 4 and Mod 5 shows that the gas absorption coefficient-Planck function correlation cannot be neglected. As discussed by Coelho [12] this correlation is positive, resulting in a large increase in the radiant fractions (Mod 5 vs. Mod 4) in the present cases. Finally, Mod 1 extends Mod 5 by considering the effects of the absorption coefficient self-correlation on the absorption term. This term contributes to enhance flame absorption leading to a non-negligible reduction in the radiant fraction (see table 1).

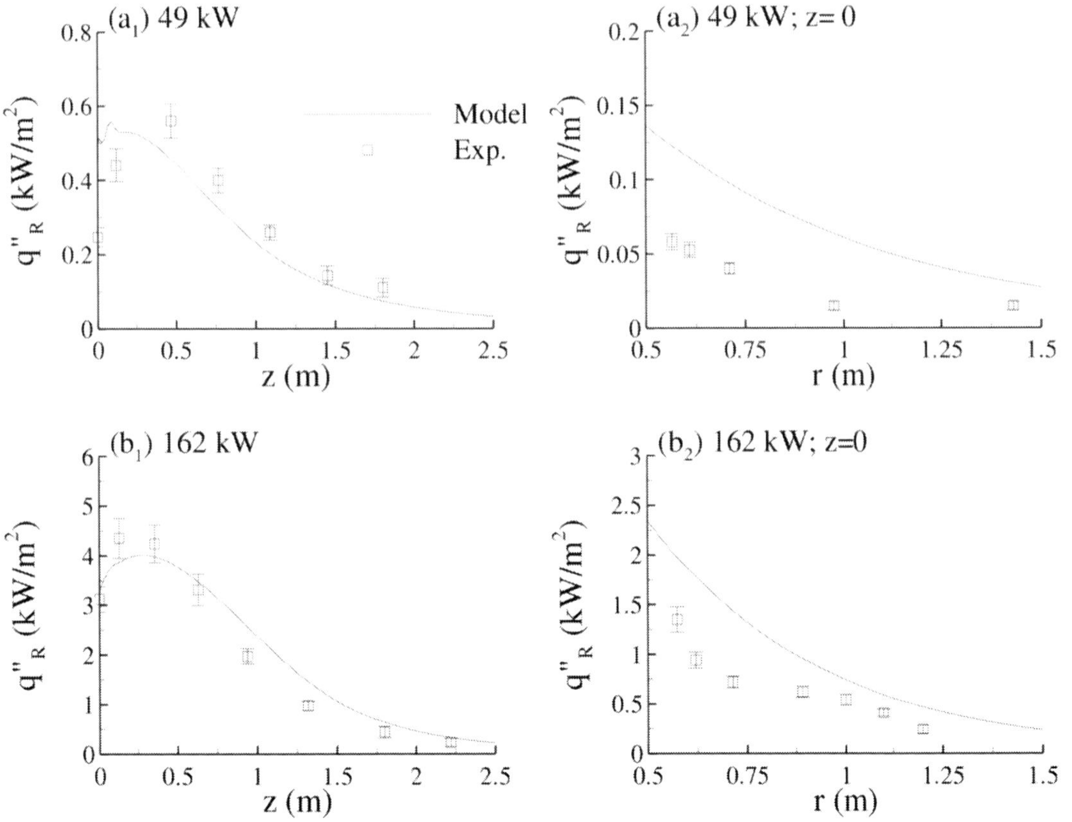

Figure 2. Predicted heat fluxes vs. experiments for (a) the 49 kW pool fire and (b) the 162 kW pool fire. The indexes 1 and 2 refer to the vertical distribution of heat flux and to the radial distribution of heat flux at z=0, respectively.

Journal of Physics: Conference Series **369** (2012) 012012 doi:10.1088/1742-6596/369/1/012012

Figure 3 shows the effects of these different closures on the vertical and radial distributions of the radiative flux. A comparison between the simulations made with Mod 1 and Mod 2 shows that heat fluxes are severely underpredicted when TRIs are neglected: the peaks in the vertical direction are lower by factors of about 2.8 and 2.5 for the 49 kW and the 162 kW pool fires, whereas the radial distributions are also strongly underestimated. The predictions obtained with the other TRI scenarios follow the same trends as those observed for the radiant fractions (see table 1). Figure 3 shows also that the different closures modify the shape of the distributions. It can be observed that both predicted vertical and radial profiles are much flatter in absence of TRIs than when Mod 1 is considered.

Table 1. Effects of the different TRI closures on the radiant fraction. The values in brackets indicate the relative error (in %) defined as $E_r = (\chi_{R,Mod1} - \chi_{R,model})/\chi_{R,model} \times 100$ Quantities denoted without an overbar refer to evaluations from mean variables, *i.e.* without accounting for TRIs.

Test Case	Absorption	Emission	49 kW	162 kW
Exp.	-	-	0.08 (0.071-0.089)[a]	0.16 (0.142-0.178)[a]
Mod 1	$\bar{k}$	$\overline{k \times aI_b}$	0.095	0.206
Mod 2	k	$k \times aI_b$	0.041 (131.7%)	0.074 (178.4%)
Mod 3	k	$k \times \overline{aI_b}$	0.048 (97.9%)	0.103 (100.0%)
Mod 4	k	$\bar{k} \times \overline{aI_b}$	0.061 (57.3%)	0.134 (53.7%)
Mod 5	k	$\overline{k \times aI_b}$	0.120 (-21%)	0.223 (-7.6%)

[a] the uncertainty on χ_R are estimated as 11%.

4. Concluding remarks

Two methane pool fires generated by a burner with a diameter of 1 m were simulated by using a buoyancy-modified k-ε model and the SLF model to model chemistry. The FSCK was applied to account for the non-grey nature of gas and soot radiation. TRIs were modelled on the basis of the OTFA; the emission term and the mean absorption coefficient were then closed by using a presumed pdf of the mixture fraction. In addition the importance of TRIs and the influence of different levels of closure for TRIs on the prediction of radiative losses and radiative fluxes on remote surfaces were quantified. The following conclusions can be drawn:

1. Predicted radiant fractions and radiative heat fluxes on remote surfaces are reasonable agreement with available experimental data when TRIs are considered.
2. TRI considerably enhance the radiative losses and the radiative heat fluxes on remote surfaces. Disregarding TRIs also significantly modifies the shape of both vertical and radial distributions of radiant flux.
3. The complete gas absorption coefficient-Planck function correlation should be considered in order to properly take into account the influence of TRIs on the emission term.
4. The effects of absorption coefficient self-correlation on the absorption term cannot be neglected. They contribute to increase the absorption and in turn to decrease the radiative losses and the heat fluxes on remote targets.

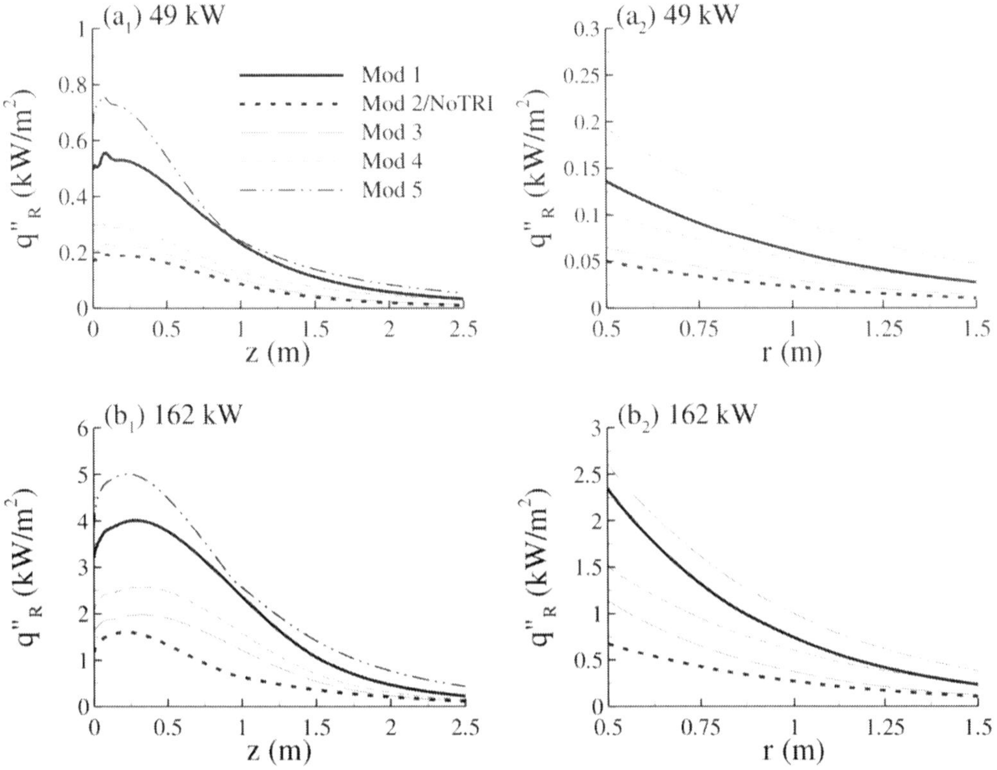

Figure 3. Effects of different TRI closures on radiative heat fluxes for (a) the 49 kW pool fire and (b) the 162 kW pool fire. The indexes 1 and 2 refer to the vertical distribution of heat flux and to the radial distribution of heat flux, respectively.

References

[1] Tieszen S R, O'Hern T J, Schefer R W, Weckman E J and Blanchat T K 2002 *Combust. Flame* **129** 378
[2] Xin Y, Filatyev S A, Biswas K, Gore J P, Rehm R G and Baum H R 2008 *Combust. Flame* **153** 499
[3] Cheung S C P and Yeoh G H 2009 *Int. J. Thermal Sci.* **48** 2187
[4] Wang A, Modest M F, Haworth D C and Wang L 2008 *J. Quant. Spectrosc. Radiat. Transf.* **109** 269
[5] Mehta R S, Haworth D C and Modest M F 2010 *Combust. Flame* **157** 982
[6] Mehta R S, Modest M F and Haworth D C 2010 *Combust. Theory Model.* **14** 105
[7] Yan Z and Holmstedt G 1999 *Combust. Flame* **117** 574
[8] Dembele S and Wen J X 2003 *Numer. Heat Transf. - Part B* **44** 365
[9] Tessé L, Dupoirieux F, and Taine J 2004 *Int. J. Heat Mass Transf.* **47** 555
[10] Modest M F and Zhang H 2002 *ASME J. Heat Transf.* **124** 30
[11] Demarco R, Consalvi J L, Fuentes A and Melis S 2011 *Int. J. Thermal Sci.* **50** 1672
[12] Coelho P J *2007 Prog. Ener. Combust.* ***Sci.*** **33** 311
[13] Coelho P J 2004 *Combust. Flame* **136** 481
[14] Mehta R S 2008 *Detailed modeling of soot formation and turbulence-radiation interaction in turbulent jet flames* Ph.D. Thesis *The Pennsylvania State University*
[15] Song T H and Viskanta R 1987 *J. Thermophys. Heat Transf.* **1** 56
[16] Gore J P, Ip U S and Sivathanu Y R 1992 *J. Heat Transf.* **114** 487

[17] Hartick J W, Tacke M, Früchtel G, Hassel E P and Janicka J 1996 *Proc. Combust. Inst.* **26** 75
[18] Xu X, Chen Y L and Wang H 2006 *Int. J. Heat Mass Transf.* **49** 2347
[19] Coelho P J, Teerling O J and Roekaerts D 2003 *Combust. Flame* **133** 75
[20] Mazumder S and Modest M F 1999 *Int. J. Heat Mass Transf.* **42** 971
[21] Li G and Modest M F 2003 *J. Heat Transf.* **125** 454
[22] Li G and Modest M F 2002 *J. Quant. Spectrosc. Radiat. Transf.* **73** 461
[23] Modest M F and Riazzi R J 2005 *J. Quant. Spectrosc. Radiat. Transf.* **90** 169
[24] Leonard B P and Drummond J E 1995 *Int. J. Numer. Method Fluids* **20** 421
[25] Chow W K and Cheung Y L 1999 *Numer. Heat Transf. – Part A* **35** 311
[26] Liu F and Wen J X 2002 *Fire Saf. J.* **37** 125
[27] Hanjalic K 1994 *Proc. of the 10th Int. Heat Transf. Conf.* p 1
[28] Davidson L 1990 *Int. J. Heat Mass Transf.* **33** 2599
[29] Peters N 1984 *Prog. Ener. Combust. Sci.* **10** 319
[30] Carbonell D, Oliva A and Perez-Segarra C D 2009 *Combust. Flame* **156** 621
[31] Qin Z, Lissianski V V, Yang H, Gardiner W C, Scott S G and Wang H 2000 *Proc. Combust. Inst.* **28** 1663
[32] Heyl A and Bockhorn H 2001 *Chemosphere* **42** 449
[33] Brookes S J and Moss J B 1999 *Combust. Flame* **116** 486
[34] Leung K M, Lindstedt R P and Jones W P 1991 *Combust. Flame* **87** 289
[35] Naggle J and Strickland-Constable R.F 1962 *Proc. 5th Conf. on Carbon* (London: Pergamon Press) p 1
[36] Fenimore C P and Jones G W 1967 *J. Phys. Chem.* **71** 593
[37] Wooley R M, Fairweather M and Yunardi 2009 *Fuel* **88** 393
[38] Consalvi J L, Demarco R, Fuentes A, Melis S and Vantelon J P 2011 *7th Mediterranean Combustion Symp.* (Sardinia) vol 7
[39] Soufiani A and Taine J 1997 *Int. J. Heat Mass Transf.* **40** 987
[40] Wang L, Modest M F, Haworth D C, and Turns S 2005 *Combust. Theory Model.* **9** 673
[41] Chui E H, Raithby G D and Hughes P M J 1992 *J. Thermophys. Heat Transf.* **6** 605
[42] Hostikka S, McGrattan K B and Hamins A 2003 *Proc. of the Int. Symp. on Fire Safety Science, vol. 4*, ed D D Evans (Boston: International Association for Fire Safety Sciences) pp 383–94
[43] Zukoski E E 1986 *Proc. of the Int. Symp. on Fire Safety Science, vol. 1*, ed C Grant and P Pagni (Washintin D C: Hemisphere) pp. 1
[44] Heskestad G 1991 *Combust. Flame* **83** 293
[45] Quintiere J G and Grove B S 1998 *Proc. Combust. Inst.* **27** 2757

Radiative cooling of a horizontal layer of participating medium

S Prasanna and SP Venkateshan
HTTP Lab, IIT Madras, Chennai, India
E-mail: spv@iitm.ac.in

Abstract. The present paper studies the role of radiative cooling on natural convection in a square cavity. Radiative cooling of the air sets up convection. The interaction depends on multitude of parameters such as Planck number, optical length and the length scale of the enclosure. Both gray and non-gray gases have been considered for the study.

1. Introduction

During night time, the ground cools faster than the surrounding air. However under certain conditions the air layers very close to the ground have been observed to be cooler than the ground, for example Ramdas phenomenon (lifted temperature minimum)[1, 2, 3], convection in a radiative fog[4, 5]. Such temperature inversions are departure from the normal and understanding these phenomena are important for weather and meteorological studies.
Radiative cooling creates instability as the top layers of the medium are cooler than the ground, a condition akin to Rayleigh Benard convection. Gille and Goody[6] studied the effect of participating radiation on onset of convection in a fluid layer enclosed between two black walls maintained at constant temperature experimentally and observed that radiation has a tendency to suppress convection and hence delay the onset of convection. The study has been well supported by numerical and experimental studies[7, 8]. However, the nature of convection radiation interaction in the lower layers of an atmosphere is characteristically different from that between two parallel plates, as radiation is the sole cause of convection. Even though radiation has a tendency to damp out the perturbations and suppress convection, the radiative cooling of air layers would promote convection. The temperature inversion would occur when the radiative cooling rate of air layers is much higher than ground cooling rate. Studying the real dynamics of such an atmosphere is beyond the scope of the present work. Nevertheless, the present paper considers a simple geometry to deconstruct the role of radiative cooling on convective heat transfer. The motivation for the present study is to design suitable experiments to study the radiation convection interaction problem. A two dimensional square enclosure with top wall being a perfect window to radiation has been considered as the physical domain with background radiation accounting for the down-welling longwave radiative fluxes. Computations have been performed for steady state conditions for gray and non gray fluid. The following sections would treat the mathematical formulation, numerical procedure and discussion of results.

Journal of Physics: Conference Series **369** (2012) 012013 doi:10.1088/1742-6596/369/1/012013

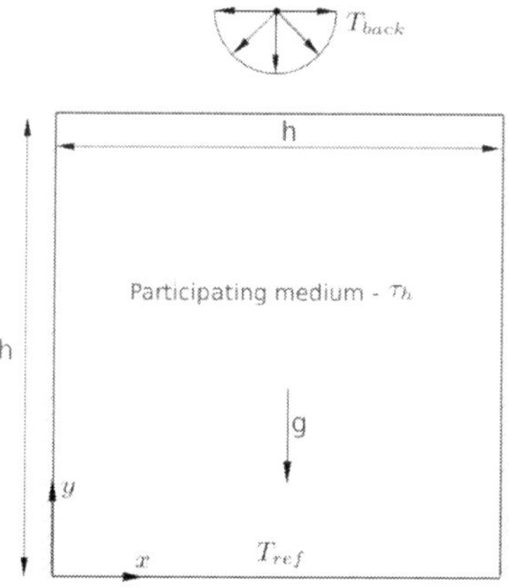

Figure 1: Physical domain of the present problem

2. Mathematical formulation

The physical domain of the present problem has been illustrated in figure 1. The bottom wall of the enclosure is maintained at a constant temperature of $T_{ref} = 300K$. The two side walls are insulated and diffuse. It has been assumed that the interaction of the enclosure with the surroundings is purely via radiation. This is unlike the real physical problem where the upper surface is a free surface continually influenced externally. However, the present study would enable understanding the influence of radiation on the state of affairs in a fluid layer.
The mass and momentum equations for the fluid are the well known continuity and Navier Stokes equations. The fluid has been assumed to be incompressible and the Boussinesq approximation has been used to account for buoyancy. Since the temperature variation within the medium is not more than $10°C$, such an assumption is justified. The presence of radiation does not have a direct effect on the mass and momentum equations. However radiation directly affects the energy equation. Even so, as radiation influences the temperature distribution and hence the flow, all the four equations have to be solved simultaneously. The non dimensionalized governing equations are

$$\nabla \cdot \mathbf{v} = 0 \tag{1}$$

$$\mathbf{v} \cdot \nabla \mathbf{v} = -\nabla p + \nabla^2 \mathbf{v} - \frac{Ra^*}{Pr}(\theta - 1)\hat{\mathbf{y}} \tag{2}$$

$$\mathbf{v} \cdot \nabla \theta = \frac{1}{Pr}\nabla^2 \theta - \frac{1}{Pr\,Pl}\nabla \cdot q_r \tag{3}$$

where Pl indicates relative strength of thermal diffusion and radiation and is equal to $\frac{k}{4\sigma T_{ref}^3 h}$. $\nabla . q_r$,the divergence of the radiant flux, indicates the radiative heating or cooling in the fluid medium which in turn drives convection.

$$\nabla \cdot q_r = \frac{1}{4\sigma T_{ref}^4} \int_{\eta=0}^{\infty} \tau_{h,\eta}(4e_{b\eta} - g_\eta)d\eta \tag{4}$$

where $e_{b\eta}$ and g_η are respectively the spectral black body emissive power given by Planck function and spectral energy density.

$$g_\eta = \int_{4\pi} I_\eta d\Omega \tag{5}$$

The equation of transfer for radiation intensity is given by

$$\mu\frac{dI_\eta}{dx} + \xi\frac{dI_\eta}{dy} = \tau_{h,\eta}(I_b - I_\eta) \tag{6}$$

The boundary conditions for the present problem are listed below

(i) $u = 0$ and $v = 0$ at the boundaries of the enclosure

(ii) $\theta = 1$, at $y = 0$

(iii) $\frac{\partial \theta}{\partial y} = 0$, at $y = h$

(iv) at the side walls, there is a balance between conduction and radiation, $\frac{\partial \theta}{\partial y} - \frac{1}{Pl} q_r = 0$

(v) $\varepsilon = 0$, $\rho = 0$ at the top wall

(vi) at side walls, $\varepsilon = 1$ (unless specified otherwise) and diffuse radiation

(vii) at bottom wall, $\varepsilon = 1$ and diffuse radiation

The background radiation is at a lower temperature compared to T_{ref}.

3. Solution Procedure

The governing equations with appropriate boundary conditions have been solved using the finite volume method. As the problem is coupled, all the variables, velocity, temperature and intensity have to be solved for simultaneously. This has been achieved by performing sequential iteration for each variable. SIMPLE algorithm has been used to couple pressure and velocity. Convection terms have been represented by first order upwinding. Staggerred grid has been used to discretize the domain, with pressure, temperature and intensities at the geometric center of the control volume and the velocities at the nodes on the boundaries of each control volume. Cosine grid has been used to discretize the physical domain where mesh is fine close to the wall and coarse away from the wall.

Weighted Sum of Gray Gas method (WSGG) has been used to model non-gray radiation. The non gray gas is represented by a weighted sum of gray gases of different absorption coefficients such that the radiative heat flux of the non gray gas is represented by an equivalent mixture of gray gases.

$$\mu \frac{dI_j}{dx} + \xi \frac{dI_j}{dy} = \tau_{h,j}(a_j \frac{\sigma T^4}{\pi} - I_j) \tag{7}$$

where I_j,a_j and τ_{hj} are the intensity, weightage and optical path of jth gray gas. The total divergence of heat flux can be calculated from spectral radiative heat flux as

$$\nabla \cdot q_r = \sum_{j}^{no\ of\ gases} \nabla \cdot q_{r,j} = \sum_{j}^{no\ of\ gases} \tau_{hj} a_j \theta^4 - \frac{1}{4} \tau_{hj} g_j^* = \overline{\tau_h} \theta^4 - \frac{1}{4} \sum_{j}^{no\ of\ gases} \tau_{hj} g_j^* \tag{8}$$

As the temperature variation within the medium is less than $10°C$, the medium is assumed to be homogeneous, thereby simplifying the calculations. The coefficients in WSGG are calculated from the line by line spectrum of the constituent gases available in HITRAN database[9] using the k-distribution method [10, 11]. The basic idea of this approach is to group the spectral intervals having the same absorption coefficient i.e. transforming from spectral distribution to absorption coefficient distribution. From the line by line data, the absorption coefficients over the entire spectral range is to be calculated which for Lorentz profile is given by

$$\kappa_\eta = \sum_{i}^{no\ of\ lines} \frac{S_i}{\pi} \frac{\gamma_i}{(\eta - \eta_i)^2 + \gamma_i^2} \tag{9}$$

where, S_i is the line strength, γ_i is the line half width and η_i is the location of line center. A cumulative distribution function of absorption coefficient weighted by the Planck function is calculated for the gas mixture. This is done by dividing absorption coefficient range into a large number of equal intervals. Due to large range of absorption coefficients, the division is performed using a logarithmic scale. The spectral range is also divided into a large number of narrow bands such that Planck function can be assumed to be constant within a given band. The weights for each absorption coefficient is determined by summing up the weighted Planck function corresponding to the absorption coefficient interval.

$$f(\kappa_i) = \frac{1}{I_b} \sum_{j=1}^{no\ of\ bands} I_{b,j} W(\kappa_j - \delta\kappa, \kappa_j + \delta\kappa)$$

$$g(\kappa_i) = \sum_{j=1}^{i} f(\kappa_j) \qquad (10)$$

During nocturnal conditions, the down-welling radiation is due to emission and scattering of radiation from the atmosphere. As the atmospheric radiation is non-gray in nature, a gray approximation may not be suitable. STREAMER[12] has been used to calculate the spectral down-welling radiation. STREAMER is a radiative transfer model that can determine radiative fluxes, longwave and shortwave, for various atmospheric and surface conditions. STREAMER performs radiative transfer calculations for water vapour, carbon di oxide, ozone and selected aerosol profiles. The model calculates spectral radiative fluxes by dividing the longwave spectrum ($20 - 2480\ cm^{-1}$) into 105 bands. Standard tropical profile provided by STREAMER, in absence of clouds and aerosols, has been used as the atmospheric profile for all the simulations. Figure 2 shows the spectral down-welling radiation obtained from STREAMER for the standard tropical atmospheric profile. The total down-welling radiative flux equals that of a black body at $T_{back} \approx 0.97 T_{ref}$ and this has been used as the background radiation for gray calculations. Figure 3 shows the cumulative distribution functions for saturated air. The weights and corresponding absorption coefficients for WSGG are determined using the Gauss Chebyshev quadrature[13].

Radiative transfer of equation for each gray gas is solved using the Discrete Ordinate method

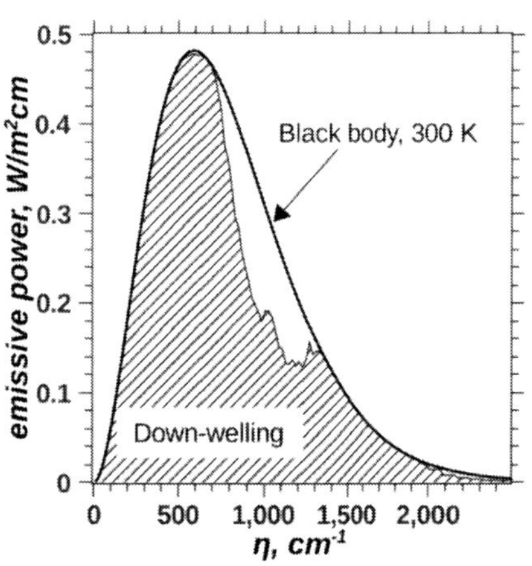

Figure 2: Downwelling radiation from STREAMER

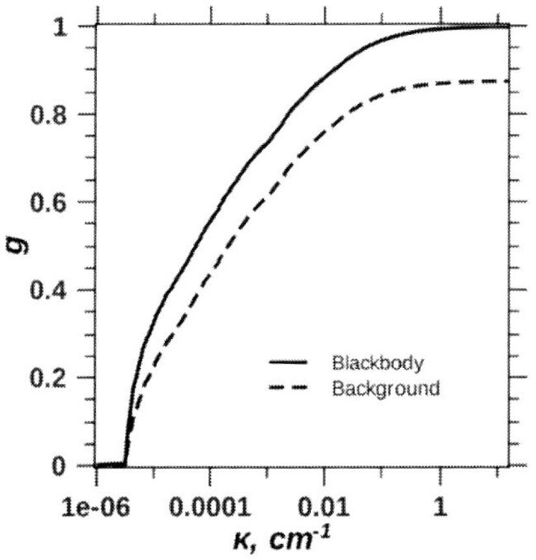

Figure 3: Cumulative distribution function for Planck distribution and background radiation

where the solid angle is discretized into discrete directions with appropriate weights. The direction and weights are to be such that the error in the computed values of fundamental radiative quantities like energy density, flux and pressure are minimum. The quadratures have to satisfy zeroth moment, first moment and the second moment. In addition, the weights should also satisfy the first moment over the half range to accurately represent the wall heat flux. As the physical domain is a square, quadrature scheme proposed by Carlson and Lathrop[14] has been used. The quadrature scheme satisfies the moment equations and also satisfies first moment over half range along the principal directions (x, y). The intensity equation for a discrete direction i is

$$I^i_{p,j} = \frac{\tau_{h,j}\Delta x\Delta y a_j I_b + |\mu|\Delta y I^i_{w,j} + |\xi|\Delta x I^i_{s,j}}{\tau_{h,j}\Delta x\Delta y + |\mu|\Delta y + |\xi|\Delta x} \tag{11}$$

where $I^i_{w,j}$ and $I^i_{s,j}$ are the intensities at the west and south (directions with respect to the ordinate direction) boundaries of the control volume. The intensity at cell center and cell boundary are related by interpolation given by

$$I_p = fI_e + (1-f)I_w = fI_n + (1-f)I_s \tag{12}$$

where f can lie any where between 0 and 1. Step scheme $f = 1$ has been used in the present numerical study. Combining equations 8 and 9, intensities are calculated iteratively by marching from one end of the boundary.
The coupling between temperature and radiation is through the divergence of radiative heat flux (equation 8). The fourth power of temperature means the coupled set of equations are non linear in nature. Newton Raphson method has been used to linearize the source term $(\overline{\tau_h}\theta^4)$ in the energy equation.

$$\overline{\tau_h}\theta^4 \equiv \overline{\tau_h}(4\theta^3_{old}\theta_{new} - 3\theta^4_{old}) \tag{13}$$

The coupled energy equation and radiative transfer equation are solved sequentially. Bi-conjugate gradient method has been used to solve the energy equation and Gauss Siedel iterations have been performed for radiative transfer equation. The procedure has been implemented using FORTRAN 90.

Table 1: Comparison of code with literature

	Present		Yucel et al.[15]	
	Total	Rad	Total	Rad
$\tau = 1.0$	38.13	29.88	38.93	31.28
$\tau = 5.0$	30.76	22.39	31.76	23.64
$\tau = 0.2$	45.14	36.89	46.11	37.40

$Ra = 5 \times 10^6$, $Pl = 0.02$

Table 2: Grid sensitivity study

Grid size	Angular quadrature	$\frac{\partial\theta}{\partial y}$
41 × 41	SN6	0.179
61 × 61	**SN6**	**0.181**
121 × 121	SN6	0.181
61 × 61	SN4	0.178
61 × 61	SN8	0.182

$\tau_h = 0.1$, $Pl = 0.001$

4. Results and Discussion

The code has been extensively validated with benchmark numerical results on convection, radiation and combined convection and radiation. Table 1 shows the comparison of numerical results from the present work and Yucel et al.[15]. A differential heated cavity with a participating medium was considered and the results of the present study compare well with the quoted results. A grid and angle sensitivity study has also been performed and is presented in Table 2. 61×61 grids with 6 discrete directions in each octant ($SN6$) are chosen for the present study.

4.1. Gray gas

Numerical experiments have been conducted for various Planck numbers and optical lengths. The length scale of the phenomenon of interest is of the order of 10 cm and hence simulations have been performed for length scale = 10 cm. Accordingly the Ra^* for a layer of air would be 3×10^7. As both optical thickness and Planck numbers are dependent on length scale of the enclosure, length scale is an important parameter in the interaction phenomenon.

The interaction of radiation on natural convection has been illustrated in figures 4 and 5. It is

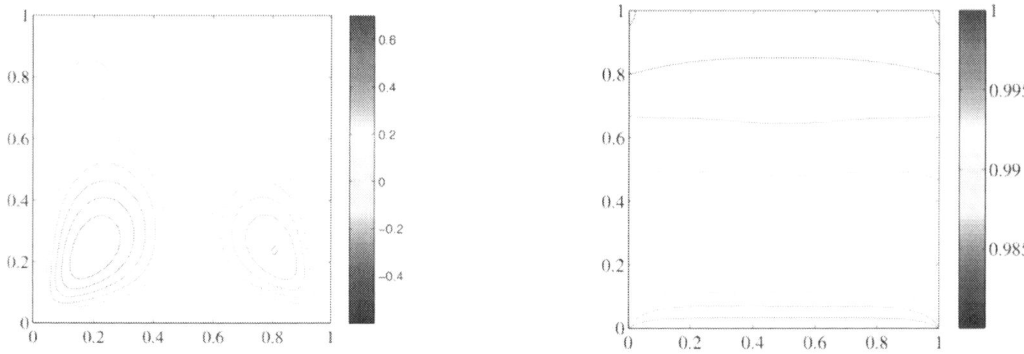

Figure 4: Streamlines and isotherms for $Pl = 0.01$ and $\tau_h = 0.1$

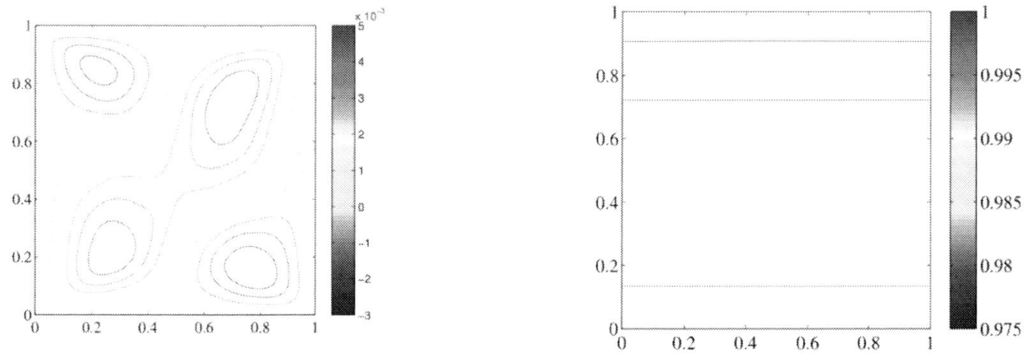

Figure 5: Streamlines and isotherms for $Pl = 0.01$ and $\tau_h = 10$

interesting to observe that radiation has contrasting roles to play in the interaction phenomena. The amount of radiative heat transfer is proportional to the optical thickness. Hence it would be reasonable to conclude that the temperature variation within the participating medium would increase with increase in optical thickness. This clearly means the Rayleigh number of the system would also increase. Therefore it would be logical to conclude convection to be stronger for a participating medium of high optical thickness. However, on the contrary it is observed that the convection is weaker for optical thickness 10 than 1. The convective strength follows a downward trend with optical thickness. However, this observation is not surprising at all, as it reaffirms the already established observation that radiation suppresses convection.
For the sake of understanding the observation, let us consider a participating medium of very high optical thickness $\tau_h \rightarrow \infty$. For such a situation, bulk of the cooling of the medium takes place close to the top wall or in the limit, from the wall itself. Accordingly, radiation can be modeled by diffusion approximation and hence the effective thermal conductivity of the system increases.

$$k_{eff} = k + k_{rad} \tag{14}$$

where k_{rad} is the "radiative conductivity" given by $\frac{16\sigma T_{ref}^3}{3\kappa}$. As a result, the effective Rayleigh number of the system reduces and would cause weak convection. Therefore, Rayleigh number defined in the conventional sense is not a good representative for a convective radiative system.
Figure 6 shows the center line vertical temperature distribution in the enclosure for $Pl = 0.01$. The temperature at the top wall clearly decreases with optical thickness. However, the temperature gradients at the bottom wall increase with decreasing optical thickness. The

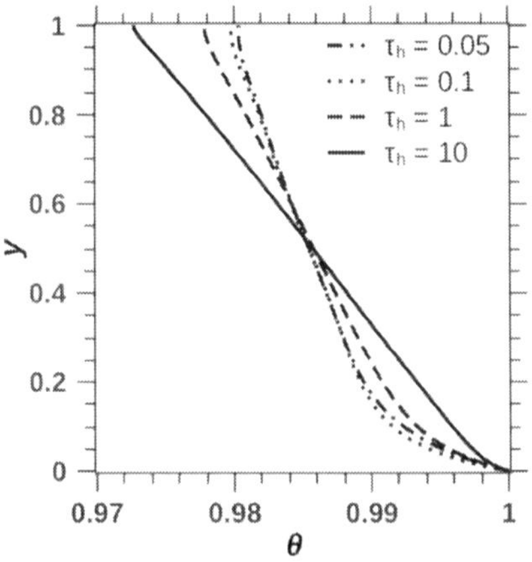

Figure 6: Center line temperature distribution for $Pl = 0.01$

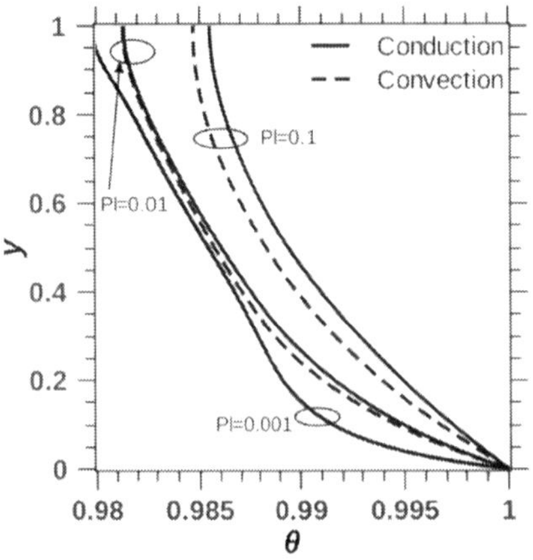

Figure 7: Center line temperature distribution for $\tau = 0.1$

possible cause of strong convection for a small optical thickness medium would be because of the strong temperature gradient close to the bottom wall. This explains the strong cell formation very close to the bottom wall. The large magnitudes of temperature gradients close to the wall is because of low values of Planck numbers. Pl signifies the relative importance of radiation and thermal diffusion. As $Pl \to 0$ the system tends towards radiative equilibrium and other modes of heat transfer would become insignificant. In absence of thermal diffusivity, there would be a temperature jump observed at the bottom wall. The temperature jump would increase with decrease in optical thickness. However, thermal diffusivity smears out the temperature jump resulting in a steep temperature gradient. As Pl increases, the thermal diffusion increases and hence the temperature variation within the medium also reduces considerably. This means the convection intensity also decreases considerably.

To illustrate the nature of interaction, comparisons have been made for cases with and without convection(diffusion only). Figure 7 shows the temperature variation for various Pl for $\tau_h = 0.1$. It is evident that there is little to choose between profiles of convection and pure conduction for $Pl = 0.01$. The difference between the two modes become significant as Pl increases. This is true, even though Ra is much higher for small Pl. This suggests that, although convection is very strong, the effective heat transfer and hence temperature profile are almost independent of convection. As Pl increases, the heat transfer through conduction from the surface becomes important and hence the temperature is affected by both convection and radiation. The typical Pl for an atmospheric layer of 10 cm and 1 m thick would be 0.01 and 0.001 respectively. This implies the presence of convection in such a scale of atmosphere would have little effect on the temperature. However, the real atmosphere is a non gray medium and it would be desirable to see the dynamics of such a medium.

4.2. Non gray medium

Water vapour is the main green house gas contributing to radiative heat transfer. Let us consider a fully saturated layer of air ($RH = 100\%$). Figure 8 shows the temperature along the center of the enclosure for different number of gray gases used in the WSGG model. In order to capture the phenomena correctly, 100 gray gases have been used to represent the non gray phenomena. The absorption coefficients and weights are calculated according to the method discussed earlier. The Pl number for such a configuration would be of the order of 0.01.

Figure 9 shows the streamline and isotherm contour plots for a air layer saturated with water

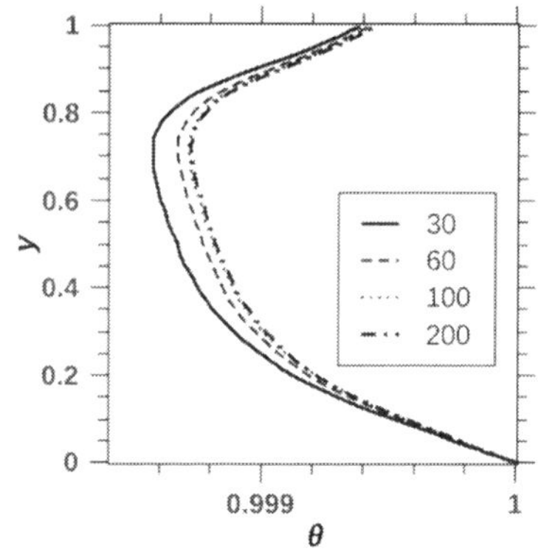

Figure 8: The sensitivity of results to number of gases

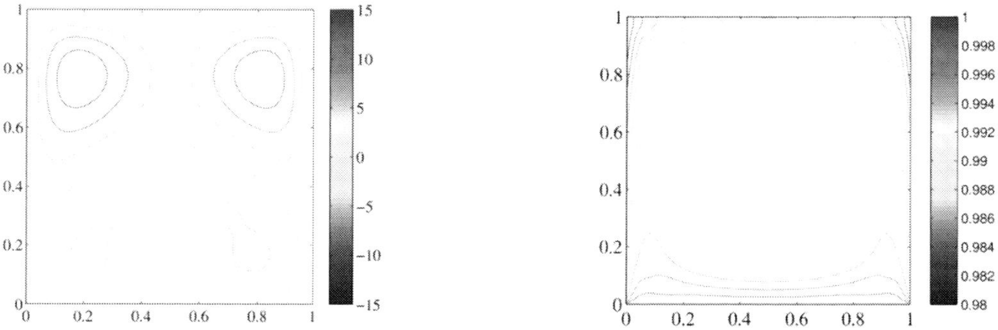

Figure 9: Streamlines and isotherms for saturated air with black side walls

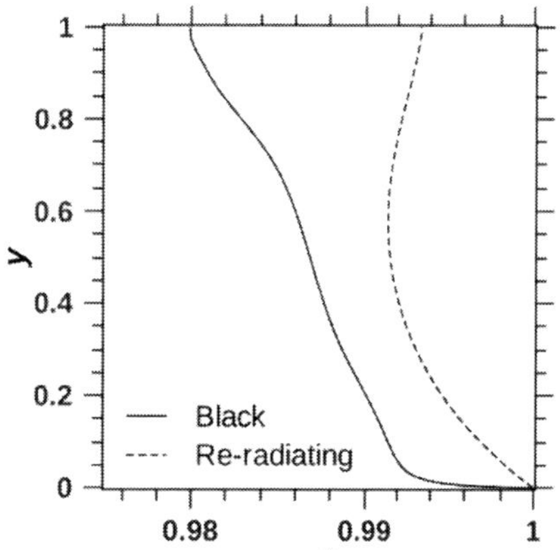

Figure 10: Temperature distribution along the side wall (black or re-radiating)

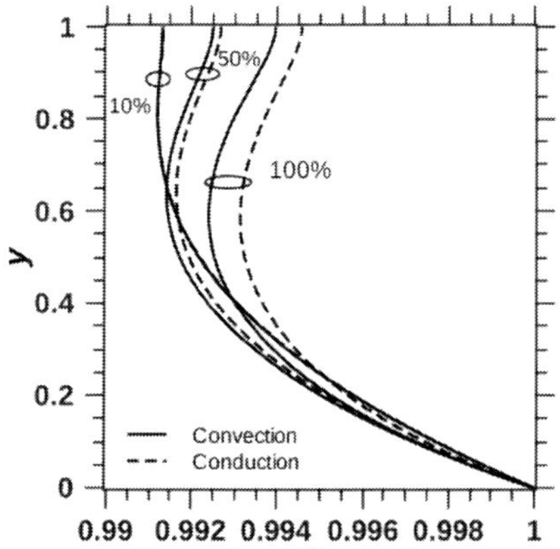

Figure 11: Center line temperature distribution for different relative humidities

vapour. The pattern is remarkably different from that for a gray medium. The side walls being perfectly black, interact with the participating medium. There would be heat loss in the window region of the participating medium, and hence the radiative losses from the side walls are much higher. As there is balance between conduction and radiation at the walls, strong temperature

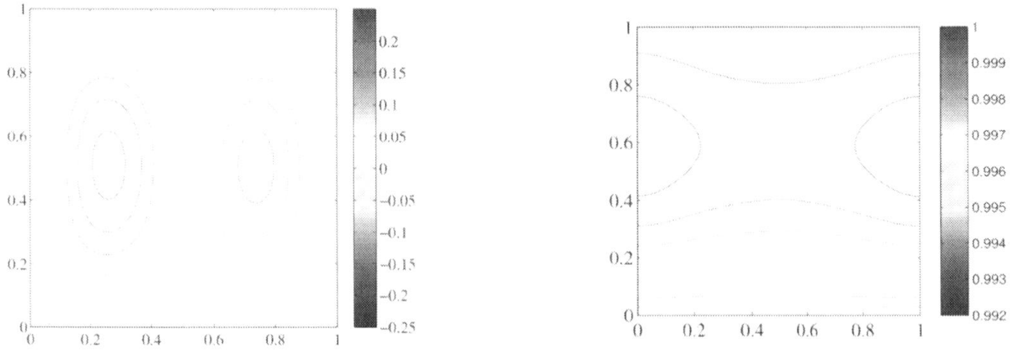

Figure 12: Streamlines and isotherms for saturated air with re-radiating side walls

gradients are developed close to the vertical walls akin to a differentially heated cavity setting up convection. Simulations have been conducted by assuming the side walls to be re-radiating i.e. the side walls are perfectly reflecting (diffuse). This means there would be no radiation heat transfer from the side walls and thus eliminating the role of side walls. Figure 10 shows the temperature variation along the side walls for black and re-radiating side wall boundary conditions which shows excessive cooling of the participating medium in the presence of black side walls. It can be concluded that radiative modeling of the boundaries are very important for accurately capturing the physical phenomena as small perturbations can change the nature of flow conditions as seen in the case above.
Figure 11 shows the temperature distribution along the center line of the enclosure for different relative humidities. Relative humidity affects the nature of convective and radiative transfer from the cell. The temperature minimum occurs at the interior of the enclosure for $RH = 50\%$ and 100%, where as for $RH = 10\%$ the lowest temperature occurs at the top. It has to be noted that the spectral radiation of humid air and the background radiation are similar. This could be a possible explanation for the maximum radiative cooling taking place from the center of the layer. There is a balance between the amount of down-welling radiation absorbed and radiation emitted. As RH increases, the absorption coefficient increases and hence the amount of absorption from the top layers of the enclosure is also higher. As a result the top layers of the enclosure are tad hotter than the central regions. However, for lower RH values, the emission from the layer would offset the role of absorption. Also, it is seen that convection alters the temperature profile to some extent as RH increases. Figure 12 shows the streamline and isotherm contour plots for saturated air. Simulations have also been performed, for both gray and non gray gases, for height of the enclosure $= 1\ m$. The solution for such cases did not converge to steady state suggesting the flow may be unstable.

5. Conclusion

Radiative cooling influences the temperature and convective heat flux to a large extent. When Pl is small, the temperature field is mainly influenced by radiative flux. It can be said that convection acts like a small perturbation to the radiation dominated phenomena. The strength of convection decreases with increase in optical thickness. The thermal characteristic of a non-gray medium is remarkably different from a gray medium. The main differences occur in the application of the radiative boundary conditions. The length scale of a Ramdas layer would be upto a meter, which from the study indicates presence of strong advection within the layer. Another observation that can be made is that the humid air can cool to a temperature below the dew point, leading to condensation and *radiative fog* (not considered in the present study). In a static environment, humid air can cause substantial cooling leading to secondary effects which would be of interest. However it has to be noted that in the presence of strong winds, radiative cooling can become less significant. The present study has considered a simple case without the complications involved in a regular atmospheric layer. However, the study has outlined the nature of interaction. The real flow and temperature field would have transient fluctuations which would be the emphasis of future works. It is also intended to design experiments to study the interaction problem.

Nomenclature

a_j	weightage of jth gas
$e_{b\eta}$	spectral blackbody emissive power
f	k-distribution function
g	cumulative k-distribution function
g_η	spectral energy density
h	height of the square cavity, m
I	non dimensional intensity
k	thermal conductivity, W/mK
p	dimensionless pressure
Pl	Planck number, $\frac{k}{4\sigma T_{ref}^3 h}$
Pr	Prandtl number
q_r	non dimensional radiative heat flux
Ra^*	Rayleigh number, $\frac{g\beta T_{ref} h^3}{\nu^2} Pr$
RH	relative humidity
S_i	line strength, cm^{-2}
T_{back}	background temperature, K
T_{ref}	reference temperature, 300 K
$\mathbf{v}$	velocity vector
x, y	dimensionless coordinates
Greek Symbols	
κ	absorption coefficient, cm^{-1}
γ	line half width,cm^{-1}
ε	hemispherical total emissivity
ρ	hemispherical total reflectivity
η	wave number,cm^{-1}
θ	non dimensional temperature, T/T_{ref}
τ_h	optical thickness, κh
ν	momentum diffusivity, m^2/s
κ	absorption index
μ, ξ	direction cosines
σ	Stefan-Boltzmann constant, $5.67 \times 10^{-8} W/m^2K^4$
subscripts	
b	blackbody
i, j	indices
η	spectral

6. References

[1] Ramdas L and Atmanathan S 1932 *Beit. Geophys* **37** 116–117
[2] Narasimha R 1994 *Current Science (Bangalore)* **66** 16–23
[3] Mukund V, Ponnulakshmi V, Singh D, Subramanian G and Sreenivas K 2010 *Physica Scripta* **2010** 014041
[4] Nishikawa T, Maruyama S and Sakai S 2004 *Boundary-Layer Meteorology* **113** 273–286
[5] Price J 2011 *Boundary-Layer Meteorology* 1–25
[6] Gille J and Goody R 1964 *J. Fluid Mech* **20** 12
[7] Arpaci V, Go D *et al.* 1973 *Physics of Fluids* **16** 581
[8] Bdéoui F and Soufiani A 1997 *Physics of Fluids* **9** 3858
[9] Rothman L, Gordon I, Barbe A, Benner D, Bernath P, Birk M, Boudon V, Brown L, Campargue A, Champion J *et al.* 2009 *Journal of Quantitative Spectroscopy and Radiative Transfer* **110** 533–572
[10] Denison M and Webb B 1993 *ASME Journal of Heat Transfer* **115**
[11] Modest M 2002 *Journal of heat transfer* **124** 30
[12] Key J 2001 *Cooperative Institute for Meteorological Satellite Studies, University of Wisconsin* **96**
[13] Boyd J 2001 *Chebyshev and Fourier spectral methods* (Dover Pubns)
[14] Carlson B and Lathrop K 1968 *Computing methods in reactor physics* 171–266
[15] Yücel A, Acharya S and Williams M 1989 *Numerical Heat Transfer* **15** 261–278

Inverse identification of intensity distributions from multiple flux maps in concentrating solar applications

Ben Erickson[1] **and Jörg Petrasch**[2,*]

[1] Department of Mechanical and Aerospace Engineering, University of Florida, Gainesville, FL 32608, USA
[2] Energy Research Center, Vorarlberg University of Applied Sciences, 6850 Dornbirn, Austria

E-mail: joerg.petrasch@fhv.at

Abstract. Radiative flux measurements at the focal plane of solar concentrators are typically performed using digital cameras in conjunction with Lambertian targets. To accurately predict flux distributions on arbitrary receiver geometries directional information about the radiation is required. Currently, the directional characteristics of solar concentrating systems are predicted via ray tracing simulations. No direct experimental technique to determine intensities of concentrating solar systems is available. In the current paper, multiple parallel flux measurements at varying distances from the focal plane together with a linear inverse method and Tikhonov regularization are used to identify the directional and spatial intensity distribution at the solution plane. The directional binning feature of an in-house Monte Carlo ray tracing program is used to provide a reference solution. The method has been successfully applied to two-dimensional concentrators, namely parabolic troughs and elliptical troughs using forward Monte Carlo ray tracing simulations that provide the flux maps as well as consistent, associated intensity distribution for validation. In the two-dimensional case, intensity distributions obtained from the inverse method approach the Monte Carlo forward solution. In contrast, the method has not been successful for three dimensional and circular symmetric concentrator geometries.

Nomenclature

$I(x,y,z,\theta,\varphi)$: Intensity at point (x,y,z) in the direction (θ,φ)

$I_0(x,y,z,\theta,\varphi)$: Intensity on the solution plane at point (x,y) in the direction (θ,φ)

$q''(x,y,z)$: Radiative flux at the point (x,y,z)

q_{jkl}: Measured radiative flux at the point corresponding to (x_j,y_k,z_l)

w_x: Weighting factor in the x-direction (equivalent for y)

x_j: x-position at the jth discretization (equivalent for r, y, z, θ, and φ)

$x_{0,jklmn}$: x-position on the solution plane corresponding to a point (x_j, y_k, z_l) and direction (θ_m,φ_n)

* To whom any correspondence should be addressed.

Δx : Discretization step in the x-direction (equivalent for r, y, z, θ, and φ)
$\Delta\Omega$: Discrete solid angle

1. Introduction

Solar concentrating systems and high flux solar simulators produce high radiative flux levels. Hirsch and Steinfeld [1] have reported fluxes in excess of 4,250 kW/m^2 for an elliptical trough based solar simulator. Petrasch et al. [2] have reported fluxes as high as 11,000 kW/m^2 from an array of elliptical reflectors coupled to Xe arc lamps. These flux levels are equivalent to blackbody stagnation temperatures of 2945 and 3730 K, respectively. Linear systems such as the EuroTrough [3] parabolic concentrator achieve actual temperatures of roughly 500 °C with concentration ratios of approximately 82:1. The LS-1 and LS-2 parabolic troughs by Luz International Ltd. [4] have geometric concentration ratios around 23:1, achieving temperatures near 400 °C. Flux mapping systems are essential for any experimental concentrating solar energy system. They provide an accurate measurement of the radiative flux distribution in the focal plane and thus form the basis for energy balances and efficiency calculations [5, 6]. Flux mapping systems consist of a diffusely reflecting flux target in the focal plane of the concentrator, a highly linear charge-coupled device (CCD) camera [5–7, 9] and a flux sensor for calibration [7, 8].

Typically, ray-tracing programs such as CIRCE2 [10], and VEGAS [11] are used to simulate flux and intensity distributions. These systems rely on idealized reflector geometries and they are adequate for the design of concentrating systems. However, they fail to predict non-ideal flux and intensity distributions due to surface imperfections introduced during manufacturing or operation [12].

In this paper, an experimentally based, inverse method to determine the intensity distribution at a desired solution plane is presented. The method uses multiple flux maps at varying distances along the concentrator's axis as an input to solve for the intensity distribution at the solution plane. There are two main approaches to solving inverse radiation problems, (i) iterative and (ii) direct methods. The iterative approach is based on an assumed parametric model of the intensity function. An optimal solution of then intensity is then found by minimizing the associated residual functionals [13]. While iterative methods are often more stable than direct methods, they are typically nonlinear and suffer from larger computation time [13]. Direct methods are based on discretizing the physical relations between measured (flux) and unknown (intensity) quantities and solving the resulting system of equations. Direct methods often result in ill conditioned linear systems of equations. A range of inverse problems involving radiation in participating medium have been studied in [14–19]. In the current paper a direct approach is chosen.

2. Methodology

Flux maps of solar concentrating systems are usually obtained using Lambertian targets and CCD cameras in conjunction with flux sensors for calibration [7]. A water-cooled Lambertian target is positioned at the focal plane. A highly linear digital camera is then used to acquire an image of the irradiated target. The brightness of each pixel of the resulting digital image is proportional to the radiative flux incident at the pixel location. Ulmer et al. detail a flux mapping system installed at the Plataforma Solar de Almeria in Spain [7]. Similar systems have been implemented at the Paul Scherrer Institute [2] and at ETH-Zurich [1]. A flux sensor is then used to establish the relation between pixel brightness and energy flux. Multiple types of flux sensors exist. Kaluza and Neumann have compiled a review of the various options. All directional information is lost in the diffuse reflection from the Lambertian target. Therefore, no predictions of the flux distribution on arbitrarily shaped receivers can be obtained via the flux maps. Furthermore, direct measurement of the directional distribution of radiation is impractical due to the very high flux levels in the focal region. If information about the flux distribution in multiple planes is available, it can be used to partially restore the directional information [9]. A schematic representation of a typical flux measurement setup is depicted in Figure 1.

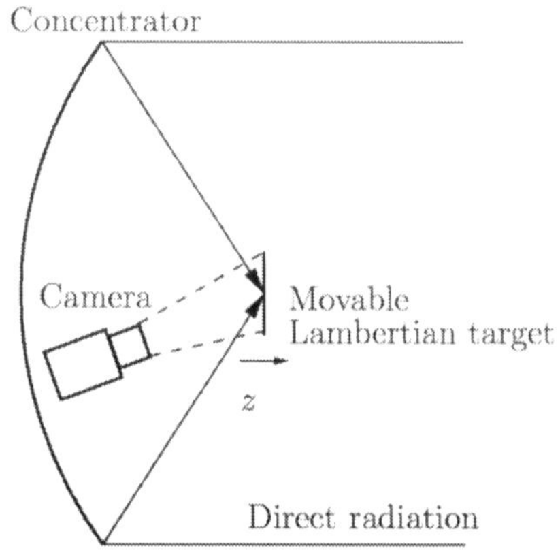

Figure 1. Flux mapping setup.

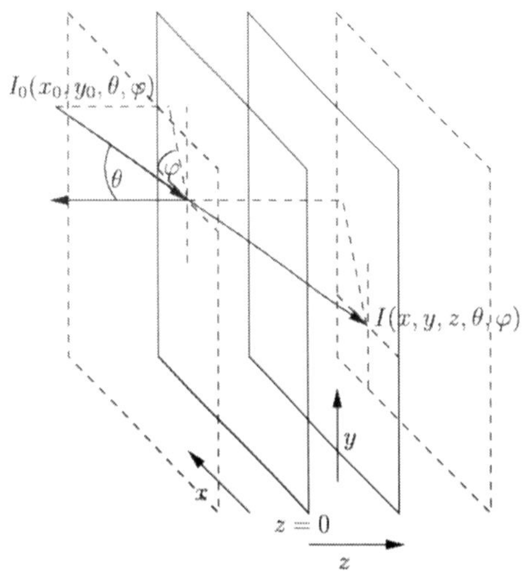

Figure 2. General case of a ray passing through multiple flux planes and the solution plane (at z=0).

2.1. Derivation for the general case

A schematic depiction of the general, three-dimensional case of a ray originating at a solar concentrator and passing through several parallel planes is shown in Figure 2.

Air is assumed to be a non-participating medium, reducing the RTE to

$$\frac{dI}{ds} = 0 \tag{1}$$

Thus, the intensity, I, is constant along any straight line. According to Figure 2, for a given direction (θ, φ) positions of equal intensity are given by,

$$I(x, y, z, \theta, \varphi) = I_o(x_o, y_o, \theta, \varphi) = I_o(x + z\cos\varphi\tan\theta,\ y + z\sin\varphi\tan\theta,\ \theta,\ \varphi). \tag{2}$$

The flux at a given position is obtained by integrating the projected intensity over the hemisphere:

$$q''(x, y, z) = \int_{\varphi=0}^{2\pi} \int_{\theta=0}^{\frac{\pi}{2}} I(x, y, z, \theta, \varphi)\cos\theta\sin\theta d\theta d\varphi \tag{3}$$

Substituting the result of equation (2) for the intensity in equation (3) one obtains

$$q''(x, y, z) = \int_{\varphi=0}^{2\pi} \int_{\theta=0}^{\frac{\pi}{2}} I_o(x + z\cos\varphi\tan\theta,\ y + z\sin\varphi\tan\theta,\ \theta,\ \varphi)\cos\theta\sin\theta d\theta d\varphi \tag{4}$$

Equation (4) can then be discretized. The measured values of radiative flux, $q''(x, y, z)$, are known at discrete locations on a Cartesian grid defined by j,k,l:

$$q_{jkl} = q''\left(x_{\text{ref}} + j\Delta x_f,\ y_{\text{ref}} + k\Delta y_f,\ z_{\text{ref}} + l\Delta z_f\right) \tag{5}$$

The unknown intensity distribution in the solution plane ($z=0$) is spatially discretized on a uniform rectangular grid in x and y while the directional distribution is uniform in elevation (θ) and azimuthal (φ) angles. Equation (4) then becomes

$$q_{jkl}=\sum_{m=1,n=1}^{N_\theta,N_\varphi}\begin{bmatrix} w_x w_y I_{o,J+1,K+1,m,n}+w_x\left(1-w_y\right)I_{o,J,K+1,m,n} \\ +\left(1-w_x\right)w_y I_{o,J+1,K,m,n}+\left(1-w_x\right)\left(1-w_y\right)I_{o,J,K,m,n}\end{bmatrix}\cos\theta_m\sin\theta_m\Delta\theta\Delta\varphi \qquad (6)$$

This is a linear system of equations of the form $\mathbf{q}=\mathbf{AI}_0$, where the radiative flux, $\mathbf{q}$, is known and the intensity at the solution plane, $\mathbf{I}_o$, is unknown. The system is further defined by equations (7-9).

$$x_{o,jklmn}=x_j+z_l\cos\varphi_n\tan\theta_m, y_{o,jklmn}=y_k+z_l\sin\varphi_n\tan\theta_m \qquad (7)$$

$$J=\text{ceiling}\left(\frac{x_{o,jklmn}-x_r}{\Delta x}\right), K=\text{ceiling}\left(\frac{y_{o,jklmn}-y_r}{\Delta y}\right) \qquad (8)$$

$$w_x=\frac{x_{o,jklmn}-x_J}{\Delta x}, w_y=\frac{y_{o,jklmn}-y_K}{\Delta y} \qquad (9)$$

Equations (7) calculate the location of intersection on the solution plane for the x and y-directions, respectively. Equations (8) calculate the index of intersection on the solution plane in the x and y directions. Equations (9) calculate the weighting factors in the x and y-directions as intersections will always occur between two nodal points. The fully three-dimensional case is presented for completeness but is not yet successfully implemented and therefore no results are available.

2.2. *Two dimensional case*

The high dimensionality of the intensity distribution on the solution plane in the most general case (two spatial and two directional coordinates) leads to very large systems of equations. Therefore an important subclass of problems, the purely two-dimensional case, is explored in detail. The 2D situation is applicable to concentrators such as parabolic, elliptical, and circular troughs and 2D compound parabolic concentrators (CPC's). Figure 3 depicts the flux distribution at varying locations on the focal axis of a 2D concentrator.

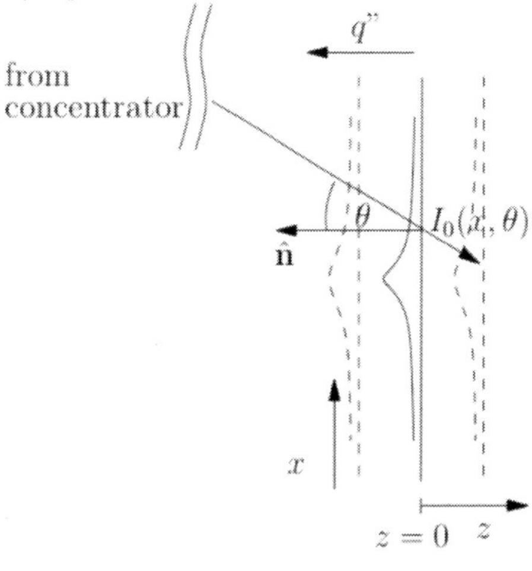

Figure 3. Two-dimensional flux maps.

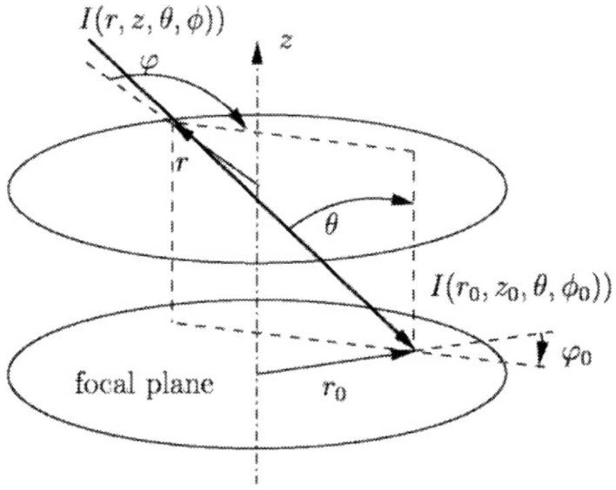

Figure 4. Schematic representation of the circular symmetric case.

The derivation of the 2D case is analogous to that of the 3D case. The intensity at an arbitrary position and direction (x, y, θ) can be related to the intensity in the solution plane according to

$$I(x,y,\theta) = I_o(x + z\tan\theta, \theta) \tag{10}$$

The radiative flux incident at a given location, $q''(x,z)$, is calculated from the intensity according to the two-dimensional equivalent of equation (4).

$$q''(x,z) = \int_{\theta=-\frac{\pi}{2}}^{\frac{\pi}{2}} I(x,z,\theta)\cos\theta\, d\theta = \int_{\theta=-\frac{\pi}{2}}^{\frac{\pi}{2}} I_o(x + z\tan\theta,\theta)\cos\theta\, d\theta \tag{11}$$

Discretization of equation (11) leads to

$$q_{jl} = \sum_{m=1}^{N_\theta}\left((1-w_x)I_{0,J,m} + w_x I_{0,J+1,m}\right)\cos\theta_m \Delta\theta \tag{12}$$

In equation (12), $q_{j,l}$ represents the radiative flux at location j on plane l. Using the 2D equivalents of Equations (7), (8), and (9), the weighting factor w_x is found. The intensity, $I_{0,J,m}$, is the intensity in direction m at location J on the solution plane.

2.3. Circular Symmetric case

Many optical concentrators are circularly symmetric. In these cases, the intensity in the focal plane depends on three independent variables: the radial position, r, and two directional angles, θ and φ. In contrast, the general case requires four independent variables. A system consisting of a circular concentrator and a plane circular target is considered.

From Figure 4 one obtains the geometrical relationship between the radial position and direction (r,θ,φ) in any plane z to a position and direction on the solution plane (r_o, θ, φ_o).

$$r_o = r^2 + (z\tan\theta)^2 - 2rz\tan\theta\cos(\pi - \varphi) \tag{13}$$

$$\varphi_o = \cos^{-1}\left(\frac{r_o^2 - r^2 + (z\tan\theta)^2}{2r_o z\tan\theta}\right) \tag{14}$$

Note that the elevation angle, θ, remains unchanged. As for the general case, the intensity along a straight line is constant leading to

$$q''(r,z) = \int_{\varphi=0}^{2\pi}\int_{\theta=0}^{\frac{\pi}{2}} I_o(r_o,\ \theta,\ \varphi_o)\cos\theta\sin\theta\, d\theta\, d\varphi \tag{15}$$

The flux distribution $q''(r,z)$ is known from measurement or ray-tracing simulations using

$$q_{j,l} = q''\left(j\Delta r_f,\ z_{ref} + l\Delta z_f\right), \tag{16}$$

discretization then yields

$$q_{j,l} = \sum_{m=1,n=1}^{N_\theta,N_\varphi}\left[\begin{matrix}(1-w_\varphi)(1-w_r)I_{o,J,m,N} + (1-w_\varphi)w_r I_{o,J+1,m,N} \\ +(1-w_r)w_\varphi I_{o,J,m,N+1} + (1-w_r)(1-w_\varphi)I_{o,J+1,m,N+1}\end{matrix}\right]\cos\theta_m \sin\theta_m \Delta\theta\Delta\varphi \tag{17}$$

where

$$J = \text{ceiling}\left(\frac{r_o - \frac{dr}{2}}{dr}\right), N = \text{ceiling}\left(\frac{\varphi_o - \frac{d\varphi}{2}}{d\varphi}\right) \tag{18}$$

$$w_r = \frac{r_o - r_J}{dr}, w_\varphi = \frac{\varphi_o - \varphi_N}{d\varphi} \tag{19}$$

2.4. Regularization

Inverse radiation problems often suffer from ill-conditioned coefficient matrices [21]. For continuous solutions, Tikhonov regularization can be employed to impose smoothness onto the solution [22]. It is implemented by appending a smoothing matrix **L** to the **A** matrix for each discretization variable.

$$\mathbf{I}_{0,\lambda} = \arg\min\left(\left\|\begin{bmatrix} \mathbf{A} \\ \lambda_x \mathbf{L}_x \\ \lambda_y \mathbf{L}_y \\ \lambda_\theta \mathbf{L}_\theta \\ \lambda_\varphi \mathbf{L}_\varphi \end{bmatrix} \mathbf{I}_0 - \begin{bmatrix} \mathbf{b} \\ 0 \end{bmatrix}\right\|_2^2\right) \tag{20}$$

Where the smoothing matrix $\mathbf{L}_x$ connects element i to its neighbour j in the x-direction:

$$\begin{aligned} \mathbf{L}_{x,ii} &= 1 \\ \mathbf{L}_{x,ij} &= -1 \end{aligned} \tag{21}$$

all other elements of $\mathbf{L}_x$ are zero.

This method is controlled via one regularization parameter per dimension of the unknown quantity (only λ_x and λ_θ in the two-dimensional case).

2.5. Monte Carlo Ray Tracing

An in-house Monte Carlo ray tracing program [11] is used to model the concentrating systems. The program features directional binning, which allows for the output of the intensity distribution at any desired location. The program provides the intensity distribution as well as the consistent, associated flux distribution, allowing for in-depth validation of the inverse method.

3. Results

3.1. Parabolic Trough

Parabolic troughs are the most common 2D concentrators. The EuroTrough 150 (ET-150) [4], which was developed under European Commission Project EuroTrough II (5th Framework Program contract number ERK6-CT-1999-00018), was modelled using the Monte Carlo ray tracing code [11]. Surface and geometry imperfections are modelled through a Gaussian distributed angular error in the direction of reflected rays around the direction of perfectly specular reflection with a standard deviation of 5 mrad.

Table 1. Parameters of the parabolic trough.

Parameter	Value
Focal length (m)	1.71
Rim angle (°)	60.0
Sun Shape (mrad)	4.649
Mirror Error (mrad)	5.0
Mirror Reflectivity (-)	0.95

The flux distributions in the focal plane and the associated intensity distributions are shown in Figures 6 and 7. The highest intensity values are found at the center of the target, where the concentration is the highest, while the intensity diminishes as the angle of incidence (θ) increases. The flux diminishes as one moves away from the focal plane (z*=0.0).

The inverse method features a range of parameters that influence the accuracy of the solution. These parameters include the number of discretization steps in both the θ and x-direction (N_{x0} and N_θ) and the length of the solution plane (L_{xo}). A range of parameter values was set based on the desired resolution of the solution. The inverse solution was generated for each possible combination of parameter values. Solutions were compared based on the RMS difference between each inverse and Monte Carlo solution with the lowest value corresponding to the optimal solution.

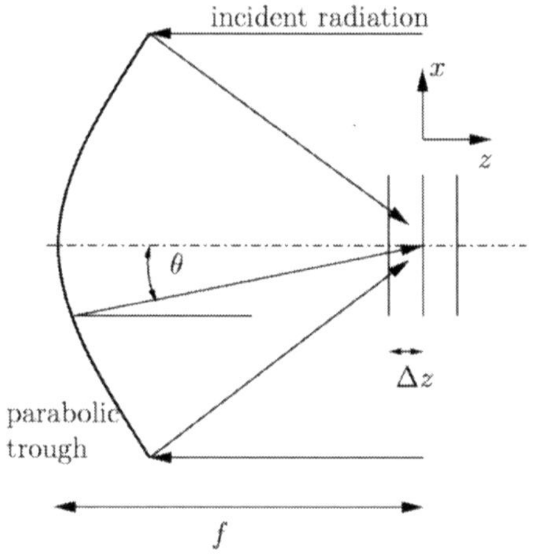

Figure 5. 2D parabolic trough as modeled in VeGaS. The useable output from the model is the intensity distribution on the target.

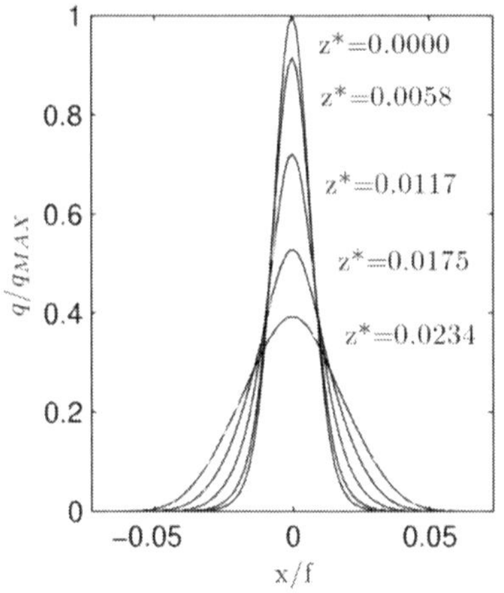

Figure 6. Parabolic trough scaled flux distribution at multiple distances ($z^*=z/f$) from the focal plane

Figure 8 shows the inverse solution based on the flux distributions from Figure 6.

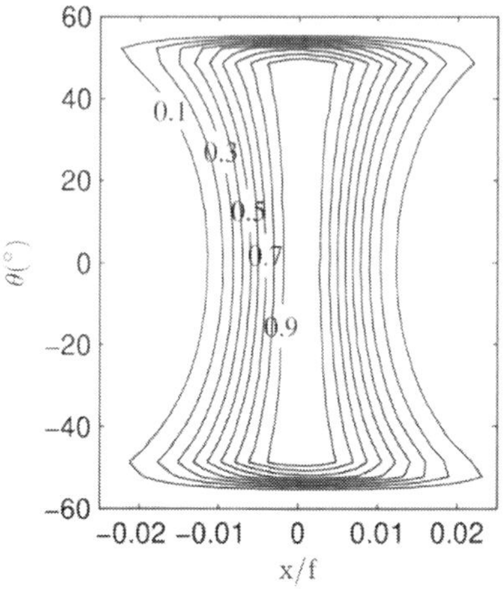

Figure 7. Scaled intensity distribution (I/I_{max}) along the focal plane of a 2D parabolic trough from Monte Carlo ray tracing.

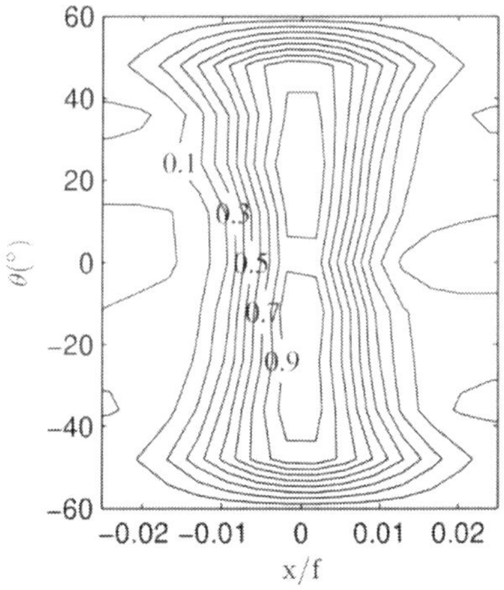

Figure 8. Scaled inverse solution results (I/I_{max}) at the focal plane (z*=0.0). N_θ = 15, N_{xo} = 20 L_{xo} = 0.12 m.

The inverse solution results closely match the Monte Carlo solutions. The discontinuity at larger angles of incidence is reproduced by the inverse solution. The largest discrepancies between inverse and Monte Carlo solutions are concentrated around areas with large gradients and discontinuities in the intensity. The optimal parameters for the parabolic trough were found to be N_θ=15, N_{xo}=20 and L_{xo}=0.12 m. The inverse solution has a maximum error of 9% with a RMS relative error of 3.1%. The linear system is of full rank with a characteristic number of 1000. Tikhonov regularization did not improve results this is attributed to the discontinuous nature of the solution.

3.2. Elliptical Trough

An elliptical trough-based Vortek-type linear concentrator set-up [1] is studied. The set-up (Table 2, Figure 9) consists of an Argon long arc lamp, an elliptical trough reflector and a cylindrical mirror on the underside of the arc. Secondary mirrors are attached at the outlet of the elliptical trough to reduce losses.

Table 2. Elliptical trough parameters

Parameter	Value
Semi-major Axis, *a* (m)	0.277
Semi-minor Axis , *b* (m)	0.190
Sun Shape (mrad)	4.649
Mirror Error (mrad)	5.0
Mirror Reflectivity (-)	0.90
Collector Length (m)	100

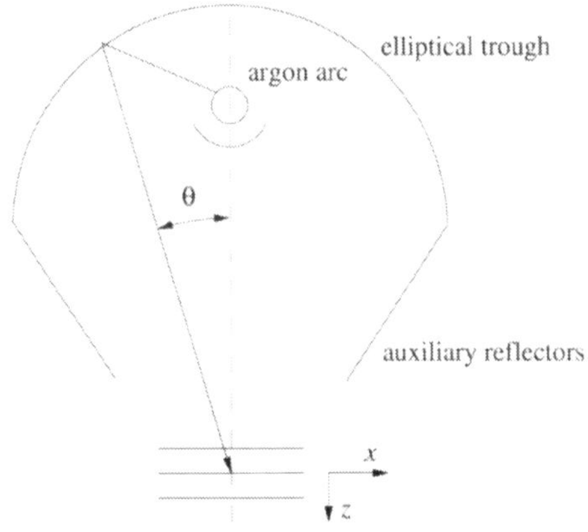

Figure 9. Schematic representation of the elliptical trough setup as implemented in Vegas.

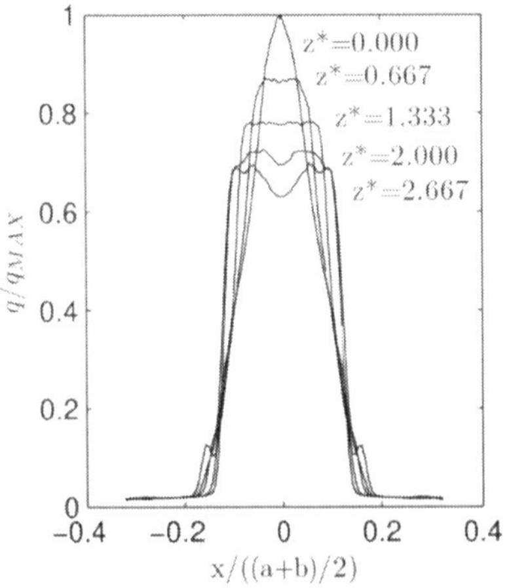

Figure 10. Scaled flux distribution at varying distances ($z^*=z/((a+b)/2)$) from the parabolic trough.

The flux and intensity distributions from Monte Carlo ray tracing are shown in Figures 10 and 11. The inverse solution is shown in Figure 12. The optimal parameters for the elliptical trough were found to be N_θ=5, N_{xo}=41 and L_{xo}=0.41 m. The solution has a maximum error of 18% with an RMS relative error of 5.1%. Tikhonov regularization significantly improves the results in the elliptical trough case. The optimal values for the Tikhonov coefficients are λ_θ=5 and λ_x=12. The condition number of the regularized linear system is 16.4.

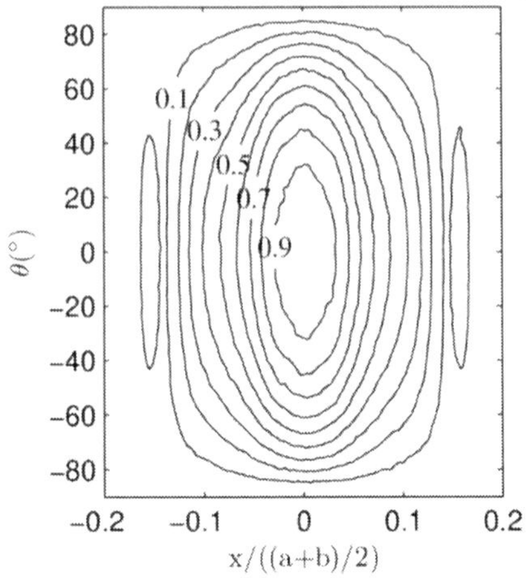

Figure 11. Scaled intensity distribution along the focal plane of a 2D parabolic trough from Monte Carlo ray tracing.

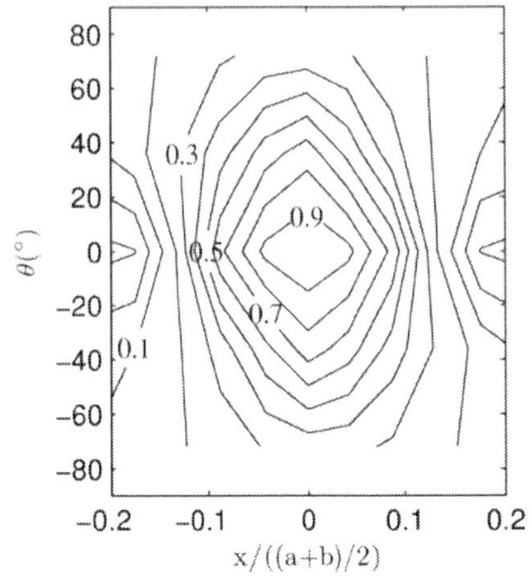

Figure 12. Scaled inverse solution results (I/I_{max}) at the focal plane (z*=0.0). N_θ = 5, N_{xo} = 41 L_{xo} = 0.41 m, λ_X = 12 λ_θ = 5.

4. Conclusions

An inverse solution method to identify intensity distributions at the focal plane of two-dimensional concentrating systems has been developed. The method uses multiple flux maps perpendicular to the concentrator axis to predict the intensity distribution in the focal plane. The mathematical derivation of the circular symmetric and general, three-dimensional cases has been presented. A parabolic trough setup was explored using an in-house Monte Carlo ray-tracing program. The optimal discretization parameters for this setup were found to be N_θ=15, N_{xo}=20, and L_{xo}/f=0.0702. The results show that the inverse solution process can be applied to a set of two-dimensional flux measurements to successfully predict the intensity distribution. The maximum relative error was less than 10% with an RMS relative error of 3.1%. Furthermore, an elliptical trough setup was also implemented. Tikhonov regularization was used to overcome ill-conditioning of the problem. The elliptical trough solution was optimal for discretization parameter N_θ=5, N_{xo} =41, and L_{xo} =1.7559 and Tikhonov parameters λ_θ =5 and λ_x =12 resulting in a relative error of less than 18% and corresponding RMS relative error of 5.1%. In the parabolic trough case Tikhonov regularization has a negative effect due to the discontinuous nature of the solution.

Acknowledgements
The authors gratefully acknowledge fruitful discussions on Tikhonov regularization with Dr. Kyle Daun at the University of Waterloo, Canada.

References
[1] Hirsch D, Zedtwitz P, Osinga T, Kinamore J and Steinfeld A 2003 *J. of Sol. En. Eng.* **125** 117-31
[2] Petrasch J, Coray P, Meier A, Brack M, Haberling P, Wuilemin D and Steinfeld A 2007 *J. of Sol.En. Eng.* **129** 405-11
[3] M. Geyer et al. 2002 *11th Int. Symp. on Concentrating Solar Power and Chemical Energy Technologies* (Zurich, Switzerland) pp 1-7
[4] Fernández-García A, Zarza E, Valenzuela L and Pérez M 2010 *Renew. and Sust. En. Rev.* **14** 1695–1721
[5] Steinfeld A, Brack M, Meier A, Weidenkaff A and Wuillemin D 1998 *Energy* **23** 803-14
[6] Z'Graggen A, Haueter P, Trommer D, Romero M, De Jesus J and Steinfeld A 2006 *Int. J. of Hyd. En.* **31** 797–811
[7] Ulmer S, Reinalter W, Heller P, Lüpfert E and Martinez D 2002 *J. of Sol. En. Eng.* **124** 182-88
[8] Cameron C P *Proc. of SPIE* Nov 1993 (San Diego, CA) 1997 pp 460-71
[9] Kaluza J and Neumann A 2001 *J. of Sol. En. Eng.* **123** 251-55
[10] Romero V J *Int. Solar Energy Conf.* Apr 1992 (Maui, HI) pp 1-9
[11] Petrasch J *ASME Conf. Proc.* Jan 2010 (Phoenix, AZ) 2010 pp. 125-32
[12] Andraka C E, Yellowhair J and Iverson B D *ASME Conf. Proc.* Jan 2010 (Phoenix, AZ) 2010 pp. 565-80
[13] Jin B and Zheng Y 2006 *Comp. Meth. in App. Mech. and Eng.* **195** 2270–88
[14] Li H and Yang C 1997 *Int. J. of Heat and Mass Trans.* **40** 1545–49
[15] Colton D and Kirsch A 1996 *Inv. Prob.* **12** 383-93
[16] Bleszynski E, Bleszynski M and Jaroszewicz T 1996 *Radio Sci.* **31** 1225-51
[17] Ho C H and Özişik M N 1988 *J. of Quan. Spec. and Rad. Tran.* **40** 553-60
[18] Subramaniam S and Menguc M P 1991 *Int. J. of Heat and Mass Tran.* **34** 253–66
[19] Park H M and Yoon T Y 2000 *Int. J. of Heat and Mass Tran.* **43** 1767-76
[20] Siegel R and Howell J R 2002 *Thermal Radiation Heat Transfer, 4th Edition* ed R H Bedford (New York: Taylor & Francis) pp 9-10
[21] Golub G H, Hansen P C and O'Leary D P 1999 *SIAM J. on Matr. Ana. and App.* **21** 185-94
[22] Neumaier A 1998 *SIAM Rev.* **40** 636–66

Inverse Optimization of Plasmonic and Antireflective Grating in Thin Film PV Cells

Shima Hajimirza and John Howell

The University of Texas at Austin
Department of Mechanical Engineering
Austin, Texas, USA
Email: Shima@ices.utexas.edu

Abstract. This work addresses inverse optimization of three dimensional front and back surface texture grating specifications, for the purpose of shaping the absorptivity spectrum of silicon thin film cells in targeted ways. Periodic plasmonic gratings with dimensions comparable or less than the incident light wavelength are known to enhance light absorption. We consider surface patterning of amorphous silicon (a-Si) thin films using front and/or back metallic nanostrips and ITO coatings, and show that wideband enhancement in unpolarized absorptivity spectrum can be achieved when back reflectors are used. The overall short circuit current enhancement using such structures is significant and can be as high as 97%. For TM-polarized wave it can be even higher as reported in previous work. In this work however, we focus on the optimization for the more realistic unpolarized radiation which is of significantly higher complexity. In addition, optimization is done with respect to two objective functions independently: spectral absorptivity and gain-bandwidth product of the absorptivity spectrum.

Keywords: Thin film solar cells, Inverse optimization, Spectral absorptivity

1. Introduction

Light trapping is very important in improving efficiencies of thin film solar cells, especially modern cells which are transparent or reflective to most of the solar radiation due to very thin absorber layers (~100nm to 1micron) and use of indirect band-gap material (Silicon). Nanoscale patterning is a light trapping technique that can greatly enhance PV efficiency in amourphous Silicon (a-Si) thin film cells [1-4], and has become practical with the aid of modern thin film deposition techniques. Precise characterization of such effects is however not trivial, as several physical mechanisms are independently responsible for absorption enhancement across the visible and near infrared spectrum. Thin antireflective coatings introduce gradual change of refractive index on the thin film interface and reduce surface reflectivity. These effects are mostly modelled by Fresnel's equations. In addition, Sub-wavelength dimensions of thin film cells result in physical mechanisms that are not modelled by conventional ray tracing radiation; Small structures on the same order of size as incident wavelengths can produce forward scattering and localize light below the diffraction limit, effects which both lead to enhanced absorption. Metallic gratings and particles of even smaller nanoscale dimensions can use optical wavelengths to trigger surface plasmon polariton (SPP) waves or localized surface plasmon resonances (LSPR) as investigated in [5-10]. These free-electron mobilizations represent focused electromagnetic fields that absorb evanescently into an adjacent a-Si medium. Long periodic metallic structures (nanostips) create waveguide modes that can resonate with the incident light, and result in peaks in spectral absorptivity at those eigen-modes. Further, nanoscale patterns (such as gratings, coating or cladding) provide different mechanisms when present on the back surface of the cell. Coatings designed to change the refractive index on the rear side of a device can reflect energy back through the silicon for an additional round of absorption.

When independent light trapping mechanisms are combined in a poly-process design, the cumulative absorption effects can be coupled and/or non-linear, complicating analytical studies. Variations of source angle, polarization and solar irradiance add to the challenge. Scant work exists in the literature that theoretically analyzes the limits of light trapping in nanoscale thin films [19, 20], and even those are subject to many simplifications and assumptions.

Numerical and experimental methods have been favourable for investigating discrete solutions to nanostructured PV cells. A few of such recent results studying the effects of plasmonic nanoparticles or gratings, and ITO are given in [2-17]. These studies were forced to use an exhaustive search method to access a small number of flexible parameters in two-dimensional device geometries. The exhaustive search approach involves calculation of the complete parameter map, which is intensely time consuming for problems with large number of flexible parameters and/or wide numerical ranges. Computational expense is a formidable challenge, with convergence times bottlenecked by extraordinarily small time steps and/or element sizes required to resolve the light/structure interaction accurately. Therefore, the addition of merely a few more flexible parameters makes an exhaustive search scheme intractable on even the fastest machines. Given the size of the available design space, the ability to conduct accurate and comprehensive analysis would be extremely valuable, precluding the need for expensive trial and error fabrication. One way to achieve this goal is through inverse optimization.

In previous work, we applied hybrid optimization techniques to solve for optimal front and back surface metallic gratings as well as indium tin oxide (ITO) antireflective layers to achieve maximum enhancement factor in the absorption of solar energy, and reported preliminary results for the case of two-dimensional simulations [18]. In this work, we generalize the previous preliminary results in several ways. We consider three-dimensional PV cells invoking more intense computational loads for numerical optimization. We allow for the front and back surface grating structures to be misaligned, giving more degrees of freedom to locations of resonated waveguide modes. In addition, we consider optimization for the case of unpolarized incident light which is an improvement over most existing numerical results. Furthermore, we consider the optimization of the texture morphology against an alternative objective function of gain-bandwidth product in addition to the short circuit current. In addition, we also analyse the sensitivity of the optimal designs to structural variations that might occur in the geometry vector as a result of numerical or fabrication error, and compare the results for different texture models.

The conclusions obtained from numerical simulations are similar to the case of 2D simulations previously studied. In particular, we observe that antireflective effect of the thin ITO layer on the absorption spectrum superimposes with the waveguide resonance created by back surface metallic grating, and creates a more productive coupling than with front surface gratings. In general, metallic structures on the back surface are less sensitive to structural and light incident angle variations and couple better with the antireflective coating than the front surface metallic gratings.

2. Structures for Metal Grating and ITO Coating

We study three different light trapping mechanisms based on periodic arrays of infinitely long metallic nanostrips and anti-reflective ITO coatings on the front and/or rear sides of a thin a-Si layer of a fixed thickness. Since these structures and the thin film absorber have sub-wavelengths dimensions, the study of light trapping in these cases extends beyond conventional ray tracing techniques, and require solving near field Maxwell's equations as will be discussed in the next section [19,21].

The studied structures are shown in Figure 1-3. The structure of Figure 1 consists of a periodic array of silver nanostrips and a thin layer of ITO mounted on top the silicon absorber, and is parameterizes with four variables: period Λ, silver height h_{Ag}, silver width w_{Ag} and ITO thickness h_{ito}. Front metallic gratings contribute to the absorption enhancement by various means; they produce forward scattering and excite evanescent surface plasmon polariton (SPP) waves that absorb in the underlying layers. This phenomenon is only present in thin film layers with smaller (or comparable) dimensions than incident photons wavelengths and is not characterized by the ray tracing models. In addition periodic nanostrips are waveguide-slots that create guided modes of electromagnetic energy. The electric field components of these modes peak at the centre of the slot, and due to the continuity of the electromagnetic field at the interface, create excitation in the absorber layer. However, the refractive index of silver is small compared to a-Si, thus a large fraction of the resonated energy is reflected. The reflection of incident energy at the interface is governed by Frensel equations [21]. A thin ITO layer in between silicon and nanostrips helps reduce reflectivity by smoothening the refractive index contrast between a-Si and Ag. ITO is a relatively transparent material with low absorption (small k index) with a refractive index between that of Ag and a-Si.

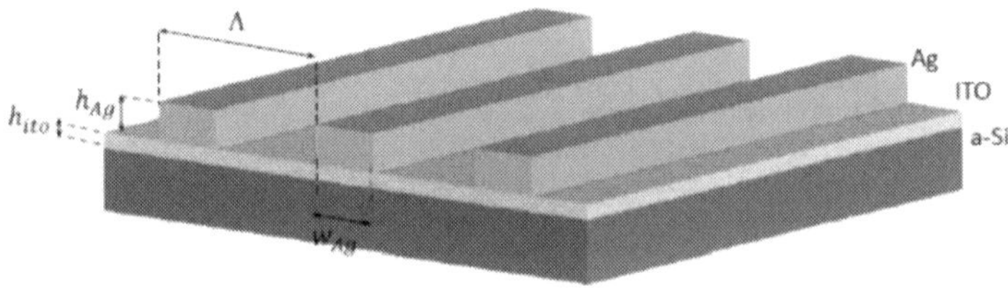

FIGURE 1: Front surface silver nanostrip grating and ITO coating structure.

The second structure is that of Figure 2, consisting of metallic nanostrip gratings on the back surface and ITO coating on the front of silicon. The parameters are the same as those of Figure 1. Unlike the front surface grating, metallic structures on the back are sought to be good reflectors, scattering light back into the cell. In addition they also create guided optic modes that couple with the incident light at certain wavelengths and create resonances, leading to increased absorptivity. Multiple peaks in the absorptivity spectrum are justified by the presence of such waveguide modes.

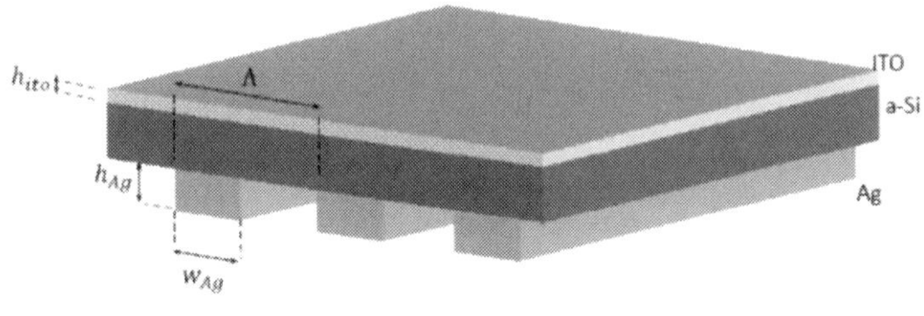

FIGURE 2: Back surface silver nanostrip grating and front ITO coating structure.

Finally, Figure 3 shows a third structure we study, which is a combination of both of the structures in Figures 1 and 2, with an additional thin ITO coating on the back. This geometry is characterized by 8 parameters: period Λ, height and width of front nanostrips h^f_{Ag}, w^f_{Ag}, height and width of back nanostrips h^b_{Ag}, w^b_{Ag}, and height of front and back ITO layers h^f_{ito} and h^b_{ito}. In addition, we allow for spacing δ between the centres of the front and back nanostrips. Such spacing potentially helps compensate the phase shifts due to light refraction on the front surface and higher of coupling with the guided modes on the back.

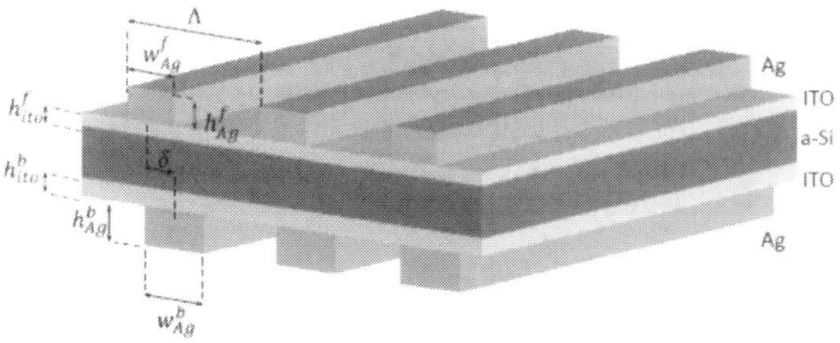

FIGURE 3: Front and back surface silver nanostrip gratings and ITO coating structure.

In a previous work [18] we reported optimization of similar structures using a simplified 2D model restricted to TM polarized sources only. An interesting observation made was that unlike front grating structures, back surface Ag gratings result in a modification of the absorptivity spectrum which superimposes with that of ITO, leading to wide-band absorption enhancement. Consequently, improvements as large as 100% in the short circuit current were reported. Furthermore, back surface structures are less sensitive to changes in the source polarization. We study these issues for the 3D nanostrip grating structures and consider unpolarized incident wave in this paper.

3. Details of Forward and Inverse Solvers

Here we discuss how numerical optimization is used to find a geometry structure with maximum absorption enhancement. Suppose that a general light trapping scheme is considered, which could be parameterized with a geometry vector **x**. This can represent the positions and dimensions of metallic grating or antireflective coating on the front or back surface of the a-Si layer (as discussed in Section 2). An inverse optimization involves finding appropriate values for the coefficients of the unknown **x**. To understand how this is done, we first discuss the forward problem of finding absorptivity characteristics for a given **x** .

3.1. Forward Problem

For a known PV structure, the most accurate way of determining the spectral absorptivity of the thin film absorber is through solving near field Maxwell's equation. Once the Maxwell's equations are solved assuming a source with 1Watt/m^2 radiation power, the spectral absorptivity of the silicon slab is found numerically by:

$$\alpha(\lambda) = \int_V -\left(\frac{\pi}{\lambda}\right) |\mathbf{E}|^2 \epsilon'' dV \tag{1}$$

where ϵ'' is the imaginary part of the electric permittivity of the absorbing material (a-Si), **E** is the electric field vector, and V is the volume of integration which encompasses a-Si slab. Furthermore, for unpolarized radiation (such as that of solar) the unpolarized absorptivity $\bar{\alpha}(\lambda)$ shall be considered:

$$\bar{\alpha}(\lambda) = \frac{2}{\pi}\int_0^{\frac{\pi}{2}} \alpha_\varphi(\lambda) d\varphi \qquad (2)$$

where $\alpha_\varphi(\lambda)$ is the absorptivity for a plane wave source with polarization angle φ[1]. The spectral reflectivity of a PV cell tells a lot about its conversion efficiency. If the irradiation air-mass spectrum is $I(\lambda)$, then the total number of absorbed photons in a spectral range Ω can be computed from a weighted expectation over the spectral absorptivity as follows:

$$N_a = \int_\Omega \lambda I(\lambda)\bar{\alpha}(\lambda) d\lambda \qquad (3)$$

Note that the above quantity is also proportional to the short circuit current of an ideal PV cell, neglecting the recombination effects within the semiconductor. Therefore, the photo-electric conversion of a cell is derived from the solution of Maxwell's equations. The forward solution therefore includes solving Maxwell's equations for sources with various polarization angles, and then numerically evaluating (2) and (3). Finite Difference Time Domain (FDTD) method is often used in solving partial differential equations such as Maxwell's.

3.2. Inverse Problem

The goal of the inverse problem is to find a PV structure that maximizes an objective function related to the spectral absorptivity. When the mapping between the geometry and material properties to the resulting objective function is not analytically tractable and is complex, numerical optimization methods should be used. Most global optimization techniques call on the forward problem many times as they try to find the best combination of free parameters in the geometry. Furthermore, forward problems can be considerably time consuming to converge, not to mention that multiple simulations are necessary to account for an unpolarized source. Therefore an inverse optimization can last a very long time for such problems. This motivates the use of very efficient global optimization algorithms.

The objective function is a goodness measure for a geometry vector **x**. A useful choice of the objective function is the *enhancement factor* in the number of absorbed photons defines by:

$$f_{EF}(\mathbf{x}) = \frac{N_a^{\text{a-}Si}(\mathbf{x})}{N_a^{\text{a-Si}}} \qquad (4)$$

where $N_a^{\text{a-}Si}(\mathbf{x})$ is the number of absorbed photons in the a-Si layer equipped with the light trapping scheme with geometry vector **x**, and $N_a^{\text{a-Si}}$ is the number of absorbed photons in the bare a-Si. For a vector, these quantities are computed by equation (3) in the forward solution. Note that $N_a^{\text{a-Si}}$ needs only be computed once, since we consider an absorber layer of fixed thickness.

In addition, we briefly discuss a alternative choice of objective function called the *gain-bandwidth product*. This is equivalent to the product of the maximum spectral absorptivity and the bandwidth, defined by:

$$f_{GBP}(\mathbf{x}) = \Delta B \times \max_{\lambda \in \Omega} \bar{\alpha}(\lambda) \qquad (5)$$

Where ΔB is the spectrum bandwidth defined as the difference between the two frequencies where the absorption drops by a factor p from its peak value. Such criterion is commonly used in the design of

[1] $\varphi = 0$ means TM polarized and $\varphi = \pi/2$ means TE polarized.

wideband amplifiers in electronics, and might not have a trivial physical interpretation for PV cells considered here. However, we suggest that gain-bandwidth product can be a measure of the stability of the light conversion efficiency of the PV cell to variations in the solar irradiance. Specifically, different realizations of solar radiation have different spectral irradiances. A cell with wideband absorptivity has therefore less sensitivity to such variations.

3.3. Optimization Algorithms

We have successfully adopted a combination of random and deterministic optimization algorithms such as *Simulated Annealing* and *Quasi-Newton*, and have created a tuneable hybrid optimization package that can solve the inverse geometry problems for thin films cells with considerable efficiency. Many of the details follow the descriptions given in previous work [15-18], so we skip them here to avoid repetition. We emphasize that our optimization algorithms are constrained, meaning that only sub-ranges of values for **x** are acceptable. This is considered with the motivation of narrowing down the search space and also for finding light trapping mechanisms that are feasible to fabricate. For a more comprehensive discussion on this, please consult with [17].

4. Results

We optimized geometry parameters of different structures using simulated annealing and an iterative coordinate-wise optimization. FDTD simulations of the forward problem were run in the Lumerical FDTD software [22]. The thickness of silicon is fixed 80nm, and spectral range of $\Omega = (300,700)\text{nm}$ is considered. Solar irradiance is assumed to be the statistically averaged AirMass1.5 [23], and radiation is normal to the surface with a polarization angle that varies uniformly between 0 and $\pi/2$. For the case of back surface gratings and front ITO (Figure 2), we tried two different objective functions: unpolarized enhancement factor (UEF) and gain bandwidth product (GBW) with 90% peak bandwidth definition. For all other cases, only UEF objective is considered. The search regions for the optimum **x** are constrained by upper and lower bounds: $50\text{nm} \leq h_{Ag}, w_{Ag} \leq 200\text{nm}$, $10\text{nm} \leq h_{ito} \leq 100\text{nm}$ and $0 \leq \delta \leq 50\text{nm}$. In addition, a minimum spacing of 50nm between adjacent nanostrips is imposed to assure that the resulting structures are feasible to fabricate. Note that even with these limitations, an exhaustive search method takes a very long time to find the best solution, especially for the case of both front and back patterns.

Numerical values of the optimization solutions are provided in Table 1. Figure 4 shows the evolution of the objective function (enhancement factor) during the course of SA optimization for the case of rear surface grating and front ITO, which has been included as an example. Note that only selected candidates are represented in this plot (for more details about SA, please refer to [15]). In addition, the unpolarized absorptivity spectrums of the optimal structures of Table 1 are plotted in Figure 5, where they are compared with the bare a-Si spectrum. Also, the enhancement factor in the number of absorbed photons of the optimized structures is plotted for varying source polarization angles in Figure 6.

Table1. Numerical values of inverse optimization solutions.

Case	Geometry(x)
Front Grating &ITO	$[\Lambda, w_{Ag}, h_{Ag}, h_{ito}] = [167,54,100,37]\text{nm}$
Back Grating & Front ITO(UEF)	$[\Lambda, w_{Ag}, h_{Ag}, h_{ito}] = [196,82,198,50]\text{nm}$
Back Grating &Front ITO(GBW)	$[\Lambda, w_{Ag}, h_{Ag}, h_{ito}] = [148,75,160,30]\text{nm}$
Back & Front Gratings and ITO's	$[\Lambda, w^b_{Ag}, h^b_{Ag}, h^b_{ito}, w^b_{Ag}, h^b_{Ag}, h^b_{ito}, \delta] =$ $[190,52,50,60,100,173,20,24]\text{nm}$

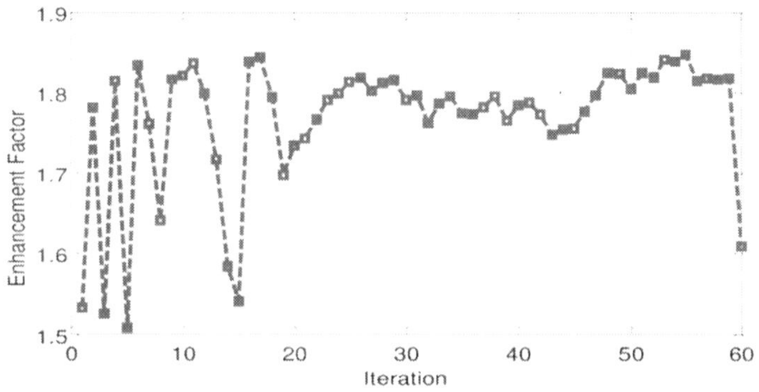

FIGURE 4: Evolution of the objective function (enhancement factor) during SA optimization for the case of rear surface gratings and front ITO layer.

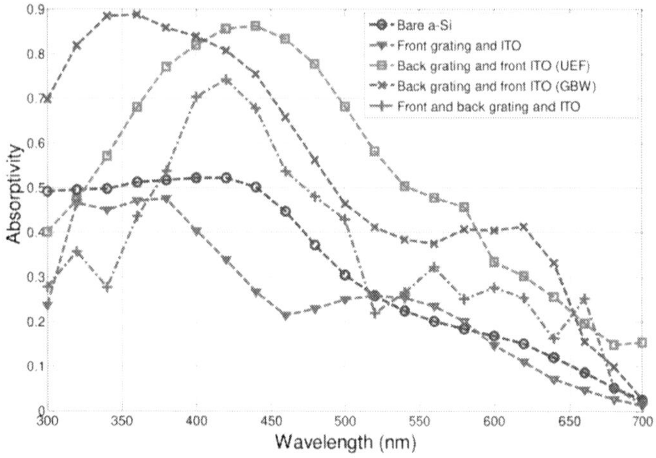

FIGURE 5: Spectral absorptivity of the inverse optimization solutions for different structures. Back grating and front ITO optimizations are done for two objective functions: Unpolarized Enhancement Faactor (UEF) and Gain Bandwidth product (GBW).

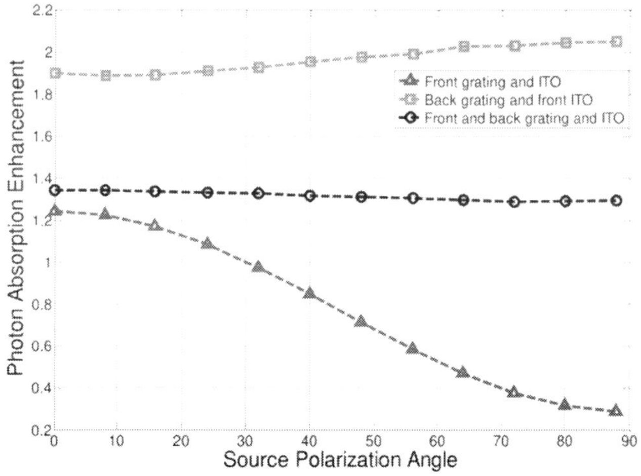

FIGURE 6: Enhancement of the total spectral absorptivity (enhancement factor) as a function of the source polarization angle for different structures.

4.1. Discussion

As indicated by Figure 5, using back surface waveguides enhancement factors as large as 1.85 in the conversion efficiency of a-Si can be achieved for unpolarized radiation. Front nanopatterns only enhance the TM polarized light absorption and are not very effective for unpolarized sources. When both back and front surface structures are used, a very stable enhancement factor over all variations of polarization angles is achieved. However, the net absorption in this case is not as high as back reflectors and front coating. These conclusions are in agreement with the results previously observed for the 2D model [18].

Note that the spectral absorptivity of the solution obtained from optimization using the GBW objective has a higher 90% bandwidth than the UEF objective, but its peak occurs at shorter a wavelength. AM1.5 spectrum peaks at around 460nm, and thus the UEF optimum solution has an overall higher photon absorption.

Referring to the numerical values of Table 1, we observe that back surface nanostrips are considerably wider and thicker than front surface gratings. Wider metallic structures are better reflectors and, when on the rear, are constructive light trapping mechanisms. Furthermore, very thick metallic gratings on the front are highly absorbent, and are thus to be avoided.

4.1.1. Sensitivity to Structural Variations

For the solutions obtained using inverse optimization, we perform a sensitivity analysis to evaluate the effect of geometry deviations on the absorptivity of the cell. This is a measure of stability of the design. In [17] a statistical analysis of efficiency robustness was reported using Monte Carlo simulations. Here, we use a different metric for measuring sensitivity. For a geometry vector $\mathbf{x}$, we define relative average sensitivity of absorption to geometry deviations as the normalized mean of the derivative of spectral absorptivity with respect to the coefficients of $\mathbf{x}$.

$$\mu(\lambda) = \frac{1}{\bar{\alpha}(\lambda)}\left(\frac{1}{n}\sum_{i=1}^{n}\left|\frac{\partial\bar{\alpha}(\lambda)}{\partial x_i}\right|\right) \quad (6)$$

We numerically calculated $\mu(\lambda)$ for the optimum structures obtained by inverse optimization (for the UEF objective). The resulting curves are plotted in Figure 7. Note that front grating structures have a highly geometry-sensitive spectrum. In contrast, rear surface structures are fairly robust to average structural deviations.

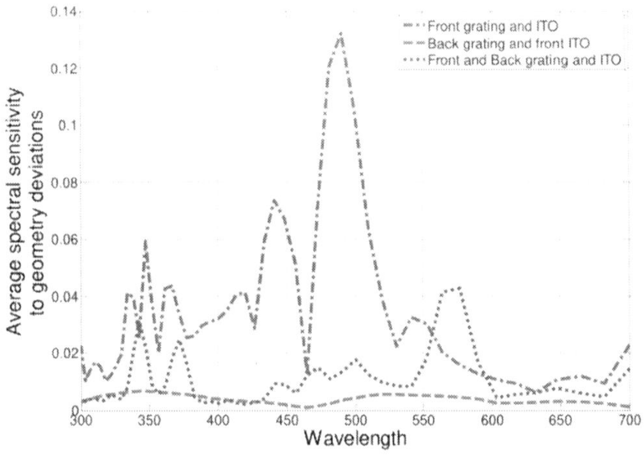

FIGURE 7: Average spectral sensitivity to geometry deviations for different structures.

5. Conclusion

We used inverse optimization algorithms to optimize combinations of back and/or front surface gratings and ITO coating to maximize photon absorption of thin film a-Si cells for unpolarized radiation. The problem is of a significant computational load and thus, locating optimal geometries for the nanoscale patterns require efficient numerical optimizations. We concluded that nanostrip back reflectors are significantly more effective in enhancing the spectral absorptivity of thin film silicon than front structures, and are less sensitive to variations in source polarization and structural deviations. Our optimization framework allows for tuneable light trapping which can maximize arbitrary objective functions. For instance, we demonstrated the gain-bandwidth product maximization of absorptivity spectrum using back surface gratings. Future work shall report experiments with other forms of physically meaningful objectives.

References

[1] Rockstuhl, C., Fahr, S. and Laderer, F. 2008 Absorption Enhancement in Solar Cells by Localized Plasmon Polaritons , *J. Appl. Phys.*, Vol. 104(12).
[2] Beck, F.J., Polman, A. and Catchpole, K. R. 2009 Tunable Light Trapping for Solar Cells Using Localized Surface Plasmons, *J. Appl. Phys.*, Vol. 105.
[3] Wang, W., Wu, S. 2010 Reinhardt, K., Lu, Y. and Chen, S.: Broadband Light Absorption Enhancement in Thin-Film Silicon Solar Cells, *Nano Lett.*, Vol. 10.

[4] Sefunc, M.A., Okyay, A.K. and Demir, H.V. 2011 Volumetric Plasmonic Resonator Architecture for Thin-Film Solar Cells, *Appl. Phys. Letts.*, Vol. 98.
[5] Schaadt, D. M, Feng B. and Yu E. T. 2005 Optical Absorption Via Surface Plasmon Excitation in Metal Nanoparticles, *Appl. Phys. Lett.*, Vol. 86.
[6] Panoiu, N.C. and Osgood, R.M. 2007 Enhanced Optical Absorption for Photovoltaics Via Excitation of Waveguide and Plasmon-Polariton Modes, *Opt. Lett.*, Vol. 32.
[7] Shchegrov, A. V., Joulain, K., Carminati, R. and Greffet, J.J. 2000 Near-Field Spectral Effects due to Electromagnetic Surface Excitations, *Phys. Rev. Letts.*, Vol. 85.
[8] Derkacs, D., Lim, S.H., Matheu, P., Mar, W. and Yu, E.T. 2006 Improved Performance of Amorphous Silicon Solar Cells Via Scattering From Surface Plasmon Polaritons in Nearby Metallic Nanoparticles, *Appl. Phys. Lett.*, Vol. 89.
[9] Pillai, S., Catchpole, K.R., Trupke T. and Green, M. A. 2007 Surface Plasmon Enhanced Silicon Solar Cells. *J. Appl. Phys.*, Vol. 101.
[10] Ferry V.E., Sweatlock L.A., Pacifici D. and Atwater H.A. 2008 Plasmonic Nanostructure Design for Efficient Light Coupling Into Solar Cells, *Nano Lett.*, Vol. 8.
[11] Munday, J. and Atwater, H.A. 2010 Large Integrated Absorption Enhancement in Plasmonic Solar Cells by Combining Metallic Gratings and Antireflection Coatings, *Nano Letters*, Vol. 11.
[12] Muller, J., Rech, B., Springer, J. and Vanecek, M. 2004 TCO and Light Trapping in Silicon Thin Film Solar Cells, *Solar Energy*, Vol. 77.
[13] Ferry, V.E., Verschuuren, M.A., Li, H.B.T., Verhagen, E., Walters, R.J., Schropp, R.E.I., Atwater, H.A. and Polman, A. 2010 Light Trapping in Ultrathin Plasmonic Solar Cells, *Opt. Express*, Vol. 18.
[14] Zhao L., Zuo Y.H., Zhou C.L., Li H.L., Diao H.W. and Wang W.J. 2010 A Highly Efficient Light-Trapping Structure for Thin-Flm Silicon Solar Cells, *Sol. En.*, Vol. 84.
[15] Hajimirza, S., El Hitti, G., Heltzel, A. and Howell, J. 2011 Specification of Micro-Nanoscale Radiative Patterns Using Inverse Analysis for Increasing Solar Panel Efficiency, *Proc. ASME/JSME* 8th *Thermal Eng. Joint Conf. AJTEC2011 (accepted ASME J. Heat Transfer).*
[16] Hajimirza, S., El Hitti, G., Heltzel, A. and Howell, J. 2011 Using Inverse Analysis to Find Optimum Nano-scale Radiative Surface Patterns to Enhance Solar Cell Performance, *Int. Conf. Ther. & Mat. NanoSci. & NanoTech. TMNN (accepted Int. J. Ther. Sci.).*
[17] Hajimirza, S. and Howell, J. 2011 Robust Nanoscale Patterns for Thin Film Solar Cells Using Inverse Optimization of Non-Uniformly Sampled Absorption Spectrum, *ASME Int. Mech. Eng. Conf. & Exp..*
[18] Hajimirza, S., Heltzel, A. and Howell, J. 2012 Broadband Absorption Enhancement in Thin Film Solar Cells Using Inverse Optimization of Light Trapping Mechanisms, *ASME Micro/Nanoscale Heat and Mass Trans. Int. Conf. MNHMT.*
[19] Yu, Z., Raman, A. and Fan, S. 2010 Fundamental Limit of Nanophotonic Light Trapping in Solar Cells, *Proc. Nat. Acad. Sci. USA, doi:10.1073/pnas.1008296107.*
[20] Han, S.E. and Chen, G. 2010 Toward the Lambertian Limit of Light Trapping in Thin Nanostructured Silicon Solar Cells, *Nano Lett.*, Vol. 10 (11).
[21] Howell, R., Siegel R. and Menguc P.M. 2011 Thermal Radiation Heat Transfer, *CRC Press* 5th Ed.
[22] http://www.lumerical.com/fdtd.php
[23] American Society for Testing and Materials, 2003, ASTM standard tables for reference solar spectral irradiances. See also URL http://www.astm.org.

Estimation of strength of source in a 2D participating media using the differential evolution algorithm

Ajit K Parwani, Prabal Talukdar[1], PMV Subbarao
Department of Mechanical Engineering, Indian Institute of Technology Delhi, New Delhi 110016, India

Phone: +91 11 26596337, Fax: +91 11 26582053, E-mail: prabal@mech.iitd.ac.in

Abstract. A numerical model is developed to estimate the time varying strength of a heat source without prior information about the unknown function. The medium is considered to be gray, absorbing and emitting and the heat transfer is assumed to take place by both radiation and conduction. All the boundaries of the 2-D enclosure are considered to be black with known temperature. The finite-volume method is used for discretization of the energy as well as the radiative transfer equation. This inverse radiation-conduction problem is solved with the minimization of a performance function, which is expressed by the sum of square residuals between calculated and observed temperature, utilizing the differential evolution algorithm. The prediction of the strength of the source by the present algorithm is found to be quite reasonable.
Keywords: Inverse radiation, finite volume method, differential evolution algorithm, participating medium.

1. Introduction

A direct heat transfer problem is about solution of the heat equation depicting a model of heat transfer - Conduction, Convection, Radiation or a combination of more than one of them. These mathematical models are governed by various input factors e.g. surface temperature, heat source strength, heat flux, dimension of the conducting body / convecting media, etc. and their goal is to predict the temperature profile of the body either independent of (steady) or depending on the time (transient).These types of problems are mathematically "well posed" and, therefore, can be solved with straight forward and well-established mathematical procedures.

But in the real life situations, where it is not possible to measure the casual factors (boundary temperature or flux) directly, however the temperature can be measured inside the enclosure by a sensor/thermocouple, a natural problem becomes to predict the unknown casual factor. This leads to the inverse problem.

The determination of the unknown source term in conduction problems have been investigated by several authors due to numerous applications in engineering and science. Huang and Ozisik [1] and Neto and Ozisik [2–4], determined the unknown time wise variation of the strength of a heat source, located inside a flat plate, using the conjugate gradient method [CGM]. Yang [5] solved 2D inverse heat source problem by linear least square error model. Jin and Marin [6] employed the method of fundamental solutions (MFS) to recover the heat source in steady-state heat conduction problems. Yan et al. [7] also applied the MFS to solve an inverse heat source problem with arbitrary geometry.

[1]Corresponding author

Girault et al. [8] used experimentally built low order model for the estimation of time-varying heat source involving thermal diffusion with convective and radiative boundary conditions.

Recently, inverse heat transfer analysis in radiation has received much attention due to its various applications [9]. Among others, especially, the inverse conduction–radiation analysis was applied to estimate the thermal properties [10-13, 18-19] or source term in a medium [14-16] or heat flux [17]. Matthews et al. [10] estimated the extinction coefficient, back scattering fraction, and thermal conductivity when the temperature and transmittance measurements are given in one-dimensional planer layer. They have used a nonlinear parameter estimation technique. Li [11] estimated the single scattering albedo, the optical thickness, the conduction-to-radiation parameter, and the scattering phase function with the exit radiation intensities in a one-dimensional plane-parallel medium using CGM in parameter estimation approach. Park and Lee [12] estimated the spatially varying heat transfer coefficient and the absorption coefficient in a radiant cooler by improved adjoint variable method. Park and Yoon [13] estimated radiative parameters using the CGM in a 3D participating media, where radiation and conduction occur simultaneously. Salinas [14] applied CGM to estimate the strength of a heat source in a two-dimensional gray media from radiative intensities exiting in some points of boundary surfaces. Park and Lee [15] and Park and Yoo [16] also estimated the strength of a heat source using modified CGM in a furnace. Kim and Baek [17] estimated total heat flux distribution utilizing Levenberg–Marquardt method on heater surface in a two-dimensional concentric cylindrical absorbing, emitting and scattering medium.

The deterministic method like CGM is most popular method for inverse problems because being an iterative regularization method the regularization procedure is performed during the iterative processes and thus the determination of optimal regularization conditions is not needed. The CGM derives from the perturbation principles and transforms the inverse problem to the solution of three problems, namely, the direct, sensitivity and the adjoint problem. Deterministic methods are in general computationally faster than stochastic methods or search-based methods like Genetic algorithm (GA) and Differential Evolution (DE), although they can converge to a local minima or maxima, instead of the global one. On the other hand, stochastic algorithms can ideally converge to a global maxima or minima, although they are computationally slower than the deterministic ones.

Stochastic methods also are applied for inverse problems for their outstanding characteristics, for instance, less dependence on initial value and non-requirement of gradient information. Lobato et al. [18] used DE approach for the estimation of radiative properties in two-layer participating media. The results obtained with this methodology are compared with other approaches, namely the simulated annealing algorithm (SA), the Levenberg–Marquardt method (LM) and the hybridization simulated annealing and Levenberg–Marquardt (SA-LM). Das et al. [19, 20] used GA to solve inverse transient conduction–radiation problem for estimation of scattering albedo, conduction–radiation parameter and boundary emissivity. They used the lattice Boltzmann method for conduction and the finite volume method (FVM) for radiation. Liu [21] used modified GA for solving inverse heat conduction problem to estimate unknown transient heat source.

The DE is a structural algorithm proposed by Storn and Price [22] for optimization problems. Besides its good convergence properties and suitability for parallelization, DE's main assets are its conceptual simplicity and ease of use. This approach is an improved version of Goldberg GA for faster optimization. There has not been extensive work done using DE approach in inverse heat transfer problems. To the knowledge of the authors, DE approach has still not been used for the estimation of strength of source in a participating medium. So, its applicability and performance for such kind of problem needs to be tested. Therefore, in this work, an inverse radiation problem is studied to estimate an unknown transient heat source using DE algorithm. The remainder of this paper is organized as follows. In Section 2 the direct and inverse heat transfer problem is described and formulated. It shows that the inverse problem can be converted into an optimization problem. Section 3 provides an overview of the DE algorithm. Some numerical results of the implementation of these algorithms for solving the inverse problem are discussed in Section 4. Section 5 summarizes this paper with certain conclusions.

2. Problem statement and formulation

2.1. Direct problem

A 2D enclosure of dimension 1m x 1m as shown in figure 1 is considered. Heat transfer within the enclosure is contributed by conduction as well as radiation. All boundaries of medium are considered as black with known temperature. Table 1 shows all the data about the thermophysical and the radiative properties taken for study. The governing equations for the temperature field are described in following subsection.

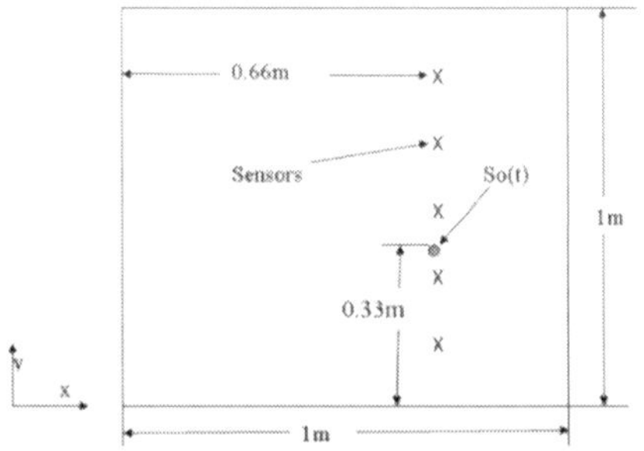

Figure 1. Problem description

Table 1. Property data of medium except stated otherwise in the main text

Property	Value	Property	Value
absorption coefficient, α	1.0 m^{-1}	heat capacity, $\rho\, c_p$	1.2 kJ/kgK
conduction-radiation parameter, $N = k\,\beta/4\sigma_b T^3$	0.008	scattering coefficient, σ_s	0.0 m^{-1}
extinction coefficient, β	1.0 m^{-1}	thermal conductivity, k	0.05 W/mK

2.1.1. Energy equation. The transient state energy conservation equation considering conduction, radiation and heat generation can be expressed as

$$\rho C_p \frac{\partial T}{\partial t} = k\nabla^2 T - \nabla . q_R + So(t)\delta(x - x^*)\delta(y - y^*) \quad (1)$$

where the ∇ operator is with respect to some physical length s, $So(t)$ denotes the strength of heat source and δ is the dirac delta function which determines the location of heat source. In the present work, a heat source at a location of $x^* = 0.66$ m and $y^* = 0.33$ m has been considered. It is assumed that the thermal conductivity k of the medium is independent of temperature T of the medium. The relevant boundary conditions are:

- At all boundaries, $T = T_w = 300$ K (2)

where T_w is wall temperature. The divergence of the radiative heat flux required in the energy equation (equation (1)) is given by

$$\nabla . q_R = \kappa(4\pi I_b - \int_{4\pi} I(\vec{r}, \hat{s}) d\Omega) \quad (3)$$

where I is the intensity, I_b is the blackbody intensity and Ω is the solid angle.

2.1.2. Radiative transfer equation (RTE). To calculate the divergence of radiative heat flux (equation (3)), the solution of RTE is required. The RTE for a gray, absorbing and emitting medium (neglecting scattering) in the direction $\hat{s}$ can be written as

$$\nabla . (\hat{s} I) = -\beta(\vec{r})\, I(\vec{r}, \hat{s}) + \kappa(\vec{r}) I_b(\vec{r}) \quad (4)$$

In Eqs. (3, 4), $\vec{r}$ and $\hat{s}$ are the position vector and the unit vector describing the radiative intensity direction respectively. β (= $\kappa + \sigma_s$) is the extinction coefficient and is equal to absorption coefficient κ since the scattering is absent ($\sigma_s = 0$) in this problem. The numerical model uses the same control volume for both conduction and radiation as both the energy equation (equation (1)) and RTE (equation (4)) are solved with the finite volume method (FVM). The step scheme is used for spatial discretization of the RTE. The readers are referred to the work of Talukdar et al. [23, 24] for the details of the discretized equations. The boundaries are isothermal (equation (2)) and black and are given by:

$$I(\vec{r},\hat{s}) = I_b = \sigma_b T_w{}^4/\pi \tag{5}$$

where σ_b is Stefan-Boltzmann constant. The direct problem is solved using the FVM. The numbers of control volumes and control angles considered are 30 x 30 and 12 x 12 respectively while using nineteen numbers of time steps. The control volume size are $\Delta x = \Delta y = 3.3$ cm for the geometry illustrated in figure 1. The total experiment duration t_f is taken as 180 s and time step size $\Delta t = 10$ s.

The computational procedure for calculation of temperature field is as follows:

1. Assume a temperature field.
2. Calculate I_b using the given temperature field.
3. Solve the RTE (equation (4)) using the FVM to obtain the radiation intensity I.
4. Calculate the divergence of the radiative heat flux from equation (3).
5. Solve equation (1) with the divergence of the radiative heat flux calculated from equation (3) as a source term to obtain the temperature field.
6. Return to step 2 with the updated temperature field.
7. Iterate until a convergence in the temperature field is obtained.

2.2 Inverse problem

In the inverse problem, strength of the source is unknown and is to be estimated. The additional information which requires for solution of this inverse problem is some temperature data inside the solution domain. In a practical problem, these temperature data are found by measuring the temperature of the enclosure at different locations using some sensors. In this work, instead of measured data, some temperature data are taken from the solution of the direct problem described in section 2.1.

The solution of present inverse problem is to be sought in such a way that the objective function $J(So)$ is minimized.

$$J\big(So(t)\big) = \sum_{t=1}^{D}\sum_{m=1}^{M}[\mathrm{T_c}(x_m, y_m, t;\ So) - T_{\mathrm{m}}(x_m, y_m, t)]^2 \tag{6}$$

where M is the number of the mounted sensors, $T_c(x_m, y_m, t; So)$ are the calculated temperatures at the measurement locations with estimated value of source and $T_m(x_m, y_m, t)$ are the measured temperatures taken at D discrete time intervals.

The stopping criteria is specified as

$$J(So) < \varepsilon \tag{7}$$

where ε is a small predefined constant. In this work ε is taken as 10^{-5}.

However, the observed temperature data may contain measurement errors. Therefore, we do not expect the functional (equation (6)) to be equal to zero at the final iteration step. We use the discrepancy principle as the stopping criterion. The following expression is obtained for stopping criteria ε:

$$\varepsilon = M\sigma^2 D \tag{8}$$

where σ is standard deviation of measurements.

3. Differential Evolution

DE is a parallel direct search method which utilizes Np D-dimensional parameter vectors [20]:

$$\vec{x}_{i,G} = [x_{i,1,G}, x_{i,2,G} \dots\dots\dots\dots\dots x_{i,D,G}],\ i = 1,2,3\dots\dots\dots Np \tag{9}$$

as a population for each generation G, i.e. for each iteration of the optimization. Each variable in the population represents strength of heat source to be estimated such as $x_{i,1,G}$, $x_{i,2,G}$, ... $x_{i,D,G}$ represents estimated strength of heat source during different time steps ($j = 1, 2, \dots\dots\dots D$) in i^{th} population and in generation G. The initial population is chosen randomly between the lower and the upper bounds and should try to cover the entire parameter space uniformly. The crucial idea behind DE is a scheme for generating trial parameter vectors. DE algorithm consists of three genetic operators: mutation, crossover and selection. Mutation and crossover operators generate new individuals and selection operator determines suitable individual which gives minimum objective function values, and in this way population gets the better individuals.

The mutation operator used in this article is:

$$\vec{v}_{i,G} = \vec{x}_{i,G} + F_1.(\vec{x}_{best,G} - \vec{x}_{i,G}) + F_2.(\vec{x}_{r1,G} - \vec{x}_{r2,G}) \tag{10}$$

where the indexes r_1 and r_2 represent the random and mutually different integers generated within range (1, *Np*) and also different from index i. F_1 and F_2 are the mutation scale factors usually less than 1. $\vec{x}_{best,G}$ is the best vector in generation G.

After mutation, the new individuals are generated using the following scheme in crossover procedure:

$$u_{i,j,G} = \begin{cases} v_{i,j,G} \text{ if rand } (0,1) \leq CR \\ x_{i,j,G} \text{ otherwise} \end{cases}, \text{ i} = 1,2,\ldots\ldots,Np \text{ and j} = 1, 2,\ldots\ldots,D \tag{11}$$

where *CR* is a crossover parameter or factor usually less than 1 and presents the probability of creating parameters for trial vector from a mutant vector.

In selection procedure for each target individual $\vec{x}_{i,G}$, the fitness value of the trial individual $\vec{u}_{i,G}$ is compared with that of the target individual $\vec{x}_{i,G}$, and the individual with the maximum fitness value is selected for the next generation. Selection is given by equation (12),

$$\vec{x}_{i,G+1} = \begin{cases} \vec{u}_{i,G} \text{ if } f(\vec{u}_{i,G}) < f(\vec{x}_{i,G}) \\ \vec{x}_{i,G} \text{ otherwise} \end{cases} \tag{12}$$

The overall computational procedure is given in flow chart (Figure 2). The parameters used for DE algorithm in this work are Np = 10, F_1 = 0.8, F_2 randomly varying in range (0, 1) and CR = 0.5. The upper and lower limit of source strength for inverse problem taken is 10,000 W/m^2 and 0 W/m^2 respectively at all time steps.

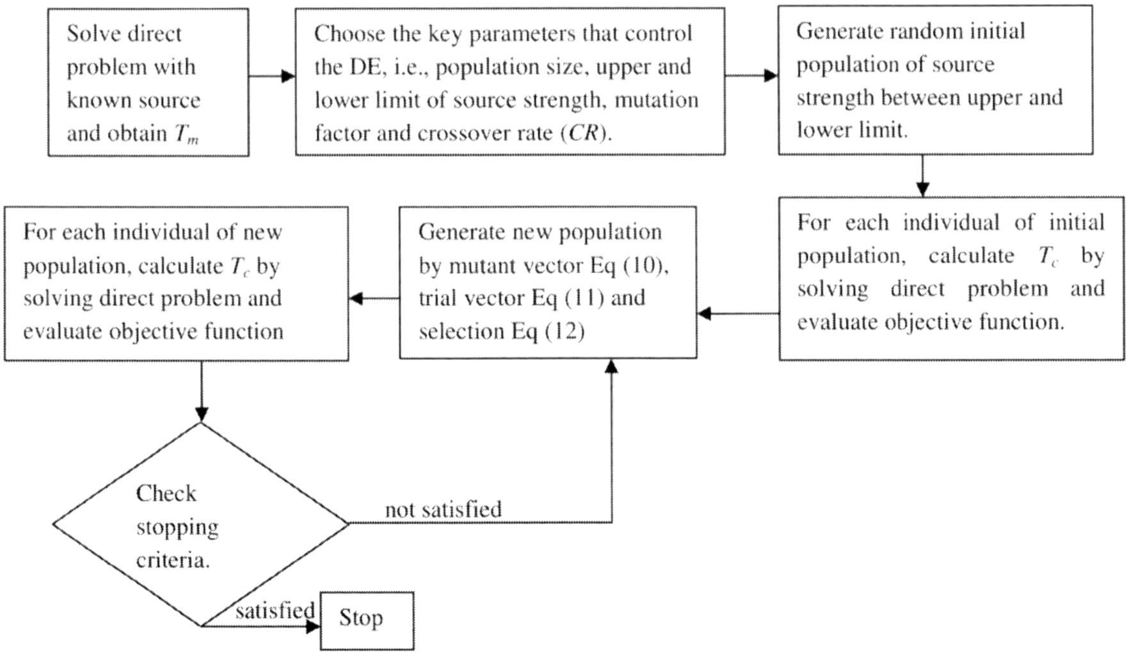

Figure 2. Computational procedure

4. Results and Discussion

The objective of this work is to estimate the strength of a transient heat source without any prior information on the functional form of the source. The unknown strength of source is assumed to be any of the three different forms of function listed below to examine the accuracy of the proposed model.

$$So(t) = 30000\sin\left(\frac{\pi t}{180}\right), \text{for } 0 \le t \le 180 \tag{13}$$

$$So(t) = \begin{cases} 30000\left(\frac{t}{180}\right), & \text{for } 0 \le t \le 90 \\ 30000\left(1 - \frac{t}{180}\right), & \text{for } 90 < t \le 180 \end{cases} \tag{14}$$

$$So(t) = \begin{cases} 1000, & \text{for } t \le 30 \text{ and } t \ge 150 \\ 30000, & \text{for } 30 < t < 150 \end{cases} \tag{15}$$

The effects of number of measurements on the accuracy of estimations are investigated. Figure 1 shows schematic of five thermocouples ($M = 5$) arrangement in y direction at $x = 20\Delta x = 0.66$m with spacing of $7\Delta y$ between each two sensors and with the first and last sensor placed at $y = 0.5\ \Delta y$ and $29.5\ \Delta y$ respectively. Similar type of arrangement at same position of $x = 20\Delta x$ is assumed for $M = 8$ with corresponding spacing of $4\Delta y$ respectively. The accuracy of estimations is also investigated with single sensor positioned at source location. It is also noted that with $M = 8$ third sensor from bottom placed exactly at source position while with $M = 5$, no sensor comes at source position.

In experiments, there could be errors in the measurement of temperature. Since in the present work, these temperature data are taken from the solution of direct problem, they are error free. In order to be more realistic, errors are added randomly in these temperature data so as to represent them more like an experimental data. Thus, the measured temperature data T_m are represented by adding random errors to the exact temperature T_e computed from the solution of the direct problem as:

$$T_m = T_e + \omega\sigma \tag{16}$$

where ω is a random variable of normal distribution with zero mean and unit standard deviation. The value of ω is calculated by the Fortran subroutine 'random_number' and chosen over the range -2.576 $< \omega <$ 2.576. Standard deviation, σ can be related to relative measurement error as follows:

$$e_{rel} = \frac{2.576 \times \sigma}{T_e} \tag{17}$$

The σ is adjusted such that e_{rel} is zero, 1%, 3% and 5% respectively.

The RMS error is calculated to compare the quality of estimation of strength of source as:

$$RMS = \frac{1}{D}\sqrt{\sum_{i=1}^{D}[So_{ex}(t) - So_{est}(t)]^2} \tag{18}$$

where So_{ex} and So_{est} are the exact and estimated strength of source values.

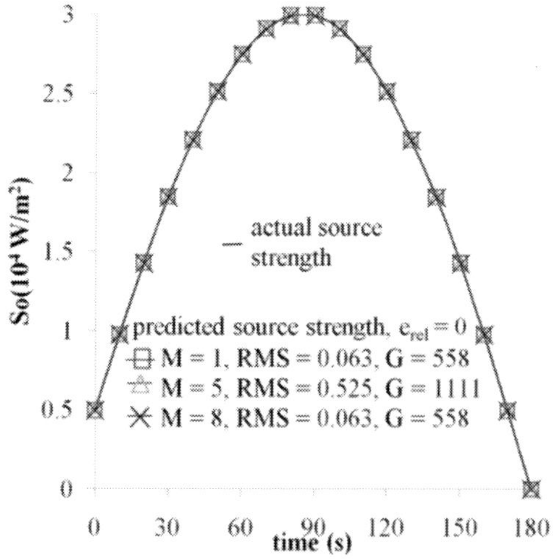

(a)

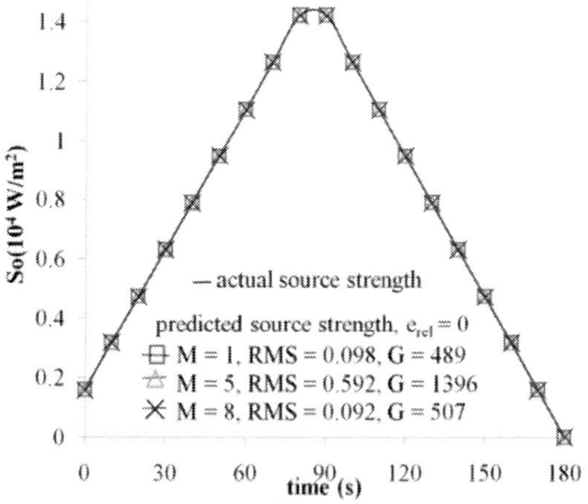

(b)

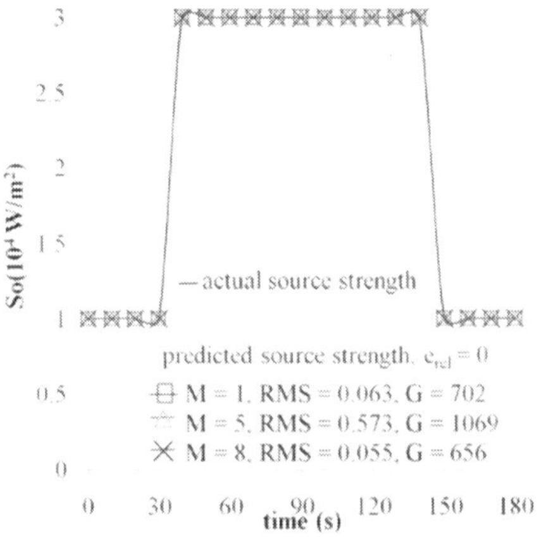

(c)

Figure 3: The exact and estimated profiles of heat source function **(a)** sinusoidal, **(b)** triangular and **(c)** step, with different number of measurements and $e_{rel} = 0$

As the first test case of the present work, an idealized situation is considered in which there are no measurement errors i.e. $e_{rel} = 0$. Figures 3(a), 3(b) and 3(c) show the exact and estimated profiles of the heat source function $So(t)$ for the three functional forms and with three different set of measurements as described above. In the same figures, the RMS error and number of iterations to satisfy stopping criterion are also indicated. The RMS error with one and eight measurements set are almost similar. With $M = 5$ sensors possesses relatively low sensitivity of heat source strength as no sensor placed at source position therefore it takes more number of iterations for stopping criteria and RMS error is also high compared with $M = 1$ and $M = 8$.

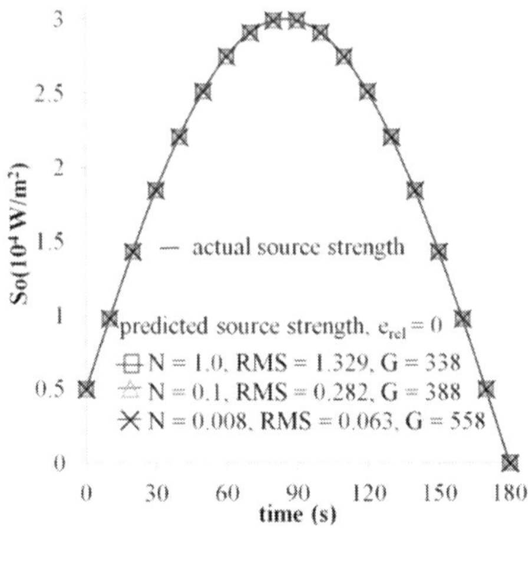

Figure 4: The exact and estimated sinusoidal profiles of heat source function with eight errorless ($e_{rel} = 0$) measurements and different values of N.

The next consideration is the effect of variation of conduction-radiation parameter N (= $k\,\beta/4\sigma_b T^4$) on the accuracy of estimation. . Figure 4 shows the exact and estimated profiles of the sinusoidal heat source function with eight errorless measurements and different values of N. In addition to the value of N (= 0.008) as given in Table 1, two other values of N (= 0.1 and 1.0) are considered in this parametric study. When the value of N increases, the conduction heat transfer gradually increases and for N = 1, the contribution of conduction heat transfer becomes much more dominant than radiation heat transfer. It is observed from the results that as the value of N increases the RMS error increases. This is due to the fact that the sensitivity value becomes smaller when the conduction heat transfer becomes more dominant than radiative heat transfer as also pointed by Kim and Baek [17].

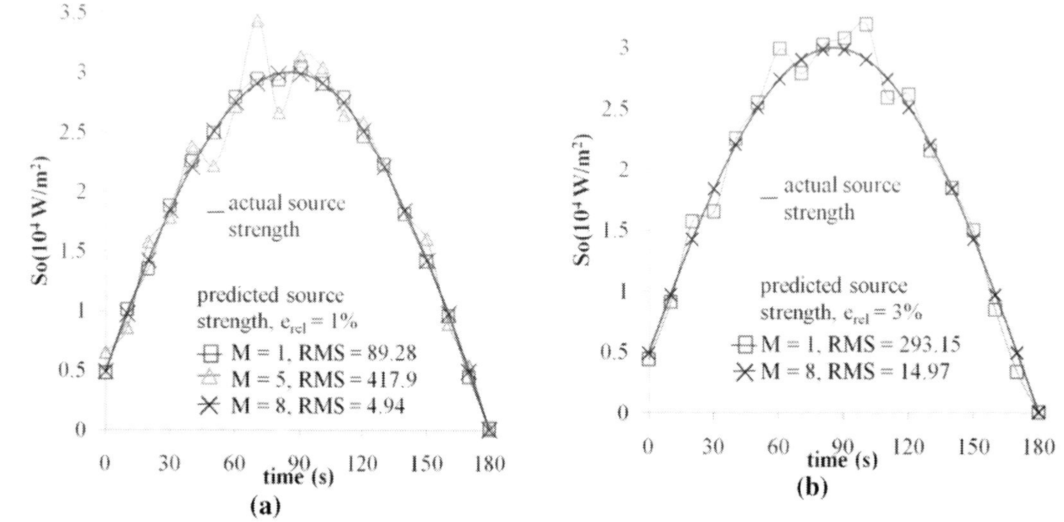

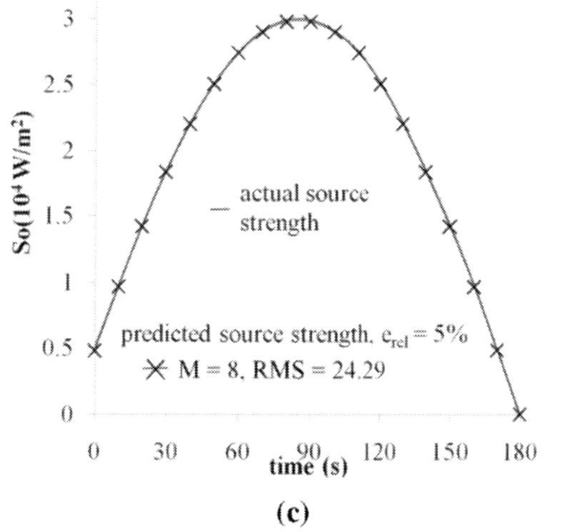

Figure 5. The exact and estimated profiles of sinusoidal heat source function with e_{rel} = 1% **(a)**, e_{rel} = 3% **(b)** and e_{rel} = 5% **(c)**.

Finally, the effect of measurement errors on the accuracy of estimation is investigated. Figure 5 shows the exact and estimated profiles of the sinusoidal heat source function *So(t)* with measurement error, e_{rel}= 1%, 3% and 5%. When e_{rel} is 1%, RMS error with M = 5 is 639.31 as shown in figure 5(a). It is also seen that the RMS error increases with increase in e_{rel}. Therefore, for M = 5, the RMS errors becomes very large when e_{rel}= 3% and 5%, and hence it is not shown in Figures 5(b)-(c) to avoid congestion. Same is the case with M = 1 when e_{rel}= 5% and is not shown in figure 5(c). In all cases of measurement errors, results show better estimation with M = 8.

5. Conclusions

The Differential evolution (DE) algorithm is successfully applied for the solution of inverse radiation problems to estimate the unknown source in a 2D participating media. Several test cases involving different functions of heat source, different numbers of thermocouples, different values of conduction-radiation parameter and introduction of measurement errors are considered. Even a single accurate sensor embedded at source location is sufficient to estimate the heat source profile by the DE algorithm. When a sensor is placed other than source position both the RMS error and number of iteration to reach stopping criterion increases as with case $M = 5$. Higher the value of conduction-radiation parameter higher is the RMS. It is also observed when the error in the measurements (direct problem temperature data) are added, eight set of measurements gives better estimation. The overall results show that the inverse solutions obtained by DE remain stable and found to be quite reasonable.

6. References

[1] Huang C H and Ozisik M N 1992 Inverse problem of determining the unknown strength of an internal plane heat source *J. Franklin Institute* **329 (4)** 751–64.
[2] Neto A J S and Ozisik M N 1993 Inverse problem of simultaneously estimating the timewise-varying strengths of two plane heat sources *J. Appl. Phys.* **73 (5)** 2132–37.
[3] Neto A J S and Ozisik M N 1993 Simultaneous estimation of location and timewise varying strength of a plane heat source *Numer. Heat Transfer* 24, 467–77.
[4] Neto A J S and Ozisik M N 1992 Two-dimensional inverse heat conduction problem of estimating the time-varying strength of a line heat source *J. Appl. Phys.* **71** 5357–62.
[5] Yang C Y 1998 Solving the two-dimensional inverse heat source problem through the linear least-squares error method *Int. J.Heat and Mass Transfer* **41 (2)** 393–8.
[6] Jin B and Marin L 2007 The method of fundamental solutions for inverse source problems associated with the steady-state heat conduction *Int. J. Numer. Methods Engg.* **69 (8)** 1570–89.
[7] Yan L, Fu C L and Yang F L 2008 The method of fundamental solutions for the inverse heat source problem *Engg Analysis with Boundary Elements* **32 (3)** 216–22.
[8] Girault M, Videcoq E and Petit D 2010 Estimation of time-varying heat sources through inversion of a low order model built with the Modal Identification Method from in-situ temperature measurements *Int. J. Heat Mass Transfer* **53** 206-19.
[9] Ozisik M N and Orlande H R B 2000 Inverse Heat Transfer *Taylor & Francis* (New York) Chapter 1.
[10] Matthews L K, Viskanta R and Incropera F P 1984 Development of inverse methods for determining thermophysical and radiative properties of high-temperature fibrous materials *Int. J. Heat Mass Transfer* **27** 487–95.
[11] Li H Y 1999 Estimation of thermal properties in combined conduction and radiation *Int. J. Heat Mass Transfer* **42**, 565–72.
[12] Park H M and Lee W J 2003 An inverse radiation problem of simultaneous estimation of heat transfer coefficient and absorption coefficient in participating media *Int. J. Numer. Meth. Engg.* **56** 787–807
[13] Park H M and Yoon T Y 2000 Solution of the inverse radiation problem using a conjugate gradient method *Int. J. Heat Mass Transfer* **43** 1767-76.
[14] Salinas C T 2010 Inverse radiation analysis in two-dimensional gray media using the discrete ordinates method with a multidimensional scheme *Int. J. of Thermal Sciences* **49** 302–10
[15] Park H M and Lee W J 2002 The solution of inverse radiation problems using an efficient computational technique *JQSRT* **73** 41–54
[16] Park H M and Yoo D H 2001 A multidimensional inverse radiation problem of estimating the strength of a heat source in participating media *Int. J. Heat Mass Transfer* **44** 2949–56.
[17] Kim K W and Baek S W 2007 Inverse radiation–conduction design problem in a participating concentric cylindrical medium *Int. J. Heat Mass Transfer* **50** 2828–37.
[18] Lobato F S, Steffen V Jr and Neto A J S 2010 Solution of inverse radiative transfer problems in two-layer participating media with differential evolution *Inverse Problems Sc. Engg.* **18(2)** 183–95.
[19] Das R, Mishra S C, Ajith M and Uppaluri R 2008 An inverse analysis of a transient 2-D conduction–radiation problem using the lattice Boltzmann method and the finite volume method coupled with the genetic algorithm *JQSRT* **109** 2060–77.

[20] Das R, Mishra S C and Uppaluri R 2010 Inverse analysis applied to retrieval of parameters and reconstruction of temperature field in a transient conduction–radiation heat transfer problem involving mixed boundary conditions *Int. Commun. Heat Mass Transfer* **37** 52–57.

[21] Liu F B 2008 A modified genetic algorithm for solving the inverse heat transfer problem of estimating plan heat source *Int. J. Heat Mass Transfer* **51** 3745–52.

[22] Storn R and Price K 1997 Differential Evolution – A simple and efficient heuristic for global optimization over continuous spaces *J. Global Optimization* **11(4)** 341–59.

[23] Talukdar P, Steven M, Issendorff F V and Trimis D 2005 Finite volume method in 3-D curvilinear coordinates with multiblocking procedure for radiative transport problems *Int. J. Heat Mass Transfer* **48** 4657-66.

[24] Talukdar P, Issendorff F V, Trimis D and. Simonson C J 2008 Conduction–radiation interaction in 3D irregular enclosures using the finite volume method *Heat Mass Transfer* **44** 695-704.

Eurotherm Conference No. 95: Computational Thermal Radiation in Participating Media IV IOP Publishing
Journal of Physics: Conference Series **369** (2012) 012017 doi:10.1088/1742-6596/369/1/012017

A massively parallel solution strategy for efficient thermal radiation simulation

P D Nguyen[1*] , V Moureau[1], L Vervisch[1] and N Perret[2]
[1] CORIA-CNRS, INSA de Rouen, Campus du Madrillet, Avenue de l'Université, BP 8, 76801 Saint Etienne du Rouvray Cedex, France. Phone: +33 (0)3 32 95 97 85
[2] Rhodia Opérations, Département Technologie & Expertise Process, Modélisation, Statistiques et CFD, 85 avenue des Frères Perret, BP 62, 69192 Saint Fons, France. Phone: +33 (0)3 04 72 89 64 20

Corresponding author's e-mail address: phucdanh.nguyen@arcelormittal.com

Abstract. A novel and efficient methodology to solve the Radiative Transfer Equations (RTE) in thermal radiation is discussed. The BiCGStab(2) iterative solution method, as designed for the non-symmetric linear equation systems, is used to solve the discretized RTE. The numerical upwind and central schemes are blended to provide a stable numerical scheme (MUCS) for interpolation of the cell facial radiation intensities in finite volume formulation. The combination of the BiCGStab(2) and MUCS methods proved to be very efficient when coupling with the DOM approach to solve the RTE. A cost-effective tabulation technique for the gaseous radiative property model SNB-FSCK using 7-point Gauss-Labatto quadrature scheme is also introduced. The whole methodology is implemented into a massively parallel unstructured CFD code where the radiative and fluid flow solutions share the same domain decomposition, which is the bottleneck in current radiative solvers. The dual mesh decomposition at the cell groups level and processors level is adopted to optimize the CFD code for massively parallel computing. The whole method is applied to simulate the radiation heat-transfer in a 3D rectangular enclosure containing non-isothermal CO_2 and H_2O mixtures. Two test cases are studied for homogeneous and inhomogeneous distributions of CO_2 and H_2O in the enclosure. The result is reported for the heat flux and radiation energy source and the comparison is also made between the present methodology BiCGStab(2)/MUCS/tabulated SNB-FSCK, the benchmark method SNB-CK (implemented at $25cm^{-1}$ narrow-band) and some other methods available in the literature. The present method (BiCGStab(2)/MUCS/tabulated SNB-FSCK) yields more accurate predictions particularly for the radiation source term. When comparing with the benchmark solution, the relative error of the radiation source term is remarkably reduced to less than 4% and the CPU time is drastically diminished.

1. Introduction

The success in the design of advanced combustion systems relies heavily on accurate and efficient CFD solvers, combustion and heat transfer modeling. In the latter, thermal radiation occupies a large portion in the total balance of heat transfer. Recent developments of Modest [1] and Liu *et al* [2] in global gas radiative modeling gave birth to the Statistical Narrow Band Full Spectrum Correlated K

* Current address: ArcelorMittal Global Research & Development, Maizières Process, Voie Romaine, BP 30320, F-57283 Maizières-lès-Metz Cedex, France. Phone: +33 (0)3 87 70 40 36.

(SNB-FSCK) model which is a compromise between the solution accuracy and computational effort if one compares with the standard band SNB-CK model (accurate but less efficient) [3] and with the global Weighted-Sum-of-Gray-Gases (WSGG) model (efficient but less accurate) [4]. The SNB-FSCK model can be implemented into radiative solvers in its expensive on-line calculation or in its pre-computation or tabulation version [2][5][6]. Nevertheless, the level of the computational accuracy and efficiency of the SNB-FSCK model is dependent on the way one couples the model with solution methods for the RTE. The tabulated SNB-FSCK model that can be used in an accurate way to gain in efficiency will be explored in the present study.

The Discrete Ordinate Method (DOM) [7] is presently one of the most widely used solution methods for discretizing the Radiative Transfer Equation (RTE) in directions because it ensures a good compromise between solution accuracy and computational cost in many practical applications [8]. This method combined with the Finite Volume (FV) discretization is particularly adapted to practical applications with complex geometry because it can be used with non-structured grids [6]. The DOM is suited to be used along with the CK and global gas radiative property models (WSGG, SLW, ADF, FSK) [8], and recently with the SNB-FSCK model. The Exponential [9], Step [10] and Diamond Mean Flux Schemes (DMFS) [11] are the conventional numerical schemes for solving the RTE, but many other schemes categorized under the Normalized Variable Diagram (NVD), Total Variation Diminishing (TVD), Essentially Non-Oscillatory (ENO), Genuinely multidimensional schemes families are also available [12,13]. However, concerning the methods to handle the RTE for unstructured grids in complex geometries, this is still a challenging task and very little work has been done to date in this direction. A sweeping optimization technique [6] was used to reorganize the cell faces numbering according to radiation wave propagation direction. The technique itself is intrinsically a serial process and the paralleling computing was carried out based on splitting the number of discrete directions. In some cases, the calculations of the emission and absorption terms had to be split into different processors. This method of parallelizing is completely decoupled from the spatial domain decomposition used in CFD and would be limited when dealing with massively parallel computation. Santos *et al.* [14] developed a methodology to couple a LES solver with a radiation code, but the two codes didn't use the same computational domain; the LES solver was parallelized by domain decomposition whereas the radiation code was based on optical rays and was parallelized by splitting in the spectral domain. An additional data exchange protocol was needed to ensure the coupling between the two codes, the computational cost was still very heavy (two-thirds of processors for radiative solver and one-third for combustion solver) and the simulation was limited to 2D cases [14].

To overcome these difficulties, a new methodology is proposed in this study to solve the RTE in the same computational grid mesh with CFD solver. In this light, the BiCGStab iterative solution method [15] is adapted to solve the non-symmetric linear equation system resulting from discretizing the RTE. It should be noted that the variant BIGSTAB was used in a parallel unstructured CFD code [16] for radiation calculation. Concerning the numerical scheme used to interpolate the cell facial radiation intensities, the Upwind scheme is blended with the Central scheme to give a stable scheme with limited dispersion and dissipation, called Mixed Upwind and Central Scheme (MUCS). According to the authors' acknowledge, it is the first time that the tabulated SNB-FSCK method, the MUCS scheme and the BiCGStab iterative solution method are combined and implemented in an unstructured code to calculate thermal radiation. The advantage of this method is that the radiative and CFD solvers share the same domain decomposition and parallelism and they are coupled in a natural manner needing no correction for load balancing or no adjustment on the number of cores devoted to each solver as done in [14]. This is paving the way for coupling radiative transfers with turbulent combustion for practical systems in the context of massively parallel computing as the solver used in the study has proved to be successful in simulating a semi-industrial swirl burner with a mesh of 2.6 billion cells [17].

The section below describes the methodology of combining the MUCS scheme and the BiCGStab iterative solution methods to solve the non-symmetric linear system of the discretized RTE with some specific care for boundary treatment. In the next section, the devised methodology is employed to calculate the thermal radiation within a 3D enclosure and the simulation results are compared with other methods presented in the work of Liu *et al.* [2] with emphasis on solution accuracy and computational efficiency of our radiative solver. Finally, conclusions will be drawn out.

2. Methodology

The discretization of the RTE proceeds through three steps: discretization in frequency, discretization along propagation directions of radiation beams, and discretization in space within computational domain. The discretization in frequency which is particularly related to the SNB-FSCK method is presented in [1,2]. The 7-point Gauss-Labatto quadrature scheme [2] is used in the study to pre-define the discretization points in the cumulative distribution g-space for the RTE. The DOM/S_4 [18] is used in the study for the direction discretization because of a good compromise between solution accuracy and computational requirements in practical calculations of combustion systems. The discretization of the RTE in space is carried out by using the Finite Volume method in the context of unstructured mesh based applications. The upwind and central schemes are blended and an iterative solver is used for inverting the non-symmetric linear system resulting from the discretization of the RTE. These methods, with special care about the specific boundary handling for this problem, are developed together in the following. The RTE in the cumulative distribution g-space to be solved can be rewritten in the form [1,2]:

$$s \cdot \nabla I_g = -k^* I_g + k^* a I_b \tag{1}$$

where I_g (W/m) is the radiation intensity in the g-space, I_b (W/m) the blackbody radiation intensity, k^* (m^{-1}) the full spectrum absorption coefficient, a the weight function. This equation will be integrated over a node-centered control volume V_i (see Figure 1.a) using the Gauss' theorem:

$$\sum_{j=1}^{N_{faces}} A_j (n_j \cdot s) I_{g,j} + k^* V_i I_{g,i} = (k^* a I_b) V_i \tag{2}$$

where n_j is the outward unit normal to the face j, A_j is the area of the face j, i denotes the node i, and V_i is the node-centered control volume at node i. Figure 1.a shows an illustration of the finite volume discretization; one can see, for example, that V_i is the volume bounded by the dash lines.

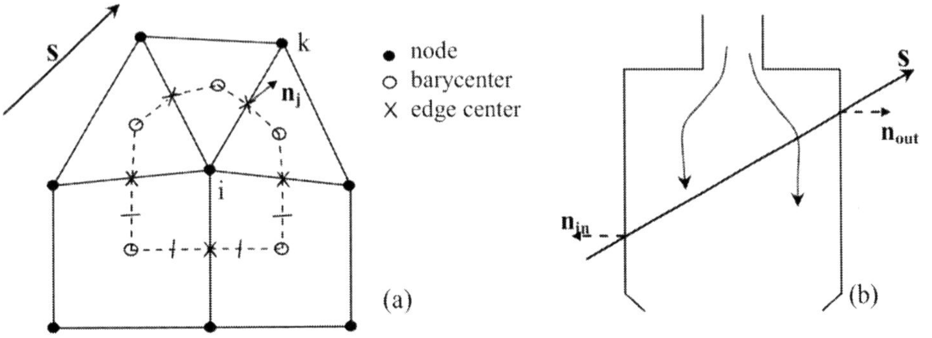

Figure 1. (a) Illustration of the MUCS scheme. (b) Illustration of the boundary treatment.

To close Equation (2), i.e, to calculate the nodal radiation intensity $I_{g,i}$, one needs to express the radiation intensities on the control volume surfaces $I_{g,j}$ in term of the nodal radiation intensity $I_{g,i}$. We propose an alternative to the conventional schemes (Exponential [9], Step [10], DMFS [11], NVD, TVD, ENO, Genuinely multidimensional [12,13]) to solve the RTE. The alternative is the BiCGStab iterative solution method combined with the stable Mixed Upwind and second-order Central Scheme (MUCS). In the framework of kinetic-energy conserving discretization in finite volume method [19], the second-order central scheme (CS) can be blended with the Upwind Scheme (US) to give a scheme which possesses the advantages of the two mother schemes such as being close to second-order accuracy, limiting the intensity diffusion and keeping the intensity bounds. The MUCS scheme is expressed in the form:

$$I_{g,j} = \begin{cases} \gamma I_{g,i} + (1-\gamma)(I_{g,i} + I_{g,k})/2 & if\ D_j > 0 \\ \gamma I_{g,k} + (1-\gamma)(I_{g,i} + I_{g,k})/2 & if\ D_j \le 0 \end{cases} \tag{3}$$

where D_j is the scalar product between the unit vector of the discrete direction s and the normal to the face j (n_j). The blending factor γ, $0 \le \gamma \le 1$, determines how much numerical diffusion of the upwind scheme is introduced. The value of $\gamma = 0.2$ led to a low-dispersion MUCS scheme and therefore is adopted in the study. The BiCGStab(2) method [15], as designed for the non-symmetric linear equation systems, is adopted in the study and consists in solving the following linear system:

$$Ax = b \tag{4}$$

The discretized RTE of Equation (2) can be transformed in the form of Equation (4). b is the Right Hand Side (RHS) of the linear system and corresponds to the emission term, i.e., $b = (k^* a I_b)_i V_i$. This emission term depends on local radiative properties and is known prior to the solving of the linear system. Ax is the Left Hand Side (LHS) of the linear system and contains two terms: *the propagation term* $\sum_{j=1}^{N_{faces}} A_j (n_j \cdot s) I_{g,j}$ and *the absorption term* $k^* V_i I_{g,i}$. x is a vector of unknowns $I_{g,i}$ (after expressing $I_{g,j}$ in term of $I_{g,i}$ using the MUCS scheme). The matrix A is always non-symmetric whatever the blending factor γ is.

The cell elements having faces lying on the boundary should be handled properly in the iterative solution method. The solution consists in considering two situations according to the sign of the scalar product between the unit vector of the discrete direction s and the normals to the face lying on the boundary (n_{in} and n_{out}), as shown in Figure 1.b. $D_{out} = s \cdot n_{out} > 0$: the boundary radiation in this direction is calculated from the radiation obtained in the cell volume adjacent to the boundary, so the boundary radiation is an unknown, therefore the *boundary propagation term* $A_j (n_j \cdot s) I_{g,j}$ is kept in Ax of Equation (4), i.e., on the LHS of the linear system. $D_{in} = s \cdot n_{in} \le 0$: the boundary radiation in this direction is calculated from the radiation known on this boundary. This corresponds to the application of the Dirichlet condition and therefore the *boundary propagation term* $A_j (n_j \cdot s) I_{g,j}$ is sent to the RHS of the linear system (Equation (4)). The whole methodology is implemented into the massively parallel unstructured CFD code called YALES2 [17], where the radiative and fluid flow solutions share the same domain decomposition, which is the bottleneck in current radiative solvers. The dual mesh splitting at the cell groups level and processors level is used to optimize the CFD code for massively parallel computing.

As the online calculation of the full spectrum absorption coefficient k^* can consume a large amount of CPU time in thermal radiation calculations, it is preferable to pre-compute k^* prior to solving the RTE [2,6]. In this way, the absorption coefficient k^*, the weight function a, and the blackbody radiation intensity I_b are calculated first in function of temperature, molar fractions of CO_2 and H_2O. This allows for the tabulation of the absorption coefficient k^* and the emission term $k^* a I_b$, for 7 Gauss-Labatto quadrature points, as functions of three indices: temperature (T), molar fraction of CO_2 and H_2O (X_{CO_2}, X_{H_2O}). A multi-dimensional linear interpolation method is implemented to interpolate k^* and $k^* a I_b$ at the actual local state (T, X_{CO_2}, X_{H_2O}) of the mixture during the simulation. It should be noted that, the method for tabulating k^* and $k^* a I_b$ used in the present study is different from the k^* pre-calculation method in the work of Liu *et al.* [2]. In fact, Liu *et al.* [2] used the 4-point Gauss quadrature scheme and the values of $k^*_{CO_2} / pX_{CO_2}$ and $k^*_{H_2O} / pX_{H_2O}$ were calculated and fit to a 9th-order polynomial function of T varying between 300 K and 2600 K for a range of reference temperatures T_0 between 500 K and 1600 K with intervals of 100 K. A 1D linear interpolation was employed to obtain the radiative properties at the actual reference T_0. The radiative property database from Soufiani and Taine [20] together with the approximate Malkmus band method [21] appear to be an appropriate strategy to calculate the mixture property, as follows:

$$k_{\Delta v,mix} = k_{\Delta v,CO_2} + k_{\Delta v,H_20}$$
$$\frac{k^2_{\Delta v,mix}}{\Phi_{\Delta v,mix}} = \frac{k^2_{\Delta v,CO_2}}{\Phi_{\Delta v,CO_2}} + \frac{k^2_{\Delta v,H_20}}{\Phi_{\Delta v,H_2O}} \quad (5)$$

where $k_{\Delta v,mix}$ (m^{-1}) is the average absorption coefficient of the mixture over a narrow band Δv, $\Phi_{\Delta v,mix}$ (dimensionless) is the average line width to spacing ratio of the mixture. $k_{\Delta v,mix}$ and $\Phi_{\Delta v,mix}$ are SNB model parameters to calculate the full-spectrum mixture absorption coefficient k^* that depends on local temperature and partial pressure, i.e., mole fractions of radiating gases and also on the reference state of the mixture. The reference state of the mixture can be chosen beforehand; Liu and Smallwood [5] proposed to choose the reference mixture consisting of $X_{CO_2,0} = 0.1$ and $X_{H_2O,0} = 0.2$ and the reference temperature can be calculated using the volume averaged temperature formulation proposed by Modest and Zhang [22] such as:

$$T_0 = T_{ref} = \frac{1}{V} \int_V T dV \quad (6)$$

3. Results and discussion

3.1. Configuration of the test case

The configuration for the test is a 3D rectangular enclosure as used in the works of Liu *et al.* [2] and of Liu and Smallwood [5]. The test case simulates a water-cooled gas-fired furnace. The enclosure has 2m in x, 2m in y and 8m in z and is divided into 17(x).17(y).40(z) = 11560 cell volumes with regular grid distribution in three directions. The number of cell volume is kept as the same used in the work of Liu *et al.* [2] for comparison purpose. The mixture contained in the enclosure consists of CO_2, H_2O and N_2 under 1atm pressure. The non-uniform distribution of temperature is specified as

$T = (T_{axis} - 800)f(r) + 800$ where T_{axis} and $f(r)$ define the temperature distribution along the centerline and radial direction, respectively, with

$$T_{axis} = \begin{cases} 400 + (2100 - 400z)/0.75 & if\ 0\,m \le z < 0.75\,m \\ 800 + (2100 - 800)(8 - z)/7.25 & if\ 0.75\,m \le z \le 8\,m \end{cases} \quad (7)$$

and $f(r) = 1 - 3r^2 + 2r^3$, where r is the distance to the centerline. In the four corner regions where $r > 1\,m$, T simply takes the value of temperature evaluated at $r = 1\,m$. Temperature is expressed in Kelvin. Equation (6) can be now applied and it yields the reference temperature $T_0 = 948.48\,K$. In the following sections, we present the result of the calculations for two cases: homogeneous and inhomogeneous distributions of CO_2 and H_2O. The result consists of wall heat flux and radiation source term. To validate our methodology, the present result will be compared to that obtained by Liu *et al.* [2] in the same configuration. The result of Liu *et al.* [2] marked by SNB-CK is used as the benchmark solution, the SNB-FSCK/M1 and SNB-FSCK/M2 methods used the on-line inversion and the pre-computation of the full spectrum absorption coefficient k^*, respectively, and the SNB-CK/9band employed the optimized 9-band model. Liu *et al.* [2] used the same 7-point Gauss-Labatto quadrature scheme for the benchmark solution while the 4-point Gauss quadrature sheme was used for other methods.

3.2. Homogeneous distribution of CO_2 and H_2O

In this case, the mole fractions of CO_2 and H_2O take values of $X_{CO_2} = 0.1$ and $X_{H_2O} = 0.2$, and are the same everywhere inside the enclosure. Figure 2 shows the contour of radiative heat source in the Oxz plane. Because the mole fraction distributions of CO_2 and H_2O are uniform, the contour of radiative source term basically follows the pattern of the temperature contour (not shown here). The mixture radiates the most in the region of high temperature around $z = 0.775\,m$ along the centerline. The negative and positive values of the radiation source term indicate the heat loss and gain, respectively, due to the radiative heat transfer. In the parietal regions, the gases absorb the heat emitted from the central region.

Figure 3 shows the evolution of the wall heat flux along the center of the side wall and the radiation source term along the centerline of the enclosure. In the same figure, the results of the present method BiCGStab(2)/MUCS/tabulated SNB-FSCK, denoted by SNB-FSCK/MUCS, are compared with those from Liu *et al.* [2]. Concerning the distribution of the wall heat flux shown in Figure 3.a, the present result (SNB-FSCK/MUCS) is generally better than the result using the SNB-FSCK/M2 method in Liu *et al.* [2]. The relative error measured between the present method (SNB-FSCK/MUCS) and the SNB-CK benchmark at the maximum absolute values of the wall heat flux reduces to 7% as compared to 15% resulting from the comparison between the SNB-FSCK/M2 method and the SNB-CK benchmark. This amelioration lies in the more efficient tabulation method, the use of the Malkmus band to handle the spectral properties of the mixture and of the 7-point Gauss-Labatto quadrature scheme in the present study. In Liu *et al.* [2], the absorption coefficients were calculated by making the assumption on the correlation of CO_2 and H_2O radiative properties. This assumption yields large error for the mixture absorption coefficient.

Another very interesting result is obtained when one observes the distribution of the radiation source term as shown in Figure 3.b. The result of the present method (SNB-FSCK/MUCS) is in very good agreement with that of the benchmark SNB-CK. At the maximum absolute values of radiation source term, the present method (SNB-FSCK/MUCS) produced less than 4% of error as compared to 14% resulted from the SNB-FSCK/M2 method. The present method (SNB-FSCK/MUCS) gives a similar result with the SNB-FSCK/M1 and SNB-CK/9band methods but with a computational time largely reduced, as shown later in the paper. It can be seen that the present method (SNB-FSCK/MUCS) exhibits an advantage over the SNB-FSCK/M1, SNB-FSCK/M2 and SNB-CK/9band methods.

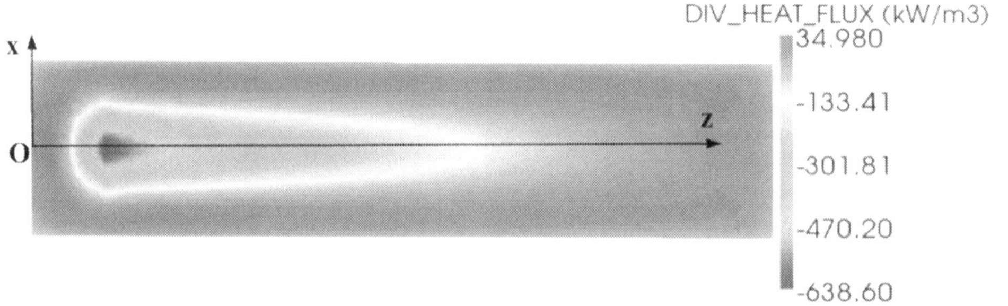

Figure 2. Homogeneous case. Contour of heat flux divergence, i.e., radiation source term (kW/m^3) in the Oxz plane (at y = 0).

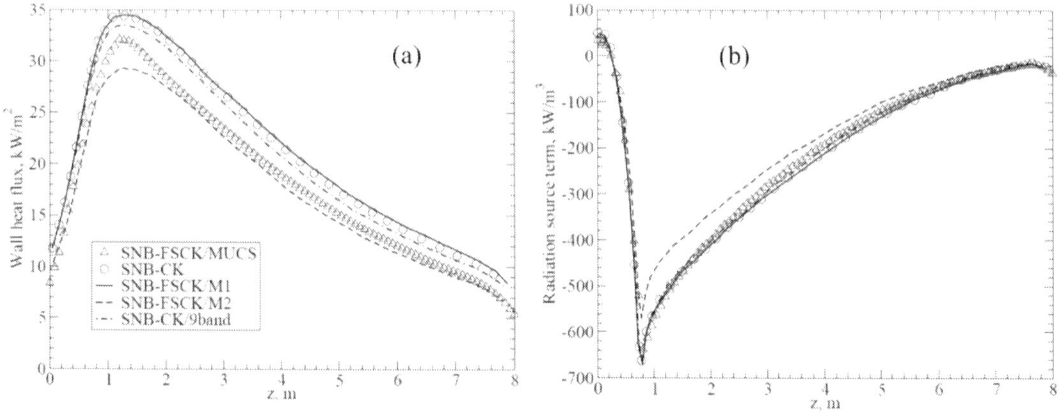

Figure 3. Homogeneous case. (a) Wall heat flux along the center of the side wall, (b) Radiation source term along the centerline of the enclosure. Triangle: present method BiCGStab(2)/MUCS/tabulated SNB-FSCK; circle: benchmark correlated K model; line: SNB-FSCK with on-line inversion of k^*; dashed-line: SNB-FSCK with pre-computation of k^*; dotted-line: optimized 9-band SNB-CK model.

3.3. Inhomogeneous distribution of CO_2 and H_2O

In the second test, the mole fraction of CO_2 and H_2O are linear in z-direction and kept uniform in x and y directions. The distributions of CO_2 and H_2O along the z-direction are given as follows:

$$X_{CO_2} = \begin{cases} 0.6z & if\ 0m \le z \le 0.5m \\ (-0.25z + 2.375)/7.5 & if\ 0.5m < z \le 8m \end{cases} \tag{8}$$

$$X_{H_2O} = \begin{cases} z & if\ 0m \le z \le 0.5m \\ (-0.45z + 3.975)/7.5 & if\ 0.5m < z \le 8m \end{cases} \tag{9}$$

Similarly to the homogeneous test case, Figure 4 shows the evolution of the wall heat flux along the center of the side wall and the radiation source term along the centerline of the enclosure. For the wall heat flux, the present method (SNB-FSCK/MUCS) shows the result which is practically equivalent to that obtained with the SNB-FSCK/M2 method (Figure 4.a). The relative error between the present method (SNBFSCK/MUCS) and the SNB-CK benchmark at the maximum absolute values of the wall heat flux is 12% as compared to 11% resulted from the comparison between the SNBFSCK/M2 method and the SNB-CK benchmark. We can see again, from the result of the radiation source term shown in Figure 4.b, that the present method (SNB-FSCK/MUCS) is in a very good agreement with the SNB-CK benchmark. At the point of maximum absolute values of the radiation source term, the relative error between the present method (SNB-FSCK/MUCS) and the benchmark method SNB-CK is reduced to less than 4% as compared to 18% resulted from the SNB-FSCK/M2 and SNB-CK comparison. The present method (SNB-FSCK/MUCS) predicts the radiation energy source much better than the SNBFSCK/M2 in a large portion of the enclosure and also better than the SNB-CK/9band in the region of high absolute values of the radiation energy source.

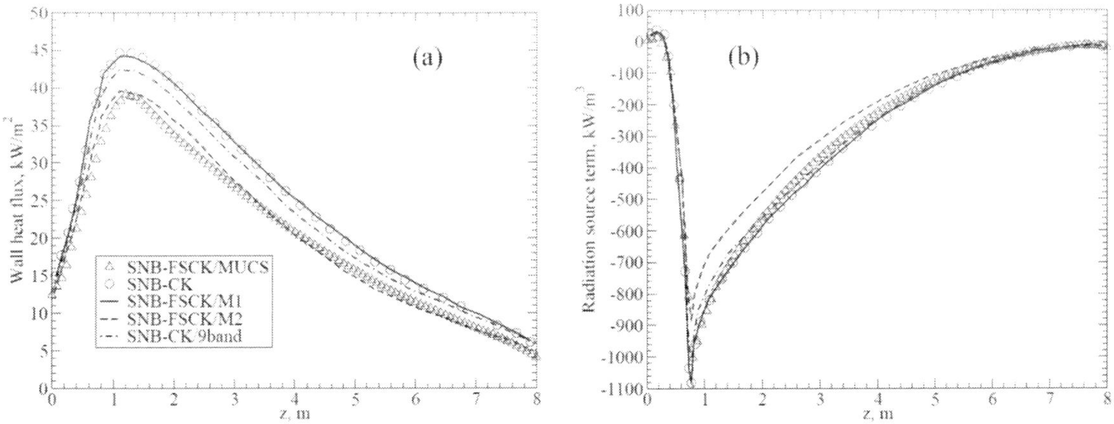

Figure 4. Inhomogeneous case. (a) Wall heat flux along the center of the side wall, (b) Radiation source term along the centerline of the enclosure. Triangle: present method BiCGStab(2)/MUCS/tabulated SNB-FSCK; circle: benchmark correlated K model; line: SNB-FSCK with on-line inversion of k^*; dashed-line: SNB-FSCK with pre-computation of k^*; dotted-line: optimized 9-band SNB-CK model.

3.4. Computational efficiency and accuracy

CPU times are compared in Table 1 to indicate the efficiency of the method developed in the present study. However, this should not be taken as an exact comparison because the present calculations and those of Liu *et al.* [2] were performed on different computers. The present calculations were carried out on IDRIS-CNRS computing resources (Vargas computer: IBM SP Power6, 3584 processors, 18 Tbytes, 68 Tflops) [23] while those of Liu *et al.* [2] on a Pentium 4 2.8 GHz PC. The computational times of runs using one processor are reported in Table 1 for comparison.

It is clearly noted that the CPU time of the present method (SNB-FSCK/MUCS) is significantly lower than that of the benchmark method SNB-CK by a factor of about 368 which is the expected CPU time reduction because we reduced the number of narrow bands from 367 to 1 global band, indicating the excellent efficiency of the new method. The CPU time of the present method (SNB-FSCK/MUCS) is systematically lower than that of the SNB-FSCK/M1 method (a factor of about 6), that of the SNB-FSCK/M2 method (a factor of about 1.2) and that of the SNB-CK/9band method (a factor of about 3.4). Three important points should be noted to understand the superiority of the SNB-FSCK/MUCS method. The first point is that the present method (SNB-FSCK/MUCS) used the 7-point Gauss-Labatto quadrature scheme for a better description of the gas radiative properties as compared to the 4-point Gauss quadrature scheme used in the SNB-FSCK/M1 and SNB-FSCK/M2. The second point is that the present method (SNB-FSCK/MUCS) is implemented into an unstructured software designed for complex geometries and it is known that an unstructured code needs about four times longer than a structured code and this constitutes a very important advantage of the present method. The third point is that the present method (SNB-FSCK/MUCS) can predict the radiation source term with more accuracy than other methods with less CPU times spent.

It should also be noted from Table 1 that, the first-order accuracy Upwind scheme (US) even reduced the calculation time down to nearly one and a half times if we compare with the MUCS scheme (i.e., 24.7s compared with 16.9s). However, this efficiency was obtained at the expense of accuracy. Figure 5 shows indeed the comparison between the MUCS and US schemes for the wall heat flux and radiation source term in the homogeneous case. It is clear from this figure that the MUCS scheme gives much accurate results than the US scheme when comparing to the SNB-CK benchmark. The radiative solver runs massively in parallel and as one example of this, 1024 processors of Vargas machine [23] were used to calculate the same configuration with 24 million of cells for the homogeneous distributions of CO_2 and H_2O. The corresponding computational time is 560.5s and the result of the parallel run is also presented in Figure 5.

Table 1. Comparison of CPU times.

Method	No. of quadrature points	No. of bands	CPU time (s)
SNB-FSCK/MUCS	7	1	24.7
SNB-FSCK/US	7	1	16.9
SNB-CK(*)	7	367	9086.3
SNB-FSCK/M1(*)	4	1	146.9
SNB-FSCK/M2(*)	4	1	30.7
SNB-CK/9band(*)	4	9	84.91

(*) Results of Liu *et al.* [2]

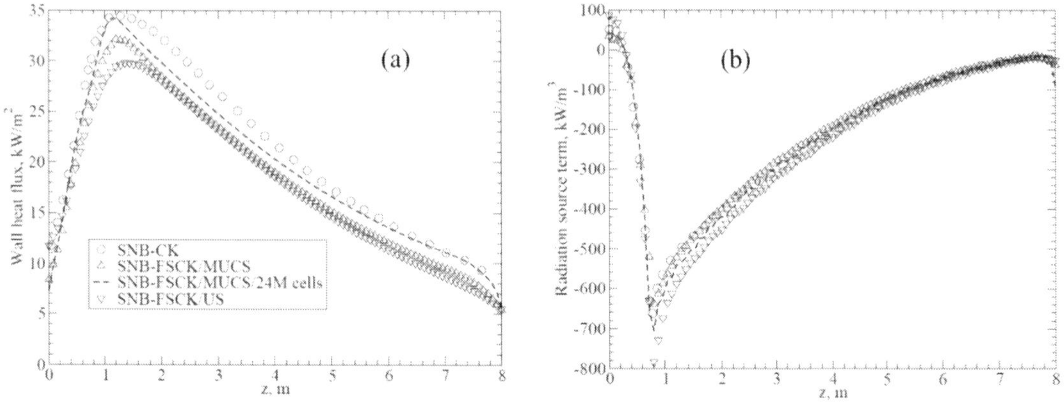

Figure 5. Homogeneous case. Comparison of accuracy between the US and MUCS schemes and results of massively parallel run (marked with dashed line). (a) Wall heat flux along the center of the side wall; (b) Radiation source term along the centerline of the enclosure. Circle: benchmark correlated K; triangle-up: BiCGStab(2)/MUCS/tabulated SNB-FSCK with 11560 cells (monoprocessor); dashed-line: BiCGStab(2)/MUCS/tabulated SNB-FSCK with 24 million cells (1024 processors); triangle-down: BiCGStab(2)/US/tabulated SNB-FSCK with 11560 cells (monoprocessor).

4. Conclusion

A new method to solve the RTE in radiation problems is introduced in the present study. The second-order central and first-order upwind schemes are blended to give a stable numerical scheme (MUCS). The BiCGStab(2) iterative solution method is used to solve the non-symmetric linear system of the discretized RTE. The BiCGStab(2)/MUCS method proved to be very efficient when coupling with the

DOM/FV discretization. An efficient tabulation technique for the gaseous radiative property model SNB-FSCK using 7-point Gauss-Labatto quadrature scheme is also introduced. The proposed methodology was implemented into a highly massive parallel unstructured code. The radiation solution benefits fully from the optimized parallelism based on the dual mesh splitting.

The method was applied to calculate the radiation in a 3D rectangular enclosure containing non-isothermal homogeneous and inhomogeneous mixtures of CO_2 and H_2O. The result was reported for the heat flux and radiation energy source and comparison was also made between the present method BiCGStab(2)/MUCS/tabulated SNB-FSCK and the benchmark method SNB-CK and some other methods presented in Liu *et al.* [2]. The present method BiCGStab(2)/MUCS/tabulated SNB-FSCK gives more accurate predictions particularly for the radiation source term. When comparing with the benchmark solution, the relative error of the radiation source term is remarkably reduced to less than 4% as compared to 14%-18% resulted from the SNB-FSCK/M2 method in [2]. The CPU time is notably reduced and proportional to the reduction number of the narrow bands and lower than the current methods available in [2], indicating the accuracy and efficiency of the present method. The developed radiative solver was also tested with success on massively parallel platforms. Based on these advantages of the present methodology that was implemented into the proven massively parallel software YALES2, it opens an interesting way to simulate large-scale combustion systems. A future work will be dedicated to a coupling between combustion and radiation using this software.

5. References

[1] Modest M F. *JQSRT*, 76:69-83 (2003).
[2] Liu F, Yang M, Smallwood G and Zhang H. Proceedings of ICHMT 2004 (4th Int. Symp. on Radiative Transfer, Istambul, Turkey, 2004), ISSN: 961-91393-0-5.
[3] Lacis A A and Oinas V. *J. of Geophysical Research*, 96:9027-9063 (1991).
[4] Hottel H C and Sarofim A F, Radiative transfer. New York: McGraw-Hill; 1967.
[5] Liu F and Smallwood G J. *JQSRT*, 84:465-475 (2004).
[6] Poitou D. Institut National Polytechnique de Toulouse, PhD dissertation, 2009.
[7] Fiveland W A. *J. Heat Transf.*, 106:699-706 (1984).
[8] Coelho P J. *Prog. Energy Combust. Sci.*, 33:311-383 (2007).
[9] Sakami M, Charette A and LeDez V. *JQSRT*, 59(1/2):117-136 (1998).
[10] Chai J C, Parthasarathy G, Lee H S, Patankar S V. *J. Thermophysics and Heat Transfer*, 9 (1995) 410-415.
[11] Ströhle J, Schnell U and Hein K N G. *In 3rd Int. Conference on Heat Transfer*, Antalya, 2001.
[12] Coelho P J. *JQSRT*, 109:189-200 (2008).
[13] Godoy W F and DesJardin P E. *J. of Computational Physics*, 229:3189-3213 (2010).
[14] Santos R G D, Lecanu M, Ducruix S, Gicquel O, Iacona E and Veynante D. *Comb. & Flame*, 152:387-400 (2008).
[15] Van der Vorst H A. AGARD Special Course on Parallel Computing in CFD, 1995.
[16] Capdevila R, Pérez-Segarra C D, Oliva A. *JQSRT*, 111:264-273 (2010).
[17] Moureau V, Domingo P and Vervisch L. doi:10.1016/j.crme.2010.12.001 (2011).
[18] Modest M F, Radiative heat transfer, 2nd ed., New York: McGraw-Hill; 2003. ISBN:0-12-503163-7.
[19] You D, Ham F and Moin P. *Physics of Fluids* 20 (2008).
[20] Soufiani A, Taine J. *Int. J. of Heat and Mass Transfer* 40:987-991 (1997).
[21] Liu F, Smallwood G J and Gülder O L. *JQSRT*, 68:401-417 (2001).
[22] Modest M F and Zhang H. *ASME, J. Heat Transfer*, 124:30-38 (2002).
[23] www.idris.fr.

Acknowledgments Financial support by RHODIA group is gratefully acknowledged.

Journal of Physics: Conference Series **369** (2012) 012018 doi:10.1088/1742-6596/369/1/012018

On the Analytical Solution of the S_N Radiative Transport Equation in a Slab for a Space-dependent Albedo Coefficient

Rubem M. F. Vargas
Pontifícia Universidade Católica do Rio Grande do Sul - FENG - PGETEMA
Porto Alegre, RS, Brazil

E-mail: rvargas@pucrs.br

Cynthia F. Segatto and Marco Túllio Vilhena
Universidade Federal do Rio Grande do Sul - PROMEC - PPGMAp
Porto Alegre, RS, Brazil

E-mail: csegatto@pq.cnpq.br; vilhena@pq.cnpq.br

Abstract. In this work, we report a genuine general analytical solution for the linearized S_N radiative-conductive transfer problem in a heterogeneous plane parallel atmosphere with the albedo coefficient depending continuously on the spatial variable. By general solution, we mean that the solution is valid for an arbitrary albedo coefficient continuous functions of the spatial variable having the property of fulfill the requirements of existence and uniqueness. The key feature of this novel approach embodies the steps: following the idea of the Decomposition method, we transform the original problem into a set of recursive problems with constant albedo coefficients, having the main feature that the sources terms takes the information of the spatial dependency of the albedo coefficient into account. This procedure allows us to solve, analytically, the resulting recursive system by the LTS_N method developed for a constant albedo coefficient. Finally, we present the error control analysis of the solution and numerical comparisons against the literature results.

1. Introduction

Radiative transfer considers problems that model the physical phenomenon of energy transfer by radiation in media. These phenomenon appears in variety of realms including engineering applications like heat transport by radiation [1]. The nonlinearity of this equation comes from the local thermal description using the Stefan-Boltzmann law that is related to heat transport by radiation which in turn is related to the radiation intensity and renders the radiative transfer problem a radiative-conductive one [2, 3]. Solutions in the literature are basically linearized and of numerical nature [4, 5, 6, 7, 8, 9]. However, it is worth mentioning that a general solution from an analytical approach for this type of problems exists only in the discrete ordinate approximation and for homogeneous media as reported by Segatto et al. in reference [10]. Its basic idea comprehends the steps: following the idea of the Decomposition method [11, 12], the authors constructed a recursive system of linearized S_N radiative-conductive equations with the main feature that the nonlinearity information is carried out by the source term, unlike the well known Decomposition method [13] which expands the nonlinearity term in polynomials series ($\hat{A}_n$ and A_n polynomials). Further, the authors also splitted the operator associated to the equation considered, as a sum of the linear and non linear operators, but not likewise the Decomposition approach, which decomposes the original problem as the summation of the higher derivative, remaining linear and non linear operators. Morever the authors constructed a recursive system of differential, but not integral equations like the Decomposition technique, whith was then solved by the authors by the LTS$_N$ method [10]. For more detais see the works of Segatto et al. [14], and Gonçalves et al. [15]. Motivated by the task of extending this of sort solution for heterogeneous media, in this work, we report a genuine general analytical solution for the linearized S$_N$ radiative-conductive transfer problem in a heterogeneous plane parallel atmosphere with the albedo coefficient depending continuously on the spatial variable. By general solution, we mean that the solution is valid for an arbitrary albedo coefficient continuous functions of the spatial variable having the property of fulfill the requirements of existence and uniqueness [16, 17]. By analytical we mean that no approximation is done along the solution derivation. We are aware of the works of Ćengel and Özişik [9], and Garcia and Siewert [18], but they are restricted for very specific albedo functions, namely polynomials and exponential functions. The key feature of this novel approach embodies the steps: procceeding in similar manner of the nonlinear problem discussed, we transform the original problem into a set of recursive problems with constant albedo coefficients, having the main feature that the sources terms takes the information of the spatial dependency of the albedo coefficient into account. This procedure allows us to solve, analytically, the resulting system by the LTS$_N$ method developed for a constant albedo coefficient. Therefore, the series solution attained, allow us to obtain numerical results with a prescribed accuracy, by controlling the number of the terms in the series solution. To our best knowledge this sort of solution is not found in literature. We complete our analysis presenting simulations and comparisons against the literature results.

2. The Analytical Solution

To construct the analytical solution of the S$_N$ radiative transfer in an inhomogeneous, emitting, absorbing and grey plane-parallel slab of optical thickness, with variable albedo coefficient with position, without source [19], let us consider the problem:

$$\mu\frac{\partial I}{\partial \tau}(\tau,\mu) + I(\tau,\mu) = \frac{\omega(\tau)}{2}\sum_{\ell=0}^{L}\beta_\ell P_\ell(\mu)\int_{-1}^{1}P_\ell(\mu')I(\tau,\mu')d\mu', \qquad (1)$$

for $0 < \tau < \tau_0$ and to the ensuing reflecting and emitting boundary condition:

$$I(0,\tau) = f_1(\mu) + 2\rho_1\int_0^1 I(0,-\mu)\mu d\mu, \qquad \mu > 0 \qquad (2)$$

and

$$I(\tau_0,\tau) = f_2(\mu) + 2\rho_2\int_0^1 I(\tau_0,\mu)\mu d\mu, \qquad \mu > 0. \qquad (3)$$

Here $I(\tau,\mu)$ is radiation intensity, is the space-dependent albedo coefficient, n is refractive index of the medium, σ is the Stefan-Boltzmann constant, T_i, ϵ_i and ρ_i are respectively the temperature, the diffuse emissivity and reflectivity at the boundary. Moreover the expansions coefficients β_ℓ and the Legendre polynomials $P_\ell(\mu)$ and $P_\ell(\mu')$ are associated with the definition of the anisotropic scattering phase-function expressed as:

$$p(\mu,\mu') = \sum_{\ell=1}^{L}\beta_\ell(\mu)P_\ell(\mu)P_\ell(\mu') \qquad (4)$$

here $a_0 = 1$. In order to determine the searched analytical solution, we now consider that the albedo coefficient has the form:

$$\omega(\tau) = \bar{\omega} + \omega_0(\tau) \tag{5}$$

From this assumption, we recast equation (1) like:

$$L(I) = \mu\frac{\partial I}{\partial \tau}(\tau,\mu) + I(\tau,\mu) - \frac{\bar{\omega}}{2}\sum_{\ell=0}^{L}\beta_\ell P_\ell(\mu')\int_{-1}^{1} P_\ell(\mu')I(\tau,\mu')d\mu' = S(I) \tag{6}$$

where the source $S(I)$ is written like:

$$S(I) = \frac{\omega_0(\tau)}{2}\sum_{\ell=0}^{L}\beta_\ell P_\ell(\mu)\int_{-1}^{1} P_\ell(\mu)I(\tau,\mu)d\mu. \tag{7}$$

Following this idea of Decomposition method [11], we assume that the radiation intensity reads like:

$$I(\tau,\mu) = \sum_{m=1}^{M} I_m(\tau,\mu) \tag{8}$$

Replacing the expression (8) in equation (6) we come out with the following recursive system:

$$\begin{aligned} L(I_0) &= 0 \\ L(I_1) &= S(I_0) \\ &\vdots \\ L(I_m) &= S(I_{m-1}) \end{aligned} \tag{9}$$

for $0 \leq m \leq M$. At this point, we must remark that the first equation of the recursive system (9) satisfy the boundary condition given by equations (2) and (3), meanwhile the remaining equations fulfill the homogeneous boundary condition mentioned. From the previous discussion, we must realize that we reduce the solution of the original problem to a set of problems with constant albedo coefficient which are then readly solved, analytically, by the well known $\mathrm{LTS_N}$ method [14]. Here we need underline that the number of recursive problems to be solved (M) is governed by the prescribed accuracy to be reached. For sake of completeness, in sequel, we report the $\mathrm{LTS_N}$ solution for the set of recursive problems discussed:

$$\begin{aligned} \mathbf{I}(\tau) = \mathbf{X}\Big(\mathbf{E}^+(\mathbf{D}(\tau-\tau_0)) + \mathbf{E}^-(\mathbf{D}\tau)\Big)\mathbf{V} + \mathbf{X}\Big(&\int_0^\tau (\mathbf{E}^-(\mathbf{D}(\tau-\xi))\mathbf{X}^{-1}\mathbf{S}(\xi)d\xi + \\ &\int_{\tau_0}^0 \mathbf{E}^+(\mathbf{D}(\tau-\xi))\mathbf{X}^{-1}\mathbf{S}(\xi)d\xi\Big)\mathbf{M} \end{aligned} \tag{10}$$

where $\mathbf{D}$ and $\mathbf{X}$ are respectively the diagonal eingenvalues matrix and the eigenvector matrix of the $\mathrm{LTS_N}$ matrix $\mathbf{A}$, whose entries are written as:

$$a(i,j) = \begin{cases} -\dfrac{1}{\mu_i} + \dfrac{\omega_j}{2\mu_i}\left[\displaystyle\sum_{\ell=0}^{L}\beta_\ell P_\ell(\mu_i)P_\ell(\mu_j)\right], & se\ i = j \\ \dfrac{\omega_j}{2\mu_i}\left[\displaystyle\sum_{\ell=0}^{L}\beta_\ell P_\ell(\mu_i)P_\ell(\mu_j)\right], & se\ i \neq j \end{cases} . \tag{11}$$

Further, from the Decomposition of the diagonal matrix $\mathbf{E}(\mathbf{D}(\tau))$:

$$\mathbf{E}(\mathbf{D}(\tau)) = \mathbf{E}^{+}(\mathbf{D}(\tau)) + \mathbf{E}^{-}(\mathbf{D}(\tau)) \tag{12}$$

we remark that the entries of $\mathbf{E}^{+}(\mathbf{D}(\tau))$ and $\mathbf{E}^{-}(\mathbf{D}(\tau))$ are respectively written like:

$$e^{+}_{ii}(\tau) = \begin{cases} e^{d_i \tau}, & \text{if } d_i > 0 \\ 0, & \text{if } d_i < 0 \end{cases}, \quad e^{-}_{i}(\tau) = \begin{cases} e^{d_i \tau}, & \text{if } d_i < 0 \\ 0, & \text{if } d_i > 0 \end{cases}. \tag{13}$$

In addition the column vetor $\mathbf{M}$ has the form:

$$\mathbf{M} = \left[\frac{1}{\mu_1} \ \frac{1}{\mu_2} \ \ldots\ldots \frac{1}{\mu_N} \right]^T \tag{14}$$

Further the integration constant vector $\mathbf{V}$ is a vector which is determined by solving the linear system resulting form the application of the boundary condition. More information and details about this solution is found in the works of Segatto et al. [10] and Vilhena et al. [1].

3. Numerical Results

In order to make comparisons with available results in literature, in the sequel, without loosing generality, we specialize the application of the proposed methodology for the following problems:

In the first problem we focus our attention to show the convergence of the proposed solution for heterogeneous media ranging the number of equations in the recursive system, from 1 to 10. Bearing in mind the proved convergence of the LTS_N method when N goes to the infinitive[20], we report numerical simulations for $N = 400$. We will show in the next problem that this value for N (400) is suitable for our purpose. To this end, let us consider the following radiative transfer problem in a heterogeneous slab of thickness $L = 1$ mfp, assuming the albedo polynomial function, $\omega(x) = 1 - 1.4x + 0.6x^2$ and the boundary condition: $\rho_1 = \rho_2 = 0$, $f_1(\mu) = 1$, $f_2(\mu) = 0$. Looking for the results displayed in table 1, we promptly realize the numerical convergence of eight rounded places for the results attained for the heterogeneous problem considered. We reinforce this affirmative noticing that all the recursive problems are solved by the LTS_{400} approach.

Next we display in table 2, numerical comparisons of our results for the reflectivity (R) and transmissivity (T) for the first and second order polynomial albedo coefficient functions for the same problem against the one achieved by Cengel and Ozişik [9] . To highlight the generality of this method we also report in table 3 numerical comparisons against the ones attained by Garcia and Siewert [18].

Table 1. Numerical convergence of the recursive system solution for the radiation intensity Iteration in the discrete directions at $\tau = 0.5$ considering $\omega(x) = 1 - 1.4x + 0.6x^2$ from j varing fron 1 to 11

μ / j	1	2	3	4	5	6	7	8	9	10	11
-1.0	.0366944770	.0236823372	.0245079378	.0243117509	.0243222678	.0243192122	.0243193413	.0243192932	.0243192946	.0243192939	.0243192939
-0.9	.0398819082	.0258490716	.0267378842	.0265266208	.0265379305	.0265346402	.0265347790	.0265347272	.0265347288	.0265347279	.0265347279
-0.8	.0436610581	.0284477145	.0294092155	.0291806061	.0291928240	.0291892638	.0291894136	.0291893575	.0291893592	.0291893583	.0291893583
-0.7	.0482062356	.0316195548	.0326649665	.0324163223	.0324295832	.0324257112	.0324258736	.0324258127	.0324258145	.0324258135	.0324258135
-0.6	.0537618820	.0355734442	.0367156621	.0364438949	.0364583500	.0364541182	.0364542951	.0364542285	.0364542305	.0364542294	.0364542294
-0.5	.0606747269	.0406298726	.0418824412	.0415843215	.0416001217	.0415954802	.0415956733	.0415956002	.0415956023	.0415956012	.0415956012
-0.4	.0694319109	.0473017675	.0486747970	.0483479882	.0483652250	.0483601379	.0483603482	.0483602680	.0483602703	.0483602691	.0483602691
-0.3	.0806620851	.0564463479	.0579324707	.0575791624	.0575976758	.0575921786	.0575924037	.0575923171	.0575923196	.0575923182	.0575923182
-0.2	.0948542979	.0695411329	.0710686056	.0707083173	.0707270904	.0707214909	.0707217185	.0707216302	.0707216327	.0707216313	.0707216313
-0.1	.1108739241	.0891284580	.0904410684	.0901526219	.0901684948	.0901640374	.0901642331	.0901641631	.0901641653	.0901641642	.0901641642
-0.001	.1266859016	.1204017233	.1215924838	.1215444920	.1215624561	.1215619337	.1215622190	.1215622144	.1215622190	.1215622190	.1215622190
0.001	.1270398499	.1212319506	.1224408249	.1223998390	.1224182279	.1224178253	.1224181192	.1224181167	.1224181215	.1224181215	.1224181216
0.1	.1528223500	.1803601613	.1847253445	.1852558686	.1853351701	.1853449674	.1853463992	.1853465790	.1853466049	.1853466082	.1853466086
0.2	.2309256998	.2831752004	.2904668502	.2914434182	.2915770505	.2915948901	.2915973208	.2915976459	.2915976900	.2915976960	.2915976968
0.3	.3266824747	.3853596428	.3933408850	.3944271298	.3945734022	.3945931999	.3945958639	.3945962241	.3945962726	.3945962792	.3945962801
0.4	.4108631731	.4687840963	.4765796555	.4776465700	.4777893745	.4778088012	.4778114032	.4778117565	.4778118039	.4778118103	.4778118112
0.5	.4798176207	.5346918061	.5420349929	.5430426909	.5431771551	.5431954939	.5431979444	.5431982778	.5431983225	.5431983285	.5431983293
0.6	.5358043287	.5870787140	.5939154502	.5948550881	.5949802393	.5949973338	.5949996149	.5949999257	.5949999672	.5949999729	.5949999736
0.7	.5816238242	.6293573872	.6357062386	.6365796772	.6366958689	.6367117555	.6367138734	.6367141622	.6367142008	.6367142060	.6367142067
0.8	.6195873124	.6640431241	.6699453286	.6707578682	.6708658648	.6708806412	.6708826099	.6708828784	.6708829143	.6708829192	.6708829198
0.9	.6514483094	.6929403454	.6984414766	.6991991734	.6992998154	.6993135927	.6993154274	.6993156778	.6993157112	.6993157158	.6993157164
1.0	.6785135233	.7173490398	.7224923798	.7232010597	.7232951438	.7233080286	.7233097437	.7233099779	.7233100091	.7233100134	.7233100139

Journal of Physics: Conference Series **369** (2012) 012018 doi:10.1088/1742-6596/369/1/012018

H

Table 2. Numerical comparisons for reflectivity and transmisivity for the albedo polynomial albedo thickness $L = 1$ and transparent boundaries ($\rho_1 = \rho_2 = 0$, $f_1(\mu) = 1$, $f_2(\mu) = 0$)

	Cengel Results [9]		LTS_{200} Results		LTS_{300} Results		LTS_{400} Results	
$\omega(x)$	R	T	R	T	R	T	R	T
τ	0.064838	0.313115	0.064840	0.313119	0.064843	0.0313121	0.064845	0.313123
$0.1 + 0.8\tau$	0.076897	0.310771	0.076900	0.310775	0.076901	0.310776	0.076902	0.310777
$0.2 + 0.6\tau$	0.089796	0.308978	0.89801	0.308981	0.089800	0.308981	0.089801	0.308982
$0.3 + 0.4\tau$	0.103594	0.307712	0.103601	0.307715	0.103599	0.307714	0.103598	0.307715
$0.4 + 0.2\tau$	0118358	0.306959	0.1183367	0.306961	0.118362	0.306960	0.118361	0.306960
0.5	0.134165	0.306709	0.134175	0.306711	0.134170	0.306710	0.134168	0.306709
$0.6 - 0.2\tau$	0.151104	0.306959	0.151116	0.306960	0,151109	0.306959	0.151107	0.306959
$0.7 - 0.4\tau$	0.169279	0.307712	0.169292	0.307713	0.169284	0.307712	0.169281	0.307712
$0.8 - 0.6\tau$	0.188806	0.308978	0.188821	0.308979	0.188812	0.308977	0.188808	0.308977
$0.9 - 0.8\tau$	0.209824	0.310771	0.209841	0.310772	0.209831	0.310771	0.209827	0.310770
$1 - \tau$	0.232491	0.313115	0.232508	0.313115	0.232498	0.313115	0.232493	0.313114
$0.3 + 0.2\tau + 0.6\tau^2$	0.081072	0.311079	0.081078	0.311083	0.811078	0.311085	0.081078	0.311087
$0.4 - 0.2\tau + 0.6\tau^2$	0.108266	0.307972	0.108275	0.307975	0.108271	0.307975	0.108270	0.307976
$0.8 - \tau + 0.6\tau^2$	0.175240	0.307972	0.175255	0.307973	0.175246	0.0307972	0.175242	0.307971
$1 - 1.4\tau + 0.6\tau^2$	0.216635	0.311079	0.216652	0.311079	0.216642	0.311078	0.216637	0.311078

Table 3. The exit distribution $I(\tau,\mu)$ for $\omega(\tau) = exp(-\frac{\tau}{s})$

$I(0,\mu)$								
μ	$s = 1$		$s = 10$		$s = 10^2$		$s = 10^3$	
	Results from [18]	LTS_{400}	Results from [18]	LTS_{400}	Results from [18]	LTS_{400}	Results from [18]	LTS_{400}
-1.0	.19055	.19055	.44517	.44517	.66146	.66146	.72904	.72904
-0.9	.20517	.20517	.46615	.46615	.67943	.67943	.74510	.74510
-0.8	.22223	.22223	.48910	.48910	.69805	.69805	.76139	.76139
-0.7	.24239	.24239	.51427	.51427	.71733	.71734	.77788	.77788
-0.6	.26656	.26656	.54197	.54197	.73730	.73730	.79455	.79455
-0.5	.29609	.29609	.57257	.57257	.75799	.75800	.81143	.81143
-0.4	.33296	.33296	.60647	.60647	.77946	.77946	.82857	.82857
-0.3	.38031	.38031	.64418	.64418	.80181	.80181	.84606	.84606
-0.2	.44328	.44328	.68632	.68632	.82523	.82523	.86410	.86410
-0.1	.53112	.53112	.73398	.73398	.85028	.85028	.88319	.88319
-0.05	.58966	.58966	.76081	.76082	.86400	.86400	.89361	.89361
$I(5,\mu)$								
μ	$s = 1$		$s = 10$		$s = 10^2$		$s = 10^3$	
0.05	.6075 (-5)	.60745(-5)	.58031(-2)	.58026(-2)	.62883(-1)	.62881(-1)	.96845(-1)	.96843(-1)
0.1	.69252(-5)	.69250(-5)	.63702(-2)	.63700(-2)	.69024(-1)	.69023(-1)	.10629	.10629
0.2	.96423(-5)	.96422(-5)	.76183(-2)	.76181(-2)	.80567(-1)	.80567(-1)	.12363	.12363
0.3	.16234(-4)	.16234(-4)	.91482(-2)	.91481(-2)	.91918(-1)	.91917(-1)	.14012	.14012
0.4	.43858(-4)	.43858(-4)	.11119(-1)	.11119(-1)	.10342	.10342	.15622	.15622
0.5	.16937(-3)	.16937(-3)	.13725(-1)	.13725(-1)	.11523	.11523	.17211	.17211
0.6	.57347(-3)	.57347(-3)	.17183(-1)	.17183(-1)	.12744	.12744	.18790	.18790
0.7	.15128(-2)	.15128(-2)	.21680(-1)	.21680(-1)	.14010	.14010	.20364	.20364
0.8	.32437(-2)	.32437(-2)	.27331(-1)	.27331(-1)	.15319	.15319	.21933	.21933
0.9	.59604(-2)	.59604(-2)	.34166(-1)	.34166(-1)	.16667	.16667	.23496	.23496
1.0	.97712(-2)	.97712(-2)	.42142(-1)	.42142(-1)	.18047	.18047	.25049	.25049

Given a closer looking for the table 1,2 and 3, we promptly realize the very good comparison with at least five rounds positions, with shows the good computational performance of the discussed method for all the problems discussed.

4. Conclusion

The main idea of the classical method to handle radiative transfer problems in a heterogeneous plane parallel atmosphere consists in the approximation of the heterogeneous media into a multilayer slab, having the albedo coefficient in each layer an constant averaged or constant value. Once the local solution for the homogeneous slab is known, the global solution for the multilayered slab is then determined imposing the boundary condition and the interface condition of continuity for the radiation intensity at the interfaces. The proposed approach posses a new philosophy. In fact, we begin constructing the global solution, by solving the radiative transfer equation in the entire domain, assuming an averaged value for the albedo coefficient. The influence of the heterogeneity is then incorporated in the solution by solving the discussed recursive system of problems in the entire domain, with the property that the sources terms carry out the information of the heterogeneity. In fact, we construct the global heterogeneous solution progressively from the homogeneous one. Besides the genuine character of the proposed analytical solution for the S_N approach of the radiative transfer equation, as well the very good comparisons of the encountered results against the ones of the literature, we are confident to affirm that this approach is an interesting and promising methodology to work out such problems in heterogeneous media with error control, once the solution for the same problem in homogeneous media is known. We reinforce this claim recalling the proved convergence of the LTS_N method whe N goes to infinitive [20]. Furthermore, we need also to emphasize the generality of this solution, in the sense it is valid for all albedo coefficient functions of the spatial variable which fulfill the mathematical requirements of existence and uniqueness. By analytical we mean that no approximation is done along the solution derivation. Therefore, in this sense we may say that the solution given by equation (10) is an analytical solution of the S_N problem (1).

So far, bearing in mind that we pave the road to extend this solution for the nonlinear radiative transfer problem in a heterogeneous plane parallel atmosphere taking the advantage of the known solution for the same problem in homogeneous media [10], we shall focus our future attention on this direction.

References

[1] Vilhena MT, Bodmann BEJ, Segatto CF. Non-Linear Radiative-Conductive Heat Transfer in a Heterogeneous Gray Plane-Parallel Participating Medium. *Heat Transfer, In: Amimul Ahsan. (Org.)*. v. 1, pp. 177-196, 2011.

[2] Özişik MN, *Radiative transfer and interaction with conduction and covection*, John Wiley & Sons Inc., ISBN:0-471-65722-0, New York, 1973.

[3] Pomraning GC, *The equations of hidrodynamics*, Dover Publ. Inc., ISBN: 0-486-44599-2, Mineola, New York, 2005.

[4] Asllanj F, Jeandel G, Roche JR: Numerical solution of radiative transfer equation coupled with non-linear heat conduction equation. *Int J of Numerical Methods for Heat and Fluid Flow*, Vol. 11, No. 5, pp 449-473, 2001.

[5] Attia MT: On exact solution of the generalized equation of radiative transfer in a two-region inhomogeneous slab. *J of Quantitative Spectroscopy and Radiative Tranfer*, Vol. 66, pp 529-538, 2000.

[6] Muresan C, Vaillon R, Menezo Ch, Morlot R: Discrete ordinates solution of Coupled Conductive Radiative Heat Transfer ia a Two Layer Slab with Fresnel Interfaces Subject to Diffuse and Obliquely Collimated Irradiation. *J of Quantitative Spectroscopy and Radiative Tranfer*, Vol. 84, No 4, pp 551-562, 2004.

[7] Siewert CE & Thomas JR: A Computational Method for Solving a Class of Coupled Conductive Radiative Heat Transfer Problem. *J of Quantitative Spectroscopy and Radiative Tranfer*, Vol. 45, No 5, pp 273-281, 1991.

[8] Krishnapraka CK, Narayana KB, Dutta B: Combined Conduction and Radiation Heat Transfer in a Gray Anisotropically Scattering Medium with Difuse-Specular Boundaries. *International Communications, in Heat and Mass Transfer*, Vol. 28, No 1, pp 77-86, 2001.

[9] Çengel YA, Özisik MN: Radiative Transfer in Plane Parallel Medium with Space-Dependent Albedo, $\omega(x)$. *Int J of Heat and Mass Transfer*, Vol. 27, No. 10, pp. 1919-1922, 1984.

[10] Segatto CF, Vargas RF, Vilhena MT, Bodmann BEJ: A Solution for the Non-linear S_N Radiative Conductive Problem in a Grey Parallel Plane-parallel Participating Medium. *Int J of Thermal Sciences*, Vol. 49, No 9, pp. 1493-1499, 2010.

[11] Adomian EM. A Review of The Decomposition Method in Applied-Mathematics. *Journal of Mathematical Analysis and Applications*, Vol. 135, No 2, pp. 501-544, 1988.

[12] Adomian G.*Nonlinear Stochastic Operator Equations*. Academic Press, Orlando, FL, 1986.

[13] Adomian G. Coupled Differential Equations and Coupled Boundary Conditions. *Journal of Mathematical Analysis and Applications*, Vol. 112, No 1, pp 129-135, 1985.

[14] Segatto CF, Vilhena MT, Gomes MG. The One-Dimensional LTS_N Solution in a Slab with High Degree of Quadrature. *Annals of Nuclear Energy*, Vol. 26, pp. 925-934, 1999.

[15] Gonçalves GA, Segatto CF, Vilhena MT. The LTS_N particular solution in a slab for an arbitrary source and large order of quadrature. *J of Quantitative Spectroscopy and Radiative Transfer*, Vol. 66, No 3, pp. 271-276, 2000.

[16] Azevedo FS, Sauter E, Thompson M, Vilhena MT. Existence theory for a one-dimensional problem arising from the boundary layer analysis of radiative flows. Progress in Nuclear Energy (New series), v. 53, p. 1105-1113, 2011.

[17] Azevedo FS, Sauter E, Thompson M, Vilhena MT. Existence Theory and Simulations for One-dimensional Radiative Flows. *Annals of Nuclear Energy*, Vol. 38, pp. 1115-1124, 2011.
[18] Garcia RDM and Siewert CC: Radiative Transfer in Finite Inhomogeneous Plane-Parallel atmospheres. *J of Quantitative Spectroscopy and Radiative Tranfer*, Vol.27, No 2, pp. 141-148, 1982.
[19] Chandrasekhar S . *Radiative Transfer*. New York: Dover. ISBN 0486605906, 1960.
[20] Pazos RP, Thompson M, Vilhena MT. Error Bounds for spectral Collocation Method For the linear Boltzmann Equation. *Int J of Computational and Numerical Analysis and Applications*, Vol 1, pp. 237-268, 2002.

Eurotherm Conference No. 95: Computational Thermal Radiation in Participating Media IV IOP Publishing
Journal of Physics: Conference Series **369** (2012) 012019 doi:10.1088/1742-6596/369/1/012019

The Simplified Spherical Harmonics Method For Radiative Heat Transfer

Michael F. MODEST(*) **and Shenghui LEI**(**)

(*) Shaffer and George Professor of Engineering, University of California, Merced 95343.
(**) Postdoctoral associate.

E-mail: mmodest@ucmerced.edu

Abstract. The *Simplified* P_N*-Approximation* was introduced some 50 years ago, as an intuitive three-dimensional extension to the one-dimensional slab P_N-formulation. The set of simplified-P_N or SP_N equations was formulated, such that they reduced to the standard P_N-approximation for a one-dimensional slab, resulting in $2(N+1)$ simultaneous first order differential equations, subject to a modified set of Marshak boundary conditions. In this article the method is developed further for radiative heat transfer applications by reducing the governing equations to a small set of simultaneous second-order, elliptic PDEs, similar to the simple P_1 equation. Each of the equations is subject to P_1-type boundary conditions. Several two-dimensional examples are presented, including scattering, strongly inhomogeneous temperatures and absorption coefficients.

1. Introduction

The radiative transfer equation (RTE) is an integro-differential equation in five independent variables (three in space and two in direction), which is exceedingly difficult to solve. Several approximate methods have been developed over time. Among these approximate methods the *Spherical Harmonics Method* (SHM), the *Discrete Ordinates Method* or the *Finite Volume Method,* and the *Monte Carlo Method* are presently used most frequently [1]. Both the *Spherical Harmonics Method* and DOM/FVM approximate the directional variation of the radiative intensity.

Like all methods, the SHM has a number of drawbacks. The lowest-order P_N-approximation, the P_1-approximation, has enjoyed great popularity because of its relative simplicity and compatibility with standard solution methods [1], but performs poorly in the optically thin limit and other nonisotropic radiative intensity fields, which to a great extent can be mitigated using the modified differential approximation approach [2, 3]. Mathematical complexity increases rapidly if higher-order P_N-approximations for multi-dimensional geometry are desired. Consequently, only a few and very limited multi-dimensional formulations of higher order have been presented [4, 5], as reviewed by Yang and Modest [6]. Recently, Modest and Yang [7] have shown how the $(N+1)^2$ first-order PDEs of the general three-dimensional P_N-approximation can be condensed into a set of $N(N+1)/2$ second-order, elliptic PDEs, which are compatible with, and can be embedded into, standard commercial codes. For two-dimensional problems the set can be reduced to $(N+1)^2/4$ elliptic PDEs. However, the set of equations is very complex, in particular if allowing for spatially varying absorption and anisotropic scattering, and includes many cross-derivatives, which makes it difficult to apply higher order P_N-methods to actual problems.

Facing these mathematical difficulties Gelbard [8] introduced the *Simplified* P_N*-Approximation* some

Published under licence by IOP Publishing Ltd

50 years ago, as an intuitive three-dimensional extension to the one-dimensional slab P_N-formulation, and their Marshak boundary conditions. Gelbard formulated his set of simplified-P_N or SP_N equations, such that they reduced to the standard P_N-approximation for a one-dimensional slab and some other narrow circumstances, but lacking any theoretical foundation, which impeded its acceptance. Theoretical justifications were found many years later by Larsen *et al.* [9] (showing SP_N to be an asymptotic correction to the diffusion approximation, and by Pomraning [10] (showing the SP_N to be asymptotically related to the P_N-equations for the slab geometry). A fine review of the SP_N-method has recently been given by McClarren [11].

While the developments of Larsen and Pomraning provide theoretical credentials to the method, they are rather tedious, and are probably instrumental in the fact that the method has to date been ignored by the heat transfer community. We will here only provide the intuitive development of Gelbard, and develop the method further for radiative heat transfer applications by reducing the governing equations to a small set of simultaneous second-order, elliptic PDEs without any cross-derivatives: 2 in the case of SP_3 (as opposed to 6 for P_3), 3 for the case of SP_5 (vs. 15 for P_5), and so on. Each of the governing equations is similar to the simple P_1 equation, connected only through simple source terms. Similarly, each of the equations is subject to P_1-type boundary conditions, also connected only through simple source terms.

Several two-dimensional examples will be presented, including scattering, strongly inhomogeneous temperatures and absorption coefficients, and a typical axisymmetric, strongly nongray combustion scenario. Results from the SP_3- and SP_5-models will be compared with various full P_N-approximations and exact Monte Carlo results, indicating that SP_3 and SP_5 perform as well as the equivalent P_N-models in many situations, but at much lower computational cost.

2. Theoretical Development

The general radiative transfer equation (RTE) is given by [1]

$$\hat{\mathbf{s}} \cdot \nabla_\tau I + I = (1-\omega) I_b + \frac{\omega}{4\pi} \int_{4\pi} I(\hat{\mathbf{s}}')\, \Phi(\hat{\mathbf{s}} \cdot \hat{\mathbf{s}}')\, d\Omega', \tag{1}$$

where space coordinates have been nondimensionalized using the extinction coefficient, i.e., $d\tau = \beta\, ds$ (as indicated by the subscript τ in ∇_τ). Here I is radiative intensity, I_b is blackbody intensity (or *Planck function*), $\omega = \sigma_s/\beta$ is the single scattering albedo and σ_s the scattering coefficient, and Φ is the scattering phase function. Equation (1) requires the outgoing intensity to be specified everywhere along the surface of the enclosure.

In the one-dimensional P_N-approximation for a plane-parallel slab the radiative intensity is expanded into a series of Legendre polynomials P_l

$$I(\tau, \mu) \simeq \sum_{l=0}^{N} I_l(\tau) P_l(\mu), \tag{2}$$

where $\mu = \cos\theta$ and θ is the polar angle for the unit direction vector $\hat{\mathbf{s}}$ (see Fig. 1). Equation (2) is approximate because the series is truncated beyond $l = N$, i.e., we assume $I_l(\tau) = 0$ for all $l > N$. The scattering phase function for such a medium may also be expanded into Legendre polynomials, or [1]

$$\Phi(\mu, \mu') = \sum_{m=0}^{M} A_m P_m(\mu') P_m(\mu), \tag{3}$$

where M is the order of approximation for the phase function. The general P_N-approximation for a one-dimensional anisotropically scattering slab may then be written as $N + 1$ simultaneous first-order partial

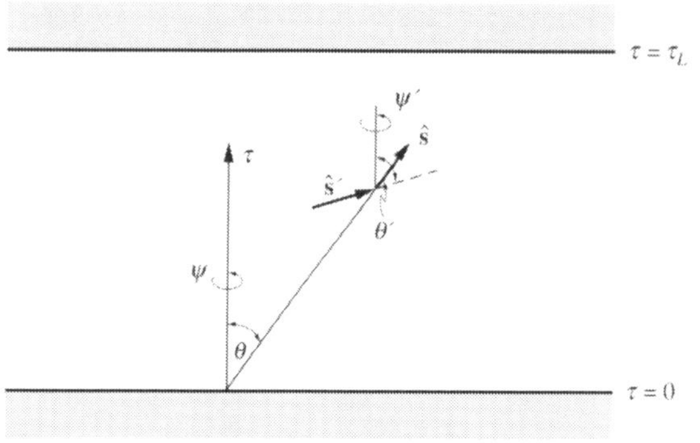

Figure 1. Coordinates for a one-dimensional plane-parallel medium.

differential equations [1]

$$\frac{k+1}{2k+3}I'_{k+1}(\tau)+\frac{k}{2k-1}I'_{k-1}(\tau)+\left(1-\frac{\omega A_k}{2k+1}\right)I_k(\tau) = (1-\omega)I_b(\tau)\delta_{0k},$$
$$k = 0,1,\ldots,N. \tag{4}$$

The Marshak boundary conditions for this set of equations, assuming diffuse surfaces, i.e., $I_w = J_w/\pi$, where J_w is the radiosity at the wall, are given by [1]

$$\sum_{l=0}^{N} I_l(0)\int_0^1 P_l(\mu)P_{2i-1}(\mu)\,d\mu = \frac{J_{w1}}{\pi}\int_0^1 P_{2i-1}(\mu)\,d\mu, \qquad i=1,2,\ldots,\tfrac{1}{2}(N+1); \tag{5a}$$

$$\sum_{l=0}^{N} I_l(\tau_L)\int_{-1}^0 P_l(\mu)P_{2i-1}(\mu)\,d\mu = \frac{J_{w2}}{\pi}\int_{-1}^0 P_{2i-1}(\mu)\,d\mu, \qquad i=1,2,\ldots,\tfrac{1}{2}(N+1). \tag{5b}$$

While the developments of Larsen [9] and Pomraning [10] provide theoretical credentials to the SP_N-method, they are rather tedious, and we will here only provide the intuitive development of Gelbard [8], as a basis for further development for radiative heat transfer applications. Depending on whether k is odd or even, Gelbard made the following substitutions in Eqs. (4) and (5):

$$k\text{ odd}: \qquad I_k(\tau)\to \mathbf{I}_k(\tau_x,\tau_y,\tau_z), \qquad I'_k=\frac{dI_k}{d\tau}\to\nabla_\tau\cdot\mathbf{I}_k, \tag{6a}$$

$$k\text{ even}: \qquad I_k(\tau)\to I_k(\tau_x,\tau_y,\tau_z), \qquad I'_k=\frac{dI_k}{d\tau}\to\nabla_\tau I_k, \tag{6b}$$

i.e., for every odd k the I_k becomes a vector and differentiation is replaced by the divergence operator, while even I_k remain scalars and their differentiation is replaced by the gradient operator. Substituting Eqs. (6) into Eq. (4) leads to

$$k=0,2,\ldots,N-1\ \text{(even)}: \qquad \frac{k+1}{2k+3}\nabla_\tau\cdot\mathbf{I}_{k+1}+\frac{k}{2k-1}\nabla_\tau\cdot\mathbf{I}_{k-1}+\alpha_k I_k=\alpha_k I_b\delta_{0k}, \tag{7a}$$

$$k=1,3,\ldots,N\ \text{(odd)}: \qquad \frac{k+1}{2k+3}\nabla_\tau I_{k+1}+\frac{k}{2k-1}\nabla_\tau I_{k-1}+\alpha_k\mathbf{I}_k=0, \tag{7b}$$

where

$$\alpha_k = 1-\frac{\omega A_k}{2k+1}. \tag{7c}$$

Solving Eq. (7b) for $\mathbf{I}_k$ and substituting the result into (7a) produces a set of simultaneous elliptic partial differential equations in the unknown scalars I_k (k even):

$$k = 0, 2, \ldots, N-1 \quad \text{(even)}:$$
$$\frac{(k+1)(k+2)}{(2k+3)(2k+5)}\nabla_\tau \cdot \left(\frac{1}{\alpha_{k+1}}\nabla_\tau I_{k+2}\right) + \frac{(k+1)^2}{(2k+3)(2k+1)}\nabla_\tau \cdot \left(\frac{1}{\alpha_{k+1}}\nabla_\tau I_k\right)$$
$$+\frac{k^2}{(2k-1)(2k+1)}\nabla_\tau \cdot \left(\frac{1}{\alpha_{k-1}}\nabla_\tau I_k\right) + \frac{k(k-1)}{(2k-1)(2k-3)}\nabla_\tau \cdot \left(\frac{1}{\alpha_{k-1}}\nabla_\tau I_{k-2}\right) = \alpha_k(I_k - I_b\delta_{0k}). \quad (8)$$

Similarly, sticking Eqs. (6) into the P_N boundary conditions, Eqs. (5), gives us a consistent set of conditions for the SP_N-equations:

$$\sum_{k \text{ even}}^{N-1} I_k \int_0^1 P_k(\mu)P_{2i-1}(\mu)d\mu + \sum_{k \text{ odd}}^{N} \hat{\mathbf{n}} \cdot \mathbf{I}_k \int_0^1 P_k(\mu)P_{2i-1}(\mu)d\mu = \frac{J_w}{\pi} \int_0^1 P_{2i-1}(\mu)d\mu,$$
$$i = 1, 2, \ldots, \tfrac{1}{2}(N+1), \quad (9)$$

or, with the definition of the Legendre polynomial half-moments $p^m_{n,j}$ given by

$$p^m_{n,j} = p^m_{j,n} = \int_0^1 P^m_n(\bar{\mu})P^m_j(\bar{\mu})d\bar{\mu}, \quad (10)$$

$$\sum_{k \text{ even}}^{N-1} p^0_{k,2i-1} I_k + \sum_{k \text{ odd}}^{N} p^0_{k,2i-1} \hat{\mathbf{n}} \cdot \mathbf{I}_k = \frac{p^0_{0,2i-1}}{\pi} J_w, \qquad i = 1, 2, \ldots, \tfrac{1}{2}(N+1). \quad (11)$$

Again, eliminating the odd $\mathbf{I}_k$ with Eq. (7b), this set of boundary conditions reduces to

$$\sum_{k \text{ even}}^{N-1} p^0_{k,2i-1} I_k - \sum_{k \text{ odd}}^{N} \frac{p^0_{k,2i-1}}{\alpha_k}\left[\frac{k}{2k-1}\hat{\mathbf{n}} \cdot \nabla_\tau I_{k-1} + \frac{k+1}{2k+3}\hat{\mathbf{n}} \cdot \nabla_\tau I_{k+1}\right] = \frac{p^0_{0,2i-1}}{\pi} J_w$$
$$i = 1, 2, \ldots, \tfrac{1}{2}(N+1). \quad (12)$$

No direct formula for intensity is derived, but one may assume a series of the form

$$I(\mathbf{r}, \hat{\mathbf{s}}) = I_0(\mathbf{r}) + \mathbf{I}_1(\mathbf{r}) \cdot \hat{\mathbf{s}} + I_2(\mathbf{r})P^0_2(\hat{\mathbf{s}}) + \ldots, \quad (13)$$

which is no longer a complete series of orthogonal functions and, therefore is not guaranteed to approach the exact answer in the limit. However, assuming this to be an orthogonal set, we can obtain incident radiation G and radiative flux $\mathbf{q}$ from their definitions as

$$G(\mathbf{r}) = \int_{4\pi} I(\mathbf{r}, \hat{\mathbf{s}})\, d\Omega = 4\pi I_0(\mathbf{r}), \quad (14)$$
$$\mathbf{q}(\mathbf{r}) = \int_{4\pi} I(\mathbf{r}, \hat{\mathbf{s}})\, \hat{\mathbf{s}}\, d\Omega = \frac{4\pi}{3}\mathbf{I}_1(\mathbf{r}) = -\frac{4\pi}{3\alpha_1}\left[\nabla_\tau I_0 + \frac{2}{5}\nabla_\tau I_2\right]. \quad (15)$$

While Eqs. (8) and (12) form a self-consistent set of $(N+1)/2$ simultaneous elliptic partial differential equations and their boundary conditions, the problem can be further simplified by recognizing that the combination of variables

$$J_k = \frac{k+1}{2k+1} I_k + \frac{k+2}{2k+5} I_{k+2} \quad (16)$$

appears repeatedly in, both, governing equations and boundary conditions. Thus we may rewrite Eqs. (8)

$$k = 0, 2, \ldots, N-1 \quad \text{(even)}:$$
$$\frac{k+1}{2k+3}\boldsymbol{\nabla}_\tau\cdot\left(\frac{1}{\alpha_{k+1}}\boldsymbol{\nabla}_\tau J_k\right) + \frac{k}{2k-1}\boldsymbol{\nabla}_\tau\cdot\left(\frac{1}{\alpha_{k-1}}\boldsymbol{\nabla}_\tau J_{k-2}\right) = \alpha_k(I_k - I_b\delta_{0k}). \tag{17}$$

A tabulation of the $p^m_{n,j}$ has been given by Modest [12]; inspection shows that $p^0_{n,j} = 0$ if $n + j$ = even, with the exception of $n = j$, and boundary conditions (12) reduce to

$$\frac{p^0_{2i-1,2i-1}}{\alpha_{2i-1}}\hat{\mathbf{n}}\cdot\boldsymbol{\nabla}_\tau J_{2i-2} = \sum_{k=0}^{\frac{N-1}{2}} p^0_{2k,2i-1} I_{2k} - \frac{p^0_{0,2i-1}}{\pi}J_w, \qquad i = 1, 2, \ldots, \tfrac{1}{2}(N+1). \tag{18}$$

The I_k on the right-hand sides may be eliminated by inverting Eq. (16), starting with $k = N-1$ (and noting that $I_{N+1} \equiv 0$). This results in individual partial differential equations for each J_k, in which J_l ($l \neq k$) occur only as source terms without derivatives. Once the J_k have been determined, incident radiation and radiative flux are obtained from Eqs. (14) and (15) as

$$G(\mathbf{r}) = 4\pi\left[J_0(\mathbf{r}) - \frac{2}{3}J_2(\mathbf{r}) + \frac{24}{55}J_4(\mathbf{r}) - + \ldots\right], \tag{19}$$
$$\mathbf{q}(\mathbf{r}) = -\frac{4\pi}{3\alpha_1}\boldsymbol{\nabla}_\tau J_0(\mathbf{r}). \tag{20}$$

We will demonstrate this by looking in more detail at the SP_1 and SP_3 approximations (even orders, such as SP_2, have also been formulated [13], but—based on the development shown here—appear to be as inappropriate as for the standard P_N-method).

2.1. SP_1-Approximation

With $N = 1$ we obtain a single equation and a single boundary condition from Eqs. (17) and (18), i.e.,

Governing equation:

$$k = 0: \qquad \frac{1}{3}\boldsymbol{\nabla}_\tau\cdot\left(\frac{1}{\alpha_1}\boldsymbol{\nabla}_\tau J_0\right) = \alpha_0(I_0 - I_b); \tag{21}$$

Boundary condition:

$$i = 1: \qquad \frac{p^0_{1,1}}{\alpha_1}\hat{\mathbf{n}}\cdot\boldsymbol{\nabla}_\tau J_0 = p^0_{0,1}(I_0 - J_w/\pi). \tag{22}$$

With $p^0_{0,1} = \frac{1}{2}$ and $p^0_{1,1} = \frac{1}{3}$, and $I_0 = J_0$ from Eq. (16), we obtain

$$\frac{1}{3}\boldsymbol{\nabla}_\tau\cdot\left(\frac{1}{\alpha_1}\boldsymbol{\nabla}_\tau J_0\right) = \alpha_0(J_0 - I_b), \tag{23}$$

with boundary condition

$$\frac{1}{3\alpha_1}\hat{\mathbf{n}}\cdot\boldsymbol{\nabla}_\tau J_0 = \frac{1}{2}(J_0 - J_w/\pi). \tag{24}$$

Not surprisingly, using $G = 4\pi I_0 = 4\pi J_0$ shows that the SP_1-approximation is identical to the P_1-method as given, for example, by Modest [1].

2.2. SP_3-Approximation

Setting $N = 3$ we get two simultaneous equations and two boundary conditions:

Governing equations:

$$k = 0: \quad \frac{1}{3}\nabla_\tau \cdot \left(\frac{1}{\alpha_1}\nabla_\tau J_0\right) = \alpha_0(I_0 - I_b) = \alpha_0\left(J_0 - \frac{2}{3}J_2 - I_b\right), \tag{25a}$$

$$k = 2: \quad \frac{3}{7}\nabla_\tau \cdot \left(\frac{1}{\alpha_3}\nabla_\tau J_2\right) + \frac{2}{3}\nabla_\tau \cdot \left(\frac{1}{\alpha_1}\nabla_\tau J_0\right) = \alpha_2 I_2 = \frac{5}{3}\alpha_2 J_2, \tag{25b}$$

or, subtracting 2×Eq. (25):

$$k = 2: \quad \frac{3}{7}\nabla_\tau \cdot \left(\frac{1}{\alpha_3}\nabla_\tau J_2\right) = \left(\frac{5}{3}\alpha_2 + \frac{4}{3}\alpha_0\right)J_2 - 2\alpha_0(J_0 - I_b). \tag{25c}$$

Boundary conditions:

$$i = 1: \quad \frac{p^0_{1,1}}{\alpha_1}\hat{\mathbf{n}} \cdot \nabla_\tau J_0 = p^0_{0,1}(I_0 - J_w/\pi) + p^0_{2,1} I_2, \tag{26a}$$

$$i = 2: \quad \frac{p^0_{3,3}}{\alpha_3}\hat{\mathbf{n}} \cdot \nabla_\tau J_2 = p^0_{0,3}(I_0 - J_w/\pi) + p^0_{2,3} I_2. \tag{26b}$$

With $p^0_{2,1} = p^0_{2,3} = \frac{1}{8}$, $p^0_{3,3} = \frac{1}{7}$, $p^0_{0,3} = -\frac{1}{8}$, and eliminating the I_k, the boundary conditions become

$$i = 1: \quad \frac{1}{3\alpha_1}\hat{\mathbf{n}} \cdot \nabla_\tau J_0 = \tfrac{1}{2}(J_0 - \tfrac{2}{3}J_2 - J_w/\pi) + \tfrac{1}{8}\tfrac{5}{3}J_2 = \tfrac{1}{2}(J_0 - J_w/\pi) - \tfrac{1}{8}J_2, \tag{26c}$$

$$i = 2: \quad \frac{1}{7\alpha_3}\hat{\mathbf{n}} \cdot \nabla_\tau J_2 = -\tfrac{1}{8}(J_0 - \tfrac{2}{3}J_2 - J_w/\pi) + \tfrac{1}{8}\tfrac{5}{3}J_2 = -\tfrac{1}{8}(J_0 - J_w/\pi) + \tfrac{7}{24}J_2. \tag{26d}$$

Unlike the regular P_3-approximation, SP_3 has only two, and nearly separated, elliptic partial differential equations: Eqs. (25a) and (26c) for J_0 and Eqs. (25c) and (26d) for J_2, the only connection being the other J_k appearing in source terms.

3. Sample Calculations

Several one- and two-dimensional radiative heat transfer problems have been reported in [6, 7, 12], in particular a nonabsorbing, isotropically scattering medium confined in a rectangular enclosure with a heated strip was studied, as well as a two-dimensional rectangular enclosure with strongly varying temperatures and absorption coefficients. We will repeat some of the previous calculations here with the SP_N equations and compare them with regular P_N and Monte Carlo results. In all case cold and black bounding surfaces are chosen: this has no impact on solution complexity, but (i) makes presentation of results easier, and (ii) is a more severe test of the method, since reflecting boundaries cause directionally smoother intensities.

Enclosure with varying properties Although scattering poses no additional difficulty for the spherical harmonics method (as opposed to the DOM or FVM), we will limit ourselves to a nonscattering medium, in order to reduce the number of variables. The following nondimensional field will is studied for a square enclosure [12, 14]:

$$I_b = 1 + 5r^2(2 - r^2), \tag{27a}$$

$$\kappa = C_k\left[1 + 5(2 - r^2)^2\right], \tag{27b}$$

with

$$r^2 = x^2 + y^2; \qquad -1 \le x \le +1, \quad -1 \le y \le +1, \tag{27c}$$

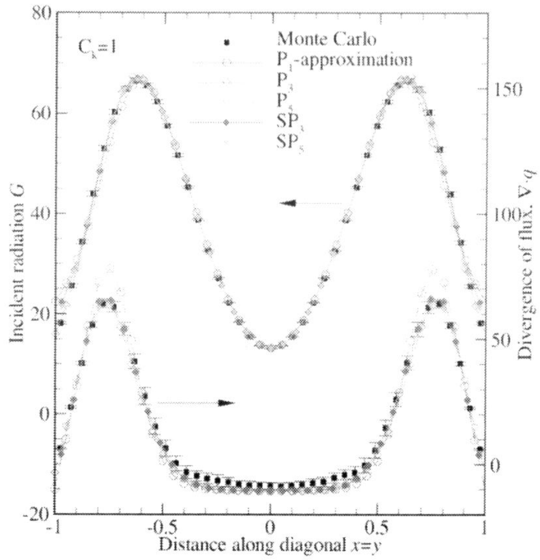

Figure 2. Incident radiation and radiative source for a square enclosure; optically thick case ($C_k = 1$).

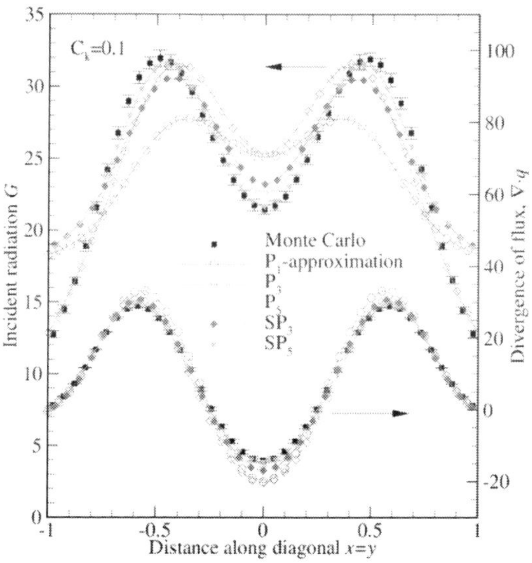

Figure 3. Incident radiation and radiative source for a square enclosure; optically intermediate case ($C_k = 0.1$).

i.e., the blackbody intensity (Planck function) is normalized with its minimum value (obtained at the center and the 4 corners), with a maximum value of $I_b = 6$ at a distance of $r = 1$ from the centerline. The absorption coefficient, normalized in terms of length units, has a maximum value of $\kappa = 16C_k$ at $x=y=0$, and rapidly diminishes away from the center to a minimum value of $\kappa = C_k$ at the four corners. Thus the problem is radially one-dimensional, except for the conditions at the four perpendicular walls, which are assumed cold and black. The optical thickness of the square enclosure is $\tau_D = 18\sqrt{2}C_k$ along a diagonal, and $\tau_D = 23.5C_k$ along an $x=0$ or $y=0$ line, respectively. Here we investigate values of $C_k = 1$ (optically thick), $C_k = 0.1$ (optically intermediate), and $C_k = 0.01$ (optically thin conditions). Results for incident radiation $G = 4\pi I_0$ and divergence of the radiative flux $\nabla \cdot \mathbf{q}$ are shown in Figs. 2 through 4, comparing results from various orders of P_N and SPN with those of a Monte Carlo simulation. All calculations were performed with a relatively fine 41×41 cell system to ensure adequate resolution of sharp gradients [especially in the case of the a zeroth-order photon Monte Carlo (PMC) method]. For the optically thick case, Fig. 2, all methods do very well, although P_1 has difficulty following the sharp peaks in both G and $\nabla \cdot \mathbf{q}$, with P_3/SP_3 being almost exact, and P_5/SP_5 even closer to the PMC results. In the optically intermediate case, Fig. 3, both P_1 and P_3 have difficulties following the peaks and valleys of the incident radiation G, doing better with $\nabla \cdot \mathbf{q}$. Only P_5 follows the variations in G well, and is essentially exact in the prediction of $\nabla \cdot \mathbf{q}$. It is seen that the SP_N schemes actually outperform their corresponding P_N counterparts. For the optically thin case, Fig. 4, P_3 and P_5 perform better than P_1, but it is clear that neither can follow the true variation in incident radiation. SP_3 and SP_5 while better than P_1 cannot follow the actual variation as well as their corresponding P_N. All methods predict $\nabla \cdot \mathbf{q}$ relatively well (i.e., the quantity of greatest importance, providing the radiative source term in the overall energy equation), but higher order methods perform only marginally better than P_1.

Figure 5 depicts radiative fluxes along a wall for the same 3 cases. It is seen that, for the optically thick case, P_1 incurs serious errors (10% at the center, 40% in the corners). This is not surprising, since it is well known that the P_1-approximation performs poorly for optically thick situations in the vicinity of temperature discontinuities [1]. P_3 and P_5, on the other hand, perform rather well. SP_3 and SP_5, while considerably better than P_1, cannot compete with their P_N counterparts. This trend becomes more

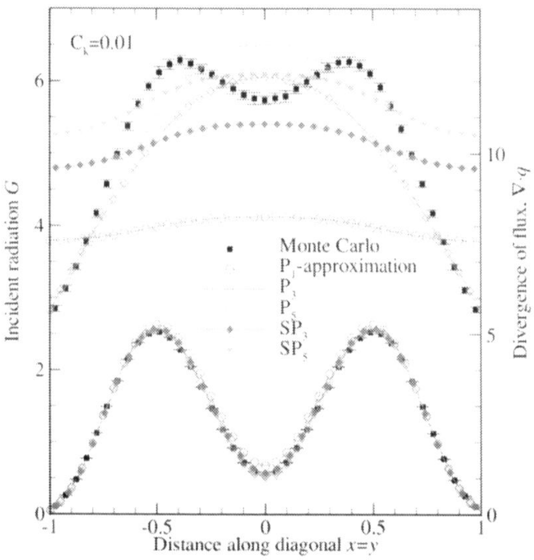

Figure 4. Incident radiation and radiative source for a square enclosure; optically thick case ($C_k = 1$).

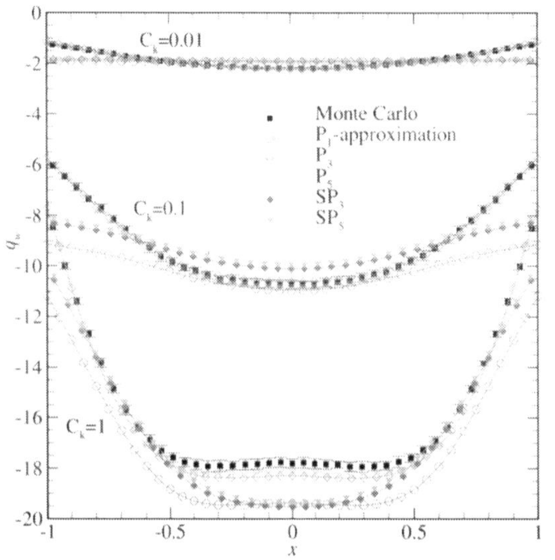

Figure 5. Radiative flux along bottom wall ($y = -1$).

obvious for optically intermediate and thin media. In the optically intermediate case P_1 does well near $x=0$ (where the problem is close to 1D), but it fails miserably toward the corners. Again, both P_3 and P_5 perform extremely well, while SP_3 and SP_5, like P_1, are not able to follow the curvature in the heat flux profile. Finally, in the optically thin case, P_1, SP_3 and SP_5 essentially predict a constant value, while P_3 and P_5 perform better, but incur maximum errors of 10–15%.

Enclosure with purely scattering medium A black-walled square enclosure with walls of width L and a constant scattering coefficient σ_s = const with one heated wall has been considered by several investigators, primarily to investigate ray effects in the discrete ordinates (DOM) and finite volume methods (FVM). Modest and Yang[7] compared the performance of P_1, P_3 and their modified versions[3] to that of various levels of the FVM, finding that *all* methods perform miserably for this problem, the FVM being better at the heated wall, and the P_N methods having the advantage along the cold walls. Figures 6 and 7 show nondimensional *incoming* irradiation $H^* = H/\sigma T^4$ (which must always be positive), comparing various levels of P_N and SP_N and their modified versions. In the optically thin case, $\tau_L = \sigma_s L = 0.1$, (Fig. 6) both P_1 and P_3 perform poorly along all 3 surfaces, albeit in a different manner, leading to physically impossible negative irradiation along the hot bottom wall. SP_3 and SP_5 perform better than P_1, but it is observed that they display P_1-type behavior, i.e., SP_3 is much closer to P_1 than to P_3. The modified versions all perform well for this wall-emission dominated configuration. Performance of P_N and SP_N becomes better with increasing optical thickness (Fig. 7), but the P_N vs. SP_N trend remains the same, i.e., while SP_3 and SP_5 show improvements over P_1, these improvements are fairly minor.

4. Summary and Conclusions

The Simplified Spherical Harmonics, or SP_N method was formulated for radiative heat transfer, resulting in $(N+1)/2$ simultaneous nearly decoupled, P_1-like elliptic partial differential equations, subject to also nearly-decoupled Marshak boundary conditions. The SP_N methods provides potentially great savings over the standard P_N approach, which requires $N(N+1)/2$ strongly coupled PDEs with cross-derivatives. Judging from the results given here, it appears that the SP_3-method constitutes a distinct improvement over P_1 for calculations of incident radiation G and flux divergences $\nabla \cdot \mathbf{q}$, but much less so as far as the

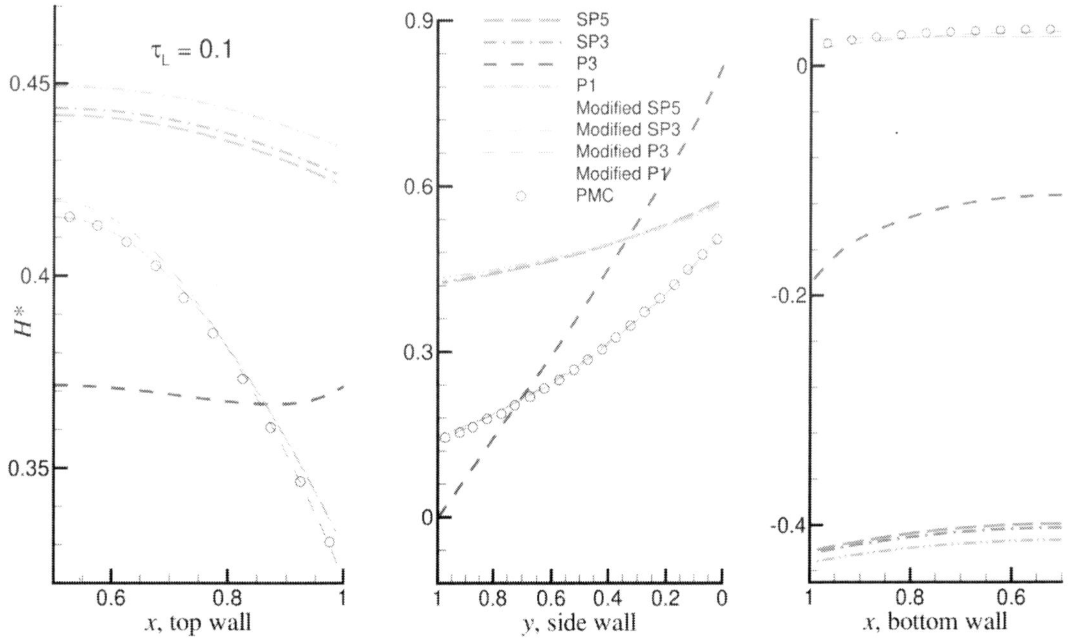

Figure 6. Comparison of various P_N and SP_N for surface heat fluxes in a purely scattering square enclosure; optically thin case $\tau_L = 0.1$.

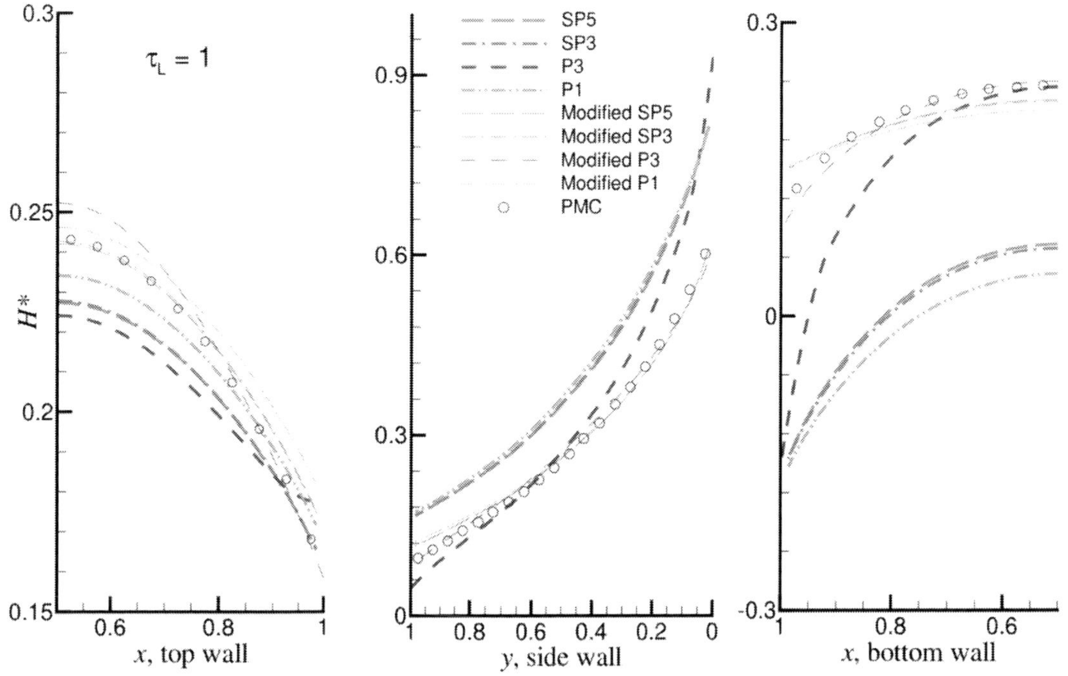

Figure 7. Comparison of various P_N and SP_N for surface heat fluxes in a purely scattering square enclosure; optically intermediate case $\tau_L = 1.0$.

evaluation of wall fluxes is concerned. The SP_3-method requires a reasonable two simultaneous PDEs as opposed to the single PDE for P_1. While SP_5 is somewhat more accurate still, the increase in accuracy is relatively small, perhaps not justifying the additional numerical effort (3 simultaneous PDEs). More sample calculations will need to be carried out to assess the overall benefit of the SP_N as opposed to more established RTE solution methods.

5. References

[1] Modest, M. F., *Radiative Heat Transfer*, Academic Press, New York 2nd edn. 2003.
[2] Olfe, D. B., A Modification of the Differential Approximation for Radiative Transfer, A.I.A.A. J. 1967, 5(4): 638–643.
[3] Modest, M. F., The Modified Differential Approximation for Radiative Transfer in General Three-Dimensional Media, J. Thermoph. Heat Transfer 1989, 3(3): 283–288.
[4] Davison, B., *Neutron Transport Theory*, Oxford University Press, London 1958.
[5] Ou, S. C. S. and Liou, K. N., Generalization of the Spherical Harmonic Method to Radiative Transfer in Multi-Dimensional Space, JQSRT 1982, 28(4): 271–288.
[6] Yang, J. and Modest, M. F., High-Order *P-N* Approximation for Radiative Transfer in Arbitrary Geometries, JQSRT 2007, 104(2): 217–227.
[7] Modest, M. F. and Yang, J., Elliptic PDE formulation and boundary conditions of the spherical harmonics method of arbitrary order for general three-dimensional geometries, JQSRT 2008, 109: 1641–1666.
[8] Gelbard, E. M., Simplified Spherical Harmonics Equations and Their Use in Shielding Problems, Technical Report WAPD-T-1182 Bettis Atomic Power Laboratory 1961.
[9] Larsen, E., Morel, J., and McGhee, J., Simplified P_N approximations to the equations of radiative heat transfer and applications, in *Proceedings of the Joint International Conference on Mathematical Methods and Supercomputing in Nuclear Applications* Portland, Oregon 1993.
[10] Pomraning, G. C., Asymptotic and variational derivations of the simplified P_n equations, Annals of Nuclear Energy 1993, 20(9): 623–637.
[11] McClarren, R. G., Theoretical aspects of the simplified P_n equations, Transport Theory and Statistical Physics 2011, 39: 73–109.
[12] Modest, M. F., Further developments of the elliptic P_N-Approximation formulation and its boundary conditions, JQSRT 2012, ??: ??
[13] Larsen, E., Thömmes, G., Klar, A., Seaïd, M., and Götz, T., Asymptotic derivation of the simplified P_N equations, J. Comp. Phys. 2002, 183: 652–675.
[14] Deshmukh, K. V., Modest, M. F., and Haworth, D. C., Higher-Order Spherical Harmonics to model radiation in Direct Numerical Simulation of Turbulent Reacting Flows, Computational Thermal Sciences 2009, 1: 207–230.

Finite volume method for radiative heat transfer in an unstructured flow solver for emitting, absorbing and scattering media

Moncef GAZDALLAH [1], Véronique FELDHEIM [1], Kilian CLARAMUNT[2] and Charles HIRSCH[2]

[1]*Faculté Polytechnique de Mons - UMONS, Thermal Engineering and Combustion Laboratory, Rue de l'Epargne 56, 7000 Mons, Belgium.*
[2]*NUMFLO, Rue Descartes, 2 - 7000 Mons – Belgium*

E-mail: moncef.gazdallah@umons.ac.be

Abstract. This paper presents the implementation of the finite volume method to solve the radiative transfer equation in a commercial code. The particularity of this work is that the method applied on unstructured hexahedral meshes does not need a pre-processing step establishing a particular marching order to visit all the control volumes. The solver simply visits the faces of the control volumes as numbered in the hexahedral unstructured mesh. A cell centred mesh and a spatial differencing step scheme to relate facial radiative intensities to nodal intensities is used.
The developed computer code based on FVM has been integrated in the CFD solver FINE™/Open from NUMECA Int. Radiative heat transfer can be evaluated within systems containing uniform, grey, emitting, absorbing and/or isotropically or linear anisotropically scattering medium bounded by diffuse grey walls. This code has been validated for three test cases. The first one is a three dimensional rectangular enclosure filled with emitting, absorbing and anisotropically scattering media. The second is the differentially heated cubic cavity. The third one is the L-shaped enclosure. For these three test cases a good agreement has been observed when temperature and heat fluxes predictions are compared with references taken, from literature.

1. Introduction

In CFD, finite volume methods are widely used because they ensure conservation laws over the control volume surrounding a node. These methods are also used for convective heat transfer. Finite volume method for radiation (FVM) [1], the numerical technique for solving the radiative transfer equation (RTE), is based on the same control volume technique as used in computational fluid dynamics. Compared with other methods such as zonal method, P_N approximation and Monte Carlo, FVM is thus known to be more compatible with the numerically discretized forms of momentum and energy in nonorthogonal structured or unstructured grids [2, 3]. It transforms the radiative transfer equation to a set of algebraic equations, one for each combination of control volume and solid angle. A differencing scheme that relates a volume facial intensity to a nodal one is required for each

To whom any correspondence should be addressed

direction. A marching order to scan all domain control volumes for each direction is required but it is memory consuming and hard to establish, especially for unstructured meshes.

In this paper, in order to avoid the establishment of this marching order when solving discrete algebraic equations, the solver only visits control volume faces as numbered in the mesh. Each face only limits two control volumes. The use of the "face by face browsing" allows dealing easily with unstructured grids.

Scattering is an important phenomenon in many scientific fields and processes. In combustion chambers, scattering of radiation is mainly due to the presence of particles (e.g. soot). To take scattering into account, anisotropic and linear anisotropic scattering phase functions have been used in the present study.

The computer code developed and integrated in the CFD solver FINE^{TM}/Open from NUMECA Int. calculates the variables needed by the coupling procedure between the radiative transfer and the CFD when solving the energy conservation equation. These variables are the energy radiative source terms and the radiative heat fluxes at the boundaries. The code estimates this information starting from the fields of concentration and temperature provided by the solution of Navier-Stokes equations using a hexahedral mesh. The radiative heat transfer can be evaluated within media containing uniform, grey, emitting, absorbing and/or isotropically or linear anisotropically scattering medium bounded by diffuse grey walls.

2. Finite volume method

2.1. Radiative Transfer Equation (RTE)

The radiation intensity for an absorbing, emitting and scattering grey medium at any position $\vec{r}$ along a path $\vec{s}$ is governed by:

$$\frac{dI(\vec{r},\vec{s})}{ds} = -\beta I(\vec{r},\vec{s}) + \kappa I_b(\vec{r}) + \frac{\sigma_s}{4\pi}\int_{\Omega'=4\pi} I(\vec{r},\vec{s}')\Phi(\vec{s}'\rightarrow\vec{s})\,d\Omega' \tag{1}$$

Where κ and σ_s are the absorption and scattering coefficients and $\beta = \kappa + \sigma_s$ is the extinction coefficient. $\Phi(\vec{s}'\rightarrow\vec{s})$ is the scattering phase function for a radiation coming from a direction $\vec{s}'$ to a scattered direction $\vec{s}$. For grey diffusely reflecting walls, the radiative intensity emitted into the medium from the boundary at a point r is given by:

$$I(\vec{r}_w) = \varepsilon(\vec{r}_w)I_b(T(\vec{r}_w)) + \frac{1-\varepsilon(\vec{r}_w)}{\pi}\int_{\vec{s}'.\vec{n}_w<0} I(\vec{r}_w,\vec{s}')\left|\vec{s}'.\vec{n}_w\right|d\Omega' \tag{2}$$

Where $\vec{n}_w$ is the local unit outward normal vector, $\varepsilon(\vec{r}_w)$ is the wall local emissivity and $I_b(\vec{r}_w)$ the blackbody radiation intensity at the local temperature of the wall.

2.2. Principle

In the FVM formulation [1, 2, 3], the domain is divided into a number of control volumes; for each one, the radiative intensity is located at the centre. The solid angle, 4π steradians, is discretized into $N=N_\theta.N_\varphi$ directions. θ is the polar angle ranging from 0 to π and φ is the azimuthal angle ranging from 0 to 2π. A set of discretized linear equations is obtained by integrating the radiative heat transfer equation over the whole domain. The resolution of the radiative transfer equation for each direction M (M=1…N) allows to obtain the directional intensities I_M. Associated with control solid angles, they are used to evaluate the radiative source terms and wall heat fluxes.

2.2.1. Angular discretization

A discrete direction $\vec{s}_i$ is defined by the polar angle θ_m and the azimuthal angle φ_n and has the following form:

$$\vec{s}_i = \sin\theta_m \sin\varphi_n \vec{i} + \sin\theta_m \cos\varphi_n \vec{j} + \cos\theta_m \vec{k} \tag{3}$$

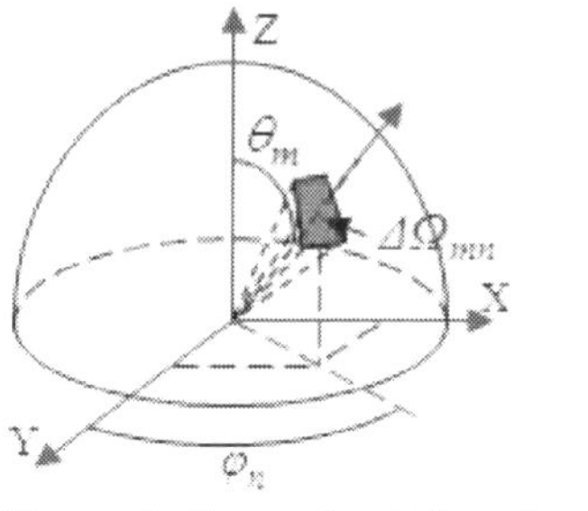

Figure 1: Control solid angle.

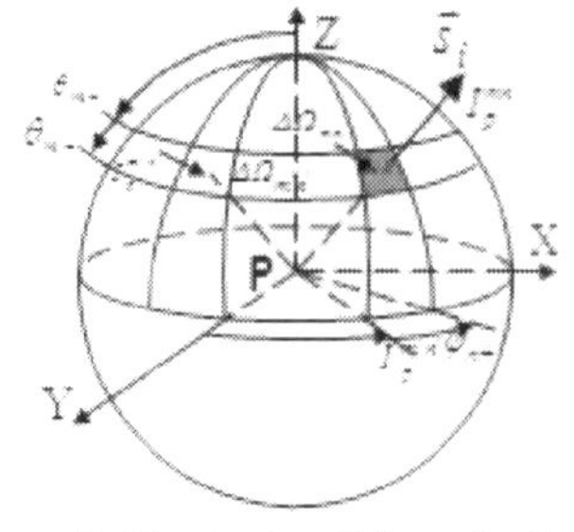

Figure 2: Control solid angle limits.

The subscripts m and n represent, respectively, the number of polar angles and azimuthal angles associated with the direction $\vec{s}_i$. Although the control solid angle can be defined arbitrarily [5] in the present study a uniform angular discretization is used:

$$\Delta\theta_m = \theta_{m+} - \theta_{m-} = \pi / N_\theta \quad , \quad \Delta\varphi_n = \varphi_{n+} - \varphi_{n-} = 2\pi / N_\varphi \quad , \quad \theta_m = \theta_{m+} - \Delta\theta_m / 2 \quad , \varphi_n = \varphi_{n+} - \Delta\varphi_n / 2 \tag{4}$$

2.2.2. Finite volume formulation

Integration of equation (1) over a given control volume V_p and control solid angle $\Delta\Omega_{mn}$ is done by assuming that the magnitude of the intensity is constant but its direction varies within the control volume V_p and control solid angle $\Delta\Omega_{mn}$. The following finite volume formulation can be obtained (N_f is the number of faces limiting the control volume and A_j the surface areas) [28]:

$$\sum_{j=1}^{N_f} I_j^{mn} A_j D_{i,j}^{mn} = \beta\left[- I_p^{mn} + S_R^{mn}\right] V_p \, \Delta\Omega_{mn} \tag{5}$$

$$S_R^{mn} = (1-\omega) I_{b,p} + \frac{\omega}{4\pi} \sum_{m'=1}^{N} I_p^{mm'} \overline{\Phi}_{m'm} d\Omega_{m'} \tag{6}$$

$$D_{i,j}^{mn} = \int_{n\Delta\Phi}^{(n+1)\Delta\Phi} \int_{m\Delta\theta}^{(m+1)\Delta\theta} (\vec{s}_i . \vec{n}_j) \sin\theta \, d\theta \, d\phi \tag{7}$$

$\omega = \frac{\sigma_s}{\beta}$ is the scattering albedo and $D_{i,j}^{mn}$ is the directional weight.

For a given control volume and a given direction $\vec{s}_i$, equation (5) represents an algebraic equation where the unknowns are the values of the intensity I_p^{mn} related to the control volume and the values of the intensities I_j^{mn} at its faces. In order to solve this equation, a spatial differencing scheme must be selected to link control volume faces intensities I_j^{mn} to a nodal one.

2.2.3. Differencing scheme

A weighted difference scheme is very useful in the case of unstructured meshes [6, 7, 8, 9, 10 and 11]. It is based on the following formulation:

$$I_p^{mn} = \alpha I_{j,out}^{mn} + (1-\alpha) I_{j,in}^{mn} \tag{8}$$

Where α is a weighting factor ranging from 0 to 1. Following a direction $\vec{s}_i$, $I_{j,in}^{mn}$ and $I_{j,out}^{mn}$ are a couple of faces intensities, respectively entering ($D_{i,j}^{mn} < 0$) and leaving ($D_{i,j}^{mn} > 0$) the control volume V_p. For any face j of a control volume V_p, the following expression can be written:

$$D_{i,j}^{mn} I_j^{mn} = \max(D_{i,j}^{mn}, 0) I_{j,out}^{mn} - \max(-D_{i,j}^{mn}, 0) I_{j,in}^{mn} = D_{i,j,out}^{mn} I_{j,out}^{mn} + D_{i,j,in}^{mn} I_{j,in}^{mn} \tag{9}$$

From equation (8), $I_{j,out}^{mn}$ can be written as:

$$I_{j,out}^{mn} = \frac{I_p^{mn}}{\alpha} + \frac{(\alpha - 1)}{\alpha} I_{j,in}^{mn} \tag{10}$$

By considering equations (5, 8, 9 and 10) and after some arrangements, the following general linear system of equations can be obtained:

$$a_p^{mn} I_p^{mn} = \sum_{j=1}^{N_f} a_j^{mn} I_{j,in}^{mn} + b_p^{mn} \tag{11}$$

Where

$$a_p^{mn} = \sum_{j=1}^{Nf} A_j \frac{D_{i,j,out}^{mn}}{\alpha} + \beta V_p \Delta\Omega_{mn} \ , \ a_j^{mn} = -A_j \left(D_{i,j,out}^{mn} \frac{(\alpha - 1)}{\alpha} + D_{i,j,in}^{mn} \right) \ , b_p^{mn} = \beta S_R^{mn} V_p \Delta\Omega_{mn} \tag{12}$$

Equation (11) is an algebraic equation relating intensity I_p^{mn} at the control volume centre to the entering intensity $I_{j,in}^{mn}$. The boundary condition in equation (2) for a diffusely emitting and reflecting wall can be derived easily as:

$$I_w^{mn} = \varepsilon_w I_{bw} + \frac{1-\varepsilon_w}{\pi} \sum_{\vec{s}_i' \cdot \vec{n}_w < 0} I_w^{m'n'} \left| D_{i,w,in}^{m'n'} \right| \tag{13}$$

Where

$$D_{i,w,in}^{mn} = \int_{\Delta\Omega_{mn}} (\vec{s}_i \cdot \vec{n}_w) d\Omega \qquad \vec{s}_i \cdot \vec{n}_w < 0 \tag{14}$$

$D_{i,w,j}^{mn}$ and I_w^{mn} are, respectively, the directional weight and the intensity at the boundary wall. Equation (11) for each node, together with boundary conditions, provides the equations set for the nodal intensities in direction $\vec{s}_i$. Similar set of equations are formed for the other directions.

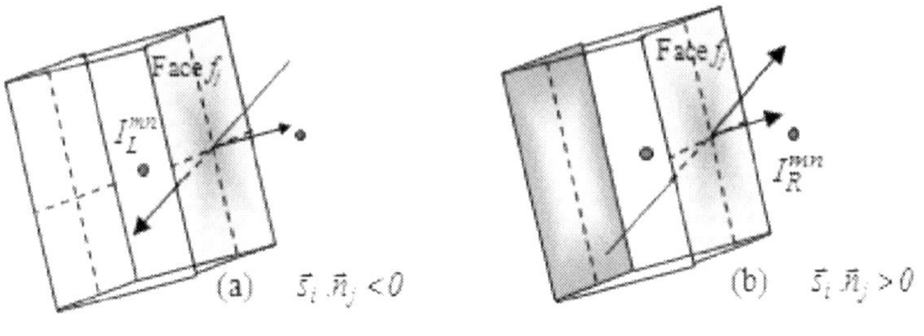

Figure 3: Inflow and outflow radiation through a control volume face.

Considering the face f_j in figure 3, where I_L^{mn} is the nodal intensity at the centre of the control volume located at the left and I_R^{mn} is the nodal intensity at the centre of the control volume located at the right, following the sign of the scalar product $\vec{s}_i \cdot \vec{n}_j$, the following can be stated:

If $\vec{s}_i \cdot \vec{n}_j < 0$: radiation transmitted from the right control volume to the left control volume through the face f_j is: $I_{j,out}^{mn} = \frac{I_R^{mn}}{\alpha} + \frac{(\alpha - 1)}{\alpha} I_{j,in}^{mn}$ if $\alpha = 1$ $I_{j,out}^{mn} = I_R^{mn}$

If $\vec{s}_i \cdot \vec{n}_j > 0$: radiation transmitted from the left control volume to the right control volume through the face f_j is $I_{j,out}^{mn} = \frac{I_L^{mn}}{\alpha} + \frac{(\alpha - 1)}{\alpha} I_{j,in}^{mn}$ if $\alpha = 1$ $I_{j,out}^{mn} = I_L^{mn}$

α = 1 corresponds to the step scheme which states that, for a given control volume Vp and direction $\vec{s}_i$, the value of the intensity at any face of the control volume is assumed equal to the value of the intensity at the centre of the nearest control volume placed upstream of the direction $\vec{s}_i$. Then, it can be equal to the value in the neighbouring control volume or to the value of the control volume Vp. By using this differencing scheme, the calculation can be done face by face.

2.2.4. Control volume faces browsing

In a control volume faces browsing method, the solver visits the faces of the control volumes as numbered in the mesh. To record the exchange between control volumes across each visited face, the equation (11), relating intensity I_p^{mn} at the control volume centre to the entering radiant intensity at its faces, is transformed as a ratio between two terms ($T_{1,p}^{mn}$, $T_{2,p}^{mn}$). From equations (11) and (12) the following expression can be written:

$$I_p^{mn} = \frac{T_{1,p}^{mn}}{T_{2,p}^{mn}} \tag{15}$$

Where

$$T_{1,p}^{mn} = T_{1,p,0}^{mn} + \sum_{j=1}^{N_f} a_j^{mn} I_{j,in}^{mn} \quad with \quad T_{1,p,0}^{mn} = S_R^{mn} \beta V_p \, \Delta\Omega_{mn} \tag{16}$$

and

$$T_{2,p}^{mn} = T_{2,p,0}^{mn} + \sum_{j=1}^{Nf} A_j D_{i,j,out}^{mn} \quad with \; T_{2,p,0}^{mn} = \beta V_p \, \Delta\Omega_{mn} \tag{17}$$

Each term $T_{1,p}^{mn}$ or $T_{2,p}^{mn}$ consists of two parts. The first one depends on the geometric and radiative properties of the control volume and angular discretization. The second one is the sum of the entering intensities across the faces of the control volume. The source term S_R^{mn} is included into the first part of $T_{1,p}^{mn}$. An iterative procedure is needed to compute the whole field of the intensities I_p^{mn} in the case of diffusely reflecting walls. Once the intensity field is obtained, the wall radiative heat flux can be obtained as follows:

$$q_w = \sum_{m=1}^{N_\theta} \sum_{n=1}^{N\varphi} I_w^{mn} (D_{i,w,in}^{mn} + D_{i,w,out}^{mn}) = \sum_{m=1}^{N_\theta} \sum_{n=1}^{N\varphi} I_w^{mn} (-\max(-D_{i,w}^{mn}, 0) + \max(D_{i,w}^{mn}, 0)) \tag{18}$$

For a given control volume, the incident radiation G_p evaluated at its centre P is calculated as follows:

$$G_p = \int_{4\pi} I(\vec{r}, \vec{s}) d\Omega = \sum_{m=1}^{N_\theta} \sum_{n=1}^{N\varphi} I_p^{mn} \Delta\Omega_{mn} \tag{19}$$

The source term S_r is calculated as follows:

$$S_r = k_a \left[4\pi . I_{p,b} - G_P \right] \tag{20}$$

2.2.5. Scattering phase functions

For isotropic scattering the phase function **Φ(** $\vec{s}\,'$, $\vec{s}$ **)** is equal to 1 while for linear isotropic scattering, the following expression can be written:

$$\Phi(\mu_p) = 1 + a\mu_p \quad with \quad -1 < a <= 1$$
$$\mu_p = cos(\Psi) \tag{21}$$

Ψ is the angle between incoming direction $\vec{s}$ and scattered direction $\vec{s}\,'$ [12].

3. Test cases: results and discussion

The code is tested for enclosures of simple geometry already studied by other authors. The first test refers to a three dimensional rectangular enclosure also called ideal furnace studied by Truelove [14,15] for emitting, absorbing and/or anisotropically or isotropically scattering media. The second is a differentially heated cavity, studied by many investigators using different methods [16, 17 and 18]. The third one is a L-shaped enclosure studied by Malalasekera et al using the discrete transfer method [19].

3.1. Three dimensional enclosure

The three-dimensional rectangular enclosure (see figure 5) is considered filled with grey absorbing emitting and/or scattering medium. Its dimensions are 2m x 2m x 4m – in the x, y and z direction respectively. As shown in figure 5, small walls are at 1200 and 400 K while the other four walls are

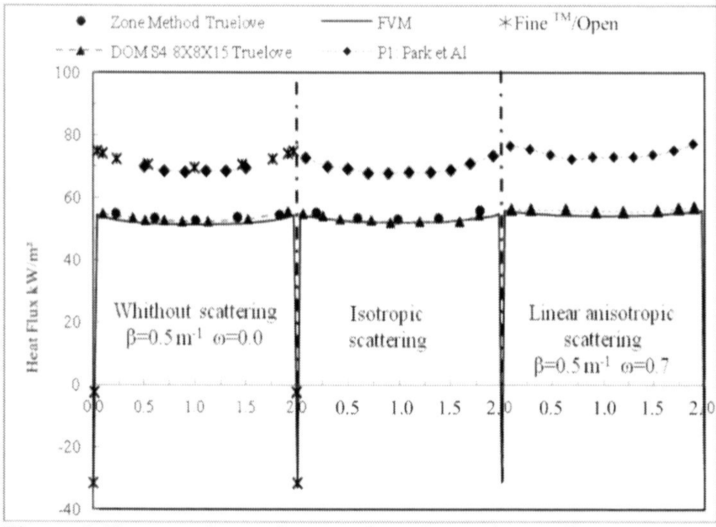

Figure 4: Rectangular enclosure - heat flux at hot wall.

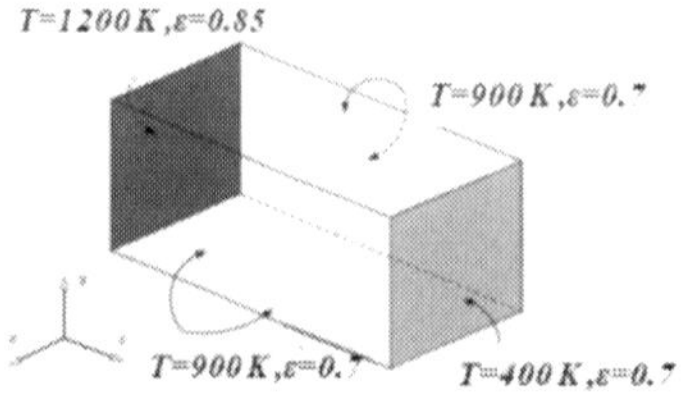

Figure 5: Boundary conditions.

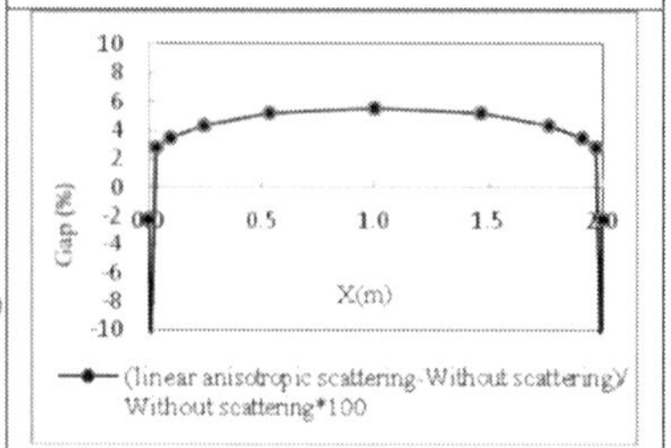

Figure 6: Heat flux increase by linear anisotropic scattering at hot wall (y=1, z=0).

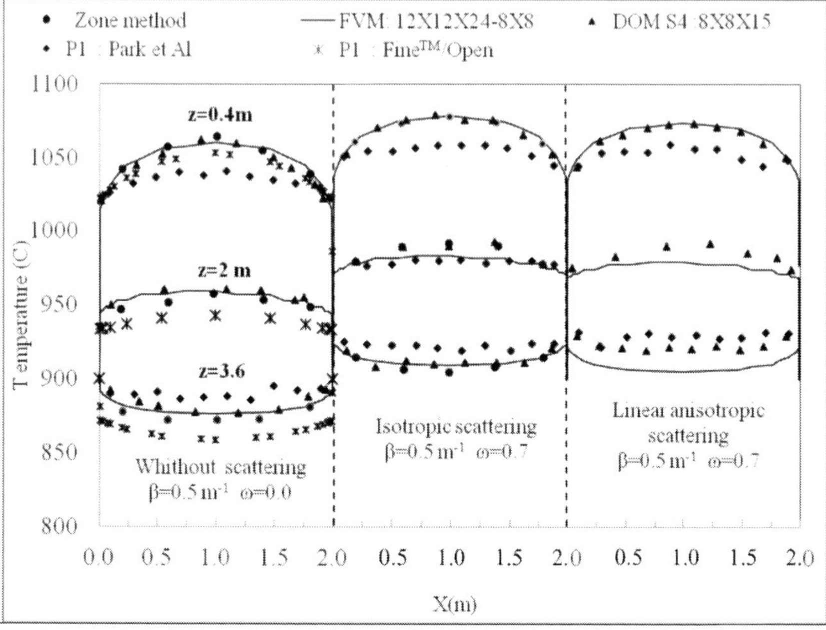

Figure 7: Comparison of temperature profiles, at y=1 and three z locations, from FVM, Zone method [14], DOM S4 [14] and P1 Park [24], P1 FINE^{TM}/Open from NUMECA Int, for three phase functions: without scattering, isotropic and linear anisotropic scattering.

maintained at 900 K. The walls are supposed to reflect and emit diffusively, and to have emissivities of 0.85 for the hottest wall and 0.7 for the remaining walls [22]. In this problem, conduction and convection effects are neglected. Hence, the medium is assumed to be at radiative equilibrium with an additional homogeneous heat source $S = 5kW/m^3$. The radiative equilibrium condition implies that the only term in the energy equation is the divergence of the radiative heat flux. In the presence of a source term S, the energy equation is simply written as:

$$\vec{\nabla}\cdot\vec{Q}_r + S = 0 \tag{25}$$

Two phase functions are tested: isotropically and linear anisotropically scattering medium corresponding to an extinction coefficient $\beta = k_a + \sigma_s = 0.5$ and a scattering albedo $\omega = 0.7$.
Results calculated by Finite Volume Method (FVM) are performed with a (12x12x24) mesh and 64 directions (N_θ=8 and N_φ=8). For isotropic scattering, comparison in figure 4 shows a good agreement of the FVM results with those of zone method and DOM S4 [14]. For linear anisotropic scattering the agreement of the FVM and DOM S4 is also good. However, results of P1 method [24] denote a significant difference with the zonal and DOM S4 methods.
As explained by Truelove [14], the effect of isotropic scattering is an increase of the emissive power σT^4 by $\omega\nabla.q/(4\beta(1-\omega))$, while the heat flux distribution is unchanged for specified radiative flux in the medium. The FVM results given in figure 4 are consistent with this explanation.

Linear anisotropic scattering increases the absolute value of the heat flux by about 5% at the walls as shown in figure 6. This effect is to be expected because the net radiative flux is directed from the hot to the cold surface (in the + z direction) and is enhanced by forward scattering (a=1). Comparison in figure 7 shows a good agreement of the temperature profiles calculated by FVM, the zone method [14], and DOM S4 [14], for the case of purely emitting and absorbing media and isotropically scattering media. However, for linear anisotropic scattering, the gap between FVM and DOM S4 is more pronounced near the cold wall. P1 results by Park [24] and by FINE^{TM}/Open from NUMECA Int. denote a significant difference with the zonal, FVM and DOM S4 methods for the three cases. Compared to the case without scattering, isotropic scattering increases temperature profiles by increasing uniformly throughout the medium the emissive power σT^4 by $\omega\nabla.q/(4\beta(1-\omega))$, while linear anisotropic scattering tends to make temperature distribution more uniform than isotropic scattering.

3.2. Differentially heated cavity

For this case, the study is focused on combined radiation and natural convection for different optical thicknesses and different Rayleigh numbers for the case of purely absorbing, emitting grey medium in a cubic enclosure with sides of unit length L=1m. Scattering is neglected according to the previous results in the literature [18, 25 and 26] which state that scattering has hardly any effect on the heat transfer in this case. As shown in figure 9, the two opposite side walls of the enclosure are kept at a different temperature. The "west" wall is at T_h, the "east" wall at T_c with $T_h > T_c$ while the others are adiabatic. All walls are treated as black bodies (ε=1). To characterize the medium, the following assumptions are considered:

- The flow is laminar, newtonian, and incompressible
- The thermophysical properties of the fluid are constant and a Boussinesq variable density formulation is used to incorporate change in density due to temperature difference. The density is also constant in all the terms of the Navier-Stokes equations except for the buoyancy term.

The strategy of the Navier Stokes equations (NSE) and radiative transfer equation (RTE) solution is stated as follows:

1. Initialize the velocity, temperature, and intensity field
2. Solve the RTE from the current temperature field for a fixed number of iteration

3. Compute the radiative source term
4. Solve the coupled velocity and temperature field with the above radiative source term
5. Update the temperature field
6. Repeat 2 to 5 until the convergence of the solution of the Navier-Stokes equations and the RTE.

Dimensionless convective (Q_C^*) and radiative (Q_R^*) heat fluxes called also, respectively, convective and radiative Nusselt number are calculated as follows:

$$Q_C^* = -\frac{L}{\Delta T}\frac{\partial T^*}{\partial x^*} \qquad Q_R^* = \frac{LQ_R}{k\Delta T} \tag{26}$$

Where L is the side length, $T_0^* = T_c / \Delta T$ the reference temperature ratio (fixed to 17) and $\Delta T = T_h - T_c$.

For this study, the Planck number $Pl = k\Delta T / (L\sigma T_c^4)$, where k is the convection coefficient and σ the Stephan-Boltzman constant, is fixed to Pl=0.016. The Prandtl number is fixed to Pr= 0.71. Tables 1 and 2 show average of dimensionless convective and radiative heat fluxes at the hot wall for three Rayleigh numbers and 3 different optical thicknesses (τ) in a range from 1 to 30 evaluated by four methods. The agreement of the FT_n-FVM [27], DOM S4 [16] and FVM is good for radiation and convection. As expected, for low optical thicknesses, the P1 method denotes a significant difference with FT_n-FVM [27], DOM S4 [16] and FVM for the evaluation of average of dimensionless radiative heat fluxes.

In FT_n FVM the polar angle is divided uniformly into an even number, n, while the azimuthal angle is uniformly divided so that the total number of the control angles, N, is given as n(n + 2), which is the same as the number of the discrete ordinate directions of the S_N DOM. More details on FT_n-FVM can be found in [27].

Table 1. Dimensionless average convective heat flux at hot wall Q_C*: meshing 33x33x33, N_θ =4, N_φ=4

	τ=1				τ=10				τ=30			
Rayleigh	FT_n - FVM	DOM	FVM	P1	FT_n - FVM	DOM	FVM	P1	FT_n - FVM	DOM	FVM	P1
Ra=10^4	2.45	2.25	2.46	3.48	2.23	2.11	2.30	2.24	*	2.08	2.23	2.11
Ra=10^5	4.04	3.92	4.10	4.69	4.46	4.21	4.27	4.40	*	4.34	4.45	4.39
Ra=10^6	*	7.63	7.67	7.83	*	8.28	8.08	8.34	*	8.68	8.50	8.69

Table 2. Dimensionless average radiative heat flux at hot wall Q_R*: meshing 33x33x33, N_θ=4, N_φ=4

	τ=1				τ=10				τ=30			
Rayleigh	FT_n - FVM	DOM	FVM	P1	FT_n - FVM	DOM	FVM	P1	FT_n - FVM	DOM	FVM	P1
Ra=10^4	5.12	4.69	5.35	8.40	1.65	1.54	1.87	1.63	*	0.79	0.86	0.66
Ra=10^5	5.88	5.44	5.98	8.81	2.99	2.80	3.06	2.97	*	1.60	1.65	1.35
Ra=10^6	*	6.25	6.74	9.43	*	4.36	4.52	4.69	*	2.86	2.80	2.48

Profiles of dimensionless average total heat flux versus dimensionless depth Z/L at the hot wall in figure 8 confirm the agreement of results of both methods; DOM S4 from Colomer [16] and FVM. They also show that the radiative heat flux at the hot wall decreases as the optical thickness increases. Isothermal surfaces in figure 9, calculated for four Rayleigh numbers and an optical thickness τ=1, show that for low Rayleigh numbers the conduction heat transfer prevails and the isotherms are nearly vertical (parallel to the gravity field). For high Rayleigh numbers, the isotherms become more horizontal, since the natural convection term, which dominates the conductive term, tends to convey the hottest fluid above the coldest. Thus the isotherms are nearly horizontal (perpendicular to gravity field) in the middle region of the enclosure [16].

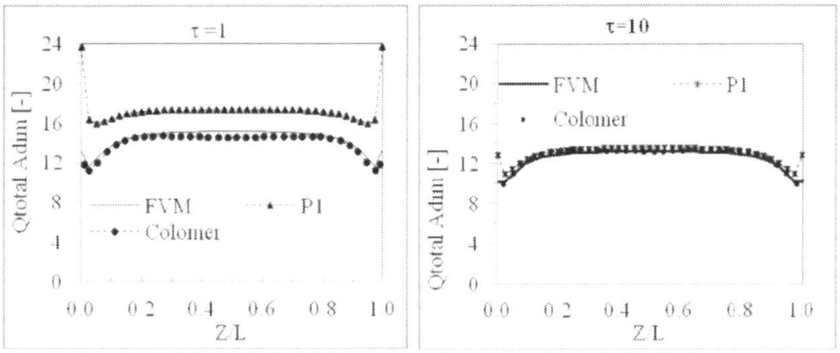

Figure 8: Average total heat flux along Y axes versus dimensionless depth Z/L at the hot wall Meshing:33x33x33, $N_\theta=4$, $N_\varphi=4$ Rayleigh $=10^6$

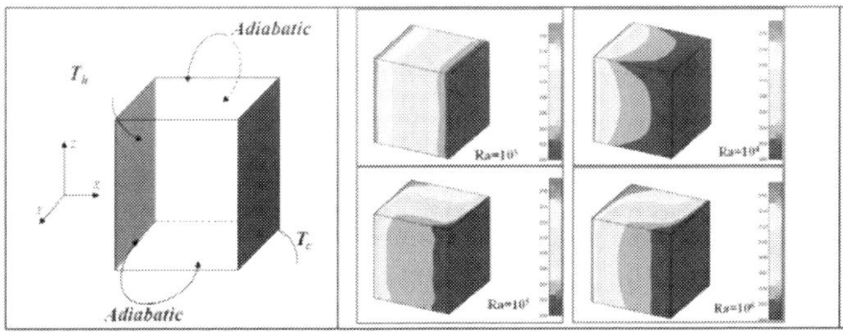

Figure 9: Isothermals surfaces distribution for $\tau=1$.

3.3. L-shaped enclosure

In order to test the code in the case of enclosures with arbitrary geometry a L-shaped enclosure [19, 20, 22 and 23] containing an emitting-absorbing medium at a temperature of 1000 K is considered (see figure (10)). The walls are black at 500 K. Figure (11) shows the evolution of the predicted net heat flux along the A-B line (marked on figure 10) for five values of the absorption coefficient of the medium. Results calculated by FVM are performed with a mesh of 16384 control volumes and 16 directions (N_θ =4 and N_φ=4).The FVM predictions are in good agreement with those of Malalasekera et al [20] who used the discrete transfer method with a body-fitted grid arrangement to discretize the enclosure and 16x16 rays for angular discretization.

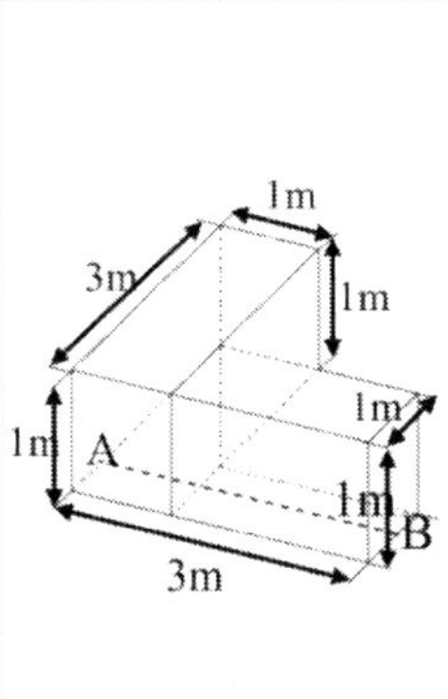

Figure 10: L-shaped enclosure.

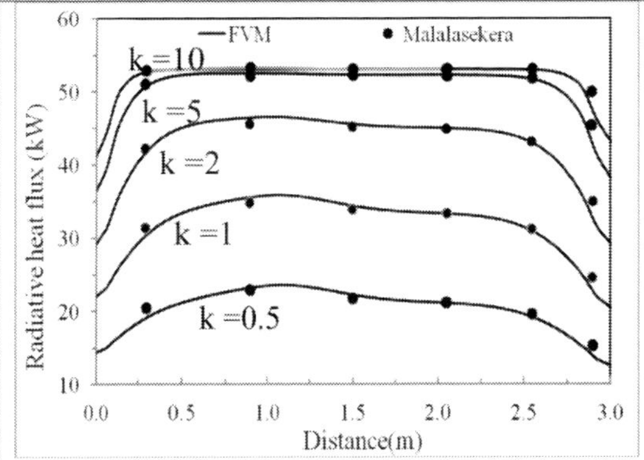

Figure 11: Radiative heat fluxes along the line A-B, shown in figure 10, for five values of absorption coefficients k.

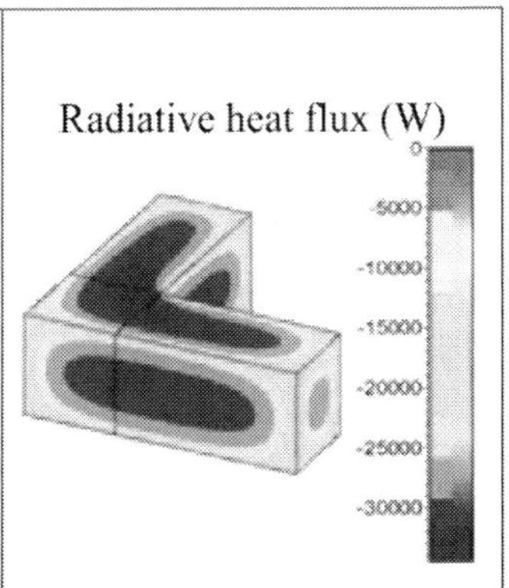

Figure 12: Radiative heat flux distribution at walls. (absorption coefficient k=1)

4. Conclusion

In this study, a finite volume method (FVM) is developed and used to predict three-dimensional radiative heat transfer within an absorbing, emitting and scattering media without a need to a pre-processing step establishing a particular marching order to visit all the control volumes. The developed computer code based on this method is integrated in the unstructured CFD solver FINE™/Open from

NUMECA Int. Then it is applied to three types of enclosures. In the first enclosure, the effects of isotropic and linear anisotropic scattering is investigated and results show that linear anisotropic scattering increases the absolute value of the heat flux. In the second enclosure, the study of combined radiation and natural convection for different optical thicknesses and different Rayleigh numbers shows that heat transfer is dominated by the radiation for lower optical thicknesses. When Rayleigh number increases, the natural convection term dominates the conductive term and isotherms become more horizontal. In the third enclosure, the effect of the shape of the enclosure is studied. For the three enclosures, calculated results are compared with the zone method, discrete ordinates method (DOM) or discrete transfer method from the literature. A good agreement has been observed for both temperature and heat fluxes predictions by the FVM. As expected, the comparison between FVM and P1 shows that the FVM is more accurate for lower optical thicknesses and as accurate as P1 for higher values. Thus, integration of the FVM in FINE^{TM}/Open from NUMECA Int. leads to accurate prediction of the variables needed by the coupling procedure between the radiative transfer and the CFD when solving the energy conservation equation.

Acknowledgments
The authors would like to acknowledge the financial support of the Ministère de la Région Wallonne, in the framework of research program Multi-Φ of the Plan Marshall. The authors express their acknowledgement to NUMECA Int. for their collaboration in carrying out this work.

References
[1] Raithby GD and Chui EH, 1990 *J. Heat Transfer*, **112** 415–423
[2] Raithby GD and Chui EH, 1993 *Num. Heat Transfer*, B 23 269–288
[3] Raithby GD, 1999 *Numer Heat Transfer*, B 35 389–405
[4] Jendoubi S and Lee HS, 1993 *J. Thermophys and. Heat transfer* 7 213-219
[5] Murthy JY and Mathur SR. 1998 *J. Thermophys and. Heat transfer* 12 313–321
[6] Borges De Miranda A and Sacadura JF, 1996 *J. Heat Transfer* **118** 650–653
[7] Cheong KB and Song TH, 1995 *Num. Heat Transfer* **B 27** 155–173
[8] Jamaluddin AS and Smith P J, 1988 *Combust. Sci. Technol.* **59** 321
[9] Lathrop KD, 1969 J. Comput. Phys. 4 475
[10] Joseph D, El Hafi M, Fournier R and Cuenot B , 2005 *Int. J. of Thermal Sciences* **44** 851–864
[11] Capdevila R , Pérez-Segarra CD and Oliva A, 2010, JQSRT, **111 (2)** 264–273
[12] Chiao MC and Churchill S W , 1955 *J. Opt. Soc. Am.* **45**, 958-961
[13] Chai JC and Patankar SV, 2000 *Advances in Numer. Heat Transfer* **vol 4** ed Taylor and Francis
[14] Truelove JS, 1988, *J. Quant. Spec. and Rad. Heat Transfer* **39** 27-31
[15] Grissa H, Askri F, Ben Salah M and Ben Nasrallah S, 2007 *JQSRT* **105(3)** 388–404
[16] Colomer G, Costa M, Consul R and Oliva A, 2004 *Int. J. of Heat and Mass Transfer* **47** 257-269
[17] Borjini MN, Aissia HB, Halouani K, and Zeghmati B, *2008 Int. J. Heat Fluid Flow*, **29** 107-118
[18] Kumar P and Eswaran V, 2010 *J. Heat Transfer*, **132** 023501-1-023501-13
[19] Malalasekera WMG and James EH, 1996 *J. Heat transfer* **118** 225-227
[20] Grissa H, Askri F, Ben Salah M and Ben Nasrallah S, 2010 *JQSRT* **111(1)** 144–154
[21] Wang CA,Sadat H, Ledez V and Lemonnier D, 2010 *IJTS* **49(12)** 2282–2288
[22] Mengüç M and Viskanta R, 1985, *J. Quant. Spec. and Rad. Heat Transfer* **33** 533-549
[23] Grissa H, Askri F, Ben Salah M and Ben Nasrallah S, 2011 *JQSRT* **112** 2661-2675
[24] Park HM and Ahluwalia RK and Im KH, 1993 *Int. J. of Heat and Mass Transfer* ***36-5***, 1181-1189
[25] Tan Z and Howell JR, 1991 *Int. J. of Heat and Mass Transfer* **34 (3)** 785-793
[26] Mondal B and Mishra SC, 2009 *Num. Heat Transfer*, **Part A 55** 18–41
[27] Kim SH and Huh KY, 2000, *Int. J. of Heat and Mass Transfer* ***43***, 1233-1242
[28] Kim MY, Baek SW and Park JH, 2001 *Num. Heat Transfer*, **Part B 39** 617–635

Journal of Physics: Conference Series **369** (2012) 012021 doi:10.1088/1742-6596/369/1/012021

Efficiency of algorithm for solution of vector radiative transfer equation in turbid medium slab

V.P.Budak[1,3], **D.S.Efremenko**[2], **O.V.Shagalov**[1]

[1]National research university "MPEI", Krasnokazarmennaya, 14, Moscow, 111250, Russia
[2]Remote Sensing Technology Institute, German Aerospace Centre, Oberpfaffenhofen, Wessling, Germany

BudakVP@mpei.ru

Abstract. The numerical solution of the vectorial radiative transfer equation (VRTE) is possible only by its discretization, which requires elimination of the solution anisotropic part including all the singularities. Discretized VRTE for the turbid medium slab has the unique analytical solution in the matrix form. Modern packages of matrix (linear) algebra allow only one possible algorithm of VRTE solution by computer. Various realizations of such an algorithm differ by the method of the elimination of the solution anisotropic part. Methods of the solution anisotropic part elimination are analysed in the paper. The codes created by the authors of these methods are analysed in simple situations in order to define its influence on the code efficiency. It is shown that the most effective method is based on the small angle modification of the spherical harmonics method (MSH). The code based on MSH is investigated in details by the influence of different properties of hard and software.

1. Boundary value problem

The comparison of different solution codes of scalar and vector radiative transfer equations (RTE, VRTE) has shown that results of calculations are almost the same [1]. This suggests that different computational methods of polarization fields in a turbid medium slab are in fact variants of the uniform VRTE solution method [2, 3]. In this paper we present VRTE general solution, analysis and optimization of the solution algorithm, and its effective computer implementation on the basis of the general expression.

We have VRTE boundary problem for the turbid medium slab of optical depth τ_0, on which a plane wave of the arbitrary polarized light incidents in the direction $\hat{\mathbf{l}}_0$ [3]:

$$\begin{cases} \mu \dfrac{\partial}{\partial \tau} \vec{\mathrm{L}}(\tau, \hat{\mathbf{l}}) + \vec{\mathrm{L}}(\tau, \hat{\mathbf{l}}) = \dfrac{\Lambda}{4\pi} \oint \vec{\mathrm{R}}(\chi) \vec{x}(\hat{\mathbf{l}}, \hat{\mathbf{l}}') \vec{\mathrm{R}}(\chi') \vec{\mathrm{L}}(\tau, \hat{\mathbf{l}}') d\hat{\mathbf{l}}', \\ \vec{\mathrm{L}}(\tau, \hat{\mathbf{l}}) \Big|_{\tau=0,\, (\hat{\mathbf{z}}, \hat{\mathbf{l}})>0} = \vec{\mathrm{L}}_0 \delta(\hat{\mathbf{l}} - \hat{\mathbf{l}}_0), \quad \vec{\mathrm{L}}(\tau, \hat{\mathbf{l}}) \Big|_{\tau=\tau_0,\, (\hat{\mathbf{z}}, \hat{\mathbf{l}})<0} = 0; \end{cases} \tag{1}$$

[3] To whom any correspondence should be addressed.

where $\vec{L}(\tau,\hat{\mathbf{l}})$ is the vector of Stokes parameters of the light field in the medium at the optical depth τ in the direction $\hat{\mathbf{l}}$. We use Cartesian coordinate system *OXYZ* where axis *OZ* is directed down perpendicularly to the slab border: $\hat{\mathbf{l}}=\left\{\sqrt{1-\mu^2}\cos\varphi,\ \sqrt{1-\mu^2}\sin\varphi,\ \mu\right\}$, $\hat{\mathbf{l}}_0=\left\{\sqrt{1-\mu_0^2},\ 0,\ \mu_0\right\}$, $\mu=(\hat{\mathbf{l}},\hat{\mathbf{z}})$, $\mu_0=(\hat{\mathbf{l}}_0,\hat{\mathbf{z}})$, $\hat{\mathbf{z}}$ is the unit vector along *OZ*. $\ddot{x}(\hat{\mathbf{l}},\hat{\mathbf{l}}')$ is the phase scattering matrix, Λ is the single scattering albedo. $\ddot{\mathrm{R}}(\chi)$ is the change matrix (rotator) of Stokes parameters at the reference plane rotation. χ is the dihedral angle between two planes $(\hat{\mathbf{z}}\times\hat{\mathbf{l}})$ and $(\hat{\mathbf{l}}\times\hat{\mathbf{l}}')$, and χ' is the angle between planes $(\hat{\mathbf{l}}\times\hat{\mathbf{l}}')$ and $(\hat{\mathbf{l}}'\times\hat{\mathbf{z}})$. The unit vectors have a symbol "^", the columns have the right arrow, the rows have the left arrow, the matrices have the double arrow above the symbol.

For a numerical VRTE solution the integrals should be replaced with the finite sums that could be done by one of two methods: the method of spherical harmonics (SH) and the discrete ordinates method (DOM). The DOM is the best method for implementation since VRTE gains a clear ray interpretation that essentially simplifies the definition of the composite boundary conditions. In this case VRTE and boundary conditions take a form of interacting streams of radiation for the fixed directions in space.

However in the initial boundary value problem (1) the replacement of the integral by the Gaussian quadrature is impossible due to the solution singularity. The singularities are an inherent feature of the ray approximation. Really, any break in the boundary conditions propagates inside the medium slab in the form of the edge between light and shadow that gives the singularity in the radiance angular distribution. Therefore, it is necessary to search the solution as the generalized function. To consider the solution singularities, the desired vector of Stokes parameters is represented as the sum of two parts: the anisotropic one that contains all the solution singularities, and the regular one – the smooth function of angular variables:

$$\vec{L}(\tau,\hat{\mathbf{l}})=\vec{L}_a(\tau,\hat{\mathbf{l}})+\vec{L}_r(\tau,\hat{\mathbf{l}}). \tag{2}$$

It is assumed that the anisotropic part can be found analytically. This allows expressing its integral also analytically – a singularity elimination method. The substitution of (2) in (1) changes the boundary value problem to the equivalent one

$$\begin{cases}\mu\dfrac{\partial\vec{L}_r(\tau,\hat{\mathbf{l}})}{\partial\tau}+\vec{L}_r(\tau,\hat{\mathbf{l}})=\dfrac{\Lambda}{4\pi}\oint\ddot{\mathrm{R}}(\chi)\ddot{x}(\hat{\mathbf{l}},\hat{\mathbf{l}}')\ddot{\mathrm{R}}(\chi')\vec{L}_r(\tau,\hat{\mathbf{l}}')\,d\hat{\mathbf{l}}'+\vec{\Delta}(\tau,\hat{\mathbf{l}});\\ \vec{L}_r(\tau,\hat{\mathbf{l}})\Big|_{\tau=0,\,\mu>0}=\vec{0};\ \vec{L}_r(\tau,\hat{\mathbf{l}})\Big|_{\tau=\tau_0,\,\mu<0}=\vec{0},\end{cases} \tag{3}$$

with the source function on the right side of the equation

$$\vec{\Delta}(\tau,\hat{\mathbf{l}})=\frac{\Lambda}{4\pi}\oint\ddot{\mathrm{R}}(\chi)\ddot{x}(\hat{\mathbf{l}},\hat{\mathbf{l}}')\ddot{\mathrm{R}}(\chi')\vec{L}_a(\tau,\hat{\mathbf{l}}')\,d\hat{\mathbf{l}}'-\mu\frac{\partial\vec{L}_a(\tau,\hat{\mathbf{l}})}{\partial\tau}-\vec{L}_a(\tau,\hat{\mathbf{l}}). \tag{4}$$

2. Discretization of VRTE

The effectiveness of DOM application in the case of VRTE is strongly reduced since in the scalar type it is determined by the possibility of the phase function presented by the series of the surface harmonics. This allows reducing the double integral on the basis of the addition theorem to the single one [4]. In the case of polarization the scattering matrix is surrounded by the rotator matrices $\ddot{\mathrm{R}}$ that disturbs the azimuthal symmetry and does not allow using the addition theorem for the surface harmonics. Kuščer-Ribarič [5] uses the circular basis to determine polarization

$$\vec{L}_{CP}=\begin{bmatrix}L_{+2}\\L_{+0}\\L_{-0}\\L_{-2}\end{bmatrix}=\frac{1}{2}\begin{bmatrix}Q-iU\\I-V\\I+V\\Q+iU\end{bmatrix}=\frac{1}{2}\begin{bmatrix}0&1&-i&0\\1&0&0&-1\\1&0&0&1\\0&1&i&0\end{bmatrix}\begin{bmatrix}I\\Q\\U\\V\end{bmatrix}=\ddot{\mathrm{T}}_{CS}\,\vec{L}_{SP},\ \ddot{\mathrm{T}}_{SC}\equiv\ddot{\mathrm{T}}_{CS}^{-1}. \tag{5}$$

It changes the rotator form and allows using generalizes surface functions $\mathrm{P}^k_{mn}(\cos\theta)$ for the scattering matrix representation that obey to the special form of the addition theorem

$$e^{-im\chi}\mathrm{P}^k_{mn}(\hat{\mathbf{l}}\cdot\hat{\mathbf{l}}')e^{-in\chi'}=\sum_{q=-k}^{k}(-1)^q\,\mathrm{P}^k_{mq}(\hat{\mathbf{l}}\cdot\hat{\mathbf{z}})\mathrm{P}^k_{qn}(\hat{\mathbf{z}}\cdot\hat{\mathbf{l}}')e^{iq\Delta\varphi},\quad \Delta\varphi=\varphi-\varphi'. \tag{6}$$

As the results, all the coefficients in VRTE become complex numbers, and this makes use of the effective numerical methods for equation system solution difficult. Therefore, after applying the addition theorem for the generalized spherical function one should return to the Stokes presentation (SP).

Let's consider the local transformation matrix of the ray at scattering in details:

$$\ddot{\mathrm{S}}(\hat{\mathbf{l}}',\hat{\mathbf{l}})\equiv\ddot{\mathrm{R}}(\chi)\ddot{x}(\hat{\mathbf{l}}',\hat{\mathbf{l}})\ddot{\mathrm{R}}(\chi')=\ddot{\mathrm{T}}_{SC}\,\ddot{\mathrm{R}}_{CP}(\chi)\ddot{x}_{CP}(\hat{\mathbf{l}},\hat{\mathbf{l}}')\ddot{\mathrm{R}}_{CP}(\chi')\ddot{\mathrm{T}}_{CS}\equiv\ddot{\mathrm{T}}_{SC}\,\ddot{\mathrm{S}}_{CP}(\hat{\mathbf{l}},\hat{\mathbf{l}}')\ddot{\mathrm{T}}_{CS}, \tag{7}$$

where $\ddot{x}_{CP}(\hat{\mathbf{l}},\hat{\mathbf{l}}')\equiv\ddot{\mathrm{T}}_{CS}\,\ddot{x}_{SP}(\hat{\mathbf{l}},\hat{\mathbf{l}}')\ddot{\mathrm{T}}_{SC}$ is the scattering matrix, $\ddot{\mathrm{R}}_{CP}(\chi)=\ddot{\mathrm{T}}_{CS}\,\ddot{\mathrm{R}}(\chi)\ddot{\mathrm{T}}_{SC}$ is the rotator in the circular polarization (CP) presentation.

Then we present the elements of scattering matrix in the form of the expansion on the generalized spherical functions

$$\left[\ddot{x}_{CP}(\cos\gamma)\right]_{rs}=\sum_{k=\max(r,s)}^{K}(2k+1)x^k_{rs}\,\mathrm{P}^k_{r,s}(\cos\gamma), \tag{8}$$

where r and s have only values ±0 and ±2; K is the number of harmonics in expansion of the scattering matrix in the generalized spherical functions.

Taking into account the addition theorem (6), one can get every element of the local transformation matrix in CP as

$$\left[\ddot{\mathrm{S}}_{CP}(\hat{\mathbf{l}},\hat{\mathbf{l}}')\right]_{r,s}=\left[\sum_{k=0}^{K}(2k+1)\sum_{q=-k}^{k}(-1)^q\,e^{iq(\varphi-\varphi')}\,x^k_{rs}\,\mathrm{P}^k_{rq}(\mu)\mathrm{P}^k_{qs}(\mu')\right]. \tag{9}$$

With the use of diagonal matrix $\ddot{\mathrm{Y}}^q_l(\mu)=\mathrm{Diag}\left(\mathrm{P}^l_{-2,q}(\mu),\mathrm{P}^l_{-0,q}(\mu),\mathrm{P}^l_{+0,q}(\mu),\mathrm{P}^l_{+2,q}(\mu)\right)$ and taking into account the reciprocal relation $\mathrm{P}^k_{m,n}(\mu)=\mathrm{P}^k_{n,m}(\mu)$ one can get the expression (7) as

$$\ddot{\mathrm{S}}(\hat{\mathbf{l}},\hat{\mathbf{l}}')=\sum_{k=0}^{K}(2k+1)\sum_{q=-k}^{k}(-1)^q\,e^{iq(\varphi-\varphi')}\ddot{\mathrm{T}}_{SC}\,\ddot{\mathrm{Y}}^q_k(\mu)\ddot{x}_k\,\ddot{\mathrm{Y}}^q_k(\mu')\ddot{\mathrm{T}}_{CS}\equiv\sum_{k=0}^{K}(2k+1)\sum_{q=-k}^{k}e^{iq(\varphi-\varphi')}\,\ddot{\mathrm{S}}^q_k(\mu,\mu'), \tag{10}$$

where $\ddot{x}_k\equiv\left[x^k_{rs}\right]$.

Then

$$\ddot{\mathrm{S}}^q_k(\mu,\mu')=(-1)^q\,\ddot{\mathrm{T}}_{SC}\,\ddot{\mathrm{Y}}^q_k(\mu)\ddot{x}_k\,\ddot{\mathrm{Y}}^q_k(\mu')\ddot{\mathrm{T}}_{CS}=\ddot{\mathrm{P}}^k_n(\mu)\ddot{\chi}_k\,\ddot{\mathrm{P}}^k_n(\mu'), \tag{11}$$

where $\ddot{\chi}_k=\ddot{\mathrm{T}}_{SC}\,\ddot{x}_k\,\ddot{\mathrm{T}}_{CS}$ is the so-called «Greek matrix»,

$$\ddot{\mathrm{P}}^l_n(\mu)=(-i)^n\,\ddot{\mathrm{T}}_{SC}\,\ddot{\mathrm{Y}}^n_l(\mu)\ddot{\mathrm{T}}_{CS}\equiv\ddot{\mathrm{P}}_R+i\,\ddot{\mathrm{P}}_I; \tag{12}$$

$\ddot{\mathrm{P}}_R$ and $\ddot{\mathrm{P}}_I$ are the real parts of matrix $\ddot{\mathrm{P}}^l_n(\mu)$ [6].

Finally, VRTE (1) could be written as

$$\mu\frac{\partial}{\partial\tau}\vec{\mathrm{L}}_r(\tau,\hat{\mathbf{l}})+\vec{\mathrm{L}}_r(\tau,\hat{\mathbf{l}})=\frac{\Lambda}{4\pi}\sum_{k=0}^{K}(2k+1)\sum_{q=-k}^{k}\ddot{\mathrm{P}}^k_n(\mu)\oint e^{iq\Delta\varphi}\,\ddot{\chi}_k\,\ddot{\mathrm{P}}^k_n(\mu')\vec{\mathrm{L}}_r(\tau,\hat{\mathbf{l}}')d\hat{\mathbf{l}}'+\vec{\Delta}(\tau,\hat{\mathbf{l}}). \tag{13}$$

The scattering integral contains complex values though the whole expression (13) is real. With the symmetry of the introduced functions $\ddot{\mathrm{P}}^n_R(\mu)=\ddot{\mathrm{P}}^{-n}_R(\mu)$, $\ddot{\mathrm{P}}^n_I(\mu)=-\ddot{\mathrm{P}}^{-n}_I(\mu)$ one can eliminate the imaginary part [6] in the equation (13). The obtained expression is correct for all scattering matrices. However, in case of the aerosol block-diagonal matrix it could be simplified [6] more. One could present the solution of problem (3) in the form [6]:

$$\mathrm{L}_r(\tau,\hat{\mathbf{l}})=\sum_{m=0}^{M}(2-\delta_{0,m})\left[\phi_1(m\varphi)\vec{\mathrm{L}}^m_1(\tau,\mu)+\phi_2(m\varphi)\vec{\mathrm{L}}^m_2(\tau,\mu)\right], \tag{14}$$

that reduces VRTE to the following form

$$\mu\frac{\partial \vec{L}_c^m(\tau,\mu)}{\partial\tau}+\vec{L}_c^m(\tau,\mu)=\frac{\Lambda}{2}\sum_{k=0}^{K}(2k+1)\vec{\Pi}_k^m(\mu)\vec{\chi}_k\int_{-1}^{1}\vec{\Pi}_k^m(\mu')\vec{L}_c^m(\tau,\mu')d\mu'+\vec{\Delta}(\tau,\mu),\ c=1,2,\tag{15}$$

where M is the term number in the Fourier series on the azimuth, $m\in\overline{0,M}$,

$$\vec{\Pi}_m^k(\mu)=\begin{bmatrix}Q_k^m(\mu)&0&0&0\\0&R_k^m(\mu)&-T_k^m(\mu)&0\\0&-T_k^m(\mu)&R_k^m(\mu)&0\\0&0&0&Q_k^m(\mu)\end{bmatrix},\quad \begin{matrix}\phi_1(\varphi)=\mathrm{diag}(\cos\varphi,\cos\varphi,\sin\varphi,\sin\varphi),\\ \phi_2(\varphi)=\mathrm{diag}(-\sin\varphi,-\sin\varphi,\cos\varphi,\cos\varphi),\end{matrix}\tag{16}$$

$$R_l^n(\mu)=0.5i^m\left(P_{m,2}^k(\mu)+P_{m,-2}^k(\mu)\right),\ T_l^n(\mu)=0.5i^m\left(P_{m,2}^k(\mu)-P_{m,-2}^k(\mu)\right),\ Q_l^n(\mu)=\sqrt{\frac{(k-m)!}{(k+m)!}}P_k^m(\mu).\tag{17}$$

In order to get the solution of the obtained equation using DOM, we present the integrals entering into the equation (15) taking into account the slab symmetry in the double Gaussian quadrature form. Accordingly we have the system of ordinary differential equations for the fixed m and c (therefore we omit them in the equation)

$$\mu_i^{\pm}\frac{d}{d\tau}\vec{L}_i^{\pm}(\tau)+\vec{L}_i^{\pm}(\tau)=\frac{\Lambda}{4}\sum_{j=1}^{N/2}w_j\sum_{k=0}^{K}(2k+1)\vec{\Pi}_k^m(\mu_i^{+})\vec{\chi}_k\left(\vec{\Pi}_k^m(\mu_j^{+})\vec{L}_i^{+}(\tau)+\vec{\Pi}_k^m(\mu_j^{-})\vec{L}_i^{-}(\tau)\right)+\vec{\Delta}_i^{\pm}(\tau),\tag{18}$$

where $\vec{L}_i^{\pm}(\tau)\equiv\vec{L}_c^m(\tau,\mu_i^{\pm})$, $\vec{\Delta}_i^{\pm}(\tau)\equiv\vec{\Delta}_c^m(\tau,\mu_i^{\pm})$, $\mu_j^{\pm}=0.5(\zeta_j\pm1)$, ζ_j, w_j are the zeroes and weights of the Gaussian quadrature of the $N/2$ order.

Now we can introduce the following matrices and vectors

$$\vec{L}\equiv\left[\vec{L}_+(\tau)\quad \vec{L}_-(\tau)\right]^T,\quad \vec{\Delta}\equiv\left[\vec{\Delta}_+(\tau)\quad \vec{\Delta}_-(\tau)\right]^T,\quad \vec{M}\equiv\mathrm{diag}\left(\mu_i^{+},-\mu_i^{+}\right),\quad \vec{W}\equiv\frac{\Lambda}{4}\mathrm{diag}\left(w_i,w_i\right),\tag{19}$$

$$\vec{A}\equiv\left[\sum_{k=0}^{K}(2k+1)\vec{\Pi}_k^m(\mu_i^{+})\vec{\chi}_k\vec{\Pi}_k^m(\mu_j^{\pm})\right],\tag{20}$$

where $\vec{L}_{\pm}(\tau)=\left[I(\tau,\mu_1^{\pm}),Q(\tau,\mu_1^{\pm}),U(\tau,\mu_1^{\pm}),V(\tau,\mu_1^{\pm}),\ \ldots,I(\tau,\mu_{N/2}^{\pm}),Q(\tau,\mu_{N/2}^{\pm}),U(\tau,\mu_{N/2}^{\pm}),V(\tau,\mu_{N/2}^{\pm})\right]^T$.
Other arrays are arranged in the same way.

Since the regular part is a smooth function of angle, all the matrices are finite. This allows rewriting the system (18) in the matrix form

$$\frac{d\vec{L}(\tau)}{d\tau}=-\vec{B}\vec{L}(\tau)+\vec{M}^{-1}\vec{\Delta}(\tau),\quad \vec{B}\equiv\vec{M}^{-1}(\vec{I}-\vec{A}\vec{W}).\tag{21}$$

3. Solution regular part

The solution of the obtained equation system has the analytic view [4, 7] in the matrix form

$$-\vec{L}(0)+e^{\vec{B}\tau_0}\vec{L}(\tau_0)=\int_0^{\tau_0}e^{\vec{B}\tau}\vec{M}^{-1}\vec{\Delta}(\tau,\mu_0)d\tau.\tag{22}$$

This expression (22) is equivalent to the solution representation that is the sum of the general solution of homogeneous equation and the particular solution of the inhomogeneous one [8]. A matrix representation is more convenient for the analytic transformation and implementation of solutions for the computer. The solution of the homogeneous equation

$$\vec{L}(\tau_0)=e^{-\vec{B}\tau_0}\vec{L}(0)\equiv\vec{P}(\tau_0)\vec{L}(0):\quad \frac{d\vec{P}(\tau)}{d\tau}=-\vec{B}\vec{P}(\tau),\tag{23}$$

connects the radiation at the slab lower boundary with an expression on the top and is called the pro-propagator [7].

The problem of the solution (22) is connected with negative and positive exponents in the expression that leads to fast worsening of the system matrix conditions. In order to eliminate this effect, we used the scale transformation [4]. The matrix exponent is represented in the form [4]

$$e^{\vec{B}\tau_0} = \vec{U}e^{\vec{\Gamma}\tau_0}\vec{U}^{-1}, \tag{24}$$

where $\vec{U}$ is the matrix of eigenvectors of the matrix $\vec{B}$ and $\vec{\Gamma} = \mathrm{diag}(\vec{\Gamma}_-, \vec{\Gamma}_+)$ is the diagonal matrix of eigenvalues sorted by ascending ordering so that $\vec{\Gamma}_- = -\vec{\Gamma}_+$.

Consequently, the equation (22) can be rewritten as

$$-\begin{bmatrix} \vec{u}_{11} & \vec{u}_{12} \\ e^{-\vec{\Gamma}_+\tau_0}\vec{u}_{21} & e^{-\vec{\Gamma}_+\tau_0}\vec{u}_{22} \end{bmatrix}\begin{bmatrix} \vec{L}_+(0) \\ \vec{L}_-(0) \end{bmatrix} + \begin{bmatrix} e^{\vec{\Gamma}_-\tau_0}\vec{u}_{11} & e^{\vec{\Gamma}_-\tau_0}\vec{u}_{12} \\ \vec{u}_{21} & \vec{u}_{22} \end{bmatrix}\begin{bmatrix} \vec{L}_+(\tau_0) \\ \vec{L}_-(\tau_0) \end{bmatrix} = \begin{bmatrix} \vec{J}_- \\ \vec{J}_+ \end{bmatrix}, \tag{25}$$

where $\vec{J} \equiv \begin{bmatrix} \vec{J}_- \\ \vec{J}_+ \end{bmatrix} = \vec{S}\int_0^{\tau_0} e^{\vec{\Gamma}\tau}\vec{U}^{-1}\vec{M}^{-1}\vec{\Delta}(\tau,\mu_0)d\tau$, $\vec{U}^{-1} \equiv \begin{bmatrix} \vec{u}_{11} & \vec{u}_{12} \\ \vec{u}_{21} & \vec{u}_{22} \end{bmatrix}$.

Note that expression (25) contains the exponent only with the positive power. Expressing in equation (25) the emerging radiation $\vec{L}_+(0)$, $\vec{L}_-(\tau_0)$ from a layer through the incident radiation $\vec{L}_-(0)$, $\vec{L}_+(\tau_0)$ we find a smooth regular part of the solution in the form of the so-called scatterers [7]:

$$\begin{bmatrix} \vec{L}_-(0) \\ \vec{L}_+(\tau_0) \end{bmatrix} = \begin{bmatrix} \vec{F}_- \\ \vec{F}_+ \end{bmatrix} + \begin{bmatrix} \vec{R}_- & \vec{T}_- \\ \vec{T}_+ & \vec{R}_+ \end{bmatrix}\begin{bmatrix} \vec{L}_+(0) \\ \vec{L}_-(\tau_0) \end{bmatrix}, \tag{26}$$

where $\begin{bmatrix} \vec{F}_- \\ \vec{F}_+ \end{bmatrix} = \vec{H}\begin{bmatrix} \vec{J}_- \\ \vec{J}_+ \end{bmatrix}$, $\begin{bmatrix} \vec{R}_- & \vec{T}_- \\ \vec{T}_+ & \vec{R}_+ \end{bmatrix} = \vec{H}\begin{bmatrix} \vec{u}_{11} & -e^{\vec{\Gamma}_-\tau_0}\vec{u}_{12} \\ e^{-\vec{\Gamma}_+\tau_0}\vec{u}_{21} & -\vec{u}_{22} \end{bmatrix}$, $\vec{H} \equiv \begin{bmatrix} -\vec{u}_{12} & e^{\vec{\Gamma}_-\tau_0}\vec{u}_{11} \\ -e^{-\vec{\Gamma}_+\tau_0}\vec{u}_{22} & \vec{u}_{21} \end{bmatrix}^{-1}$.

The obtained system (26) is the desired solution that makes definition of the radiance distribution of the reflected and transmitted radiations possible. Note that the expression (26) is the rigorous analytical solution of the boundary value problem for VRTE discretized by DOM. The transfer from VRTE (1) to the equation system (26) is possible due to the singularity elimination (2) and to the change of the scattering integral (discretization) by the Gaussian quadrature (18).

The solution obtained in the form of scatterers (26) has a functional type and allows entering a photometric concept of the layer radiance factor by the reflection or transmission. This immediately gives the possibility to formulate a matrix-operator method to replace the two adjacent layers by the equivalent one described by the expression identical to (32), but with effective parameters [3]. Consequently, the solution in the form of scatterers possesses the property of invariance.

Equation (26) can be reorganized in a form similar to the propagator (23):

$$\vec{L}(\tau_0) = \begin{bmatrix} \vec{T}_+ - \vec{R}_-\vec{T}_-^{-1}\vec{R}_+ & \vec{R}_-\vec{T}_-^{-1} \\ -\vec{T}_-^{-1}\vec{R}_+ & \vec{T}_-^{-1} \end{bmatrix}\vec{L}(0) + \vec{F}. \tag{27}$$

This kind of transformation has been called the "star product" [7]. Since one knows the differential equation (23) for the propagator, one is able to get the one-point problem with initial conditions of the matrix Riccati equation for the matrix elements of the scatterers [7], e.g. the reflection:

$$\frac{d\vec{R}_-}{d\tau} = \vec{b}_2 + \vec{b}_1\vec{R}_- + \vec{R}_-\vec{b}_1 + \vec{R}_-\vec{b}_2\vec{R}_-;\quad \frac{d\vec{R}_+}{d\tau} = -\vec{b}_2 + \vec{b}_1\vec{R}_+ - \vec{R}_+\vec{b}_1 - \vec{R}_+\vec{b}_2\vec{R}_+;\quad \vec{B} \equiv \begin{bmatrix} \vec{b}_1 & \vec{b}_2 \\ -\vec{b}_2 & -\vec{b}_1 \end{bmatrix}, \tag{28}$$

and similarly for the transmission $\vec{T}_\pm$.

It is easy to see that (28) is an equivalent to the equations of Ambartsumyan-Chandrasekhar [8], derived from the principle of invariant embedding. Thus, after eliminating the solution anisotropic part and the equation discretization we reach a unique analytical solution in the matrix form (26). In this sense the method of successive orders of scattering is an iterative method for finding the inverse matrix $\vec{A}$ in (26), and the Monte Carlo is a stochastic method.

Implementing the suggested solution (26) as the algorithm includes the calculation of the following components: zeroes and weights of the quadrature formula for the VRTE discretization (18), source function (4), eigenvectors and values of the system matrix (24), and products of matrices in (26).

The practical implementation of the indicated algorithm depends mainly on the size of matrices in (26). Note that (26) is obtained for the regular part, therefore, its size is determined mostly not by the medium optical parameters and the boundary conditions, but by the elimination of the solution anisotropic part. In general, $M \approx N \approx K$. However, in case of correct elimination of the anisotropic part solution, i.e. the smooth regular part is near to the isotropic angular distribution, it is possible that $M \ll N \ll K$. This improves considerably the algorithm performance.

4. Solution anisotropic part elimination

Now let's analyse and compare all the main known methods of the anisotropic part elimination from the point of view of matrix size M and N reduction and the enhancement of calculation efficiency of the solution (26).

4.1. Elimination of direct non-scattering radiation

For the first time Eddington proposed the elimination of the anisotropic part by subtracting the direct non-scattered radiation on the basis of the Bouguer law, Milne developed the theory and it was fully crystallized in the works of Chandrasekhar [8]. Thus, the expression for the direct non-scattered radiation in the plane of reference coincided with the plane of incidence is given by:

$$\vec{L}_a(\tau,\mu,\varphi) \equiv \vec{L}_0(\tau,\mu,\varphi) = e^{-\tau/\mu_0}\,\ddot{R}(\varphi)\vec{L}^0\,\delta(\hat{\mathbf{l}}-\hat{\mathbf{l}}_0),\ \ \vec{L}^0 = E_0\begin{bmatrix}1 & p\cos\varphi_0 & p\sin\varphi_0 & q\end{bmatrix}^T, \tag{29}$$

where $\vec{L}^0$ is the vector of Stokes parameters of incident radiation, E_0 is the normal to beam irradiance, p, q are degrees of linear and circular polarization of the beam, and φ_0 is the azimuth of polarization.

In this case the source function in the equation for the regular part takes the form

$$\vec{\Delta}(\tau,\hat{\mathbf{l}}) = \frac{\Lambda}{4\pi}e^{-\tau/\mu_0}\,\ddot{R}(\varphi)\ddot{x}(\hat{\mathbf{l}},\hat{\mathbf{l}}_0). \tag{30}$$

Under these conditions the radiation scattered into small angles slightly differs from the direct radiation, and this method becomes inefficient. Strongly anisotropic scattering leads to a significant increase in the number of terms in the expansion of the scattering matrix in a series of generalized spherical functions K, which correspondingly increases the size of matrices N and M in the solution (26).

4.2. Delta-M method

Some works were done to solve the anisotropy problem. Various algorithms based on the truncation of scattering phase function were used. Delta-M method is the most efficient among them; it could be generalized in the vector case [9]. As in the scalar one the scattering phase function is represented as a sum of a delta-function and a smooth part

$$\ddot{x}(\hat{\mathbf{l}},\hat{\mathbf{l}}') = 4\pi f\,\ddot{1}\,\delta(\hat{\mathbf{l}}-\hat{\mathbf{l}}') + (1-f)\ddot{x}^*(\hat{\mathbf{l}},\hat{\mathbf{l}}') \tag{31}$$

Here f is a part of anisotropy scattering; $\ddot{1}$ is an identity matrix.

Equation (31) for Greek matrix (11) can be written down as

$$\ddot{\chi}_k = f\ddot{1} + (1-f)\ddot{\chi}_k^*,\ \ \ddot{\chi}_k^* = (\ddot{\chi}_k - f\ddot{1})/(1-f). \tag{32}$$

This phase matrix transformation leads to the scale transformation of optical depth and single scattering albedo

$$\tau^* = (1-\Lambda f)\tau,\ \ \Lambda^* = \Lambda(1-f)/(1-\Lambda f). \tag{33}$$

Thus, delta-M method could decrease K significantly. However such an approach distorts initial boundary problem and leads to the error in a small sighting angle (intrinsically we do not consider the coarsest fraction of aerosol) or to oscillations in the angular radiance distribution.

4.3. TMS and IMS-methods

To eliminate the problems in VRTE solution mentioned above, Nakajima and Tanaka [10] returned to the idea of determination of the anisotropic part in the solution on the base of approximate analytical presentation of angular distribution of Stokes parameters for the first and second order of scattering:

$$\vec{L}(\tau,\hat{\mathbf{l}}) = \underbrace{\vec{L}_0(\tau,\hat{\mathbf{l}}) + \vec{L}_1(\tau,\hat{\mathbf{l}}) + \vec{L}_2(\tau,\hat{\mathbf{l}})}_{\vec{L}_a(\tau,\hat{\mathbf{l}})} + \vec{L}_r(\tau,\hat{\mathbf{l}}) . \quad (34)$$

Here the first two orders satisfy the equations:

$$\mu\frac{\partial\vec{L}_1(\tau,\hat{\mathbf{l}})}{\partial\tau} + \vec{L}_1(\tau,\hat{\mathbf{l}}) = \frac{\Lambda}{4\pi}e^{-\tau/\mu_0}\vec{R}(\varphi)\vec{L}^0\,\vec{x}(\hat{\mathbf{l}}_0,\hat{\mathbf{l}}), \quad (35)$$

$$\mu\frac{\partial\vec{L}_2(\tau,\hat{\mathbf{l}})}{\partial\tau} + \vec{L}_2(\tau,\hat{\mathbf{l}}) = \frac{\Lambda}{4\pi}\oint\vec{R}(\chi)\vec{x}(\hat{\mathbf{l}},\hat{\mathbf{l}}')\vec{R}(\chi')\vec{L}_1(\tau,\hat{\mathbf{l}}')d\hat{l}' . \quad (36)$$

Equations (35) and (36) are solved analytically. However the solution of equation (36) is a triple integral, and the run-time of its calculations can exceed the run-time of the initial problem solution (26). For this reason an approximation describing Stokes parameters distribution near the incident direction is generally used. The most efficient way is to use the small angle approximation, where the dispersion of scattered and non-scattered rays is not taken into account. This approximation is equivalent to the change of μ in (35), (36) by μ_0. The most known codes using both delta-M and TMS methods are DISORT [11] in the scalar case and Pstar [12] in the vector one.

4.4. Small angle modification of spherical harmonic method

It could be shown that the natural evolution of TMS-method is the inclusion in $\vec{L}_a(\tau,\hat{\mathbf{l}})$ of all orders of scattering at small angle approximation [6]. This approach is described in [6] on the basis of small angle modification of spherical harmonics method (MSH).

In this case the anisotropic part of angular distribution of Stokes parameters in CP-representation is expanded with spherical functions with reference to the incident direction $\hat{\mathbf{l}}_0$:

$$\vec{L}_a(\tau,\hat{\mathbf{l}}) = \sum_{m=-\infty}^{\infty}\sum_{k=0}^{\infty}\frac{2k+1}{4\pi}\vec{Y}_m^k(\nu)\vec{f}_m^k(\tau)\exp(im\psi), \quad (37)$$

here $\nu = (\hat{\mathbf{l}},\hat{\mathbf{l}}_0)$, ψ is azimuth $\hat{\mathbf{l}}$ in the system toward $\hat{\mathbf{l}}_0$.

For the strong anisotropic angular distribution its spectrum is a smooth function of harmonic index k. For this reason we can get the differential equation [6] for expansion coefficients

$$\mu_0\frac{\partial\vec{f}^m(\tau,k)}{\partial\tau} + (\vec{1} - \Lambda\vec{x}_k)\vec{f}^m(\tau,k) = \vec{0}, \quad (38)$$

the solution of which is simple

$$\vec{f}^m(\tau,k) = \exp\left(-(\vec{1} - \Lambda\vec{x}_k)\tau/\mu_0\right)\vec{f}^m(0,k). \quad (39)$$

After the conversion in the solution to SP–representation MSH [6] could be presented in the coordinate system about the normal to the layer boundary:

$$\vec{L}_a(\tau,\hat{\mathbf{l}},\hat{\mathbf{l}}_0) = \sum_{c=1,2}\sum_{m=0}^{M}(2-\delta_{m,0})\vec{\phi}_c(m\varphi)\sum_{k=0}^{K}\frac{2k+1}{4\pi}\vec{\Pi}_k^m(\mu)\vec{Z}_k(\tau)\vec{\Pi}_k^m(\mu_0)\vec{D}_c\vec{L}^0, \quad (40)$$

where $\vec{D}_1$=diag(1 1 0 0), $\vec{D}_2$=diag(0 0 1 1), $\vec{Z}_k(\tau) = \exp\{-(\vec{1} - \Lambda\vec{\chi}_k)\tau/\mu_0\}$.

By substituting expression (40) in (4), and after some transformations [6] we get an expression for the source function

$$\mu_0\vec{\Delta}_c^m(\tau,\mu_i) = \sum_{k=m}^{K}(2k+1)(\mu_i - \mu_0)\vec{\Pi}_m^k(\mu_i)\vec{d}_k\vec{Z}_k(\tau)\vec{\Pi}_m^k(\mu_0)\vec{D}_c\vec{L}_0$$
$$+\vec{A}_{K+1}^m\left[\vec{\Pi}_m^K(\mu_i)\vec{d}_{K+1}\vec{Z}_{K+1}(\tau)\vec{\Pi}_m^{K+1}(\mu_0) - \vec{\Pi}_m^{K+1}(\mu_i)\vec{d}_K\vec{Z}_K(\tau)\vec{\Pi}_m^K(\mu_0)\right]\vec{D}_c\vec{L}_0, \quad (41)$$

where $\ddot{\mathrm{d}}_k = \ddot{1} - \Lambda\ddot{\chi}_k$, $\left[\ddot{\mathrm{A}}_m^k\right]_{rs} = \frac{1}{k}\sqrt{(k^2 - m^2)(k^2 - s^2)}\delta_{rs}$.

This approach makes the regular part of the solution almost an isotropic function. So M can be small, and N does not practically depend on K. We implemented this method of the solution anisotropic part elimination as two codes: MDOM in the scalar case and MVDOM in the vector one. Both codes are realized by languages FORTRAN as well as Matlab [13].

5. Numerical implementation

First of all we note that the numerical implementation of the solution (26) is based on solving the basic problems of linear algebra. There are some libraries where the necessary subroutines are collected: LAPACK, IMSL, NAG etc. Math Kernel Library (MKL) is a package that is specially optimized for Intel processors (MKL has versions both for Windows and Linux systems). Great advantages result from the use of the supporting multicore and multiprocessor systems and automatic parallelization. This feature can increase computing speed significantly. It has vector implementations of element operations (e.g. square root from elements of an array) in a parallel mode. Special mathematical program packages like Matlab, Maple, Mathematica use MKL by playing the role of library cover.

A perspective way to accelerate the code is a general-purpose graphics processing units (GPUs). Compute Unified Device Architecture (CUDA) is the computing engine in nVIDIA GPUs. It gives access to the instruction set of computing elements of video card. Standard LAPACK routines can be accelerated by CUDA as well. Instead of reprogramming LAPACK subroutines, it is advisable to apply the special library CULA. Also MATLAB 2010b supports nVIDIA CUDA-capable GPUs, and no knowledge and experience in CUDA is required to use GPU-computing features.

The significant acceleration and saving the memory can be reached by using sparse matrices. The basic idea is the following: while working with matrix that has a lot of zero elements, it is wise to store only non-zero elements and the additional information that can be used to restore the indices of nonzero elements or indices themselves. There are some formats to store the sparse matrices: compressed sparse row (CSR) format, compressed sparse column format (CSC), and Coordinate Format. The choice of format depends on type of operation; for instance, CSC format is convenient when using sparse matrix since its elements are arranged by each column. The use of sparse matrices can be efficient for computing the anisotropic part (49) and dealing with a modified source function.

6. Results and discussions

We analyzed the effect of different hardware-software resources for the efficiency of implementation of solution (26) with the anisotropic part elimination on a basis of MSH by example of codes MDOM and MVDOM. The comparison of run-time for different modes of compiling and computing is represented in Table 1 for two tests (test1: $N = 101$, K=500, M=32; test 2: $N = 101$, K=1000, M=32). The tests were run on system Ubuntu 10.04, Intel Core 2 Duo 3GHz, 2 Gb RAM, Intel Fortran Compiler 11.1 with MKL 10.2. Two compilers gfortran and ifort, optimization and sparse matrix technique were used.

Note that MKL uses all available computing cores of the system (2 cores in this case) and provides the results in half of the time. The sparse matrix technique used for computing the anisotropic part reduces run-time significantly. Due to sparse matrices, the two-dimensional arrays are reduced to one-dimensional array. Thus, the run-time is proportional to K instead of to K^2.

We increased acceleration by about 20% due to the matrix multiplication on nVIDIA GeForce 480 GTX GPU. GPU computations provide advantages only for big size arrays, otherwise CPU computing is more preferable. The profiler tool shows that the calculation of eigenvectors and values take the half of the run-time. Unfortunately, the subroutines for these problems have not been implemented for Matlab GPU tool. The sizes of matrices are significantly reduced due to the solution anisotropic part elimination by MSH. In practice, N is less than 300. However in the case computations of spectra the parallelization of wavelength loop also can be implemented within CUDA technology. It is a possible way to use CUDA advantages, but not for a single wavelength problem.

Table 1. Comparison of calculating time for two tests.

Features	Run-time I, sec	Run-time II, sec
gfortran+LAPACK	240	530
gfortran+LAPACK + optimization	230	505
ifort+LAPACK	210	490
ifort+MKL	115	250
ifort+MKL+optimization	105	230
ifort+MKL+optimization + sparse matrix	33	44
Matlab 2010b	27	45
Matlab 2010b + CUDA	22	33

Now let's consider the effectiveness of different methods of the solution anisotropic part elimination by the numerical comparison of the radiance angular distribution calculation for the same task by different codes: DISORT and MDOM in the scalar case and Pstar and MVDOM in the vector one. It is worth mentioning that the speed of DISORT calculation can be significantly increased by using the sparse matrix algorithm in subroutine ZEROIT, which zeros a given matrix. In this case the speed for quite smooth phase functions (g<0.9) of MDOM and DISORT is of the same order. For the phase function Henyey-Greenstein g=0.98, the MSH method of anisotropic part elimination appears to be better. Figure 1 provides the numerical comparison of the reflectance calculations for the medium scattering anisotropy, and figure 2 for the strong anisotropy.

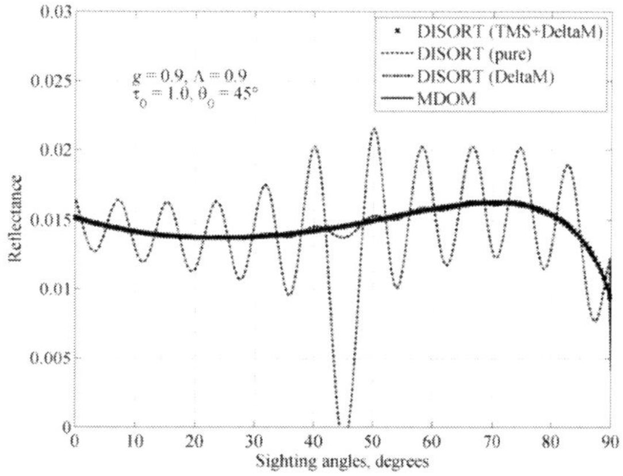

Figure 1. Comparison of MDOM and DISORT in the case of medium scattering anisotropy.

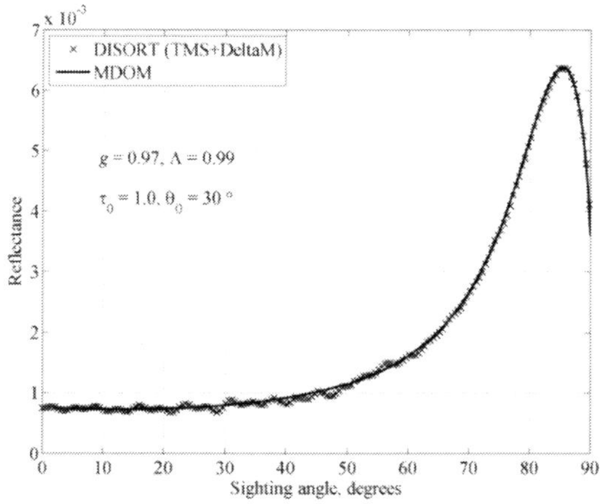

Figure 2. Comparison of MDOM and DISORT in the case of strong scattering anisotropy.

Note that even an average degree of scattering anisotropy mandatory requires the use of TMS method. Isolation of the solution singularities on the basis of MSH operates independently from the degree of scattering anisotropy. The running time in the first case is approximately the same t ~ 0.5 sec, but it differs greatly in the second case: MDOM t ~ 2 seconds, and DISORT t ~ 14 sec.

Quite similar results were obtained in the vector case. We compared Pstar and MVDOM in the cases of the phase matrix for log-normal distribution with parameters r_0=5, s=0.4, Λ=0.99; and geometry of the observation scheme with θ_0=45°, τ_0=1.0, φ=30°. Figure 3 shows a comparison of calculations of the angular distribution of polarization transmitted by slab, and figure 4 provides a comparison of the reflected radiation.

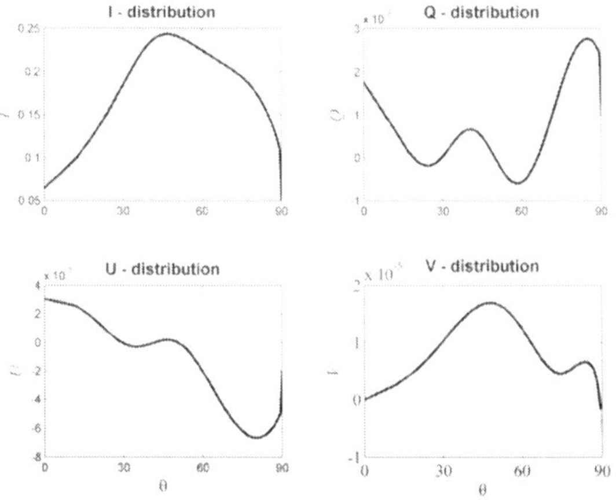

Figure 3. Polarization of the transmitted radiation.

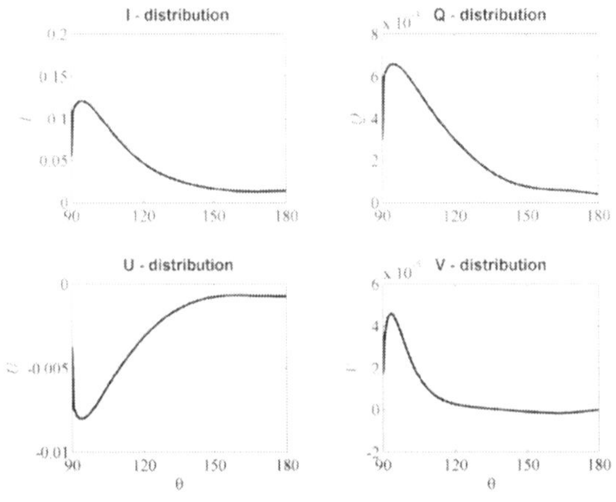

Figure 4. Polarization of the reflected radiation.

For the best fit, we used the parameters: N = 71, K = 171, M = 20 in the code MVDOM and 30 streams in the code Pstar. The comparison of results showed the perfect coincidence at the computer accuracy level, but the computation time was 11 seconds for MVDOM and for the program Pstar more than 180 seconds.

7. Conclusions

The discretized VRTE for a slab has a unique analytic solution in the matrix form. The high level of optimization of linear algebra packages permits a single algorithm implementation on a computer for this solution. Various implementations differ in the method of the solution anisotropic part elimination. Taking into account our algorithm analysis, the analytical anisotropy elimination on the basis of MSH is supposedly a more precise algorithm than TMS. At the algorithm implementation it is necessary to use the algorithm of sparse matrices, and when using the Intel processor the application of MKL library is very essential.

References

[1] Kokhanovsky AA, Budak VP, Cornet C, Duan M, Emde C, Katsev IL, Klyukov DA, Korkin SV, C-Labonnote L, Mayer B, Min Q, Nakajima T, Ota Y, Prikhach AS, Rozanov VV, Yokota T, Zege EP 2010 *J. Quant. Spectrosc. Radiat. Transf.* **111** 1931.
[2] Doicu A and Trautmann T 2009 *J. Quant. Spectrosc. Radiat. Transf.* **110** 159
[3] Budak VP, Klyuykov DA, and Korkin SV 2011 *J. Quant. Spectrosc. Radiat. Transf.* **112** 1141.
[4] Karp AH, Greenstadt J, and Fillmore JA 1980 *J. Quant. Spectrosc. Radiat. Transf.* **24** 391.
[5] Kuščer I and Ribarič M 1959 *Opt Acta* **6** 42.
[6] Budak VP and Korkin SV 2008 *J. Quant. Spectrosc. Radiat. Transf.* **109** 220.
[7] Flatau PJ and Stephens GL 1988 *J. Geophys. Res.* **93** 11037.
[8] Chandrasekhar S 1950 *Radiative transfer* (London: Oxford University Press)
[9] Wiscombe WJ 1977 *J. Atmos. Sci.* **34** 1408.
[10] Nakajima T and Tanaka M 1988 *J. Quant. Spectrosc. Radiat. Transf.* **40** 51.
[11] ftp://climate1.gsfc.nasa.gov/wiscombe/Multiple_Scatt
[12] http://www.ccsr.u-tokyo.ac.jp/~clastr/
[13] http://www.svet-mpei.org

Finite elements parameterization of optical tomography with the radiative transfer equation in frequency domain

O. Balima[1,2] and Y. Favennec[3] and F. Dubot[1] and D. Rousse[1]

[1] Industrial research chair in technologies of energy and energy efficiency (t3e). École de Technologie Supérieure, 1100, rue Notre-Dame Ouest, Montral (Qc), Canada H3C 1K3.
[2] Département des Sciences Appliquées. Université du Québec à Chicoutimi, QC, Canada G7H 2B1.
[3] LTN UMR CNRS 6607 Polytech' Nantes - La Chantrerie, Rue Christian Pauc, BP 50609, 44 306 Nantes Cedex 3, France.

E-mail: obalima@uqac.ca

Abstract. Optical tomography is a technique of probing semi-transparent media with the help of light sources. In this method, the spatial distribution of the optical properties inside the probed medium is reconstructed by minimizing a cost function based on the errors between the measurements and the predictions of a numerical model of light transport (also called forward/direct model) within the medium at the detectors locations. Optical tomography with finite elements methods involves generally continuous formulations where the optical properties are constant per mesh elements. This study proposes a numerical analysis in the parameterization of the finite elements space of the optical properties in order to improve the accuracy and the contrast of the reconstruction. Numerical tests with noised data using the same algorithm show that continuous finite elements spaces give better results than discontinuous ones by allowing a better transfer of the information between the whole computational nodes of the inversion. It is seen that the results are more accurate when the number of degrees of freedom of the finite element space of the optical properties (number of unknowns) is lowered. This shows that reducing the number of unknowns decreases the ill-posed nature of the inverse problem, thus it is a promising way of regularizing the inversion.

1. Introduction

Among the new imaging modalities expected to be available in the future, optical tomography is one of the most promising although numerous difficulties still exist. It is used in flow diagnostics, medical imaging, food processing, etc. This laser-based probing technique may be divided into direct imaging in which the emerging signal is directly used for projection and reconstruction imaging based on inverse problems also called optical tomography. For both of them, recent research tends to show that the use of the long term photons, which have travelled for a long time in the whole sample to be probed, generates more information to the image reconstruction.

Recently, an increasing interest has been devoted to the finite elements formulations when solving the radiative transfer equation. This tendency in favor of the finite elements method follows from its simplicity, flexibility and property of being able to handle complex geometries and advection type equations. From a standard Galerkin formulation [1], a number of studies have been done to improve the accuracy of these computational methods such as the Streamline Upwind Petrov Galerkin [2, 3], the Least Square formulation [4, 5] and the Discontinuous Galerkin formulation [6, 7], to name but a few.

Generally, optical tomography with finite elements methods involves continuous formulations [8, 9, 10, 11]. However, continuous finite element formulations suffer from the lack of local conservativity compared to the discontinuous formulation which allows the use of numerical fluxes to achieve local conservativity [12, 13]. In addition, high order accuracy can be achieved with the Discontinuous Galerkin formulation by using higher order polynomial approximation than the finite volume method. The Discontinuous finite elements formulation can be viewed as an all-in-one formulation as with the same formulation, the finite volume method is obtained when constant polynomial elements are used, while using a continuous function yields the standard finite element method [14].

This study presents an analysis of optical tomography where a parameterization of the optical properties is used to improve the accuracy of the reconstruction. Numerical tests of image reconstruction are performed to gauge the reconstruction accuracy in relation with the specific features of the explored finite elements models and the related finite elements space of the optical properties.

2. Physical model

In optical tomography, the forward model aims at computing the prediction of the boundary reading once the optical properties are known. Below, we present the forward model equation based on the frequency domain radiative transfer equation and the used solution method.

2.1. Radiative transfer in frequency domain

In this study, the retained forward model is the Fourier transform of the transient radiative transfer equation which writes in a given direction $\vec{\Omega}$ and for each spatial position $r \in \mathcal{D}$ by [15]:

$$\vec{\Omega} \cdot \nabla I(r, \vec{\Omega}, \omega) + \left(\frac{i\omega}{c} + \kappa + \sigma_s \right) I(r, \vec{\Omega}, \omega) = \frac{\sigma_s}{4\pi} \int_{4\pi} I(r, \vec{\Omega}', \omega) \Phi(\vec{\Omega}', \vec{\Omega}) d\Omega' \tag{1}$$

where $i = \sqrt{-1}$, c is the light speed in the medium, ω is the modulation frequency, $I(r, \vec{\Omega}, \omega)$ is the radiant power per unit solid angle per unit area at the spatial position r in direction $\vec{\Omega}$, $\kappa = \kappa(r)$ and $\sigma_s = \sigma_s(r)$ are respectively the absorption and the scattering coefficients, and $\Phi(\vec{\Omega}', \vec{\Omega})$ is the scattering phase function. Generally, the scattering phase function in tissues is given by the Henyey-Greenstein phase function [16]:

$$\Phi(\vec{\Omega}', \vec{\Omega}) = \frac{1 - g^2}{\left(1 + g^2 - 2g\cos(\Theta)\right)^{3/2}} \tag{2}$$

where Θ is the scattering angle between directions $\vec{\Omega}$ and $\vec{\Omega}'$ and g is the anisotropic factor. The boundary condition for Eq.(1) is a collimated external radiation that penetrates into the medium with the direction $\vec{\Omega}_c$, at position r_0. The boundary is taken as non reflecting with the corresponding boundary condition:

$$I(r_0, \vec{\Omega}) = q_0 \delta(r - r_0) \delta(\vec{\Omega} - \vec{\Omega}_c) \quad \vec{\Omega} \cdot \vec{\mathbf{n}} < 0 \quad \forall r \in \partial\mathcal{D} \tag{3}$$

where q_0 is the total heat flux, $\vec{\mathbf{n}}$ is the outward normal unit vector and δ is the Dirac function, $\partial\mathcal{D}$ denotes the boundary of the domain $\mathcal{D}$.

The radiative intensity is separated into two components $I = I_c + I_s$ where I_c and I_s are respectively the collimated and scattered intensities within the medium as treated in [5]. It is pointed out in [17], that this separation technique removes strong discontinuities due to Dirichlet-type boundary condition (Eq.(3)) and consequently improves the accuracy of the finite element solution. The collimated component obeys the extinction law [16] such that

$$\vec{\Omega}_c \cdot \nabla I_c(r,\omega) + \left(\frac{i\omega}{c} + \kappa + \sigma_s\right) I_c(r,\omega) = 0 \tag{4}$$

where the boundary condition is given by

$$I_c(r_0,\omega) = q_0(r,\omega)\delta(r - r_0) \quad \vec{\Omega}_c \cdot \vec{\mathbf{n}} < 0, \quad \forall r \in \partial D \tag{5}$$

The solution of the collimated part is used as a source term for the scattered component I_s such that

$$\vec{\Omega} \cdot \nabla I_s(r,\vec{\Omega},\omega) + \left(\frac{i\omega}{c} + \kappa + \sigma_s\right) I_s(r,\vec{\Omega},\omega) = \frac{\sigma_s}{4\pi}\int_{4\pi} I_s(r,\vec{\Omega}',\omega)\Phi(\vec{\Omega}',\vec{\Omega})d\Omega' + S_c(r,\vec{\Omega},\omega) \tag{6}$$

where $S_c(r,\vec{\Omega},\omega)$ is the source term induced by the scattering of I_c within the medium. S_c is given by

$$S_c(r,\vec{\Omega},\omega) = \frac{\sigma_s}{4\pi}\int_{4\pi} I_c(r,\omega)\delta(\vec{\Omega}' - \vec{\Omega}_c)\Phi(\vec{\Omega}',\vec{\Omega})d\Omega' = \frac{\sigma_s}{4\pi} I_c(r,\omega)\Phi(\vec{\Omega}_c,\vec{\Omega}) \tag{7}$$

Let us recall that Eq.(6) uses only a vacuum boundary condition which writes

$$I_s(r,\vec{\Omega},\omega) = 0, \quad \vec{\Omega} \cdot \vec{\mathbf{n}} < 0, \quad \forall r \in \partial\mathcal{D}. \tag{8}$$

For optical tomography applications, the measurable quantity used is the exitance on the boundary obtained by

$$P(r) = \int_{\vec{\Omega} \cdot \vec{\mathbf{n}} > 0} I_s(r,\vec{\Omega},\omega)\vec{\Omega} \cdot \vec{\mathbf{n}} d\Omega \quad \forall r \in \partial\mathcal{D}. \tag{9}$$

Let us point out that here the forward model is a system of two equations (Eq. (4) and Eq.(6)) that makes possible to handle the collimated direction compared to other studies [10, 18, 19, 20]. As shown in [9], the collimated source direction acts in the adjoint equation for the computation of the objective function gradient.

2.2. Solution method

The solution method is based on the Discontinuous Galerkin formulation to solve both the collimated and the scattered Eq. (4) and Eq.(6), whereas the discrete ordinates method is used to handle the angular dependency.

2.2.1. Discrete ordinates method In the Discrete Ordinates Method, integrals over solid angles are replaced by a numerical quadrature [21]. Thus, Eq.(6) is rewritten as a spatial differential equation for each discrete direction $\vec{\Omega}_m$:

$$\vec{\Omega}_m \cdot \nabla I_s^m(r,\omega) + \left(\frac{i\omega}{c} + \kappa + \sigma_s\right) I_s^m(r,\omega) = \frac{\sigma_s}{4\pi}\sum_{m'=1}^{M} I_s^{m'}(r,\omega)\Phi\left(\vec{\Omega}_{m'},\vec{\Omega}_m\right) w_{m'} + S_c^m(r,\omega) \tag{10}$$

where M is the number of directions of the quadrature, w_m and I_s^m are respectively the quadrature weight and the radiative intensity in the direction $\vec{\Omega}_m$. The corresponding boundary condition writes

$$I_s^m(r,\omega) = 0 \quad \vec{\Omega}_m \cdot \vec{\mathbf{n}} < 0, \quad \forall r \in \partial D. \tag{11}$$

The phase function in Eq.(10) is re-normalized in order to avoid normalization errors. As in [22, 23]. The re-normalized phase function is given by

$$\tilde{\Phi}\left(\vec{\Omega}_{m'},\vec{\Omega}_m\right) = f_{m'}\Phi\left(\vec{\Omega}_{m'},\vec{\Omega}_m\right) \tag{12}$$

where $f_{m'}$ is the renormalization factor defined by

$$f_{m'} = \left(\frac{1}{4\pi}\sum_{i=1}^{M} w_i\Phi\left(\vec{\Omega}_{m'},\vec{\Omega}_i\right)\right)^{-1}. \tag{13}$$

2.2.2. Discontinuous Galerkin formulation Equations Eq. (4) and Eq.(6) for each discrete direction m, are advection-type equations that are to be solved. There are many method that can be used to solve these equations, among which the finite element methods are well suited for handling complex geometries. Finite elements formulation of the radiative transfer equation has been used in optical tomography recently [10, 8, 9]. In these works, only standard or continuous finite elements are used. Here we consider the Discontinuous Galerkin formulation [24].

For an easy presentation of the formulation, let us rewrite the equations of the collimated intensity and the scattered intensity for each direction of the quadrature as

$$\begin{aligned} &\beta.\nabla u + bu = f && \forall x \in D \\ &u = h && \forall x \in \partial D^- \end{aligned} \tag{14}$$

where β is the advection direction ($\beta = \vec{\Omega}_c$ or $\beta = \vec{\Omega}_m$), $u = u(x)$ is the complex value field, b is the complex extinction coefficient, f the complex source term, $\partial D^- = \{x \in \partial D, \beta\cdot\vec{\mathbf{n}} < 0\}$ is the inflow boundary according to the advection direction β and h is the inflow boundary complex function that is applied to the system.

The variational formulation of (14) with the Discontinuous Galerkin writes [17]:

$$\begin{aligned} &\text{Find } u_h \in V_h \text{ such that} \\ &\mathcal{B}(u_h, v_h) = \mathcal{F}(v_h) \qquad \forall v_h \in V_h \end{aligned} \tag{15}$$

where V_h is the finite element space where the solution u_h is searched, $\mathcal{B}(u_h, v_h)$ and $\mathcal{F}(v_h)$ are respectively the bilinear and linear forms given by

$$\begin{aligned} \mathcal{B}(u_h, v_h) &= \sum_{k\in\mathcal{D}}(-\beta u_h, \nabla v_h)_k + \sum_{k\in\mathcal{D}}(\beta.\mathbf{n}\ u_h, v_h)_{\partial k^+} + \\ &\qquad \sum_{k\in\mathcal{D}}\left(\beta.\mathbf{n}\ u_h^-, v_h\right)_{\partial k^-\setminus\partial\mathcal{D}^-} + (bu_h, v_h)_{\mathcal{D}} \\ \mathcal{F}(v_h) &= (f, v_h)_{\mathcal{D}} - (\beta.\mathbf{n}\ h, v_h)_{\partial\mathcal{D}^-} \end{aligned} \tag{16}$$

and where u_h^+ and u_h^- are respectively the values of u_h in cell k and in the neighboring cells. In the following, the space of piecewise linear discontinuous elements is chosen for the solutions of the collimated and the scattered component of the light intensity. The solution method of the forward model equation is done by solving first the collimated component solution. The resulting source term (Eq. (7)) is used for the solution of the scattered part through an iterative procedure where the initial scattered field is null. The iterative algorithm stops when the maximum absolute relative error between the current iteration and the previous one on the scattered field is lower than a user-defined value, *i.e.* when

$$\max_{i,m}\left|\frac{\mathbf{I}_{\mathbf{s,i}}^{\mathbf{m,k-1}} - \mathbf{I}_{\mathbf{s,i}}^{\mathbf{m,k}}}{\mathbf{I}_{\mathbf{s,i}}^{\mathbf{m,k}}}\right| \leq 10^{-6} \tag{17}$$

where $\mathbf{I}_{\mathbf{s,i}}^{\mathbf{m,k}}$ is the discretized scattered component solution in direction m at iteration k and i is a degree of freedom of the finite element solution. In all computations, linear equations are solved with the SuperLu solver [25]. The criterion Eq. (17) is chosen in order to make sure that both the global convergence and the local convergence are achieved. The accuracy of the method is given in [17].

3. Inversion

In the inversion procedure, the aim is to recover the optical properties of the medium through the minimization of an objective function. Here, a gradient-type optimization is used where the objective function gradient is deduced through an additional adjoint state formulation.

3.1. Objective function and gradient computation

The objective function to be minimized in the inversion procedure is based on the errors between the measurements and the predictions at the detectors positions with a numerical forward model. Here, our model is based on a Discontinuous Galerkin formulation which leads us to use an integral form of objective function in order to be consistent with discontinuous fields as the values of the solution is not well defined at boundaries due to the discontinuity of the finite element space but their integral can be computed. Hence, the objective function writes

$$J(\theta) = \frac{1}{2}\sum_{n=1}^{N_s}\sum_{d}^{N_d}\int_{\partial\mathcal{D}_d} \|P(r,\omega_n,I_s) - M(r_d,\omega_n)\|^2 ds \tag{18}$$

where $\partial\mathcal{D}_d$ is the surface of the dth detector such that $\partial\mathcal{D}_d \subset \partial\mathcal{D}$, $\partial\mathcal{D}$ being the boundary of the domain whose properties are to be recovered. $\theta = (\kappa,\sigma)$ is the vector of parameters, *i.e.* the absorption and the scattering coefficients. The complex values $P(r,\omega_n,I_s)$ and $M(r_d,\omega_n)$ are the predictions and the measurements whom differences are integrated over all the detectors for all collimated sources. Next, N_s is the number of collimated sources, and N_d is the number of detectors, and $\|z\|^2 = z\bar{z} \quad \forall z \in \mathbb{C}$ where $\bar{z}_i$ is the conjugate of z_i. We suppose that κ and σ belong to the same finite element space such that $\theta = (\kappa,\sigma) \in \mathbf{R}^{2N_c}$ where N_c is the number of degrees of freedom of the chosen finite element space. In the following, the space of piecewise polynomial functions of degree 1 is chosen for κ and σ fields.

The gradient of the objective function is computed with the adjoint method with the following expressions [8]:

$$\begin{aligned} \nabla J(\theta)\cdot\delta\kappa &= (\lambda_s|I_s\delta\kappa)_s + (\lambda_c|I_c\delta\kappa)_c \\ \nabla J(\theta)\cdot\delta\sigma_s &= (\lambda_s|I_s\delta\sigma_s)_s + (\lambda_c|I_c\delta\sigma_s)_c - \\ &\left(\lambda_s|\left(\frac{1}{4\pi}\int_{4\pi} I_s\left(r,\vec{\Omega}',\omega\right)\Phi(\vec{\Omega}',\vec{\Omega})d\Omega'\right)\delta\sigma_s\right)_s \end{aligned} \tag{19}$$

where $(.|.)_s$ and $(.|.)_c$ are inner products associated to the solution space respectively of I_s and I_c and $\lambda_s = \lambda_s(r,\vec{\Omega},\omega)$, $\lambda_c = \lambda_c(r,\omega)$ are the corresponding complex vectors which represent the solution of the following adjoint equation system:

$$\left[-\vec{\Omega}\cdot\nabla - \frac{i\omega}{c} + \kappa_a + \sigma_s\right]\lambda_s(r,\vec{\Omega},\omega) - \frac{\sigma_s}{4\pi}\int_{4\pi}\lambda_s(r,\vec{\Omega}',\omega)\Phi(\vec{\Omega}',\vec{\Omega})d\Omega' = 0 \tag{20}$$

and

$$\left[-\vec{\Omega}\cdot\nabla - \frac{i\omega}{c} + \kappa_a + \sigma_s\right]\lambda_c(r,\omega) - \frac{\sigma_s}{4\pi}\int_{4\pi}\lambda_s(r,\vec{\Omega}',\omega)\Phi\left(\vec{\Omega}',\vec{\Omega}_c\right)d\Omega' = 0 \tag{21}$$

where $\lambda_c = 0$ and $\vec{\Omega}\cdot\vec{\mathbf{n}}\lambda_s + \dfrac{\partial J(\theta)}{\partial I_s} = 0$ for $\vec{\Omega}\cdot\vec{\mathbf{n}} > 0,\ \forall r \in \partial D$. We refer to references [8, 9, 26, 27, 28] for more details on how to deduce the gradient of the objective function through a Lagrangian formulation .

3.2. Reconstruction algorithm

Gradient-based algorithms have shown to be efficient for large scale optimization problem in optical tomography [18, 29]. Here, our reconstruction scheme is based on the limited-memory version of the BFGS method [30] where a scaling of both the objective function and its gradient are used to handle round-off errors and the low level of the boundary measurements due to the high extinction in the medium [9].

4. Tests and results

4.1. Test description

A 2 cm × 2 cm domain which contains two inclusions is chosen where the heterogeneity in the optical properties of the medium is represented by two inclusions. The optical properties of the medium are given in Table 1. It is assumed that the medium is forward-scattering where the phase function is given by the Henyey-Greenstein function with an anisotropic factor g= 0.9 . Four collimated source with zero-phased, modulated at the frequency of 600 MHz are placed at mid-centres of each side of the square and the measurements are done with eight detectors of 0.8 cm of extension regularly placed around the boundaries beyond the source positions.

The experiments measurements data is computed numerically. The above forward model is used with a regular fine triangular mesh of 8142 elements and 24 angular discrete ordinates (S_6 quadrature). The inverse problem is based on a coarser mesh of 3786 elements coupled to 24 discrete directions. In this study, the parameters (κ and σ_s) and the state variables (I_c and I_s) are computed on the same mesh, but they belong to different finite elements spaces for parameterization purpose. The obtained complex-valued intensities with the original properties (Table 1) are noised with a Gaussian distribution such that the level of the noise in dB unit is given by

$$SNR = 10\log_{10}\left(\frac{M_d}{\sigma_{M_d}}\right) \tag{22}$$

where σ_{M_d} is the standard deviation. The inversion computations are performed with the limited memory quasi-Newton method of Brodyen-Fletcher-Goldfarb-Shanno [30, 31]. The quality of the reconstruction is measured with errors ϵ_1 and ϵ_2 defined by:

$$\epsilon_1 = \frac{1}{N_c}\sum_{i=1}^{N_c}\left(\frac{\theta_i^r - \theta_i^o}{\theta_i^o}\right)^2,\quad \epsilon_2 = \left(\int_D (\theta_i^r - \theta_i^o)^2\,dx \Big/ \int_D (\theta_i^o)^2\,dx\right)^{1/2} \tag{23}$$

where N_c is the number of degrees of freedom related to the parameters finite element space, and superscripts r and o refer to the reconstructed and original images, respectively. ϵ_1 represents the mean quadratic errors per degree of freedom and ϵ_2 is the relative error of the reconstruction with respect to the original distribution.

Table 1: Optical properties of the test medium

	Background	Bottom inclusion	Top inclusion
κ	0.25 cm^{-1}	0.35 cm^{-1}	0.15 cm^{-1}
σ_s	20 cm^{-1}	30 cm^{-1}	10 cm^{-1}

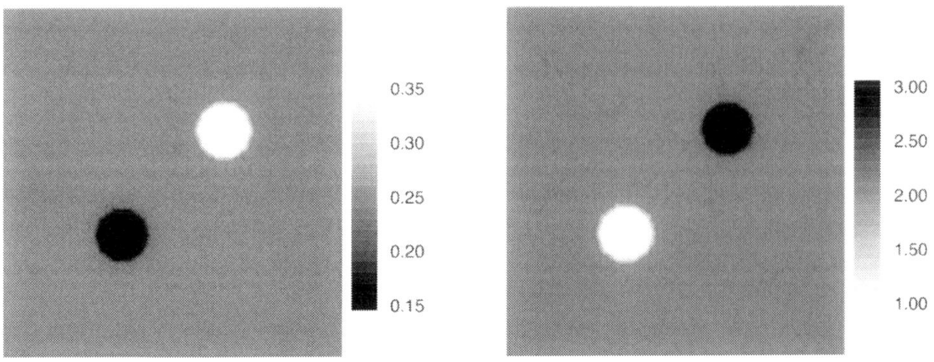

(a) (b)

Figure 1: Original distribution of the optical properties. (a) is the absorption coefficient, (b) is the reduced scattering coefficient.

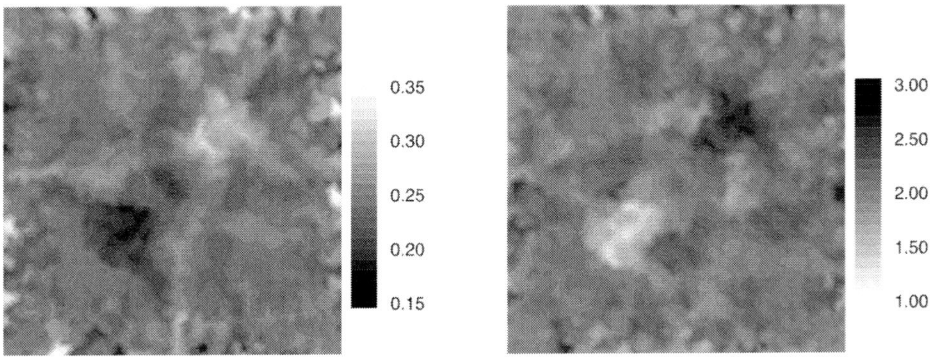

(a) (b)

Figure 2: Estimated distributions where the optical properties (κ and σ) belong to the piecewise constant finite elements space ($P0$). (a) is the estimated absorption coefficient, (b) is the estimated reduced scattering coefficient.

Table 2: Comparative accuracy of the reconstruction. ϵ_1 and ϵ_2 are the corresponding errors, $ndof$ is the number of degrees of freedom of the finite element space.

		$P0$	$P1dc$	$P1$	$P2$
$ndof$	–	3790	11370	1976	7741
ϵ_1	$\epsilon_{1,\kappa}$	0.0015	0.0008	0.0018	0.0009
	$\epsilon_{1,\sigma}$	0.0018	0.0010	0.0021	0.0010
ϵ_2	$\epsilon_{2,\kappa}$	0.0932	0.0895	0.0651	0.0720
	$\epsilon_{2,\sigma}$	0.1005	0.0964	0.0715	0.0770

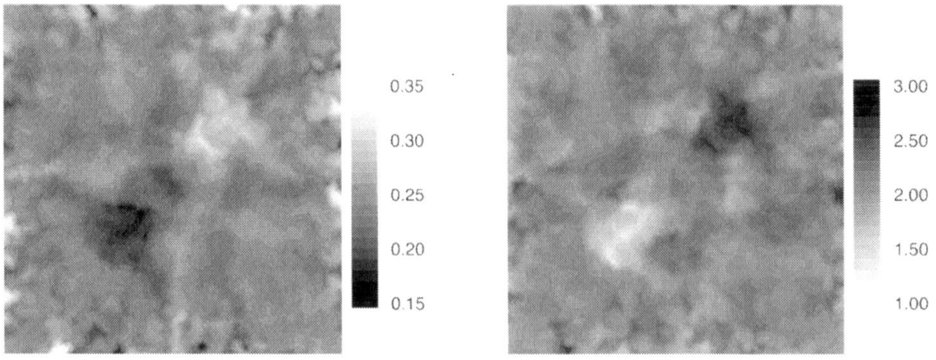

(a) (b)

Figure 3: Estimated distributions where the optical properties (κ and σ) belong to piecewise linear discontinuous finite elements space ($P1dc$). (a) is the estimated absorption coefficient, (b) is the estimated reduced scattering coefficient

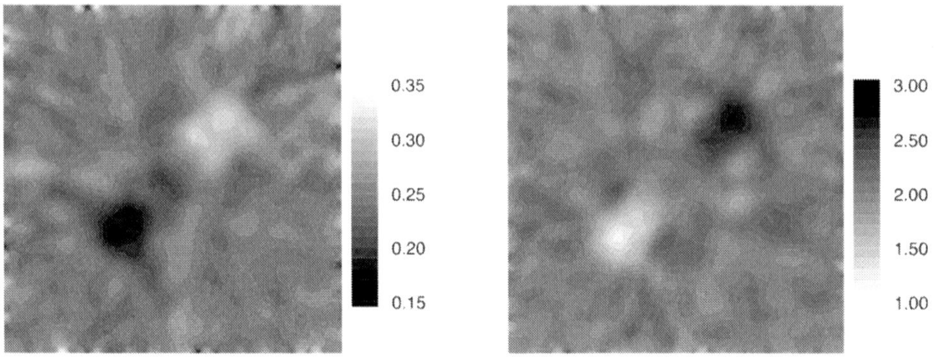

(a) (b)

Figure 4: Estimated distributions where the optical properties (κ and σ) belong to piecewise linear continuous finite elements space ($P1$). (a) is the estimated absorption coefficient, (b) is the estimated reduced scattering coefficient

4.2. Results and analysis

A comparative test is carried out with different finite element spaces for the optical properties where the intensity fields are supposed to be in the space of linear discontinuous elements ($P1dc$). The original distribution of the optical properties are presented in Fig. 1 where the reduced scattering coefficient $\sigma_s^{'} = (1 - g)\sigma_s$ is used. A comparative estimation is carried out with the computed noised data. The reconstruction is done by taking the parameters respectively in the space of piecewise constant ($P0$), piecewise linear discontinuous ($P1dc$), piecewise linear continuous ($P1$) and piecewise quadratic ($P2$) elements. The recovered distribution are reported

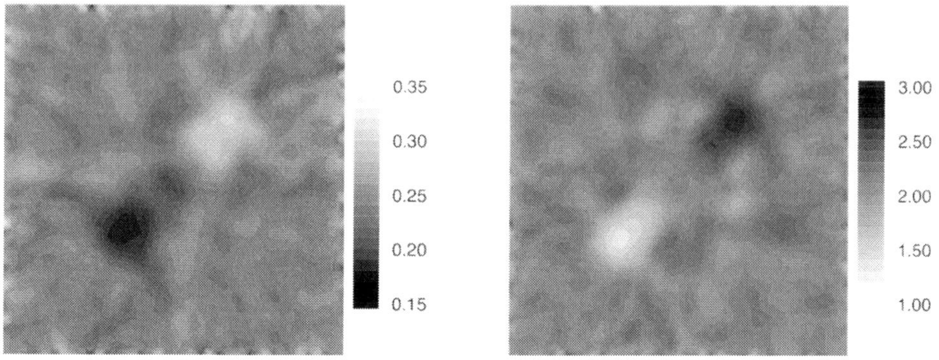

(a) (b)

Figure 5: Estimated distributions where the optical properties (κ and σ) belong to piecewise quadratic finite elements space ($P2$). (a) is the estimated absorption coefficient, (b) is the estimated reduced scattering coefficient

for piecewise constant (Fig.2), piecewise linear discontinuous (Fig.3), piecewise linear continuous (Fig.4) and piecewise quadratic (Fig.5) elements.

It is seen that continuous finite elements spaces give better results than the discontinuous one. The results show that continuous finite elements approximations of the optical properties introduce some implicit regularization on the inversion by smoothing the results. For each type of finite element space (continuous or discontinuous), it is seen that increasing the number of degrees of freedom do not improve the quality of the reconstruction. The smallest errors defining the quality of the reconstruction is given by the $P1$ finite element space whose number of degrees is the lowest (see ϵ_2 in Table 2). The errors given by ϵ_1 take into account the number of degrees of freedom and show that increasing the number of degrees of freedom lower the errors ϵ_1 as expected. However, taking into account the total number of degrees of freedom, the piecewise linear elements ($P1$) solution remains the more accurate.

From the inverse analysis point of view, high order parameterization may increase the ill-posed nature of the inverse problem such as the existence of correlations, the number of possible solutions and the sensitivity to round errors and noise. And one may end-up with less accurate solutions. Then, reducing the number of unknowns is a way of reducing the number of possible solutions *i.e* a way of regularizing the inversion. Physically, with continuous parameterization, information is well conducted among the whole computational nodes which improves the reconstruction as seen in the results.

5. Conclusion

In this study, numerical tests of reconstruction of optical properties of scattering and absorbing medium are done. The forward model is a finite element model based on the discrete ordinates method and the Discontinuous Galerkin method. A comparative study of the parameterization of the finite element space of the optical properties is done with piecewise constant, linear (continuous and discontinuous), and quadratic elements. The results show that better results are obtained when the finite element space of the unknown coefficients have a low number of degrees of freedom through a decrease of the ill-posed nature of the inverse problem as a regularization technique. The analysis shows that continuous finite element parameterization gives better results because the information is well conducted among the whole computational nodes which improves the reconstruction as seen in the results.

In a future step, we hope to speed up the forward and the adjoint model by using multiprocessing solvers in order to use high order quadrature. Also, new optimization schemes based on the finite elements flexibilities will be worked to improve the reconstruction.

Acknowledgment
This work was supported by the National Sciences and Engineering Research Council of Canada (NSERC).

References

[1] An W, Ruan L, Qi H and Liu L 2005 *Journal of Quantitative Spectroscopy & Radiative Transfer* **96** 409–422
[2] Brooks A N and Hughes T J R 1982 *Computer Methods in Applied Mechanics and Engineering* **32** 199–259 ISSN 0045-7825
[3] Tarvainen T, Vauhkonen M and Arridge S 2008 *Journal of Quantitative Spectroscopy & Radiative Transfer* **109** 2767–2778
[4] Pontaza J and Reddy J 2004 *Journal of Quantitative Spectroscopy & Radiative Transfer* **95** 387–406
[5] Balima O, Pierre T, Charette A and Marceau D 2010 *Journal of Quantitative Spectroscopy and Radiative Transfer* **111** 280–286
[6] Ciu X and Li B Q 2005 *Journal of Quantitative Spectroscopy & Radiative Transfer* **96** 383–407
[7] Gao H and Zhao H 2009 *Transport Theory and Statistical Physics* **38** 149–192
[8] Balima O, Boulanger J, Charette A and Marceau D 2011 *Journal of Quantitative Spectroscopy and Radiative Transfer* **112** 1229–1234
[9] Balima O, Boulanger J, Charette A and Marceau D 2011 *Journal of Quantitative Spectroscopy and Radiative Transfer* **112** 1235–1240
[10] Tarvaimen T, Vauhkonen M and Arridge S R 2008 *Journal of Quantitative Spectroscopy and Radiative Transfer* **109** 2767–2778
[11] Mohan P, Tarvainen T, Schweiger M, Pulkkinen A and Arridge S 2011 *Journal of Computational Physics* **230** 7364–7383
[12] Hughes T J R, Engel G, Mazzei L and Larson M 2000 *A comparison of discontinuous and continuous Galerkin methods based on error estimates, conservation, robustness and efficiency* (Springer-Verlag)
[13] Chung T J 2010 *Computational fluid dynamics* (Cambridge University press)
[14] Li B Q 2006 *Discontinuous Finite Elements in Fluid Dynamics and Heat Transfer* (Springer-Verlag)
[15] Arridge S R 1999 *Inverse Problems* **15** R41–R93
[16] Modest M F 2003 *Radiative Heat Transfer* 2nd ed (Academic Press)
[17] Balima O, Charette A and Marceau D 2010 *Journal of Computational and Applied Mathematics* **234** 2259–2271
[18] Charette A, Boulanger J and Kim H K 2008 *Journal of Quantitative Spectroscopy and Radiative Transfer* **109** 2743–2766
[19] Ren K, Ball G and Hielscher A H 2006 *Society for Industrial and Applied Mathematics* **28** 1463–1489
[20] Kim H K and Charette A 2007 *Journal of Quantitative Spectroscopy and Radiative Transfer* **104** 248–256
[21] Fiveland W 1984 *Journal of Heat Transfer* **106** 699–706
[22] Kim T and Lee H 1989 *Journal of Quantitative Spectroscopy & Radiative Transfer* **42** 225–238
[23] Liu L H, Ruan L M and Tan H P 2002 *International Journal of Heat and Mass Transfer* **45** 3259–3262
[24] Cockburn B and Shu C 1989 *Mathematics of Computation* **52** 411–435
[25] Demmel J W, Eisenstat S C, Gilbert J R, Li X S and Liu J W H 1999 *SIAM J. Matrix Analysis and Applications* **20** 720–755
[26] Favennec Y, Labbé V and Bay F 2003 *International Journal of Computational Physics* **187** 68–94
[27] Rouizi Y, Favennec Y, Ventura J and Petit D 2009 *J. Comput. Phys.* **228** 2239–2255 ISSN 0021-9991
[28] Favennec Y, Le Masson P and Jarny Y 2011 Optimization for function estimation *Thermal Measurements and Inverse techniques : Eurotherm CNRS advanced spring shool* ed of Nantes U URL `http://www.sft.asso.fr/`
[29] Hielscher A H, Klose A and Hanson K M 1999 *IEEE Transactions on Medical Imaging* **18** 262–271
[30] Liu D and Nocedal J 1989 *Mathematical Programming: Series A and B* **45** 503–528
[31] Cardona A and Idelsohn S 1986 *International Journal for Numerical Methods in Engineering* **23** 1023–1042

Journal of Physics: Conference Series **369** (2012) 012023 doi:10.1088/1742-6596/369/1/012023

The modeling of radiation transfer in highly-porous composite materials with strong scattering

O.M. Alifanov, V.V. Cherepanov and A.V. Zaytseva
Moscow Aviation Institute,
Volokolamskoe Hgw. 4, 125993, Moscow, Russia
Email: o.alifanov@yandex.ru

Abstract. In this report the problems, which arise during the radiative transfer of highly-porous thermal protective material with high spectral albedo of scattering, were studied. A three-step iterative numerical method based on splitting the problem operator by "physical processes" was developed. The method is useful and does not need any restrictions in initial approach and optical thickness of a layer. Results of computational modeling are given below.

1. Introduction

There are usually [1-4] the diffusion and transport-diffusion approximations, radiative thermal conductivity approximation or their combination for estimation of values integrated by radiative spectrum such as radiation component of thermal conductivity. However, if the material fragments are anisotropic and their sizes are comparable to the wavelength, such as the fragments of highly-porous thermal protective materials or the material layer is optically thin in the spectral range, then the radiations for these wavelengths can be far different from isotropic radiation. So it is necessary to describe the radiation by means of kinetic theory. For example, in the case of flat layer with thickness d we come to the known integral-differential boundary problem [3]

$$(\mu\frac{\partial}{\partial z}+\gamma(z))I_\nu(z,\mu)=\alpha(z)I_{b\nu}(z)+\beta(z)\int_{-1}^{1}\sigma(z,\mu,\mu_1)I_\nu(z,\mu_1)d\mu_1 \quad , \tag{1}$$

$$I_\nu(0,\mu)=I_{b\nu}(0)\,,\ \mu>0\ , \qquad I_\nu(d,\mu)=I_{b\nu}(d)\,,\ \mu<0\ , \tag{2}$$

in which I- the spectral intensity of radiation $\sigma(z,\mu,\mu_1)=\frac{1}{4\pi^2}\int_0^{2\pi}\int_0^{2\pi}p(z,\mathbf{\Omega}\cdot\mathbf{\Omega}_1)d\varphi d\varphi_1$ - is an average phase function for which the normalization requirement is met, $\int_{(4\pi)}p(\mathbf{r},\mathbf{\Omega},\mathbf{\Omega}_1)d\mathbf{\Omega}_1=1$, $n,\alpha,\beta,\gamma=\alpha+\beta,p$ – the local spectral coefficients of absorption, scattering and extinction respectively, $\mathbf{\Omega}$ – the direction unitary vector , $d\mathbf{\Omega}=sin\theta d\theta d\varphi$ - an elementary solid angle (θ,φ -polar and azimuth angles of spherical coordinate system with the polar axis Oz directed across layer), $\mu=cos\theta$, I_b - the intensity of equilibrium radiation defined by the Planck formula.

There are various methods for the numerical solution to the problems, which are similar to (1-2) [4-9]. The Monte-Carlo and «ray tracing» methods are time-consuming; they have a set of peculiarities, which are connected to their practical realization. These methods are not considered in this report as well as the semianalytic methods, which use different approximations of phase function and scattering intensity (moments methods, spherical harmonic methods and etc.)

It is very simple to solve the problem (1-2) if we introduce the "dummy" time so that we can determine the stationary solution during stabilization of the non-stationary solution, which was received with the help of the discrete ordinates method (DOM). However, the use of stable implicit finite difference approximations in the equations with integral operator leads to the linear systems with high dimension. It is difficult enough to make the explicit and comparatively low-cost DOM variants stable. Hereafter we will show that the intense scattering in media exacerbates this problem by making the one- and two-step methods unstable. It will also be demonstrated that problem is easily solvable in the frame of three-step method.

2. Scaling and the explicit one-step method of stabilization

Let us introduce the "dummy" time and use the scale natural system for realization of the residual methods during the stationary problem (1-2) solution

$$M_\gamma = M_\alpha = M_\beta = M_z^{-1} = a\gamma_{\max}, cM_t M_\gamma / n_{ef} = 1, a \in (0,1). \qquad (3)$$

As a result we come to the equation

$$(\frac{\partial}{\partial t} + \mu\frac{\partial}{\partial z} + \gamma(z))I(z,\mu) = \alpha(z)I_{b\nu}(z) + \beta(z)\int_{-1}^{1}\sigma(z,\mu,\mu_1)I_\nu(z,\mu_1)d\mu_1 \quad , \qquad (4)$$

in which the same designations are used for both the corresponding dimensional and for dimensionless variables. There will not be overmuch value of optical coefficients in (4) with the chosen way of scaling. The choice of constant a in (3) should be realized so that the change range of dimensionless variable z would not be too large, because in this case we need to work with the grids, which have a great amount of knots on z. Let us reduce the equation to the divergent form in order to define the conservative method (W – potential function)

$$\begin{gathered} \frac{\partial}{\partial t}I(z,\mu,t) + \frac{\partial}{\partial z}[V_z(z,\mu,t)I(z,\mu,t)] = 0 \,,\, V_z(z,\mu,t) = \mu + W(z,\mu,t)I^{-1}(z,\mu,t), \\ W(z,\mu,t) = \int_z^0 f(z_1,\mu,t)dz_1 \,,\mu > 0 \,,\, W(z,\mu,t) = \int_z^d f(z_1,\mu,t)dz_1 \,,\mu < 0 \\ f(z,\mu,t) = \alpha(z)I_b(z) - \gamma(z))I(z,\mu,t) + \beta(z)\int_{-1}^{1}\sigma(z,\mu,\mu_1)I(z,\mu_1,t)d\mu_1 \end{gathered} \quad . \qquad (5)$$

For the approximation of equation (5) we use the «upstream» scheme [10]. This scheme is very simple, conservative, conditionally stable and has accuracy of the first order of approximation, which leads to the appearance of the numerical diffusion. In order to decrease the diffusion influence on solution quality it is possible to introduce the additional step realizing the Flax Corrected Transport method (FCT) [11,12]. Such an approach slightly increases the amount of calculations and practically raises the approximation order in Δz of equation (5) to the second one. Also, it is possible to realize the scheme [13].

During the use of scheme [10] the most serious problems arise near to the intensity sharp changes and areas, where the intensity values are relatively close zero. It essentially increases the effective "speeds" V_z that leads to the time step reduction in this method up to the unacceptably small values.

Besides, the integral character of potential function (5) can have the great influence on the method stability. That's why we will carry out the detailed research in approximation. For simplicity we will consider a homogeneous media case α, σ=0, γ=*const* and assume μ>0 for clarity. Because the effective speeds are positive in this case, then the approximation (5) becomes

$$\begin{gathered} I_{k,j}^{n+1} = I_{k,j}^n + \frac{\Delta t}{\Delta z}[\frac{\mu_j}{\gamma}(I_{k-1,j}^n - I_{k,j}^n) - \int_{\Delta z(k-2)}^{\Delta z(k-1)} I_{,j}^n(z)dz] + o(\Delta t^2, \Delta z^2) \\ \rho_{k,j}^n \equiv \rho(z_k,\mu_j,t_n) \,,\, t_n = t_{n-1} + \Delta t_n \,, n \geq 1 \,,\, t_0 = 0 \,,\, \Delta t_n < \Delta z / \max_{k,j}|V_{k,j}^{n-1}|, \\ \mu_j = \cos((j-0.5)\pi / n_\mu),\ j = 1 \ldots n_\mu \,,\, n_\mu = 2m \,,\, z_k = (k-1)\Delta z \,,\, \Delta z = d/(n_z - 1) \,, \end{gathered} \quad ,$$

It is received after uncomplicated transformations and removals of subscripts and superscripts from functions:

Journal of Physics: Conference Series **369** (2012) 012023 doi:10.1088/1742-6596/369/1/012023

$$\frac{\partial I}{\partial t}+\left(\frac{\mu}{\gamma}-\frac{\Delta z}{2}\right)\frac{\partial I}{\partial z}+\frac{\Delta t}{2}\frac{\partial^2 I}{\partial t^2}+o(\Delta t^2)=\frac{\mu\Delta z}{2\gamma}\frac{\partial^2 I}{\partial z^2}-I+o(\Delta z^2)\,,\ \mu>0\ . \tag{6}$$

Thus, in the members of the first order grid steps of the approximating equation (6) we can see three additional effects in comparison with the equations (5):

1. The «convectional» speed changes and this change can be meaningful in case of big γ.
2. The additional "diffusive" component appears and its influence on the solution increases with reduction of γ and with the increase of μ (the instabilities process spreads over from areas with intense scattering and small μ magnitude).

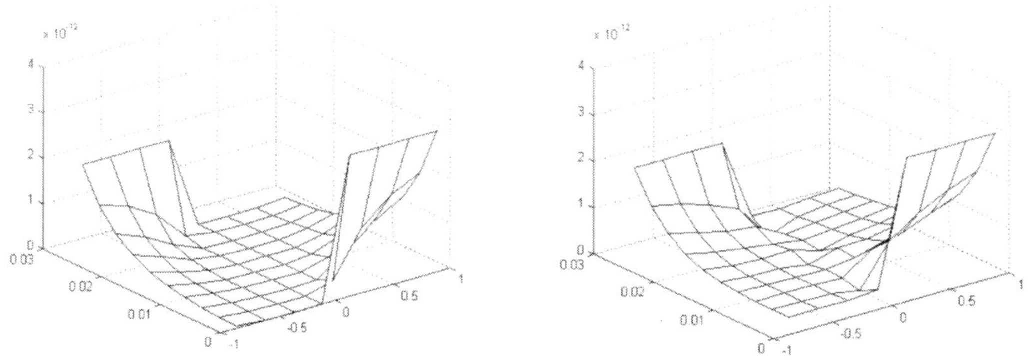

(a) (b)

Figure 1: (a) - the exact solution to (1), (b) - the intermediate solution to (5).
$d=0.026$, $\alpha=\sigma=0$, $\gamma=130$, $n_z\times n_\mu=10\times 8$, $T(C)\in[800,1200]$.

3. The additional wave disturbances of solution appear. And it is easy to show that the unstable components are *always* contained in them (see fig.1)

The reason for such unstable behavior of scheme [10] in the range of its «traditional stability», which is determined by Courant condition, is the presence of small multiplier μ near the spatial derivative of intensity. The equations with such structure usually «behave badly» during the numerical solution and it is necessary to take special steps for increasing the computing procedure stability.

The research on steady approximation of the problem can be organized in different ways. As it has been marked above such problems arise during the implicit approximation of the equations with the integral operator. At the same time it is possible to decrease their influence if we use the problem operator splitting and construct a multistage algorithm [14,15].

3. Two-stage scheme "predictor-corrector"

Let us write equation (1) in the form

$$\frac{\partial I}{\partial t}+L_1 I=\alpha I_b+L_2 I\ ,\ L_1 I(z,\mu,t)=(\mu\frac{\partial}{\partial z}+\gamma(z))I(z,\mu,t)\,,$$
$$L_2 I(z,\mu,t)=\beta(z)\int_{-1}^{1}\sigma(z,\mu,\mu_1)I(z,\mu_1,t)d\mu_1\ , \tag{7}$$

and use the following "predictor-corrector" scheme

$$\frac{I^{n+1/2}-I^n}{\tau}+L_1 I^{n+1/2}=\alpha I_b\ ,\quad \frac{I^{n+1}-I^{n+1/2}}{\tau}=L_2 I^{n+1}\ , \tag{8}$$

which is implicit on each "fractional" step. However, the scheme (8) approximates the equation (7) explicitly and with the first order of accuracy. Let us investigate the (8) more clearly.

Journal of Physics: Conference Series **369** (2012) 012023 doi:10.1088/1742-6596/369/1/012023

2.1 The analysis of "predictor" work

The first out of the equations (8) is not complicated for exact solution and we calculated the follows result:

$$I^{n+1/2}(z,\mu) = I_b(0)e^{-\int_0^z f(z_1,\mu)dz_1} + \int_0^z dz_1 g(z_1,\mu)\mathrm{e}^{-\int_{z_1}^{z} f(z_2,\mu)dz_2} , \quad \mu > 0 ,$$

$$I^{n+1/2}(z,\mu) = I_b(d)e^{\int_z^d f(z_1,\mu)dz_1} - \int_z^d dz_1 g(z_1,\mu)\mathrm{e}^{\int_{z_1}^{z} f(z_2,\mu)dz_2} , \quad \mu < 0 , \quad , \tag{9}$$

$$f(z,\mu) = \frac{1}{\tau\mu}(1+\tau\gamma(z)) , \; g(z,\mu) = \frac{1}{\tau\mu}(I^n + \tau\alpha I_b) .$$

Let us note, that functions $f(z,\mu)$ and $g(z,\mu)$ always coincide with sign of μ and that's why in exponent dimension there are always negative values. Functions $f(z,\mu)$ and $g(z,\mu)$ hyperbolically increase with decreasing of τ, μ so that the question about boundedness appears. Let us note that the fulfillment of the continuity condition

$$I^{n+1/2} \rightarrow I^n \; , \; \tau \rightarrow 0 \; , \tag{10}$$

contains, in particular, the answer to boundedness question. Let's investigate the question on solutions boundedness, assuming that the layer is optically homogeneous, i.e. setting $\alpha, \gamma = const$ and suppressing the inappreciable dependences from μ. Because of

$$\frac{1}{\tau\mu}\int_0^z dz_1 \rho(z_1,\tau)\mathrm{e}^{-(\frac{1}{\tau}+\gamma)\frac{z-z_1}{\mu}} \equiv \frac{1}{1+\tau\gamma}\int_0^{\frac{z}{\mu}(\frac{1}{\tau}+\gamma)} \rho(z-\mu\frac{u\tau}{1+\tau\gamma},\tau)\mathrm{e}^{-u}\,du \xrightarrow[\tau\to 0,\,\mu>0]{} \rho(z,0) \; ,$$

then from this as $\rho(z,\tau) = I^n(z) + \tau\alpha I_b(z)$ the condition (10) follows.

Besides it is possible to be confirmed in this equation

$$I^{n+1/2}(z,\mu) \rightarrow I_b(0)e^{-\frac{\gamma z}{\mu}} + \frac{\alpha}{\mu}\int_0^z dz_1 I_b(z_1)\mathrm{e}^{-\frac{\gamma(z-z_1)}{\mu}} \; , \quad \tau \rightarrow \infty \, , \; \mu > 0 \; ,$$

in which the solution to stationary equation (1) for scattering media is contained in right part. Analogical equations are realized for $\mu<0$. And it is possible to generalize these limitary equations in case of heterogeneous media.

It is clearly shown that in analytic realization the "predictor" step of scheme (8) works normally and leads to the necessary limitary consequences. Moreover the computing realization of equations (9) is associated with major problems.

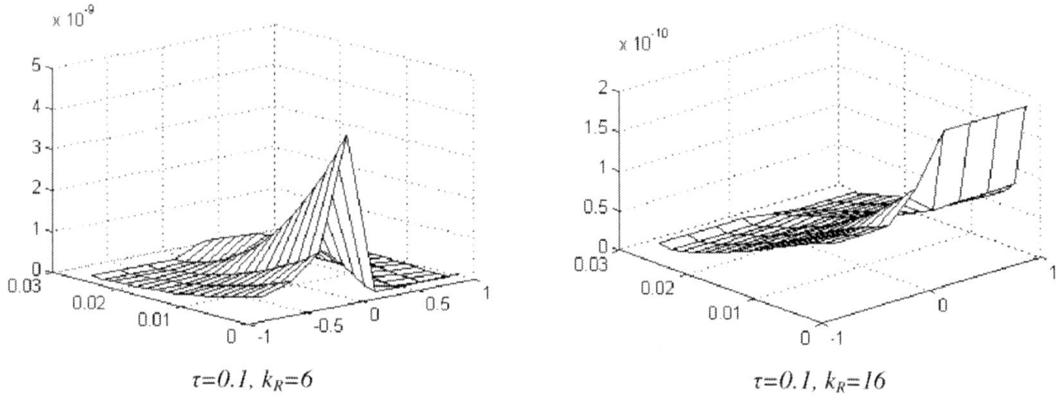

$\tau=0.1$, $k_R=6$ $\tau=0.1$, $k_R=16$

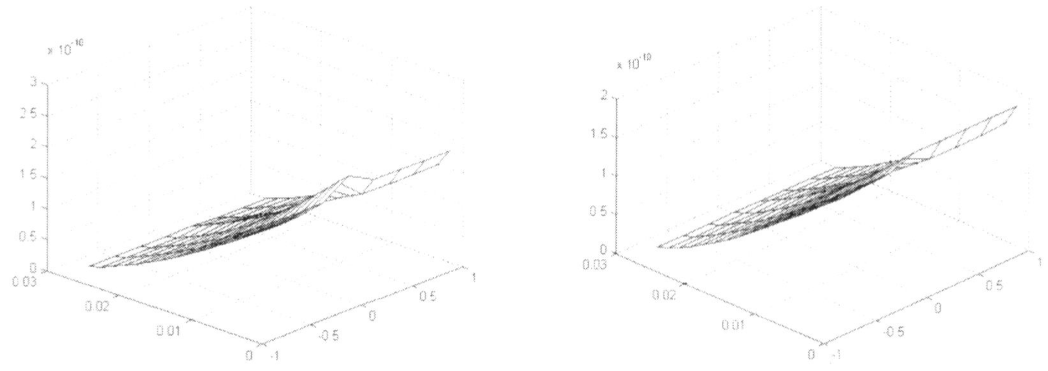

$\tau=1,\ k_R=6$ $\tau=1,\ k_R=8$

Figure 2. The exposure on equilibrium radiation intensity "predictor"-step of method (8) for different accuracy of integration at d=0.026, $\alpha=\sigma=0$, γ=130, $n_z{\times}n_\mu$=10×8, $T\in[800,1200]^{\circ}$C.

In Figure 2 the results of step-predictor exposure on equilibrium intensity at τ=0.1 and τ=1 for as the same set of parameters as for Figure1 are shown. The numerical integration was carried out by Romberg method with different accuracies. The order of main inaccuracy component is also given in figures. It is obvious from figures, that at the "predictor" step the meaningful errors of integration arise, which appear and spread over from small μ magnitude range. They can be reduced only with the rise in integration accuracy. With the small τ the accuracy conditions arise so that it excludes the practical use of method. On the *whole, the continuity condition (10) violates at the "predictor" step of scheme (8) because of current integration errors with small* $\tau\mu$.
With $\tau\mu$~1 and above such problems don't arise or they are easy to reduce. However, the condition of $\tau\mu$>1 leads to such high values of τ, that it violates the approximation of (1) by scheme (8). And arising problems in this method are not limited only by this.

2.2 Analysis of "corrector" work. Unremovable problems of two-step method
The equation of "corrector"(8) step has the form

$$I^{n+1}(z,\mu) = I^{n+1/2}(z,\mu) + \tau\beta(z)\int_{-1}^{1}\sigma(z,\mu,\mu_1)I_\nu^{n+1}(z,\mu_1)d\mu_1 \ . \tag{11}$$

Let us use the iterative process for its solution

$$I^{n+1,p+1}(z,\mu) = I^{n+1/2}(z,\mu) + \tau\beta(z)\int_{-1}^{1}\sigma(z,\mu,\mu_1)I^{n+1,p}(z,\mu_1)d\mu_1\,,\ p=0,1,\dots, \qquad I^{n+1,0}(z,\mu) =_\nu^{n+1/2}(z,\mu)\ . \tag{12}$$

Iteration convergence of "corrector" step is valid only with the small time steps submitting to $\tau\beta_{max}$<1 condition, because

$$\left\|\delta^{p+1}\right\| \equiv \max_{z,\mu}\left|I_\nu^{n+1,p+1} - I_\nu^{n+1,p}\right| \le \tau\max_z \beta(z)\int_{-1}^{1}\sigma(z,\mu,\mu_1)\left\|\delta^p\right\|d\mu_1 = \tau\beta_{\max}\left\|\delta^p\right\|.$$

The main problem of two-step method is existence of contradictory conditions: $\tau\mu$>1for "predictor" and $\tau\beta_{max}$<1 for "corrector". These conditions are not consistent during the extension of scattering effects. At this step in case of decisive "predictor" we face with the unremovable problems, which were discussed in part 2.
It turns out that the problem of ill-posed condition can be removed with the help of right organization of "fractional" steps.

4. Three-stage splitting «by physical processes»
Results from part 2 and 3 show, that the problems of the numerical solution to the equation (1) are connected with the presence of small-value coefficient near to the spatial derivative. They arise when the hard component of the equation (1) is included into any equations for "fractional steps" along with

the other components of the original equation (perhaps, except the time derivative). The following algorithm of splitting by «physical processes» is offered:

1. First of all the "fractional" step of "convection" is carried out and it only processes the hard component of the equation (1). At this step the «upstream» scheme is used. The condition of stability, which defines a method time step, is formed and spatial boundary conditions are considered:

$$I^{n+1/3}(z,\mu)-I^{n}(z,\mu)+\tau\frac{\partial}{\partial z}(\mu I^{n}(z,\mu))=0\ ,\ \tau<\Delta z/max\ \mu_j \tag{13}$$

Let us note that if the following step of similar *explicit* algorithm considers other not integrated components of the equation (1), in other words it will be connected with the consideration of physical effects of radiation extinction and secondary radiation after absorption, then as a result there is two-step method, whose first step is similar to (8).

2. That's why for the second step it is necessary to consider the action of equation (1) integrated components, with the help of connection of this step in the process of unusual "hashing" of the radiation generated by scattering:

$$I^{n+2/3}(z,\mu)=I^{n+1/3}(z,\mu)+\tau\beta(z)\int_{-1}^{1}\sigma(z,\mu,\mu_1)I^{n+1/3}(z,\mu_1)d\mu_1\ , \tag{14}$$

3. The step of taking into account secondary radiation effects and attenuation:

$$I_{\nu}^{n+1}(z,\mu)=I_{\nu}^{n+2/3}(z,\mu)+\tau(\alpha(z)I_{b\nu}(z)-\gamma(z)I_{\nu}^{n+2/3}(z,\mu))\ . \tag{15}$$

Practical use shows the simplicity and efficiency of this method. The iterations (13-15) converge at performance of the stability condition (13). The method convergence practically does not depend on the choice of an initial condition, as well as the structure of the stationary solution, which, how it must be for these problems, is attractor state dependent only on problem parameters. Naturally, in similar calculations the continuous control of stationary equation discrepancy in the knots of finite-difference grids should be also carried out .

5. Some modeling results

As the example of practical use of developed method we consider the radiation transfer with wavelength λ=2.4 μm in foamed glassy carbon RVC ETTI-CF-ERG with the following parameters: thickness d=26 mm, temperature of boundaries T_{min}=800°C, T_{max}=1200°C. The temperature profile in the layer of material is needed for solution to problems (1) and (2). This profile defines the local values of Planck's component in equation (1), the values of spectral coefficients of scattering and attenuation, and phase function of the material. Before [16] the temperature dependencies of full thermal conductivity $\lambda(T)$ and volumetric heat capacity $C(T)$ were experimentally determined (see Figure3).

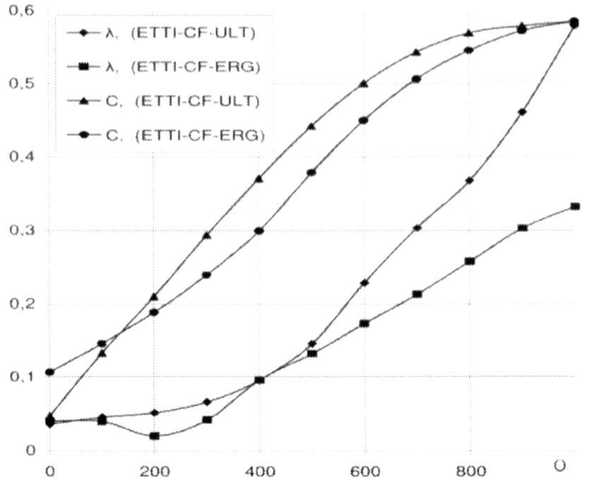

Figure 3. The temperature dependencies of full thermal conductivity and volumetric heat capacity of foam glass carbon. Experiment [16].

As the full thermal conductivity contained radiation component, the energy equation in the problem of radiation-conductive heat transfer for flat layer with known dependences of $C(T)$ and $\lambda(T)$ had the divergence form, closed under temperature and its stationary solution determine the necessity temperature profile in the material layer:

$$C(T)\frac{\partial T}{\partial t}=\frac{\partial}{\partial z}\lambda(T)\frac{\partial T}{\partial z}\ ,\ T(0,z)=T_{max}+\frac{T_{min}-T_{max}}{d}z\ ,\ T(t,0)=T_{max}\ ,\ T(t,d)=T_{min}\ , \tag{16}$$

The one-dimensional variant of implicit scheme [17] was used for solution to non-linear problem (16). There were organized additions inner iterations at each time step for account of nonlinearity in coefficients of "predictor-corrector" scheme.

The spectral dependencies of the absorption coefficient α_λ and the scattering coefficient β_λ, which were calculated for one of the representative elements [18] of carbon foam ETTI-CF-ERG shone in the direction of axis Oz, are shown on Figure 4. Values of wavelength on horizontal axes are specified in meters. The scanning of area on Figure 4 (a) was spent with step $\Delta\lambda$ =2•10-6m, on Figure 4 (b) with step $\Delta\lambda$ =2•10-8m. Diameters of knot and crosspieces are noted on horizontal axes. A smooth line, which is the best visible on Figure 4 (b), corresponds to the spectral weight function at T=500K multiplied by 0.03, which enters into expression [1] for radiating heat conductivity in a diffusive approach

$$\psi(\tau)=\tau^{-6}e^{-1/\tau}(1-e^{-1/\tau})^{-3}\ ,\ \tau=\lambda kT/hc=10^{3}\lambda T/14.39\ .$$

The bottom pulsing line, which has narrow and relatively high peaks of a resonance, corresponds to the spectral coefficient of absorption α_λ in all pictures. The pulsing line located above displays the spectral scattering coefficient β_λ, which is reduced in 10^3 times.

The calculations show, that the highest resonance peaks of absorption are located in the spectral area, which ingenuously contains the values of diameters of knot and crosspieces. The width of resonant peaks increases and the height, especially in absorption resonances, decreases in a long-wave part of a spectrum (Figure 4(a)). The especial coherency of resonant effect arises in the processes of absorption and scattering. The coherency was observed in the spectrums, which are shone on a normal of homogeneous flat layers and this coherency is well-known [19]. In this spectral range the representative element (as well as the material) starts to behave itself like a homogeneous media, because the scales of its spatial heterogeneity become insignificant in comparison with wavelength.

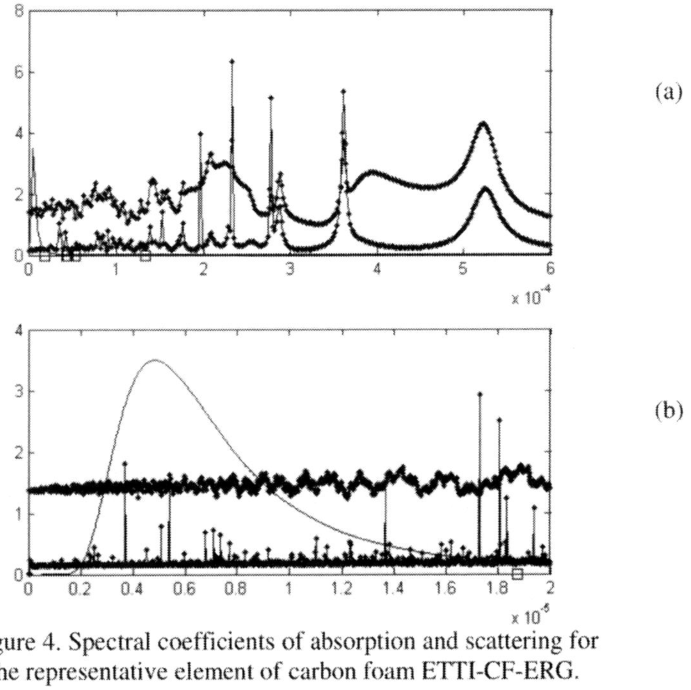

Figure 4. Spectral coefficients of absorption and scattering for the representative element of carbon foam ETTI-CF-ERG.

The resonant effects weaken in the process of the further increase in wavelength. Absorption and scattering become neglectfully small, the material becomes optically transparent. The resonant effects in the absorption are not so significant in the short wavelengths range, containing the carrier of weight function at temperature T=500K. In this range the change of spectral coefficients has a character of small fluctuations in amplitude near some average values, the absorption has little significant and the considered material behaves itself practically like the conservative media with constant properties in spectrum. Values of similar characteristics of a material can be received with the help of averaging corresponding values for representative elements. However, the scattering and absorption spectrums of representative elements of carbon foam in the temperatures range between 500 and 1000C practically do not vary [2]. It is not right for the weight function ψ, whose carrier displaces the area of shorter wavelengths in the process of temperature rising. Thus, the accuracy of the description with use of average radiating characteristics at a considered material with temperature growth rises.

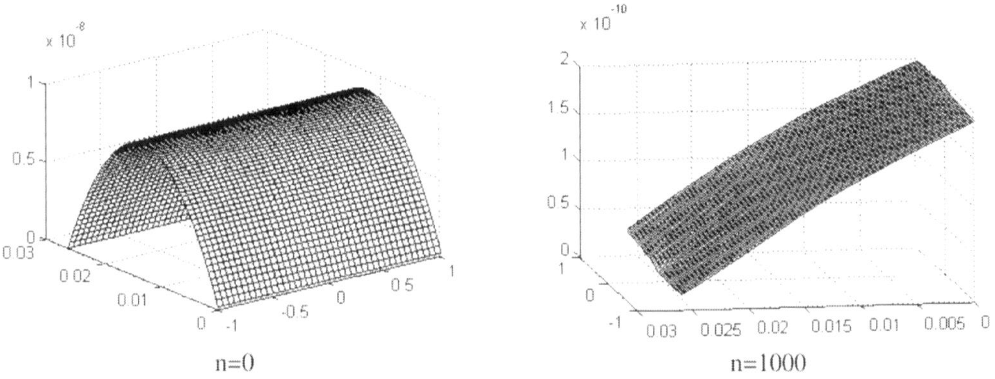

Figure 5. Residual of radiation intensity at the first and last step of approximation $I^n(z,\mu)$ to stationary condition in optically thick layer of RVC ETTI-CF-ULT

Scattering by foam glass carbon is practically isotropic and its phase function was practically constant. The grid with dimension $n_z{\times}n_\mu$=17×8 and M_γ=130 was used in calculations. The structure of intensity at the first and last iterations is shown in Figure 5. The evolution of equation (1) residual during the iterations (13)-(15) is shown in Figure 6. The modulus of difference between left and right parts is regarded as residual. The z- variable initial approach was parabolic and hard enough. Numbers of corresponding iterations are specified in pictures. As it is visible from Figure 6, at separate stages of solution the maximum of residual can rise. But it is always accompanied by cardinal reorganization of the solution form to residual. After the finish of "sweeping changes" process the relaxation to a stationary stage with minimum residual becomes monotonous. In conclusion the convergence is obtained only in three-step method for such high values of scattering coefficient.

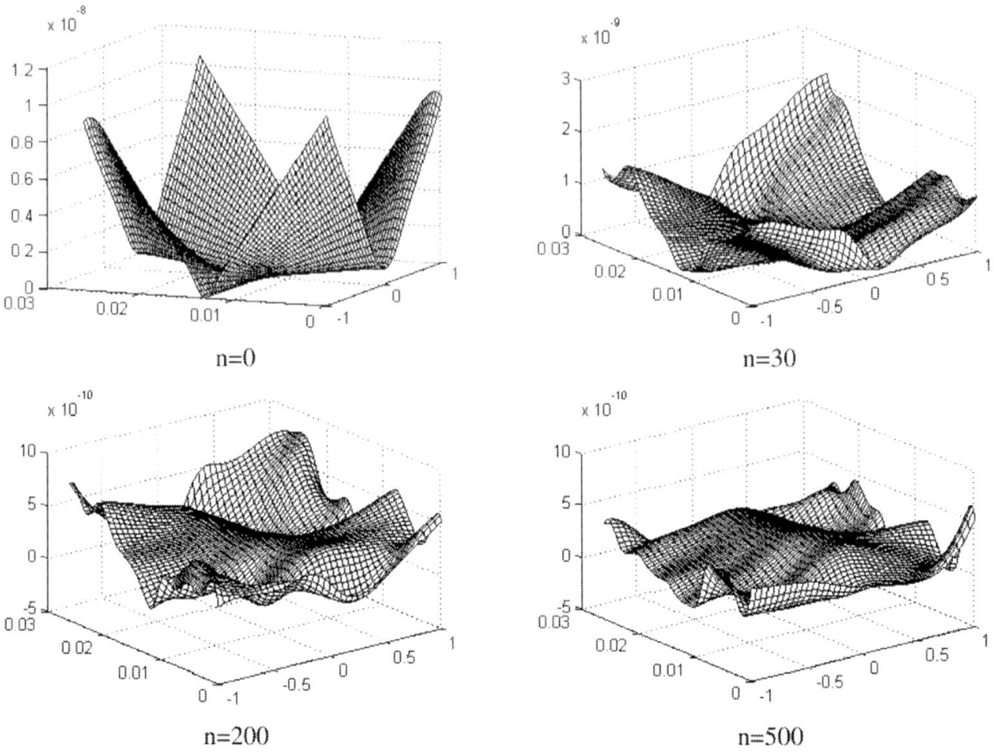

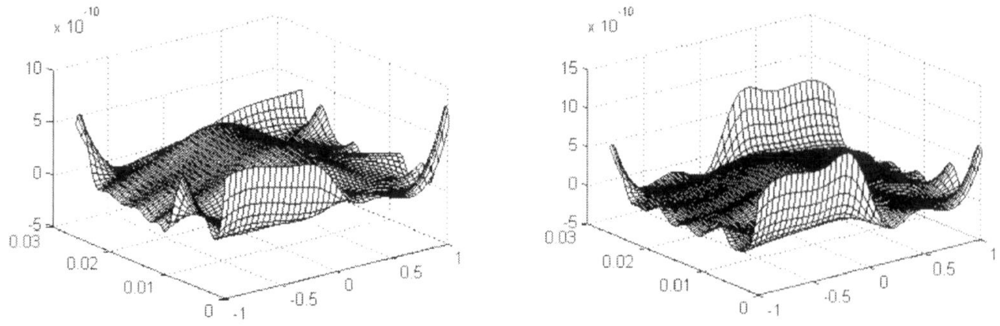

n=800 n=1000

Figure 6. Change of the equations (1) residual depending on iteration number at a radiation relaxation in optically thick layer of RVC ETTI-CF-ULT.

6. Conclusion

The method, which has great stability factor in comparison to traditional two-step method, was developed. The method is quick enough, is easily realized and doesn't lead to any restrictions on optical thickness of a layer, scattering character, etc.

7. Acknowledgments

This work was supported by ESTEC/ESA and EADS (former Astrium GmbH) in the framework of the International Science and Technology Centre grant No 3871.

8. References

[1] Alifanov O.M., Cherepanov V.V. Mathematical simulation of High-Porousity Fibrous Materials and Determination of Their Physical Properties. // High Temperature, 2009, v.47, №3, p.438-447.
[2] Dombrovskiy L.A. Quartz-fiber thermal insulation: Infrared radiative properties and calculation of radiative-conductive heat transfer. // ASME J.Heat Transfer, 1996, v.118, № 2, p.408-414.
[3] Özisik M.N. Radiative transfer and interections whith conduction and convection. - NY: John Wiley&Sons, 1974.
[4] Bass L.P., Voloshenko A.M., Germogenova T.A. The methods of discrete ordinates in problems of radiation transfer (rus.) - Moscow: Pre-print IAM AS USSR, 1986.
[5] Kotov D.V., Surdjikov S.T. Local estimation of the directed radiating ability of light-scattering volumes by the Monte-Carlo method (rus.). // TVT, 2007, v.45, №6, p. 885-895.
[6] Kalitkin N.N. Numerical methods (rus.). // M: Nauka, 1978.
[7] Ocisik M.N. Complex heat transfer (rus.).//M: Mir, 1976, p.616.
[8] Surzhikov S.T. Termal radiation of gases and plasms (rus.).//M.: MSTU n.a. N.E.Bauman Publisher, 2004, p.544.
[9] Nagirner D.I. Lection on theories of radiative transfer(rus)–//St.Petersbourg: St.Petersbourg University Publisher, 2001, p.207.
[10] Potter D. Computational methods in physics. //M: Mir, 1975, p.392.
[11] Killing J. Controlled thermonuclear fusion. // M: Mir, 1980, p.480.
[12] Cherepanov V.V. Flat walled prob in thermo-dynamical unbalanced continuous plasma (rus). // Dep. AUISTI, 1984, №1089-84 Dep. P.22.
[13] Bogomolov C.V., Zvenkov D.S. Non-smoothing gas-dynamical abruption implicit method of fractions (rus.). // Mathematical modeling, 2006, V.19, №3, p.74-86.
[14] Marchuk G.I. Methods of computing mathematic (rus.).// M: Nauka, 1977, p.456.
[15] Kovenya V.M., Yanenko N.N. Splitting technique in gas dynamic problem (rus). // Novosibirsk: Nauka, 1981, p.304.
[16] Alifanov O.M., Budnik S.A., Cherepanov V.V., Nenarokomov A.V. Experimentally-Theoretical Research of Heat Transfer in High-Porous Materials (rus.) //Thermal Processes in Engineering, 2011, v.3, №2, p.53-65.
[17] Bishop K.A. An Application of the Hybrid Alternating Direction Implicit Procedure to the Simulation of an Unconfined Aquifer. //Annales de 1 Association International Pour le Calcul Analogique, 1974, №3, p.9-14.

[18] Cherepanov V.V. Mathematical modeling of spectral and thermophysical properties of carbon foam (rus.) // Thermal Processes in Engineering, 2011, v.3, №9, p. 386-399.
[19] Boren K., Halfmen D. Absorption and scattering of light by small fractions. // M: Mir, 1986, p.664.

The Multi-Spectral Reordering (MSR) technique for the narrow band modeling of the radiative properties of nonuniform gaseous paths: the correlated/uncorrelated approximations

F André[1] and R Vaillon
Université de Lyon, CNRS, CETHIL, INSA-Lyon, UMR5008, F-69621, France

E-mail address: frederic.andre@insa-lyon.fr

Abstract. A general path averaging strategy is proposed to calculate the equivalent parameters to be used with the narrow band modeling of the radiative properties of gases in nonuniform media. Approximate formulas are in particular derived for the cases of correlated and uncorrelated gas spectra. These approximations are applied to the Narrow Band MSR-SLMB (Spectral-Line Moment-Based model together with the Multi-Spectral Reordering technique) model and assessed by comparison with LBL data. Both approximations are accurate with an absolute error less than 0.05 for transmission functions. This demonstrates the improvement that MSR modeling provides when compared to usual narrow band models, for which larger errors are observed.

1. Introduction

Radiative heat transfer in molecular gases is significant in many situations encountered in combustion or planetary atmosphere studies. Due to the excessive computational costs of the most reliable approximate model available, the so-called Line-By-Line (LBL) method, more efficient modeling techniques have to be used to get radiative properties estimates. But choosing the right approach among the existing ones can be a delicate problem especially when the gaseous medium is strongly nonuniform. Indeed, many gas radiative properties models frequently used in moderately nonuniform media, and more particularly those based on the assumption of scaled or correlated spectra, may be inappropriate for this task and their application can lead to incorrect outputs.

One of the trickiest problems consists in calculating IR emission spectra of flames after their transmission through cold gaseous atmospheres containing the same molecular species (mainly CO_2 and H_2O) as those produced by the flames. Such a situation is frequently encountered as it is representative of a large number of remote sensing problems, ranging from the tracking of missiles or aircrafts for military applications, to the sensing of exhaust gases from industrial plants or volcano for environmental purposes. It is also one of the most difficult modeling situations. Indeed, due to the so-called spectral correlation effects that may or may not exist between the radiative emission by the hot gases and its absorption by the atmosphere, the use of a scaled/correlated-*k* assumption alone may produce very large errors, both over narrow and wide bands. For this reason, several sophisticated techniques were proposed during the past decades to handle highly nonuniform situations. Among

[1] To whom any correspondence should be addressed.

them we can cite the fictitious gas approach [1], the multi-group method [2] by Modest and co-workers or mapping techniques such as those described, for example, in West et *al* [3]. Those methods were introduced in the 90s-mid 2000s to improve radiative models for nonuniform calculations, a goal that they succeeded to achieve at the cost of higher computational requirements than with the standard models. More recently [4], we have proposed a new and efficient modeling, called the multi-spectrum reordering (MSR) technique, of radiative heat transfer in nonuniform media. Fundamentally, it shares many similarities with the mapping approaches, to the extent that both methods are based on the multiple reordering of gas spectra in distinct thermophysical conditions. One main difference is that the MSR technique is used together with the Spectral-Line Moment-Based (SLMB) modeling, which enables to get very accurate results without requiring many successive reorderings as with the usual mapping schemes. In fact, "standard" mapping techniques involve the reordering of spectra over each uniform sub-paths encountered along nonuniform paths until nearly constant absorption coefficients, viz. first order moments, are obtained over each of the MSR bands and for each uniform sub-path. This process can be computationally consuming for large nonuniformities. With the Narrow Band MSR-SLMB model, on the other hand, the use of higher order *k*-moments lowers the number of successive reorderings but it requires, as a set-off, path-averaging strategies. Two original ones are introduced in the present paper.

In particular, the aim of the present work is to investigate the impact of using a correlation or uncorrelation assumption of gas spectra in various thermophysical states when they are applied over multi-spectrally reordered spectral intervals. On one hand, the uncorrelation assumption is known (see Ref. [5] for example) to provide inaccurate results when applied over narrow and wide bands. In this case, correlated models, viz. the narrow band correlated-*k* as well as the usual full spectrum models such as ADF, SLW or FSK, were demonstrated to give better results. Nevertheless, the uncorrelation assumption has never been studied together with the NB MSR-SLMB modeling. This is also one purpose of the present work. On the other hand, the correlation assumption is more usual for this kind of calculations. An approach based on *k*-moments was proposed in Ref. [6].

In this paper, we derive the general formula for the calculation of the path-averaged *k*-moments in nonuniform gaseous media. Starting from this formula, two simpler ones, obtained under the correlated and uncorrelated assumptions, are given. The uncorrelation assumption provides a nonlinear formula: it is to our knowledge the first one of this kind proposed for radiative transfer applications. Then, the MSR modeling to which we apply the previous equations is described. The building of Narrow Band MSR-SLMB model databases for CO_2 and H_2O diluted in N_2 is also briefly discussed. Finally, results for several test cases are given to assess the nonuniform models against LBL reference calculations.

2. Nonuniform narrow band *k*-moment modeling

2.1. Preliminary comments about the correlation/uncorrelation assumptions

We consider a narrow spectral interval $\Delta\nu$ over which: A1/ the Planck function can be assumed constant, A2/ in a thermophysical condition chosen arbitrarily as a reference, the absorption coefficient is nearly constant; a nonuniform optical path and two locations along this path, s' and s'', with distinct spectral absorption coefficients $\kappa_\nu(s') \neq \kappa_\nu(s'')$. For the sake of simplicity, the spectral absorption coefficient $\kappa_\nu\left[\underline{\phi}(s')\right] = \kappa_\nu\left[x(s') \quad P(s') \quad T(s')\right]$ will be written $\kappa_\nu(s')$. Similar simplifications will be used for all local quantities along the path.

Following Ref. [7] (Eq. (19.133) p. 625), gas spectra are correlated (resp. scaled) if there exists a monotonously increasing (resp. linear) function h such that $\kappa_\nu(s'')$ can be written as $h\left[s', s'', \kappa_\nu(s')\right]$. In this case, the Pearson product-moment correlation coefficient:

$$\Gamma(s',s'')=\frac{1}{\sqrt{\sigma^2(s')\sigma^2(s'')}}\left[\frac{1}{\Delta\nu}\int_{\Delta\nu}\kappa_\nu(s')\kappa_\nu(s'')d\nu-\bar{k}(s')\bar{k}(s'')\right]\approx 1 \tag{1}$$

where $\bar{k}(s')$ and $\sigma^2(s')$ are the mean and variance of the k-distribution at abscissa s'. The dependency of $\Gamma(s',s)$ on the spectral interval $\Delta\nu$ is not explicitly written here to abridge the notations. The strict equality in Eq. (1) is achieved when gas spectra are linearly correlated, which is almost the case if assumption A2 is verified.

Contrariwise, if gas spectra are uncorrelated, $\Gamma(s',s'')=0$.

In general, gas spectra are neither fully correlated nor fully uncorrelated over any spectral interval containing several lines. This is illustrated (very schematically, for pedagogical purposes) on Figure 1: high resolution spectra are shown for pure CO_2 at three temperatures, 500K (bottom), 1,500 K (center) and 2,500 K (top). From this figure, it is clear that over this narrow band, gas spectra cannot be rigorously assumed correlated or uncorrelated. Nevertheless, over smaller spectral intervals, these assumptions can be well thought-out as valid. Indeed, the correlation assumption seems acceptable over spectral intervals Z1, which correspond to wavenumbers that are located near the center of spectral lines observed at each selected temperature. Contrariwise, uncorrelation seems more accurate over intervals Z2, since some lines that appear in the two high temperature spectra (the so-called "hot lines") are missing in the spectrum of the coldest one. The rigorous processing of this problem by the k-moment approach is presented in the next two sections.

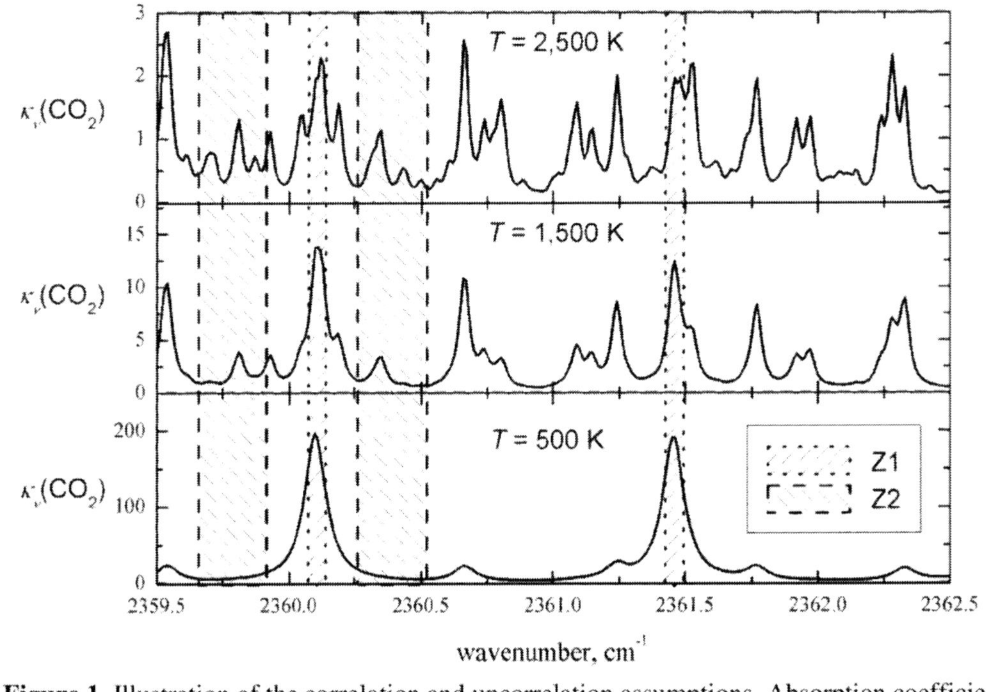

Figure 1. Illustration of the correlation and uncorrelation assumptions. Absorption coefficients are in $cm^{-1}.atm^{-1}$.

2.2. *The moment problem in nonuniform media*

For an homogeneous isothermal gaseous column of length l between two abscissa 0 and s (see Fig.2), the transmission function averaged over $\Delta\nu$ is given by [5,7]:

$$\bar{\tau}^{\Delta\nu}(0,s)=\frac{1}{\Delta\nu}\int_{\Delta\nu}\exp\left[-\kappa_\nu\left(\underline{\phi}\right)l\right]d\nu=\int_0^\infty f(k)\exp(-xPl\,k)\,dk \tag{2}$$

in which $f(k)$ is the k-distribution, defined as:

$$f(k)dk=\int_{\nu\in\Delta\nu/\kappa_\nu(\underline{\phi})\in[k,k+dk]}\frac{d\nu}{\Delta\nu} \tag{3}$$

$f(k)dk$ represents the fraction of wavenumbers inside $\Delta\nu$ such that the absorption coefficient of the gas in the thermophysical state $\underline{\phi}$ is between k and $k+dk$. It can be obtained as the inverse Laplace transform of the transmission function but can also be reconstructed from the knowledge of the k-moments using the so-called k-moment method introduced in Ref. [9]. Some theoretical links between the inverse Laplace transform and the moment problem was studied for instance in Ref. [10].

In theory, applying the moment approach to recover the actual k-distribution requires to solve an infinite set of moment equations of the following form:

$$\forall n\in\mathbb{Z},\ \int_0^\infty f(k)\,k^n\,dk=\frac{1}{\Delta\nu}\int_{\Delta\nu}\left[\kappa_\nu\left(\underline{\phi}\right)\right]^n d\nu=\mu_n\left(\underline{\phi}\right) \tag{4}$$

In practice, considering n= 1, 2 values is enough to provide accurate estimates of the transmission function (see for example Ref. [9]) in uniform media. μ_1 corresponds to the band averaged absorption coefficient $\bar{k}$.

For nonuniform paths, Eq. (2) becomes:

$$\bar{\tau}^{\Delta\nu}(0,s)=\frac{1}{\Delta\nu}\int_{\Delta\nu}\exp\left[-\int_0^s x(s')P(s')\kappa_\nu(s')\,ds'\right]d\nu=\frac{1}{\Delta\nu}\int_{\Delta\nu}\exp\left[-\kappa_{\nu,e}(0,s)u_e(0,s)\right]d\nu \tag{5}$$

where we have introduced the two path-averaged quantities (noted with the index "e"):

$$\begin{cases}u_e(0,s)=\int_0^s x(s')P(s')\,ds'\\ \kappa_{\nu,e}(0,s)u_e(0,s)=\int_0^s x(s')P(s')\kappa_\nu(s')\,ds'\end{cases} \tag{6}$$

It is possible to write Eq. (5) in a k-distribution form. In this case, the distribution is associated with the path-averaged absorption coefficient $\kappa_{\nu,e}(0,s)$:

$$\bar{\tau}^{\Delta\nu}(0,s)=\int_0^\infty f_e(k)\exp\left[-k\,u_e(0,s)\right]dk \tag{7}$$

where as a generalization of Eq. (3):

$$f_e(k)dk=\int_{\nu\in\Delta\nu/\kappa_{\nu,e}(0,s)\in[k,k+dk]}\frac{d\nu}{\Delta\nu} \tag{8}$$

Associated moment equations are:

$$\int_0^{\infty} f_e(k)\,k^n\,dk = \frac{1}{\Delta\nu}\int_{\Delta\nu}\left[\kappa_{\nu,e}(0,s)\right]^n d\nu = \mu_{n,e}(0,s) \tag{9}$$

Applying Eqs. (9) to infer f_e requires to define strategies to estimate the path-averaged *k*-moments, which is the aim of the next section, for n limited to the 1st and 2nd orders.

2.3. Nonuniform path-averaging strategies for k-moments

The processing of the first-order moment equation is obvious and provides:

$$\bar{k}_e(0,s)u_e(0,s) = \int_0^s x(s')P(s')\bar{k}(s')\,ds' \tag{10}$$

This equation, together with the definition of $u_e(0,s)$ given by Eq. (6), is common to all nonuniform narrow band models, including the usual Curtis-Godson and Lindquist-Simmons ones [5].

The case of the second-order moment equation is a little bit more complicated. Indeed, Eq. (9) provides:

$$\mu_{2,e}(0,s)\left[u_e(0,s)\right]^2 = \frac{1}{\Delta\nu}\int_{\Delta\nu}\left[\int_0^s x(s')P(s')\kappa_\nu(s')\,ds'\right]\left[\int_0^s x(s'')P(s'')\kappa_\nu(s'')\,ds''\right]d\nu \tag{11}$$

If we separate the second path-integral into two components, $T(s') = \{s'' \in [0,s] \,/\, \forall \nu \in \Delta\nu,\ \kappa_\nu(s'') = \kappa_\nu(s')\}$ and $U(s') = \{s'' \in [0,s] \,/\, \exists \nu \in \Delta\nu,\ \kappa_\nu(s'') \neq \kappa_\nu(s')\}$, with $T(s') \cup U(s') = [0,s]$, we obtain:

$$\mu_{2,e}(0,s)\left[u_e(0,s)\right]^2 = \int_{s'=0}^{s}\left(\int_{\substack{s''=0\\ s''\in T(s')}}^{s} + \int_{\substack{s''=0\\ s''\in U(s')}}^{s}\right)\left[\frac{1}{\Delta\nu}\int_{\Delta\nu} x(s')P(s')\kappa_\nu(s')x(s'')P(s'')\kappa_\nu(s'')\,d\nu\right]ds'ds'' \tag{12}$$

Let's introduce the variance of the *k*-distribution f_e defined as:

$$\sigma_e^2(0,s) = \frac{1}{\Delta\nu}\int_{\Delta\nu}\left[\kappa_{\nu,e}(0,s) - \bar{k}_e(0,s)\right]^2 d\nu = \mu_{2,e}(0,s) - \left[\bar{k}_e(0,s)\right]^2 \tag{13}$$

Then, together with Eq. (1), Eq. (12) can be written as:

$$\begin{aligned}\sigma_e^2(0,s)\left[u_e(0,s)\right]^2 = &\int_{s'=0}^{s}\int_{\substack{s''=0\\ s''\in T(s')}}^{s} x(s')P(s')x(s'')P(s'')\underbrace{\Gamma(s',s''=s')}_{=1}\sqrt{\sigma^2(s')\sigma^2(s''=s')}\,ds'ds''\\ &+\int_{s'=0}^{s}\int_{\substack{s''=0\\ s''\in U(s')}}^{s} x(s')P(s')x(s'')P(s'')\,\Gamma(s',s'')\sqrt{\sigma^2(s')\sigma^2(s'')}\,ds'ds''\end{aligned} \tag{14}$$

It can be put in the following compact form:

$$\sigma_e^2(0,s) = \sigma_{e,U}^2(0,s) + \Omega(0,s)\left[\sigma_{e,C}^2(0,s) - \sigma_{e,U}^2(0,s)\right] \tag{15}$$

in which we have introduced the path-averaged correlation coefficient $\Omega(0,s)$ defined as:

$$\Omega(0,s)=\frac{\omega(0,s,\Gamma)}{\omega(0,s,1)},\ \omega(0,s,X)=\int_{s'=0}^{s}\int_{\substack{s''=0\\ s''\in U(s')}}^{s} x(s')P(s')x(s'')P(s'')\ X(s',s'')\sqrt{\sigma^2(s')\sigma^2(s'')}ds'ds'' \tag{16}$$

where X is a function of abscissa s' and s''. Quantities $\sigma^2_{e,C}(0,s)$ and $\sigma^2_{e,U}(0,s)$ are defined hereafter. If over the nonuniform path [0,s], gas spectra are assumed uncorrelated (index "U") then if $s''\in U(s')$, $\Gamma(s',s'')=0$. From Eq. (16), $\Omega(0,s)=0$ and:

$$\sigma_e^2(0,s)=\sigma^2_{e,U}(0,s)=\left[u_e(0,s)\right]^{-2}\int_{s'=0}^{s}\int_{\substack{s''=0\\ s''\in T(s')}}^{s} x(s')P(s')x(s'')P(s'')\sqrt{\sigma^2(s')\sigma^2(s''=s')}ds'ds'' \tag{17}$$

Contrariwise, if over [0,s], gas spectra are assumed correlated (index "C") then for any $s''\in U(s')$, $\Gamma(s',s'')=1$ and $\Omega(0,s)=1$ that provides:

$$\sigma^2_{e,C}(0,s)=\left[u_e(0,s)\right]^{-2}\left[\int_{s'=0}^{s} x(s')P(s')\sqrt{\sigma^2(s')}\ ds'\right]^2 \tag{18}$$

Eq. (18) is exactly the same as the one obtained in one of our previous works, using similar assumptions [6]. Eqs. (17,18) can be rewritten in a discrete form (see Figure 2 for details) by noticing that condition $\kappa_\nu(s_j^-)=\kappa_\nu(s_i^-)$ is equivalent to consider that $T(s_j^-)=T(s_i^-)$ and $\gamma_L(s_j^-)=\gamma_L(s_i^-)$ for Lorentz lines, assumed to have the same Lorentz half width at half maximum:

$$\begin{cases}\text{correlation: } u_e(0,s)\sqrt{\sigma^2_{e,C}(0,s)}=\sum_{i=0}^{N}x(s_i^-)P(s_i^-)\sqrt{\sigma^2(s_i^-)}\delta s_i\\ \text{uncorrelation: } \left[u_e(0,s)\right]^2\sigma^2_{e,U}(0,s)=\sum_{i=0}^{N}x(s_i^-)P(s_i^-)\sigma^2(s_i^-)\Pi(s_i^-)\delta s_i\\ \text{with: } \Pi(s_i^-)=\sum_{\substack{j\text{ such that}\\ T(s_j^-)=T(s_i^-)\\ \text{and } \gamma_L(s_j^-)=\gamma_L(s_i^-)}} x(s_j^-)P(s_j^-)\delta s_j\end{cases} \tag{19}$$

The correlation assumption provides a linear path-averaging equation in terms of parameters δs_i, $i=1,..,N$, like usual nonuniform approximations such as the Curtis-Godson one. The uncorrelation formulation yields a nonlinear relation.

Eq. (15) cannot be used in many cases because the path-averaged correlation coefficient defined by Eq. (16) is usually not known, but is required in a general frame. Thus, the simplest case corresponding to uncorrelated ($\Omega(0,s)=0$) or correlated ($\Omega(0,s)=1$) spectra barely represent the actual spectral correlations. Nevertheless, they are simple enough to be compared and assessed against LBL reference calculations for radiative transfer applications. This will be detailed after some elements about the multi-spectral modeling are described in section 3.

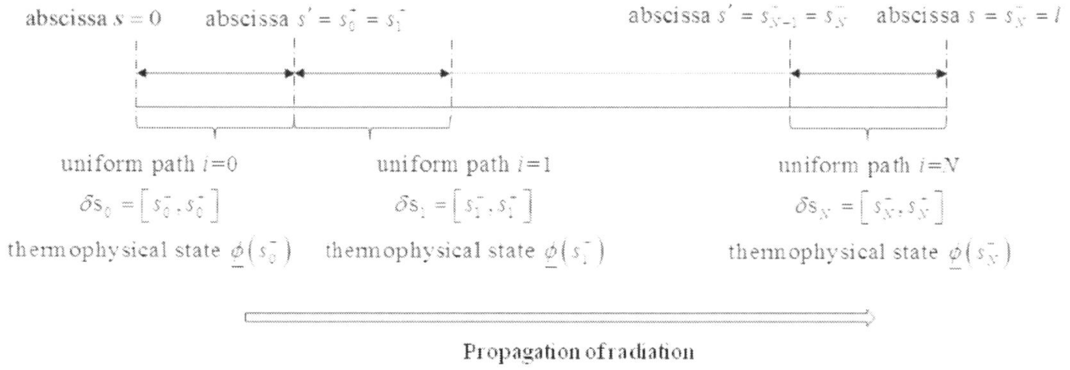

Figure 2. Notations used for the nonuniform gas path modeling

3. Application to the narrow band multi-spectral reordering technique

Path-averaging equations given in Section 2 can be used over any kind of spectral interval, as long as it is narrow enough to assume that the Planck function is constant over it. In this section, $\Delta\nu$ is defined using the multi-spectral reordering technique for the SLMB modeling of the radiative properties of the gas, whose principles can be found in Ref. [4].

The MSR technique can be formulated as follows. Let's consider two (arbitrary) reference thermophysical conditions and define, for a set of real positive parameters k_1, k_2, the following two-dimensional distribution:

$$f(k_1,k_2)dk_1dk_2 = \int_{\Delta\nu(k_1,k_2)} \frac{d\nu}{\Delta\nu} = \frac{\Delta\nu(k_1,k_2)}{\Delta\nu} \tag{20}$$

where:

$$\Delta\nu(k_1,k_2) = \left\{\nu \in \Delta\nu \,/\, \kappa_\nu\left(\underline{\phi}_1^{ref}\right) \in [k_1, k_1+dk_1] \text{ and } \kappa_\nu\left(\underline{\phi}_2^{ref}\right) \in [k_2, k_2+dk_2]\right\} \tag{21}$$

Eq. (20) extends Eq. (2) to more than one "reference". It should be noticed that usual narrow band *k*-distribution models do not explicitly make use of "reference" conditions. Indeed, these models assume that all spectra are correlated, so that any thermophysical state is implicitly a reference.

Using this two-dimensional *k*-distribution, we can rewrite the transmission function in Eq. (6) as:

$$\overline{\tau}^{\Delta\nu}(0,s) = \int_{k_1=0}^{+\infty}\int_{k_2=0}^{+\infty} f(k_1,k_2)\left\{\frac{1}{\Delta\nu(k_1,k_2)}\int_{\Delta\nu(k_1,k_2)} \exp\left[-\kappa_{\nu,e}(0,s)u_e(0,s)\right]d\nu\right\}dk_1dk_2 \tag{22}$$

The term in brackets can be estimated by using the *k*-moment method which provides, assuming an inverse Gaussian *k*-distribution over $\Delta\nu(k_1,k_2)$ [11]:

$$\frac{1}{\Delta\nu(k_1,k_2)}\int_{\Delta\nu(k_1,k_2)} \exp\left[-\kappa_{\nu,e}(0,s)u_e(0,s)\right]d\nu \approx \exp\left\{-\frac{\left[\overline{k}_e(0,s)\right]^2}{\sigma_e^2(0,s)}\left[\sqrt{1+2\frac{u_e(0,s)\sigma_e^2(0,s)}{\overline{k}_e(0,s)}}-1\right]\right\} \tag{23}$$

$\overline{k}_e(0,s)$ and $\sigma_e^2(0,s)$ are quantities spectrally averaged over $\Delta\nu(k_1,k_2)$. They can be obtained from the application of Eqs. (9,19), using the mean and variance of the absorption coefficient on each uniform sub-path along the nonuniform one.

NB MSR-SLMB databases were built to easily compute the mean and variance of the absorption coefficient using parameters (see Ref. [12] for details) estimated from LBL spectra for 25 temperatures between 300 K and 2,700 K by 100 K steps for CO_2 and H_2O. For H_2O, a second discretization as a function of H_2O molar fraction is made. Details about these calculations can be found in Ref. [4]. Their application for any thermophysical condition consists in using a simple linear interpolation formula [12]. The multi-spectral reordering strategy proposed in Ref. [4], with T_1^{ref} =500 K and T_2^{ref} =1,500 K as reference temperatures, was chosen. k-discretizations use 8 intervals of k values which means that with the bi-spectral reordering, up to 64 spectral intervals (over each narrow band) can result from the reordering. Nevertheless, most of them are empty and thus are not involved in the calculations. A pure gas at atmospheric pressure was selected as the reference composition for CO_2 and 10 % H_2O in nitrogen was chosen for water vapor. Finally, the narrow spectral bands proposed in Ref. [13] were used, which corresponds to 96 bands for CO_2 and 367 for H_2O. This discretization covers the [150 cm^{-1} ; 9,300 cm^{-1}] spectral range. NB transmission functions for mixtures of CO_2-H_2O-N_2 were calculated as the product of those of the single gases.

Using the NB MSR-SLMB model does not require excessive computational storage: CO_2 database occupies 4 Mo and H_2O 87 Mo. The calculation of around 80 path-integrals per narrow band is required. It is thus more demanding than the usual narrow band model (the Curtis-Godson approximation involves 3 path integrals only) but remains around 30 times less expensive than LBL calculations (2,500 path integrals per narrow band at 10^{-2} cm^{-1} resolution).

4. Results

In order to assess the MSR modeling and the various nonuniform approximations described in this work, we chose two test cases representative of engineering applications. The NB MSR-SLMB model with the correlated and uncorrelated approximations were used for the calculations, as well as the usual NBKM/SNB model with the Curtis-Godson approximation [5,12] and the NB MSR-SLMB together with the following formula (see Ref. [8]):

$$u_e(0,s)\frac{\sigma^2_{e,2OKM}(0,s)}{\bar{k}_e(0,s)} = \sum_{i=0}^{N} x(s_i^-)P(s_i^-)\frac{\sigma^2(s_i^-)}{\bar{k}(s_i^-)}\delta s_i \tag{24}$$

The model based on Eq. (24) will be referred as 2OKM from now on. The 1st order MSR-SLMB model, which assumes a Dirac k-distribution function instead of the Inverse Gaussian one was also considered for the comparisons. In the case of the NBKM/SNB model, nonuniformities are processed by applying the Curtis-Godson approximation:

$$\frac{u_e(0,s)\left[\bar{k}_e(0,s)\right]^3}{\sigma^2_{e,CG}(0,s)} = \sum_{i=0}^{N} x(s_i^-)P(s_i^-)\frac{\left[\bar{k}(s_i^-)\right]^3}{\sigma^2(s_i^-)}\delta s_i \tag{25}$$

As a first test we considered the temperature and molar fraction profiles corresponding to an oxy-fuel combustion scenario described in Ref. [14]. This profile was chosen because it corresponds to a high molar fraction of CO_2 (higher than 0.6) for which the optically thin approximations cannot be used *a priori*. It should be noticed that the influence of the variance is more sensitive at the optically thick than at the thin limit which justifies this choice. Results are depicted on Figure 3. In this case, all models provide similar errors that do not exceed 0.05 for the MSR models but can reach 0.27 with the usual NBKM/SNB approach. The correlated and 2OKM models are in this case the most accurate.

The second test case is taken from Ref. [15] (see Fig. 10 in this reference for a description of the specie and temperature profiles). It is representative of a plume signature situation, for which radiation emitted by the hot gases is measured after its transmission through a long path of cold air. As CO_2 and

H_2O are both in the exhaust gases and the air, but at different temperatures and concentrations, spectral correlation effects are very important in this kind of situation. Results are given in Figure 4. In this case, it is clear that the MSR modeling performs better than the standard NBKM/SNB model. The uncorrelated model provides here the best accuracy.

From these two test cases, it is difficult to conclude which model, correlated or uncorrelated, is the best one in a general frame. In fact, results indicate that some information about the correlation between spectra along the path, represented by parameter $\Omega(0,s)$, is likely to partially correct errors due to these assumptions. From a practical point of view, calculating $\Omega(0,s)$ from LBL data is possible, but would require to build specific databases, including the parameters to estimate $\Gamma(s',s'')$. Such an approach would increase the computation costs drastically as it would also require the computation of both the correlated and uncorrelated solutions involved in Eq. (15). Nevertheless, for radiative sensing and spectrum analysis purposes, such a full modeling might be useful.

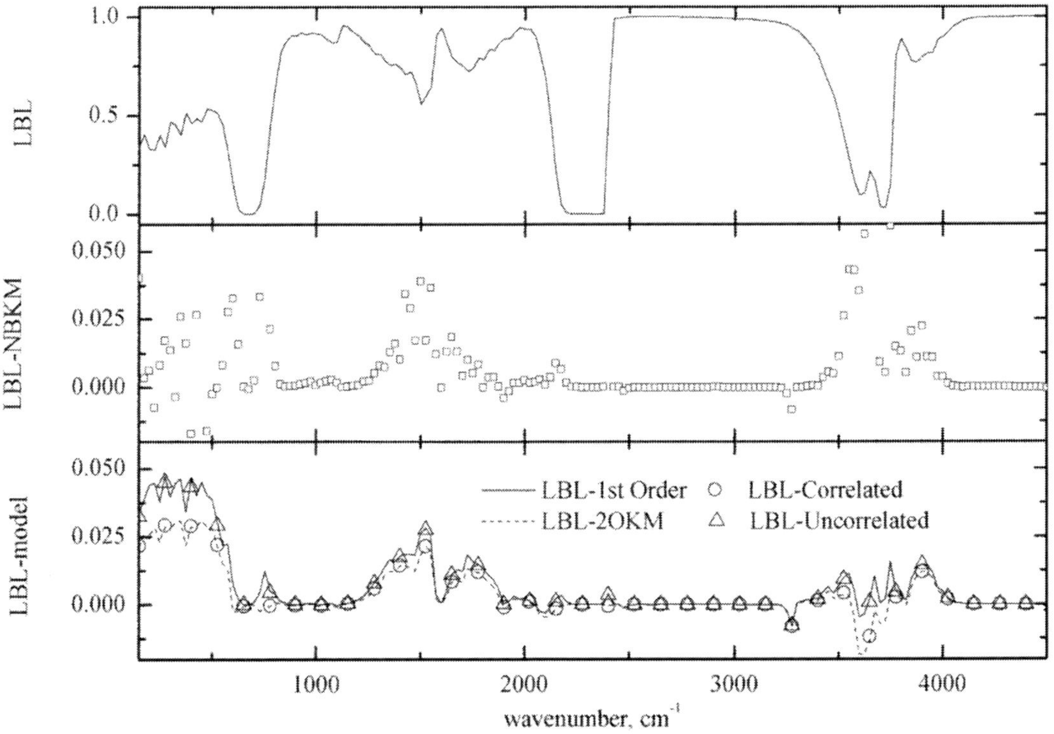

Figure 3. Narrow band transmission functions for the nonuniform oxy-fuel temperature and specie concentration profiles (from Ref. [14]).

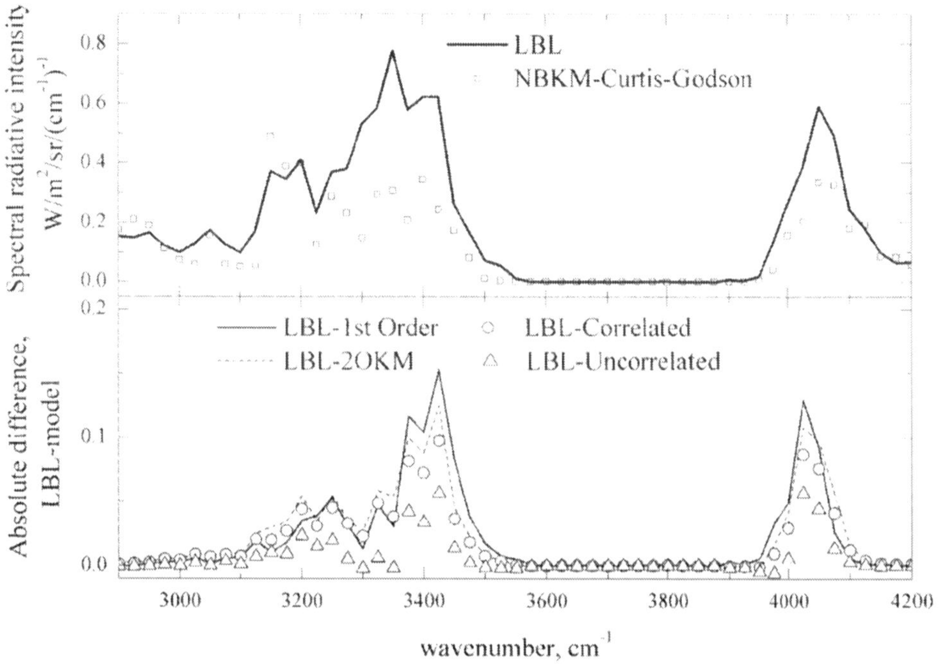

Figure 4. Narrow band intensities for an IR signature test case, taken from Ref. [15].

4. Conclusion

In this paper, a general nonuniform formula for the path-averaged narrow band model parameters has been proposed, as well as two simpler ones based on the correlation/uncorrelation assumptions. They have been applied to a NB MSR-SLMB model and assessed against LBL reference data for CO_2-N_2 or H_2O-N_2 mixtures. Several comparisons with a standard NBKM/SNB model and the Curtis-Godson approximation have also been discussed. From this study, it follows that all MSR models based on second order *k*-moments provide a similar accuracy, better than those resulting from standard narrow band models, at the cost of extra but reasonable calculation costs. Correlated/uncorrelated approaches combined with the MSR technique are acceptable for radiative transfer applications, whereas the full model might be required for fine radiative analysis. This statement will be refined in future works.

References

[1] Levi Di Leon R, Taine J 1986 *Rev. Phys. Appl.* **21** 825-31
[2] Zhang H, Modest M F 2003 *Int. J. Heat Mass Transfer* **46** 3593-603
[3] West R, Crisp D, Chen L 1990 *J. Quant. Spectr. Rad. Transfer* **43** 191-199
[4] André F, Vaillon R, Galizzi C, Guo H, Gicquel O 2011 *J. Quant. Spectr. Rad. Transfer* **112** 394-411
[5] Taine J, Soufiani A 1999 *Adv. Heat Transfer* **33** 295-414
[6] André F, Vaillon R 2010 *J. Quant. Spectr. Rad. Transfer* **111** 1900-11
[7] Modest M F 2003 *Radiative heat transfer, 2nd ed.*, Academic Press
[8] André F, Vaillon R 2008 *J. Quant. Spectr. Rad. Transfer* **109** 258-68
[9] André F, Vaillon R 2007 *J. Quant. Spectr. Rad. Transfer* **108** 1-16
[10] Widder D V 1934 *Transactions of the American Mathematical Society* **36** 107-200
[11] André F, Vaillon R 2011 submitted to *J. Quant. Spectr. Rad. Transfer*
[12] André F, Vaillon R 2010 *J. Quant. Spectr. Rad. Transfer* **111** 325-330
[13] Soufiani A, Taine J 1997 *Int. J. Heat Mass Transfer* **40** 987-991
[14] Chu H, Liu F, Zhou H 2011 *Int. J. Heat Mass Transfer* **54** 4736-4745
[15] Rivière Ph, Soufiani A, Taine J 1992 *J. Quant. Spectr. Rad. Transfer* **48** 187-203

A Multi-Source Full Spectrum K-Distribution Method for 1-D Inhomogeneous Media

John Tencer, John R. Howell

Dept of Mechanical Engineering, The University of Texas at Austin, Austin, TX 78712-0292, USA

E-mail: JTTencer@gmail.com

Abstract. The k-distribution method for treating the spectral properties of an absorbing-emitting medium represents an alternative to line-by-line calculations which reduces the number of RTE evaluations from the order of a million to the order of ten without any significant loss of accuracy. For problems where an appropriate reference temperature can be defined, the k-distribution method is formally exact. However, when no appropriate reference temperature can be defined, the method results in errors. There have been several attempts to implement corrections to the k-distribution method to extend its application to inhomogeneous media by modelling the effects of temperature, pressure, and concentration gradient. Some of these techniques are very effective at reducing these errors, but none successfully extend the k-distribution method to inhomogeneous media in a way that retains the analytical guarantee of line-by-line accuracy. The Multi-Source Full Spectrum K-Distribution Method (MSFSK) introduced here manages to attain this goal for 1-dimensional domains with piecewise constant temperature distributions by applying the superposition principle to the original RTE before applying the k-distribution transformation to decompose the problem into a set of sub-problems each of which is able to be solved exactly via the ordinary or modified full spectrum k-distribution method.

1. Introduction

Heat transfer problems in general involve solving the conservation of energy equation. If convection is important, there is an associated fluid dynamics problem that is coupled to the energy equation that must be solved in order to determine the flow field. The general energy equation for a single-component translucent fluid including all three heat transfer modes is

$$\rho c_p \frac{DT}{Dt} = \beta T \frac{DP}{Dt} + \nabla \bullet (\kappa \nabla T - q_r) + q''' + \Phi_d \qquad (1.1)$$

where β is the thermal coefficient of volume expansion of the fluid, q_r is the radiant flux vector, q''' is the local energy source, and Φ_d is the energy production due to viscous dissipation [1].

In this paper, we are only concerned with calculating the radiant flux vector and its divergence. This term is neglected in many engineering applications. However, since the radiative term scales as T^4, it becomes increasingly significant in high temperature applications. These applications could include analysis of the heat transfer in furnaces, glass melting and cooling, and spacecraft reentry. As advances in materials continue to be made, higher temperatures are becoming more common in a wider array of applications in turbo machinery as well. In addition to high temperature applications, radiation is dominant in atmospheric heating and cooling making it of special interest to meteorologists. Regardless of the application, in order to find a solution to Equation (1.1) an expression for q_r is needed.

The integral formulation of the radiant flux divergence is found by analyzing the local radiative flux. The divergence of the radiative flux vector is equal to the rate of energy emitted locally by the medium minus the energy absorbed by the medium. This relation is valid at any location within the medium. Specifically, this quantity is defined by Equation (1.2).

$$\nabla \cdot q_r(S) = 4\pi \int_{\lambda=0}^{\infty} \kappa_\lambda(\lambda,S) I_{\lambda b}(\lambda,S) d\lambda - \int_{\lambda=0}^{\infty} \int_{\Omega=0}^{4\pi} \kappa_\lambda(\lambda,S) I_\lambda(\lambda,S,\Omega) d\Omega d\lambda \tag{1.2}$$

Here, κ_λ is the local absorption coefficient as a function of wavelength, $I_{\lambda b}$ is the blackbody intensity as a function of wavelength, and I_λ is the local incoming radiative intensity as a function of wavelength and solid angle, Ω. The subscript λ indicates the dependence of the quantity upon wavelength. The use of the absorption coefficient and the blackbody intensity in the definition of the emission term is dependent upon the assumption of local thermodynamic equilibrium or LTE. For the purposes of this review, this assumption will be presumed valid in every situation.

The intensities are both intrinsically dependent upon temperature. The local black-body intensity present in the emission term is dependent upon the local temperature. The incident intensity in the absorption term is dependent upon the global temperature field. The local incoming radiative intensity is also a function of scattering (if present) and is given by Equation (1.3). The spectral absorption coefficient includes a temperature dependence as well.

$$I_\lambda(\tau_\lambda,\Omega) = I_\lambda(0,\Omega) e^{-\tau_\lambda} + \int_0^{\tau_\lambda} \hat{I}_\lambda(\tilde{\tau}_\lambda,\Omega) e^{-(\tau_\lambda - \tilde{\tau}_\lambda)} d\tilde{\tau}_\lambda \tag{1.3}$$

where $\hat{I}_\lambda(\tau_\lambda,\Omega)$ is the source function and τ_λ is the opacity or optical depth and is defined in Equation (1.4). This is the way in which scattering affects the incident intensity. While it does not directly account for emission or absorption it does decrease the effective mean free path within the medium.

$$\tau_\lambda(S) = \int_0^S \left[\kappa_\lambda(s) + \sigma_{s\lambda}(s)\right] ds \tag{1.4}$$

In some situations, scattering can be neglected. If this is the case, the source function in Equation (1.3) is given simply as the local blackbody intensity, $I_{\lambda b}$. In general however, the source function is more complicated and is given by

$$\hat{I}(\tau_\lambda,\Omega) = (1-\omega_\lambda) I_{\lambda b}(\tau_\lambda) + \frac{\omega_\lambda}{4\pi} \int_{\tilde{\Omega}=0}^{4\pi} I_\lambda(\tau_\lambda,\tilde{\Omega}) \Phi_\lambda(\Omega,\tilde{\Omega}) d\tilde{\Omega} \tag{1.5}$$

ω_λ is the scattering albedo. The albedo is a generalized reflectivity of the medium and is defined as the ratio of the scattering coefficient ($\sigma_{s\lambda}$) to the extinction coefficient ($\kappa_\lambda + \sigma_{s\lambda}$). Basically, the scattering albedo defines the portion of attenuated radiation that is scattered rather than absorbed. This is an important quantity because it helps to determine which solution methods might be appropriate. Scattered energy continues to be tracked through the system until it is absorbed, so problems with a high scattering albedo generally require additional computational expense unless special methods developed just for those cases are utilized.

2. Solving the Spectral Radiative Transport Equation in Isothermal Media

Our goal is to determine the total radiative flux divergence. As is suggested by the definition, the radiative flux divergence may be found by solving Equation (1.3) for the spectral intensity and then integrating over all wavelengths. There are many methods that have been developed for solving Equation (1.3), each with its own advantages and disadvantages. Which method is used to solve the transport equation for any given application is not relevant to this paper. Here, we are focused on the integration of the spectral intensity over all wavelengths and how to evaluate this integral numerically in a way that is both accurate and computationally efficient.

2.1. Line-by-line method

Calculation of the radiative flux divergence within a medium requires integration of the absorption coefficient and intensities over all wavelengths. The most direct and accurate method is line-by-line calculations. In a line-by-line scheme, the spectrum is divided into small spectral intervals over which the absorption coefficient and local black-body intensity may be assumed constant. In real gasses these intervals must be extremely small in order to capture the many resonance peaks in the absorption coefficient spectrum. The RTE is then solved for local intensity within each wavelength interval. This is time consuming and expensive.

The spectral RTE along a given path neglecting scattering is given by Equation (2.1). The absorption coefficient is potentially a function of S since in inhomogeneous media the absorption coefficient varies spatially.

$$\frac{dI_\lambda}{dS} = \kappa_\lambda (I_{b\lambda} - I_\lambda) \tag{2.1}$$

The total intensity is found by integrating the spectral intensity over all wavelengths as shown in Equation (2.2).

$$I = \int_0^\infty I_\lambda d\lambda \tag{2.2}$$

Evaluating this integral numerically can be very computationally expensive. In real gasses hundreds of thousands or millions of spectral intervals may be required in order to accurately calculate the total intensity.

2.2. The k-distribution method

The k-distribution method represents an alternative to line-by-line calculations which reduces the number of RTE evaluations from the order of a million to the order of ten without any significant loss of accuracy for homogeneous media [2]. This requires the formulation of a weighted cumulative distribution function of the absorption coefficient where the weight function is the Planck function.

$$g(\kappa) = \frac{\pi}{\sigma T^4} \int_{\lambda \text{ s.t. } \kappa_\lambda \le \kappa} I_{b\lambda} d\lambda \tag{2.3}$$

Figure 1 demonstrates graphically the regions over which the Planck distribution would be integrated in order to compute g for a particular value of the absorption coefficient, κ (or equivalently, the absorption cross section, C) for a narrow section of a representative spectrum.

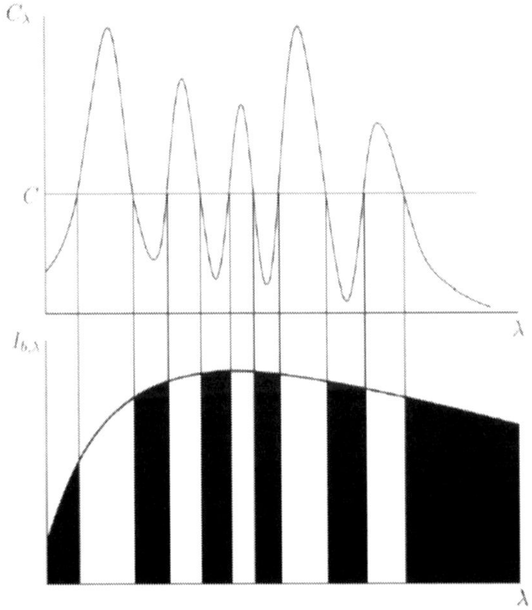

Figure 1: k-distribution Demonstration [3]

C is directly proportional to κ. The constant of proportionality is the number density, N. Assuming ideal gas behavior, this relationship is expressed in Equation (2.4).

$$\kappa = C \cdot N \approx C \cdot \frac{P}{RT} \tag{2.4}$$

where P is the pressure, T is the temperature and R is the universal gas constant expressed in consistent units.

The cumulative distribution function, g, is a smooth monotonically increasing function of κ. In fact, g is bijective. Figure 2 shows an example of a k-distribution generated for a 10% mixture of CO_2 at 600K and 1atm. Since g(κ) is a bijection, it is invertible and a unique function κ_g or C_g may be defined.

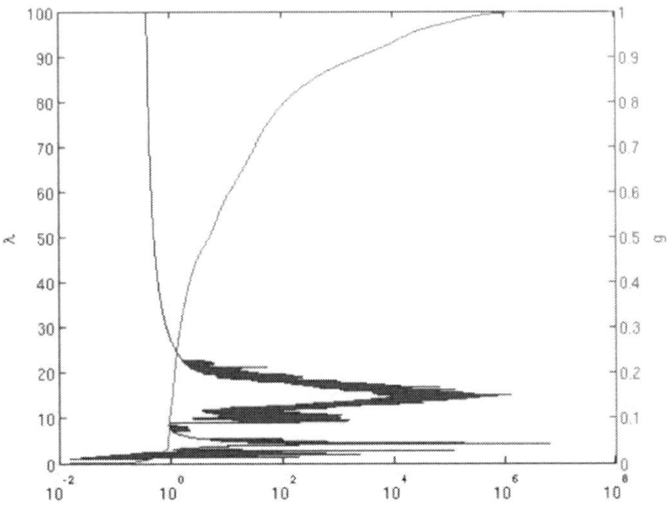

Figure 2: Example k-distribution for 10% CO_2 at 600K, 1 atm

We can then rewrite the radiative transport equation as a function of g rather than λ.

$$\frac{dI_g}{dS} = \kappa_g (I_b - I_g) \tag{2.5}$$

Additionally, we can evaluate the total intensity by integrating over g rather than λ.

$$I = \int_0^1 I_g dg \tag{2.6}$$

As can be seen from Figure 2, κ_g is a smoothly varying function of g and consequently, I_g is a smoothly varying function of g as well. This allows for a much smaller number of quadrature points to achieve near-line-by-line accuracy. The integration over g may be accomplished using any available quadrature rule. An 8 point Gaussian quadrature is found to provide good results in most cases although increased accuracy may be achieved by using higher order quadrature schemes.

In addition, it should be noted that for each value of g we find the intensity by solving the RTE for a (potentially inhomogeneous) gray gas. This solution may be found by a number of methods ranging from highly accurate but computationally expensive to extremely fast with only first order accuracy. Which solution method is appropriate may be determined based upon the optical thickness for a given value of g.

There is an inherent problem with the method just described when analyzing non-isothermal media. The Planck distribution in Equation (2.3) is not only a function of wavelength, but of temperature as well. Choosing the appropriate temperature for use in defining the k-distribution function a priori is difficult for non-isothermal problems. For cases where temperature variations across the domain are relatively small, this temperature dependence is negligible and the full spectrum k-distribution method provides good results for any reasonable choice of reference temperature including but not limited to the maximum temperature in the system, the minimum temperature in the system, the volume averaged temperature, the Planck mean temperature, and the emission weighted temperature [4].

As the temperature variations become large, significant errors can develop as a result of this approximation [5]. One way to solve this temperature dependence problem is to model the absorption coefficient's dependence on the temperature as was done in the Multi-Group Full Spectrum k-distribution Method (MGFSK) [6].

The MGFSK method provides good results for inhomogeneous media and the accuracy may be increased by the inclusion of an increasing number of groups. However, the method remains approximate. A more accurate but more computationally expensive alternative is presented here. The multi-source full spectrum k-distribution method is capable of providing exact results for one dimensional geometries with piecewise constant temperature and absorption coefficient in the same way that the full spectrum k-distribution method is able to provide exact results for homogeneous media problems.

3. Solving the Spectral Radiative Transport Equation in Non-Isothermal Media

For treatment of non-isothermal media, various approximations must be made. It is not feasible to provide a continuous definition of the medium's spectral properties as a function of temperature. As a substitute, the continuous temperature distribution within the slab is modeled as piecewise constant in what has been called a multilayer approach. Solovjov and Webb applied this method to the SLW and CW methods in 2008 and demonstrated that a relatively small number of layers is actually required in order to achieve good predictions of the radiative heat flux and flux divergence [7].

The non-isothermal slab is divided into K layers. Each layer has a reference temperature, T_k, a spectral absorption coefficient, κ_λ^k and a width, $\Delta x_k = x_k - x_{k-1}$. Within each layer, the total integrated heat flux and flux divergence are sought. The multi-layer approximation is similarly applicable to inhomogeneous media with concentration gradients. If concentration gradients exist, layers should be sufficiently thin so that the spectral absorption coefficient may be modeled as constant within each layer.

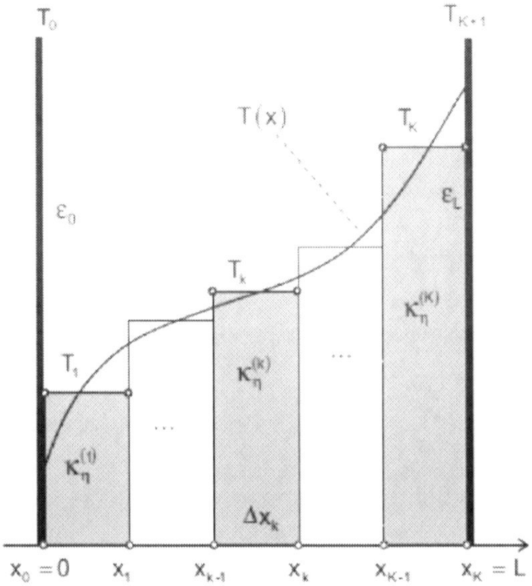

Figure 3: Schematic illustration of multilayer system and geometric nomenclature [7]

For isotropic scattering and diffuse boundary conditions, the directionally integrated spectral heat flux is only a function of optical depth and is given by Equation (3.1).

$$q_{r\lambda}(\tau_\lambda) = 2\pi \Big\{ I_\lambda^+(0)E_3(\tau_\lambda) - I_\lambda^-(\tau_{L\lambda})E_3(\tau_{L\lambda} - \tau_\lambda) + \cdots + \int_0^{\tau_\lambda} \hat{I}_\lambda(\tilde{\tau}_\lambda)E_2(\tau_\lambda - \tilde{\tau}_\lambda)d\tilde{\tau}_\lambda - \int_{\tau_\lambda}^{\tau_{L\lambda}} \hat{I}_\lambda(\tilde{\tau}_\lambda)E_2(\tilde{\tau}_\lambda - \tau_\lambda)d\tilde{\tau}_\lambda \Big\} \quad (3.1)$$

Neglecting scattering, the source function is simply the black-body intensity and is piecewise constant. The spectral heat flux then becomes

$$q_{r\lambda}(x) = 2\pi \left\{ \begin{array}{l} I_\lambda^+(0)E_3\left(\tau_\lambda(x)\right) - I_\lambda^-(L)E_3\left(\tau_\lambda(L) - \tau_\lambda(x)\right) + \cdots \\ + \sum_{i=1}^{k-1} I_{b\lambda}(T_i)\left[E_3\left(\tau_\lambda(x) - \tau_\lambda(x_i)\right) - E_3\left(\tau_\lambda(x) - \tau_\lambda(x_{i-1})\right)\right] + \cdots \\ + \sum_{i=k+1}^{K} I_{b\lambda}(T_i)\left[E_3\left(\tau_\lambda(x_i) - \tau_\lambda(x)\right) - E_3\left(\tau_\lambda(x_{i-1}) - \tau_\lambda(x)\right)\right] + \cdots \\ + I_{b\lambda}(T_k)\cdot\left[E_3\left(\kappa_\lambda^k(x - x_{k-1})\right) - E_3\left(\kappa_\lambda^k(x_k - x)\right)\right] \end{array} \right\} \quad (3.2)$$

Using the multilayer approach, the definition of the optical depth from Equation (1.4) may be simplified.

$$\tau_\lambda(x) = \int_0^x \kappa_\lambda(\tilde{x})d\tilde{x} = \sum_{i=1}^{k-1} \kappa_\lambda^i \Delta x_i + \kappa_\lambda^k\left(x - x_{k-1}\right) \quad (3.3)$$

The k-distribution method may be applied within each layer in order to reduce the number of spectral evaluations required. In order to do this, a reasonable reference temperature should be established. It has been shown [8] that in general non-isothermal problems this is not always feasible and that errors associated with this choice are difficult if not impossible to quantify.

The form of Equation (3.2) suggests an alternative to the establishment of a global reference temperature. The problem may be recast as 3K sub-problems, K of which includes emission from exactly one layer at exactly one temperature. The other 2K sub-problems involve purely absorbing media with a (complex) spectral boundary condition. The spectral heat flux at the layer boundaries is found by performing a line-by-line solution on a significantly coarser spatial grid with only K points. There is no additional assumption introduced here. The radiative transport equation is linear and we are simply representing the solution as a superposition of solutions to simpler problems.

For simplicity of explanation, we will discuss the contributions to the spectral heat flux at layer boundaries as being due to contributions from each individual layer. The total spectral heat flux is found by summing the contributions from each of the individual layers. In the 1-dimensional problem, these contributions may be written explicity.

The contribution to the spectral heat flux due to emission from the kth layer is given by Equation (3.4).

$$q_{r\lambda}^{k}(x)=2\pi I_{b\lambda}(T_k)\begin{cases}E_3\left(\tau_\lambda(x)-\tau_\lambda(x_k)\right)-E_3\left(\tau_\lambda(x)-\tau_\lambda(x_{k-1})\right) & x_k\le x\\ E_3\left(\tau_\lambda(x_k)-\tau_\lambda(x)\right)-E_3\left(\tau_\lambda(x_{k-1})-\tau_\lambda(x)\right) & x\le x_{k-1}\\ E_3\left(\kappa_\lambda^k(x-x_{k-1})\right)-E_3\left(\kappa_\lambda^k(x_k-x)\right) & x\in(x_{k-1},x_k)\end{cases} \tag{3.4}$$

In order to use the k-distribution method to simplify the numerical spectral integration, an appropriate weighting function for the generation of the absorption line blackbody distribution function (ALBDF) must be defined. Usually, a Planck distribution at a set reference temperature is used as this weighting function. However, the method may be generalized by using a general weighting function which can be defined for each problem.

$$g(\kappa)=\frac{\int\limits_{\lambda\ s.t.\ \kappa_\lambda^k<\kappa} w(\lambda)d\lambda}{\int\limits_{\lambda=0}^{\infty} w(\lambda)d\lambda} \tag{3.5}$$

$w(\lambda)$ is a weighting function that may be different for each sub-problem. Consider the K sub-problems representing the flux within a given layer due to emission from within that same layer. For the kth sub-problem of this type, the weighting function is simply the Planck distribution and is given by Equation (3.6).

$$w(\lambda)=I_{b\lambda}(T_k) \tag{3.6}$$

A database of k-distribution functions using these weighting functions has previously been developed by Modest for CO_2 and H_2O. There is no geometry information imbedded in Equation (3.6) so these functions may be evaluated and inverted offline.

There are an additional K sub-problems representing the heat flux within a given layer due to emission from layers to the left of that layer. For these sub-problems the appropriate weighting function is given by Equation (3.7). Note that the weighting functions presented here only equal the spectral heat flux up to a multiplicative constant since any such constant is eliminated in the normalization of g performed in Equation (3.5).

$$w(\lambda)=\sum_{l=1}^{k-1}\kappa_\lambda^l I_{b\lambda}(T_l)\left\{E_3\left[\tau_\lambda(x_k)-\tau_\lambda(x_l)\right]-E_3\left[\tau_\lambda(x_k)-\tau_\lambda(x_{l-1})\right]\right\} \tag{3.7}$$

The remaining K sub-problems represent the heat flux within a given layer due to emission from layer to the right of that layer. The appropriate weighting function in these sub-problems is given by Equation (3.8).

$$w(\lambda)=\sum_{l=k+1}^{K}\kappa_\lambda^l I_{b\lambda}(T_l)\left\{E_3\left[\tau_\lambda(x_{k-1})-\tau_\lambda(x_l)\right]-E_3\left[\tau_\lambda(x_{k-1})-\tau_\lambda(x_{l-1})\right]\right\} \tag{3.8}$$

The k-distribution method has the advantage that any standard quadrature rule may be employed to evaluate the spectral integral which has been transformed into an integral over g. While it is theoretically possible to vary the quadrature rule used at different spatial points within the slab, it is generally not very reasonable to do so. From this point forward, our notation will assume that the same N-point quadrature rule is used at every spatial point.

Within each layer, Equation (3.5) is inverted to define a unique function κ_g which is constant within that layer in the same manner as κ_λ and T_k. The modified wavenumber, g is in $[0,1]$ and the choice of quadrature rule defines a set of N transport equations that should be solved within each layer. Any appropriate solution technique may be used to determine the solution within each sub-problem. The solution method may be chosen on the basis of layer optical thickness.

The significant cost savings provided by this method relative to a full line-by-line solution stems from the significant computational savings and efficiency gained by the use of the k-distribution method. This decrease in cost comes at virtually no cost in accuracy besides that incurred due to the use of the multilayer approximation.

4. Computational Cost Comparison

We will compare the computational cost of the multi-source k-distribution method with the cost of a line-by-line solution and with a standard k-distribution method. The line-by-line solution requires N_{lbl} solutions of the gray radiative transport equation at M spatial locations. The cost of a single gray RTE solution scales as M^α where $\alpha \in [2,3]$ depends upon the RTE solution technique as well as the linear algebra package employed. The case where $\alpha = 2$ corresponds to the evaluation of the analytical solution expressed in (3.1). $\alpha = 3$ corresponds to the solution of a general dense matrix.

The cost of a line-by-line solution is therefore given by Equation (4.1).

$$N_{lbl} M^\alpha \tag{4.1}$$

By contrast, the cost of a solution by the standard k-distribution method is given by Equation (4.2)

$$N_{kdist} M^\alpha \tag{4.2}$$

where $N_{kdist} \ll N_{lbl}$ is the number of quadrature points used in the k-distribution spectral integration. This commonly represents an up to 5 or 6 orders of magnitude improvement in computational cost over the line-by-line method. However, for non-isothermal media, this improvement comes at a cost in accuracy which can be large in some cases and has yet to be effectively bounded in general.

The cost of the multi-source k-distribution method is between the two and is given by Equation (4.3).

$$N_{lbl} K^\alpha + 3 N_{kdist} K \left(\frac{M}{K} \right)^\alpha \tag{4.3}$$

where K is the number of layers used in the multilayer approximation. For simplicity, it is assumed that the number of spatial points per layer is uniform. For this method to be viable, $K \ll M$. Ideal applications for this method involve the desire for a high order of accuracy on a fine spatial grid.

5. Results and Discussion

Solovjov and Webb established the convergence and viability of the multilayer approach in their paper [7]. This establishes the legitimacy of approximating a slab with a continuous temperature distribution as having a piecewise constant temperature distribution as depicted graphically in Figure 3. The question remains how well the k-distribution method and the multi-source k-distribution method predict the radiative heat flux and flux divergence in slab geometries with piecewise constant temperature distributions.

As a limiting case, consider a large step change in temperature. The left half of the slab (50 cm) is at 1000 K. The right half is at 300 K. The medium is 10% CO_2 by mole and is at a uniform pressure of 1 atm. Spectral line data (centers, half-widths, and intensities) are acquired from the HITRAN 2004 database [9]. The line data is used to generate a high resolution absorption spectrum. Lorentz profiles are used for all line shapes. The same spectrum is used in both the line-by-line solution and in the generation of the k-distribution functions.

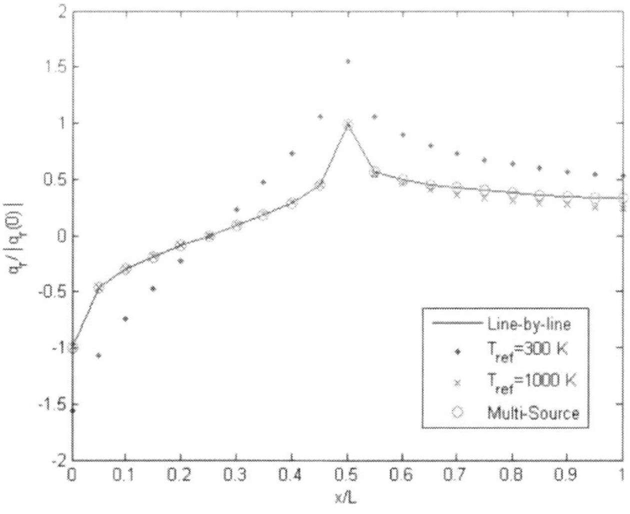

Figure 4: Heat Flux Distribution in a Non-Isothermal Medium Bounded by Cold Black Walls

Figure 4 shows the calculated heat flux using the full spectrum k-distribution method with a single reference temperature and using the multi-source k-distribution method. These results are compared to those generated using line-by-line calculations. Using a reference temperature of 300 K gives poor results. The majority of the radiative energy is emitted in the hotter left half of the slab. Using a reference temperature of 300 K is totally incorrect for this spatial region. For this reason, the heat flux is poorly represented in every part of the domain.

Using a reference temperature of 1000 K, the heat flux prediction is quite accurate in the left half of the domain. The fact that this model does a relatively poor job of predicting the behavior of energy emitted from the right half of the medium has little effect since the vast majority of the flux is due to the high temperature emission from the left half of the slab. However, this model also does a poor job of predicting the absorption of energy in the right half of the slab. This leads to potentially large relative errors in predicting the heat flux exiting the right boundary. The thicker the cold region becomes, the more significant this error becomes.

Modest and Zhang analyzed the effect of the length of the cold region on the error in the heat flux at the right wall for several different reference temperature models [4]. They used a higher temperature for the hot zone and a slightly different gas mixture. However, the general trends should hold true. Five different reference temperatures were used in this study and none were capable of achieving the kind of accuracy reported here using the Multi-Source method. Admittedly, this improvement in accuracy comes at a computational cost. However, for applications for which accuracy is paramount, the multilayer multi-source k-distribution method provides a more efficient alternative to line-by-line calculations.

This example highlights the difficulty of using the k-distribution method in non-isothermal media. The appropriate choice for a reference temperature is virtually impossible to determine a priori. Furthermore, the error associated with an incorrect choice of reference temperature is unknown. The correct value of the heat flux at any given spatial location is not guaranteed to be bounded by the heat flux calculated using the maximum system temperature as the reference temperature and the heat flux calculated using the minimum system temperature as the reference temperature. To the authors' knowledge, there has been no effective bound developed for the model error introduced in this way. As can be seen from Figure 4, this error can be substantial.

6. Conclusions

The multilayer approach as presented in this paper presents a significant increase in accuracy over other k-distribution based approaches while remaining significantly less computationally expensive than a standard line-by-line solution. It will require modification to be applicable in more general higher dimensional geometries. Additionally, the definitions of the appropriate weighting functions for the multi-source k-distribution method provided implicitly assume a slab geometry. It is unclear at this time what errors would be introduced when generalizing this method for use in two or three dimensional domains, and this is a topic undergoing further research.

7. Acknowledgments

This material is based in part upon work supported by the Department of Energy [National Nuclear Security Administration] under Award Number [DE-FC52-08NA28615].

References

[1] R. Siegel and J. R. Howell, Thermal Radiation Heat Transfer, 4th ed., New York: Taylor & Francis, 2002.

[2] M. F. Modest and R. S. Mehta, "Full spectrum k-distribution correlations for CO2 from the CDSD-1000 spectroscopic databank," *International Journal of Heat and Mass Transfer,* pp. 2487- 2492, 2004.

[3] A. Maurente, F. H. R. Franca, K. Miki and J. R. Howell, "Application of Joint Cumulative k-Distributions to FSK Radiation Heat Transfer in Multi-component High Temperature Non-LTE Plasmas," *AIAA J. Thermophysics and Heat Transfer.*

[4] M. F. Modest and H. Zhang, "The Full-Spectrum Correlated-k Distribution for Thermal Radiation From Molecular Gas-Particulate Mixtures," *ASME Journal of Heat Transfer,* 2002.

[5] G. Pal, M. F. Modest and L. Wang, "Hybrid Full-Spectrum Correlated k-Distribution Method for Radiative Transfer in Strongly Nonhomogeneous Gas Mixtures," *ASME Journal of Heat Transfer,* 2008.

[6] H. Zhang and M. F. Modest, "Scalable Multi-Group Full-Spectrum Correlated-k Distributions for Radiative Transfer Calculations," *ASME Journal of Heat Transfer,* 2003.

[7] V. P. Solovjov and B. W. Webb, "Multilayer Modeling of Radiative Transfer by SLW and CW Methods in Non-Isothermal Gaseous Medium," *Journal of Quantitative Spectroscopy and Radiative Transfer,* pp. 245-257, 2008.

[8] J. T. Tencer, "The K-Distribution Method for Radiation Heat Transfer in Non-Isothermal Real Air-Gas Plasmas," *Master thesis, Mechanical Eng., University of Texas, Austin,TX,* 2011.

[9] L. S. Rothman, D. Jacquemart, A. Barbe, D. C. Benner, M. Birk, L. R. Brown, M. R. Carleer, C. Chackerian Jr., K. Chance, L. H. Coudert, V. Dana, V. M. Devi, J. M. Flaud, R. R. Gamache, A. Goldman, J. M. Hartmann, K. W. Jucks, A. G. Maki, J. Y. Mandin, S. T. Massie, J. Orphal, A. Perrin, C. P. Rinsland, M. A. H. Smith, J. Tennyson, R. N. Tolchenov, R. A. Toth, J. V. Auwera, P. Varanasi and G. Wagner, "The HITRAN 2004 molecular spectroscopic database," *Journal of Quantitative Spectroscopy & Radiative Transfer,* vol. 96, p. 139–204, 2005.

Absorption coefficient regression scheme for splitting radiative heat sources across phases in gas-particulate mixtures

Jian Cai* **and Michael F. Modest****

(*) Postdoctoral research associate, School of Engineering, University of California, Merced.
(**) Shaffer and George Professor of Engineering, University of California, Merced. 5200 North Lake Road, Merced, CA 95343
Corresponding author's e-mail address: mmodest@ucmerced.edu

Abstract. In a gas-particle mixture particles of different makes and sizes may have different local temperature and are, therefore, modeled with separate energy equations. If the important radiative heat transfer contributions are determined with sophisticated spectral models such as k-distributions, spectral information is lost during the reordering process, making it difficult to separate the resulting radiative heat source across the various energy equations. It is shown here how radiative sources can be determined accurately with the full-spectrum k-distribution assembled from a narrowband database and combined with particle properties on a narrowband level. Proper distribution of the radiative source across the various phases is achieved using regression analysis. Numerical examples demonstrate the accuracy of the model.

1. Introduction

Thermal radiation makes major contributions to the heat transfer in fluidized bed and pulverized coal combustors, due to strong particle radiation [1,2]. A high-fidelity radiation model is necessary to improve numerical predictions of the overall heat transfer.

In gas-particle flows, polydisperse particles are grouped into phases by their size and material because of distinguishing characteristics of momentum transfer [3]. Particles of each phase also have individual dynamical responses to heat transfer, which requires a dedicated energy equation and each energy equation requires an individual radiative heat source. On the other hand, the radiation field is described by a single radiative transfer equation (RTE), which determines the total radiative heat source. Therefore, an accurate scheme is needed to split the total heat source across phases in gas-particle mixtures.

Solving the RTE requires spectral models for radiatively participating gases commonly occurring in coal combustion, such as CO_2, CO and H_2O. Their absorption coefficients have strong spectral dependency. Common particles in coal combustion include soot, coal, char, ash and limestone [4]. Their radiative properties are not gray even for gray complex refractive indices due to dependency on the size parameter [5].

While the line-by-line (LBL) spectral model [6] provides the most accurate results of radiative heat transfer, its large computational demands prevent its use in practical engineering applications. Recently, k-distribution methods have been developed for gases [7–9]. A k-distribution is a spectrally reordered absorption coefficient over a narrow-band or the full spectrum. Using k-distributions, radiative heat transfer can be solved with excellent accuracy, but at a small fraction of the effort compared to LBL calculations [9].

Although the total heat source can be accurately predicted using k-distribution methods, the reordering process makes splitting the total heat source across phases challenging because of the loss of spectral information. For example as shown in Fig. 1, for a nongray solid gas-particle mixture, multiple solid absorption coefficient values may correspond to one total absorption coefficient value. After reordering, due to the loss of spectral information, the individual solid phase absorption of radiation is indeterminate.

In this paper a regression scheme is introduced to extend the validity of the k-distribution scheme to multi-phase gas-particulate mixtures with multiple emission temperatures, and to accurately distribute the resulting radiative source across the phases. The new method is tested with several examples, demonstrating its validity and accuracy.

2. Theoretical Analysis

Consider a gas-particle mixture with particles grouped into M different phases according to their material and/or sizes with $m = 1, 2, \ldots, M$ and in particular let $m = 0$ be the gas phase. Particles of different size possess different radiative properties and may have different temperatures due to different reaction rates and responses to heat transfer. The spectral radiative transfer equation (RTE) of radiation intensity I_η at wavenumber η including absorption, scattering and multiple temperature emission for a gas-particle mixture may be stated as [5]

$$\frac{dI_\eta}{ds} = \sum_{m=0}^{M} \kappa_{m\eta} I_{b\eta}(T_m) - \sum_{m=0}^{M} \kappa_{m\eta} I_\eta - \sum_{m=0}^{M} \sigma_{sm\eta} I_\eta + \sum_{m=0}^{M} \left[\sigma_{sm\eta} \frac{1}{4\pi} \int_{4\pi} I_\eta(\hat{s}') \Phi_{m\eta}(\hat{s}, \hat{s}', \eta) d\Omega \right] \quad (1)$$

where $\kappa_{m\eta}$ are the absorption coefficient, $I_{b\eta}$ the Planck function, $\sigma_{sm\eta}$ the scattering coefficient and $\Phi_{m\eta}$ the scattering phase function of phase m. Both absorption and scattering coefficients are functions of state variable arrays $\underline{\phi}$ and $\underline{\phi}_s$, i.e., ϕ contains the gas properties, such as temperature, pressure and species concentration, and ϕ_s contains information of local particle properties such as particle number density, radius and species.

We now apply reordering according to the total absorption coefficient at reference state (κ_η^0) over the full spectral range [5, 10]:

$$\int_\eta [\bullet] \delta(k - \kappa_\eta^0) d\eta \quad (2)$$

The superscript ^{0} indicates a reference state, i.e., $\kappa_\eta^0 = \kappa_\eta(\underline{\phi}^0, \underline{\phi}_s^0)$ indicates the absorption coefficient at a reference state specified by state variable arrays $\underline{\phi}^0$ and $\underline{\phi}_s^0$. Similar to the full spectrum k-distribution method [5], after performing the integration indicated by Eq. (2), each term in Eq. (1) is transformed (reordered) as follows:

The path derivative (term on the left) becomes

$$\int_\eta \frac{dI_\eta}{ds} \delta(k - \kappa_\eta^0) d\eta = \frac{d}{ds} \int_\eta I_\eta \delta(k - \kappa_\eta^0) d\eta = \frac{d}{ds} I_k \quad (3)$$

where

$$I_k = \int_\eta I_\eta \delta(k - \kappa_\eta^0) d\eta \quad (4)$$

The emission term (first on the right) becomes

$$\int_\eta \sum_{m=0}^{M} \kappa_{m\eta} I_{b\eta}(T_m) \delta(k - \kappa_\eta^0) d\eta$$
$$= \sum_{m=0}^{M} \int_\eta \kappa_{m\eta} I_{b\eta}(T_m) \delta(k - \kappa_\eta^0) d\eta$$

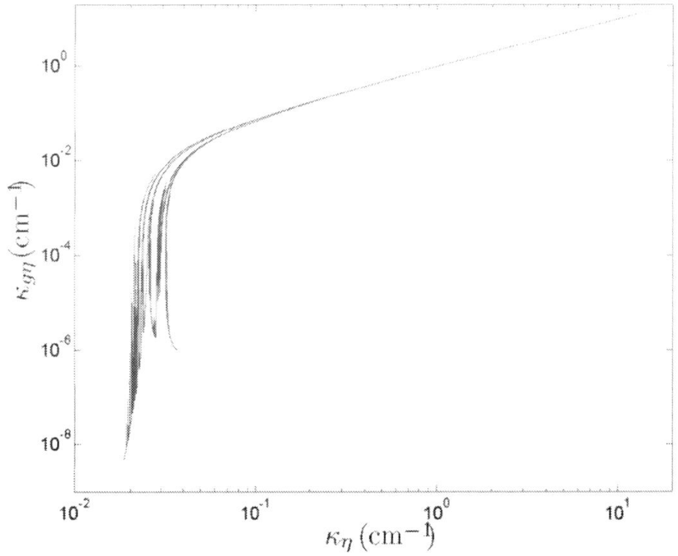

Figure 1. Line-by-line scatter plot of gas phase absorption coefficient ($\kappa_{g\eta}$) vs total absorption coefficient (κ_η). The mixture has 10% CO_2 and 90% N_2 by volume. The particles are char with a complex refractive index $m = 2.2 - 1.12i$ and a uniform diameter of 400 μm. The volume fraction of particle is 0.001. The mixture has a temperature of 600K.

$$
\begin{aligned}
&= k_0^*(k)\int_\eta I_{b\eta}(T_0)\delta(k-\kappa_\eta^0)d\eta + \sum_{m=1}^{M}\hat{k}_m(k;T_m)\int_\eta I_{b\eta}(T_m)\delta(k-\kappa_\eta^0)d\eta \\
&= k_0^* I_b(T_0) f_{FS}(T_0,\underline{\phi}^0,\underline{\phi}_s^0,k) + \sum_{m=1}^{M}\left[\hat{k}_m(k;T_m)I_b(T_m)f_{FS}(T_m,\underline{\phi}^0,\underline{\phi}_s^0,k)\right]
\end{aligned} \tag{5}
$$

where the gas absorption coefficients are assumed to be correlated with total absorption coefficients, i.e., there exists a map K_0 [10]

$$\kappa_{0\eta}(\underline{\phi},\underline{\phi}_s,\eta) = K_0(\underline{\phi},\underline{\phi}_s,\kappa_\eta(\underline{\phi}^0,\underline{\phi}_s^0,\eta)) \tag{6}$$

which gives

$$k_0^* = K_0(\underline{\phi},\underline{\phi}_s,k;\underline{\phi}^0,\underline{\phi}_s^0) \tag{7}$$

The correlation assumption for a mixture of a gas with nongray particles may be examined by plotting line-by-line gas absorption coefficients against the total absorption coefficients as shown in Fig. 1, for a typical CO_2-N_2-char mixture with properties given in the figure caption. The gas absorption coefficients cluster around a monotonic curve, suggesting the existence of the map K_0 and the validity of correlation assumption. The correlation between gases at varying temperature and concentration has been found to be good in early work on nonluminous reacting flows [9]. Such correlation is not generally available for solid absorption coefficients as shown in Fig. 2 for the same mixture. For a given total absorption coefficient, several different solid absorption coefficients are possible. However, an apparent (or relevant average) solid absorption coefficient may be determined from a regression as

$$\hat{k}_m(k;T) = \frac{\int_\eta \kappa_{m\eta} I_{b\eta}(T)\delta(k-\kappa_\eta^0)d\eta}{\int_\eta I_{b\eta}(T)\delta(k-\kappa_\eta^0)d\eta} \tag{8}$$

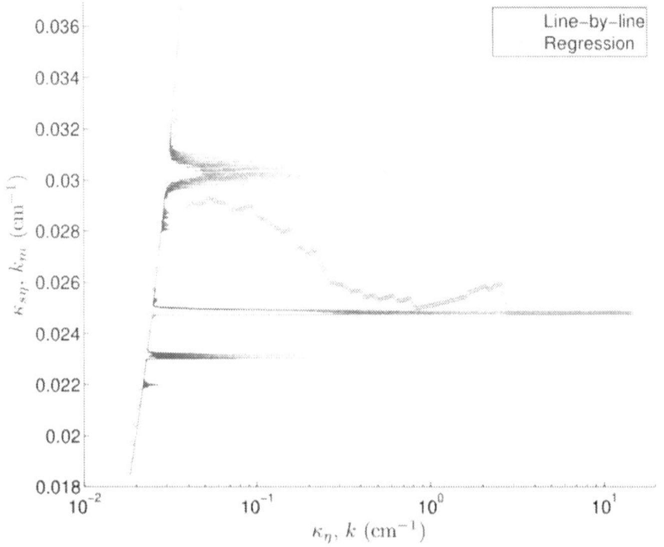

Figure 2. Line-by-line scatter plot of solid phase absorption coefficient ($\kappa_{s\eta}$) vs total absorption coefficient (κ_η). The regression of solid absorption coefficient against total absorption coefficient is shown by red circles. Mixture conditions are the same as Fig. 1

This allows us to find a proper average value of $\kappa_{m\eta}$ that occurs in the presence of a total absorption coefficient κ_η^0. The definition of the full spectrum k-distribution for a multi-temperature mixture remains the same as its definition for a single-temperature medium:

$$f_{\mathrm{FS}}(T, \underline{\phi}^0, \underline{\phi}_s^0, k) = \frac{1}{I_\mathrm{b}(T)} \int_\eta I_{\mathrm{b}\eta}(T)\delta(k - \kappa_\eta^0)d\eta \tag{9}$$

The absorption term (second on the right) becomes

$$\begin{aligned}
&\int_\eta \sum_{m=0}^{M} \kappa_{m\eta} I_\eta \delta(k - \kappa_\eta^0) d\eta \\
&= \int_\eta \kappa_{0\eta} I_\eta \delta(k - \kappa_\eta^0) d\eta + \sum_{m=1}^{M} \int_\eta \kappa_{m\eta} I_\eta \delta(k - \kappa_\eta^0) d\eta \\
&= k_0^* \int_\eta I_\eta \delta(k - \kappa_\eta^0) d\eta + \sum_{m=1}^{M} \left\{ \frac{\int_\eta \kappa_{m\eta} I_\eta \delta(k - \kappa_\eta^0) d\eta}{\int_\eta I_\eta \delta(k - \kappa_\eta^0) d\eta} \right\} \int_\eta I_\eta \delta(k - \kappa_\eta^0) d\eta \\
&\approx k_0^* I_k + \left[\sum_{m=1}^{M} \hat{k}_m \right] I_k
\end{aligned} \tag{10}$$

where the term in the curly braces is approximated as

$$\frac{\int_\eta \kappa_{m\eta} I_\eta \delta(k - \kappa_\eta^0) d\eta}{\int_\eta I_\eta \delta(k - \kappa_\eta^0) d\eta} \approx \frac{\int_\eta \kappa_{m\eta} I_{\mathrm{b}\eta}(T_m) \delta(k - \kappa_\eta^0) d\eta}{\int_\eta I_{\mathrm{b}\eta}(T_m) \delta(k - \kappa_\eta^0) d\eta} = \hat{k}_m(k; T_m) \tag{11}$$

This approximation is equivalent to assuming that $I_\eta \propto I_{b\eta}(T_m)$ over the spectrum. The proportionality factor may vary with space and direction. For example, if space and spectral dependence are separable

$$I_\eta(\underline{x}, \hat{s}, \eta) = p(\underline{x}, \hat{s}) I_{b\eta}(T_m) \tag{12}$$

Eq. (11) is still exact when p is not a function of wavenumber. Eq. (12) is valid in an isothermal homogeneous mixture bounded by cold black walls, or for a medium with a scaled absorption coefficient. Deviations may arise near a wall with a different temperature, in the presence of strong temperature gradients and if phase temperatures are very different.

The treatment of scattering will follow Modest and Riazzi [9], in which they demonstrated that gray scattering coefficients ($\sigma_{sm\eta} = \sigma_{sm}$) and phase functions ($\Phi_{m\eta} = \Phi_m$) are valid approximations even for extremely nongray scatterers. By dropping spectral dependence in scattering properties and reordering according to Eq. (2), the scattering terms become

$$\begin{aligned}\int_\eta \left\{ \sum_{m=0}^{M} \sigma_{sm\eta} I_\eta + \sum_{m=0}^{M} \left[\sigma_{sm\eta} \frac{1}{4\pi} \int_{4\pi} I_\eta(\hat{s}') \Phi_{m\eta}(\hat{s}, \hat{s}', \eta) d\Omega \right] \right\} \delta(k - \kappa_\eta^0) d\eta \\ = \sum_{m=0}^{M} \sigma_{sm} I_k + \sum_{m=0}^{M} \left[\sigma_{sm} \frac{1}{4\pi} \int_{4\pi} \Phi_m(\hat{s}, \hat{s}') I_k(\hat{s}') d\Omega \right]\end{aligned} \tag{13}$$

Assuming gas phase scattering to be negligible, the summation over phases starts from $m = 1$.

Combining the above terms, the reordered RTE becomes

$$\begin{aligned}\frac{dI_k}{ds} = k_0^*(k) I_b(T_0) f_{FS}(T_0, \underline{\phi}^0, \underline{\phi}_s^0, k) + \sum_{m=1}^{M} [\hat{k}_m(k; T_m) I_b(T_m) f_{FS}(T_m, \underline{\phi}^0, \underline{\phi}_s^0, k)] \\ - \left[k_0^*(k) + \sum_{m=1}^{M} \hat{k}_m(k; T_m) \right] I_k - \sum_{m=1}^{M} \sigma_{sm} I_k + \sum_{m=1}^{M} \left[\sigma_{sm} \frac{1}{4\pi} \int_{4\pi} \Phi_m(\hat{s}, \hat{s}') I_k(\hat{s}') d\Omega \right]\end{aligned} \tag{14}$$

Defining the cumulative k-distribution as

$$g(k; T^0, \underline{\phi}^0, \underline{\phi}_s^0) = \int_0^k f_{FS}(T^0, \underline{\phi}^0, \underline{\phi}_s^0, k') dk' \tag{15}$$

and

$$I_g(g(k), \underline{r}, \hat{s}) = I_k(k, \underline{r}, \hat{s}) / f_{FS}(T^0, \underline{\phi}^0, \underline{\phi}_s^0, k) \tag{16}$$

where T^0 is the reference temperature and T_0 the gas phase temperature. The new RTE for I_g becomes

$$\begin{aligned}\frac{dI_g}{ds} = k_0^*(g) a(T_0, T^0, g) I_b(T_0) + \sum_{m=0}^{M} \hat{k}_m(g; T_m) a(T_m, T^0, g) I_b(T_m) - \left[k_0^*(g) + \sum_{m=1}^{M} \hat{k}_m(g; T_m) \right] I_g \\ - \sum_{m=1}^{M} \sigma_{sm} I_g + \sum_{m=1}^{M} \left[\sigma_{sm} \frac{1}{4\pi} \int_{4\pi} \Phi_m(\hat{s}, \hat{s}') I_g(\hat{s}') d\Omega \right]\end{aligned} \tag{17}$$

where the nongray stretching function a is defined as

$$a(T_m, T^0, g) = \frac{f_{FS}(T_m, \underline{\phi}^0, \underline{\phi}_s^0, k(g))}{f_{FS}(T^0, \underline{\phi}^0, \underline{\phi}_s^0, k(g))} \tag{18}$$

To split the heat source across phases, one may use the same assumption as in the derivation of the absorption term in the RTE for solid phases as given by Eq. (11). The heat source for solid phase m ($m \geq 1$) then becomes

$$
\begin{aligned}
S_m = -\nabla \cdot \boldsymbol{q}_m &= -\int_\eta \nabla \cdot \boldsymbol{q}_{m\eta} d\eta \\
&= \int_{4\pi} \int_\eta \kappa_{m\eta} I_\eta d\eta d\Omega - 4\pi \int_\eta \kappa_{m\eta} I_{\mathrm{b}\eta} d\eta \\
&= \int_{4\pi} \int_\eta \kappa_{m\eta} I_\eta \int_{k=0}^{\infty} \delta(k - \kappa_\eta^0) dk d\eta d\Omega - 4\pi \kappa_{Pm} I_{\mathrm{b}}(T_m) \\
&= \int_{4\pi} \int_{k=0}^{\infty} \int_\eta \kappa_{m\eta} I_\eta \delta(k - \kappa_\eta^0) d\eta dk d\Omega - 4\pi \kappa_{Pm} I_{\mathrm{b}}(T_m) \\
&\approx \int_{4\pi} \int_{k=0}^{\infty} \hat{k}_m(k; T_m) \int_\eta I_\eta \delta(k - \kappa_\eta^0) d\eta dk d\Omega - 4\pi \kappa_{Pm} I_{\mathrm{b}}(T_m) \\
&= \int_{k=0}^{\infty} \hat{k}_m(k; T_m) \int_{4\pi} I_k d\Omega dk - 4\pi \kappa_{Pm} I_{\mathrm{b}}(T_m) \\
&= \int_{k=0}^{\infty} \hat{k}_m(k; T_m) G_k dk - 4\pi \kappa_{Pm} I_{\mathrm{b}}(T_m) \\
&= \int_{g=0}^{1} \hat{k}_m(k(g); T_m) G_g dg - 4\pi \kappa_{Pm} I_{\mathrm{b}}(T_m)
\end{aligned}
\tag{19}
$$

The gas phase heat source is, assuming the gas to be correlated,

$$
\begin{aligned}
S_0 = -\nabla \cdot \boldsymbol{q}_0 &= -\int_\eta \nabla \cdot \boldsymbol{q}_{0\eta} d\eta \\
&= \int_{4\pi} \int_\eta \kappa_{0\eta} I_\eta d\eta d\Omega - 4\pi \int_\eta \kappa_{0\eta} I_{\mathrm{b}\eta} d\eta \\
&= \int_{4\pi} \int_\eta \kappa_{0\eta} I_\eta \int_{k=0}^{\infty} \delta(k - \kappa_\eta^0) dk d\eta d\Omega - 4\pi \kappa_{P0} I_{\mathrm{b}}(T_0) \\
&= \int_{4\pi} \int_{k=0}^{\infty} \int_\eta \kappa_{0\eta} I_\eta \delta(k - \kappa_\eta^0) d\eta dk d\Omega - 4\pi \kappa_{P0} I_{\mathrm{b}}(T_0) \\
&= \int_{4\pi} \int_{k=0}^{\infty} k_0^* \int_\eta I_\eta \delta(k - \kappa_\eta^0) d\eta dk d\Omega - 4\pi \kappa_{P0} I_{\mathrm{b}}(T_0) \\
&= \int_{k=0}^{\infty} k_0^*(k) \int_{4\pi} I_k d\Omega dk - 4\pi \kappa_{P0} I_{\mathrm{b}}(T_0) \\
&= \int_{k=0}^{\infty} k_0^*(k) G_k dk - 4\pi \kappa_{P0} I_{\mathrm{b}}(T_0) \\
&= \int_{g=0}^{1} k_0^*(k(g)) G_g dg - 4\pi \kappa_{P0} I_{\mathrm{b}}(T_0)
\end{aligned}
\tag{20}
$$

The total heat source is obtained by integrating Eq. (17) over all solid angles and all g, and is conserved because

$$\begin{aligned}
\sum_{m=0}^{M} S_m &= \int_0^1 \left[k_0^*(k(g)) + \sum_{m=1}^{M} \hat{k}_m(k(g);T_m) \right] G_g dg - \sum_{m=0}^{M} 4\pi\kappa_{Pm} I_b(T_m) \\
&= \int_0^1 \left[k_0^*(g) + \sum_{m=1}^{M} \hat{k}_m(g) \right] G_g dg \\
&\quad - 4\pi \int_{g=0}^1 k_0^*(g) a_0(g) I_b(T_0) dg - 4\pi \sum_{m=1}^{M} \int_0^1 \hat{k}_m(g) a_m(g) I_b(T_m) dg \\
&= \int_0^1 \left\{ \left[k_0^*(g) + \sum_{m=1}^{M} \hat{k}_m(g) \right] G_g - 4\pi \left[k_0^*(g) a_0(g) I_b(T_0) + \sum_{m=1}^{M} \hat{k}_m(g) a_m(g) I_b(T_m) \right] \right\} dg \\
&= -\int_0^1 \nabla \cdot \boldsymbol{q}_g dg = -\nabla \cdot \boldsymbol{q}
\end{aligned} \tag{21}$$

Regression

To calculate the full-spectrum k-distribution of a gas-particle mixture, we use databased narrowband k-distributions [11] of the combustion gases and mix them based on the Modest-Riazzi model [9] to obtain the narrowband k-distribution of the gas mixture. Over each narrowband, the solid absorption coefficient of each phase may be approximated by a constant value equal to its average absorption coefficient over the narrowband. Then the narrowband k-distribution of the gas-particle mixture is obtained by adding the narrowband solid absorption coefficients of all solid phases to the narrowband gas k-distribution over each narrowband [9]. The full-spectrum k-distribution of the gas-particle mixture is then compiled from the narrowband k-distributions of the gas-particle mixture.

The regression of solid absorption coefficient in Eq. (8) is calculated with narrowband k-distributions of the mixture. Suppose that the spectrum is partitioned into $N_{\rm NB}$ narrowbands and the particle absorption coefficient of phase m is approximated by a constant value $k_{m,n}$ over the nth narrowband. Over this narrowband, the cumulative k-distribution of the gas-particle mixture is $g_{n,i}$ for a sequence of increasing k_is ($i = 1, 2, \ldots, N_i$, where N_i is the total number of k values). The regression $\hat{k}_m$ is expressed as a weighted average of $k_{m,n}$ from all narrowbands with the weights proportional to the narrowband k-distributions and the narrowband integral Planck function, i.e.,

$$\hat{k}_m(k_i) = \frac{\int_\eta I_{b\eta}\kappa_{m\eta}\delta(k_i - \kappa_\eta)d\eta}{\int_\eta I_{b\eta}\delta(k_i - \kappa_\eta)d\eta} \approx \frac{\int_{(k_i+k_{i-1})/2}^{(k_i+k_{i+1})/2}\int_\eta I_{b\eta}\kappa_{m\eta}\delta(k - \kappa_\eta)d\eta dk}{\int_{(k_i+k_{i-1})/2}^{(k_i+k_{i+1})/2}\int_\eta I_{b\eta}\delta(k - \kappa_\eta)d\eta dk} \tag{22}$$

The numerator may be expressed as

$$\begin{aligned}
&\int_{(k_i+k_{i-1})/2}^{(k_i+k_{i+1})/2}\int_\eta I_{b\eta}\kappa_{m\eta}\delta(k - \kappa_\eta)d\eta dk \\
&= \int_\eta \int_{(k_i+k_{i-1})/2}^{(k_i+k_{i+1})/2} I_{b\eta}\kappa_{m\eta}\delta(k - \kappa_\eta)dk d\eta \\
&= \sum_{n=1}^{N_{\rm NB}} \int_{\eta_n}^{\eta_n+\Delta\eta_n} \int_{(k_i+k_{i-1})/2}^{(k_i+k_{i+1})/2} I_{b\eta_n}\kappa_{m\eta_n}\delta(k - \kappa_\eta)dk d\eta \\
&= \sum_{n=1}^{N_{\rm NB}} I_{b\eta_n}\kappa_{m\eta_n} \int_{\eta_n}^{\eta_n+\Delta\eta_n} \int_{(k_i+k_{i-1})/2}^{(k_i+k_{i+1})/2} \delta(k - \kappa_\eta)dk d\eta
\end{aligned}$$

$$= \sum_{n=1}^{N_{\mathrm{NB}}} I_{\mathrm{b}\eta_n} \Delta\eta \kappa_{m,n} \int_{(k_i+k_{i-1})/2}^{(k_i+k_{i+1})/2} \frac{1}{\Delta\eta} \int_{\eta_n}^{\eta_n+\Delta\eta_n} \delta(k-\kappa_\eta) d\eta dk$$

$$= \sum_{n=1}^{N_{\mathrm{NB}}} I_{\mathrm{b}n} \kappa_{m,n} \int_{(k_i+k_{i-1})/2}^{(k_i+k_{i+1})/2} \frac{1}{\Delta\eta} \int_{\eta_n}^{\eta_n+\Delta\eta_n} \delta(k-\kappa_\eta) d\eta dk$$

$$= \sum_{n=1}^{N_{\mathrm{NB}}} I_{\mathrm{b}n} \kappa_{m,n} \int_{(k_i+k_{i-1})/2}^{(k_i+k_{i+1})/2} f_{\mathrm{NB}}(k;n) dk$$

$$= \sum_{n=1}^{N_{\mathrm{NB}}} I_{\mathrm{b}n} \kappa_{m,n} \Delta g_{i,n} \tag{23}$$

where $\kappa_{m,n}$ is the constant solid phase absorption coefficient of phase m over narrowband n, $I_{\mathrm{b}n} = \int_{\eta_n}^{\eta_n+\Delta\eta_n} I_{\mathrm{b}\eta} d\eta$ is the fraction of Planck function over narrowband n, $f_{\mathrm{NB}}(k;n)$ is the narrowband k-distribution at k over narrowband n, $\Delta g_{i,n} = \int_{(k_i+k_{i-1})/2}^{(k_i+k_{i+1})/2} f_{\mathrm{NB}}(k;n)dk$ is the change of narrowband k-distribution around k_i over narrowband n. Similarly, the denominator reads,

$$\int_{(k_i+k_{i-1})/2}^{(k_i+k_{i+1})/2} \int_\eta I_{\mathrm{b}\eta} \delta(k-\kappa_\eta) d\eta dk = \sum_{n=1}^{N_{\mathrm{NB}}} I_{\mathrm{b}n} \Delta g_{i,n} \tag{24}$$

Combining Eqs. (23) and (24), the regression may be evaluated as

$$\hat{k}_m(k_i) = k_{m,i} = \frac{\sum_{n=1}^{N_{\mathrm{NB}}} I_{\mathrm{b}n} \kappa_{m,n} \Delta g_{n,i}}{\sum_{n=1}^{N_{\mathrm{NB}}} I_{\mathrm{b}n} \Delta g_{n,i}} \tag{25}$$

A sample regression is included in Fig. 2, in which 256 k values are used, with the corresponding $\hat{k}_m$ marked by circles. Note that for small k values, gas absorption is negligible, and the $\hat{k}_m$ takes the same value as k. For large k values ($k > 3\mathrm{cm}^{-1}$), the total absorption coefficient is dominated by gas absorption at the line centers, and $\hat{k}_m$ approaches the solid absorption coefficient at the corresponding wavenumber.

3. Sample calculations

First consider the radiation in a homogeneous gas-particle field bounded by two parallel black cold walls. The gas phase has 10% CO_2 and 90% N_2. The particles are char with a complex refractive index $m = 2.2 - 1.12i$ [12] and a uniform diameter of 400μm. The volume fraction of the particles is 0.001. The mixture slab has a temperature of 600K and a thickness of 7cm. The gray scattering coefficient is evaluated at the wavenumber where the Planck function peaks and scattering is assumed to be isotropic. The gas absorption coefficient is calculated from databased narrowband k-distributions and the solid absorption coefficient is approximated as constant across each narrowband [9].

Equation (17) is solved with the P_1 approximation using 64 quadrature points for g. The results of total, gas and solid phase radiative heat sources are compared with line-by-line calculations in Fig. 3. The k-distribution method is capable of predicting all of the heat sources accurately (to within 1%).

To demonstrate the accuracy of the regression method in inhomogeneous media, a one-dimensional slab of 10cm width, bounded by two parallel black cold walls is considered. The slab has two homogenous gas-particle mixture layers with equal thickness.The left layer has a gas temperature of 600K, the composition is 5% CO_2 diluted by O_2 and N_2. The particles have uniform diameter of 200μm, volume fraction of 1×10^{-3} and temperature of 500K. The right layer has a gas temperature of 1200K, the composition is 10% CO_2 diluted by O_2 and N_2. The particles have uniform diameter of 100μm, volume fraction of 2.5×10^{-4} and temperature of 1300K. In both layers the particles are char with a complex refractive index $m = 2.2 - 1.12i$. This configuration approximates both unburned and burning

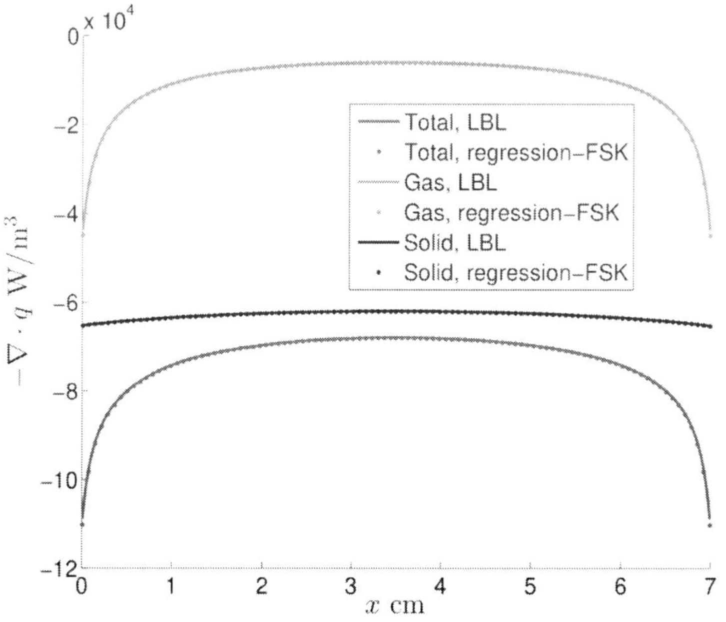

Figure 3. Heat source calculation of an inhomogeneous slab between line-by-line (solid lines) and full-spectrum k-distribution with regression (dots). Line colors are blue (total heat source), red (gas phase heat source) and black (solid phase heat source).

gas-particle environments in a pulverized coal combustor. CO_2 is assumed to be the only radiatively participating gas species, which reduces errors from k-distributions since only temperature levels (as opposed to concentration) contribute to uncorrelation. The inaccuracy hereby primarily comes from the approximation in Eq.(11).

Equation (17) is solved with the P_1 approximation using 64 quadrature points for g. The results of total, gas and solid phase radiative heat sources are compared with line-by-line calculations in Fig. 4. The k-distribution method is capable of predicting major trends of total and phase-split heat source terms. In particular, the gas phase heat source is predicted accurately (mostly within 5%) in spite of the fact that it is one order less in magnitude than the total heat source. This is because gas radiation primarily comes from strong bands, and the proposed regression scheme successfully picks the solid absorption coefficients at the corresponding wavenumbers (e.g. large k values in Fig. 2). For the solid phase heat source, the largest deviation occurs near the interface between the two layers. Equation (11) is greatly challenged by the sharp temperature gradient at the interface. In the left (cold) layer, the spectral incident radiation deviates significantly from the emission spectrum at the local temperature because of the emission from the hot layer. Away from the interface, such deviation is reduced and accuracy improves. However, this effect is hardly observed in the hot layer because emission from the cold layer is negligible. In the hot layer, the error is less than 1%, which is due to the temperature difference between gas and solid phases.

4. Conclusion

In a gas-particle mixture particles of different makes and sizes may have different local temperature and, therefore, are modeled with separate energy equations. To model the radiative heat transfer in such a mixture the full-spectrum k-distribution model was extended to the case of multi-phase mixtures with individual temperature for each phase. Full spectrum k-distributions were assembled from a narrowband database for gases, by assuming particles to be gray across narrowbands, and by using a regression to

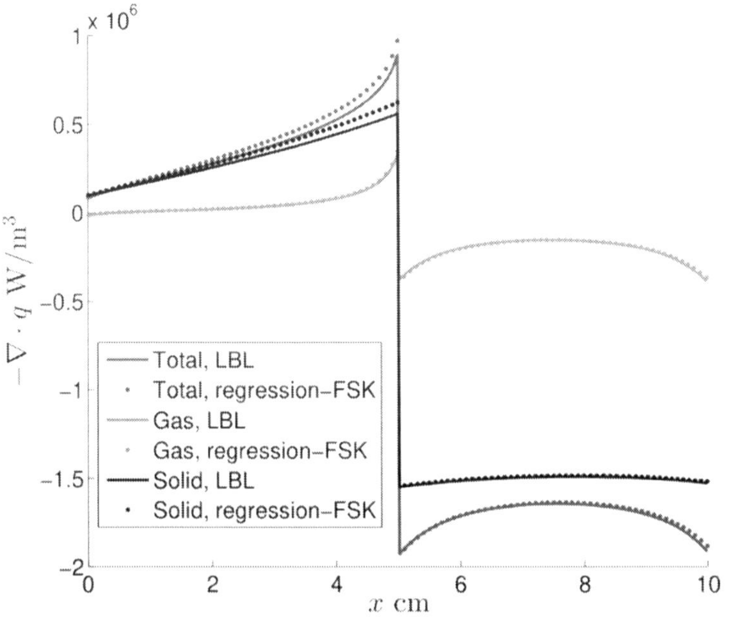

Figure 4. Heat source calculation of an inhomogeneous slab between line-by-line (solid lines) and full-spectrum k-distribution with regression (dots). Line colors are blue (total heat source), red (gas phase heat source) and black (solid phase heat source).

determine "effective" absorption coefficients for particles.

It was shown how properly distributed radiative source terms can be determined for the individual energy equations, by using a similar regression analysis. The accuracy of the method was demonstrated with several examples.

References

[1] Hua Y, Flamant G, Lu J and Gauthier D 2005 *International Journal of Heat and Mass Transfer* **48** 1145–1154
[2] Alagoz D E, Kulah G and Selçuk N 2008 *Combustion Science and Technology* **180** 910–926
[3] Syamlal M and O'Brien T 1988 *International Journal of Multiphase Flow* **14** 473–481
[4] Im K H and Ahluwalia R K 1993 *International Journal of Heat and Mass Transfer* **36** 293–302
[5] Modest M F 2003 *Radiative Heat Transfer* 2nd ed (New York: Academic Press)
[6] Taine J 1983 *Journal of Quantitative Spectroscopy and Radiative Transfer* **30** 371–379
[7] Zhang H and Modest M F 2002 *Journal of Quantitative Spectroscopy and Radiative Transfer* **73** 349–360
[8] Wang L and Modest M F 2005 *ASME Journal of Heat Transfer* **127** 740–748
[9] Modest M F and Riazzi R J 2005 *Journal of Quantitative Spectroscopy and Radiative Transfer* **90** 169–189
[10] Modest M F 2003 *Journal of Quantitative Spectroscopy and Radiative Transfer* **76** 69–83
[11] Wang A and Modest M F 2005 *Journal of Quantitative Spectroscopy and Radiative Transfer* **93** 245–261
[12] Mengüç M P and Viskanta R 1985 *Combustion Science and Technology* **44** 143–159

Radiative shielding by water mist : comparisons between downward, upward and impacting injection of droplets

S. Lechêne(*), Z. Acem(**), G. Parent(**), A. Collin(**) and P. Boulet(**)

(*)CSTB - Centre Scientifique et Technique du Bâtiment
84 avenue Jean Jaurès - Champs sur Marne - 77447 Marne-La-Vallée cedex 2 - France
(**)LEMTA, Nancy-Université, CNRS
2 av. de la Forêt de Haye - BP 160 - 54504 Vandoeuvre Cedex, France.

E-mail: pascal.boulet@lemta.uhp-nancy.fr

Abstract. Radiative shielding with water curtain has been studied numerically, investigating three different possibilities of droplet injection : downward, upward and impacting on a wall to be protected. The efficiency has been evaluated based on radiation attenuation predicted considering a given incident flux attenuated when crossing the area where water is injected. For upward and downward injection, a simple water curtain is considered. For the impacting spray case, a water film streaming on the wall is considered in addition to the spray (an idealized film with constant and fixed thickness for the moment). The dynamics has been imported from an Eulerian-Lagrangian simulation and radiative transfer has been addressed with a Monte Carlo method. Results show that the upward injection performs better than the downward injection due to a favoring dynamics that increases the residence time of droplets. The impacting spray could be even more efficient owing to the possible high attenuation efficiency of films, but present results still make use of simplifications on the water film falling on the wall and present promising observations require further verification.

1. Introduction

Water droplets injected in air may be used to build a radiative shield which can attenuate radiation coming from a radiative source like a fire or a high temperature process for example. The attenuation mechanisms are well known : absorption by droplets and vapour in air, plus scattering due to the droplets solely. This has been studied for decades (see [1–5] among others) and in particular by present authors during the last ten years. A numerical code - BERGAMOTE - has been developed, which simulates the two phase flow made of droplets in wet air with an Eulerian-Lagrangian approach and which predicts the radiation attenuation through this medium using a Monte Carlo method. The code has been validated in some situations of a single spray and further extended to a series of sprays in classical configurations of a downward injection of the droplets [6, 7]. Some preliminary predictions also indicated that a better use of the water - a better radiation attenuation - could be achieved if droplets are injected in a different manner in order to increase the residence time of droplets in air. This idea is further extended here with a shield aimed at protecting a given surface, comparing different injection possibilities: classical downward injection, or upward injection, or impacting sprays on the surface to be protected. The guiding idea with the upward injection has been already discussed in [8]: the double use of the droplets with their up and down motion leads to a stronger attenuation ability. The drawback is a weaker stability of the water curtain however, which can be dispersed by an air flow due to wind or a ventilation device for example. With the impacting spray, the goal is to take benefit from a first spray attenuation which is completed by the influence of a streaming film on the surface which also weakens the incident radiation on the surface. This direct protection of a surface could be more efficient than a simple use of a water curtain in front of the surface to be protected. Such idea was also suggested by Hald and Buchlin [9]. These different solutions of water injection for radiation attenuation have been studied numerically in the present work. Water with given droplet size distribution and flow rate is injected in air, trajectories are computed involving evaporation, turbulence and heat transfer phenomena. Then, radiative properties of the spray are calculated in the whole computational domain, combining the Mie theory and a C-k model, in addition to a direct computation of transmission through a water film for the impacting spray. A Monte Carlo method is used to calculate the radiative fluxes in order to predict the potential radiation attenuation of the resulting medium by dividing the radiative flux reaching a given target, with the reference value obtained in the same configuration with the spray off. With the impacting spray, some assumptions are required regarding the film thickness which makes the study still exploratory, but the high attenuation potential of such a film makes this method very attractive. The following sections will include a short description of the spray dynamics with some details regarding the droplet trajectories, which will help to understand the radiative properties of the medium. Then the focus will be put on radiative transfer results in a second step. They will be discussed in terms of transmission and attenuation ability.

2. Spray dynamics

Details regarding the Eulerian-Lagrangian process used for the present simulation have been already presented and validated in [6–8]. Without describing the various submodels, the involved physical phenomena are recalled here : drag and gravity effects, convection and evaporation influence on the heat transferred to the droplet, influence of the turbulence on the droplet trajectories through a dispersion model, influence of the dispersed phase on the gas phase through dedicated source terms. Finally two subroutines are run for the Eulerian simulation of the gas phase with mass, momentum, energy balances plus a k-ϵ model on the one hand, then balances for mass, momentum, position and energy for the droplets on the other hand. Interactions between phases are taken into account through an iterative process.

The strong coupling between dynamics and radiative transfer has been already demonstrated to be of importance in problems close to the present one [8]. Actually, this is mainly treated as a one-way coupling in the present work, because the droplet distributions are affecting the radiative properties of the medium and consequently the radiation propagation. However, the radiation absorption is not expected to be as strong as to modify the droplet size due to enforced evaporation process for example. The scenario considers a curtain aimed at protecting a given target from radiation coming from a high temperature source but not submitted itself to a high temperature increase (the curtain is aside the emission source, just building a shield against radiation). A two-way coupling which would consider the alteration of the spray characteristics due to strong radiation absorption would not complexify the resolution. An iterative solution between subroutines for dynamics and radiative transfer would solve the problem. For the present application the absence of coupling will allow us to save time. One has also to mention that further validation would be achieved at laboratory scale with a reduced incident flux and little impact on the spray. This would be simulated as a one-way coupling problem too.

For the present exercise the following conditions have been chosen for the spray : the case of nozzles located on the same pipe, regularly spaced, fed with the same pressure, but possibly oriented in various directions (as illustrated by Figure 1, red circles show the location of the nozzles in the different configurations). Numerical conditions mimic a nozzle referred as TP400067 by Spraying systems and Co. The pressure feed is 4 bars, the consecutive water flow rate is 0.32 L/min, the droplet distribution can be described by a Log-normal law with mean diameter set to 123 μm and dispersion coefficient equal to 0.4 at the nozzle exit. The injections provide a series of conic sprays with angles equal to 48 and 18.5°. The injection velocity is 24 m/s. All conditions are assumed to be identical whatever the injection directions: vertical upward, downward, or inclined (an assumption which could be discussed as gravity effects may alter the way droplets are formed and consequently the size distribution at the injection point in particular). In the following sections, results will consequently concern the same water consumption, yet the attenuation ability will strongly vary with the injection type as it will be seen. Supplementary choices must be done regarding the location of the injection points. For the present qualification study a complete sensitivity study cannot be conducted on the role of the various possible injection locations. Arbitrarily (and with the idea of validating the present results experimentally with a device available in our laboratory), the vertical position of the nozzles is 1.50 m in the downward case, 30 cm in the upward case and finally 1 m in the impacting case with an horizontal distance of 1 m between the nozzles and the wall). In this last case the injection angle with respect to the horizontal direction may also have an influence. As a tentative study, an angle of 45° is chosen here, a parameter which could be of course studied and optimized. A series of nozzles in parallel are considered as to generate a water curtain. Nozzles are separated by a regular space of 10 cm, assumed to be fed by a horizontal pipe. In order to describe the spray numerically, a polydispersion is considered with twenty classes between 20 and 300 μm. The classes are automatically generated in order to concern the same water flow rate amount (one twentieth of the whole water quantity for each class).

The computational domain corresponds to a box with the following dimensions $3 \times 0.05 \times 3$ m^3, which is extended to a height of 6 m for the upward injection case since a dedicated sensitivity study showed us that small droplets may reach such high positions that the upper wall at 3 m would affect their motion. Symmetry planes are defined considering the series of nozzles and restricting the analysis to the central part of the domain. The width of the computational domain is reduced to 5 cm corresponding to the half space between two neighbouring nozzles. The bottom right subfigure on Figure 1 gives a schematic view of the problem in a case of a downward injection. The emission panel is on the foreground, the target is the rear surface and radiation crosses the spray along the x-direction. Typical mesh involves in that case $48 \times 5 \times 40$ cells for this $3 \times 0.05 \times 3$m^3 domain, with cells tightened near the injection point and in the spray area. For the upward injection, the same configuration is considered except that droplets are injected in the positive z-direction (the mesh becomes $60 \times 5 \times 50$ cells). Finally for the impacting spray, droplets are injected at 45° in the positive x- and z- directions in the original $3 \times 0.05 \times 3$ m^3 domain. The other subplots of Figure 1 give a sketch of the droplet trajectories in the different cases (as commented below). For a better vizualisation several injections have been plotted using the symmetries, but only one half of a single injection has been computed.

For the sake a brievity, complete numerical conditions are not detailed, but they can be found in [10].

For the understanding of the water curtain characteristics the first key element is the description of the droplet trajectories which affect their residence time and the consecutive absorption and scattering ability of the water curtain. As shown on Figure 1, the droplets are following very different trajectories. Two classes have been chosen for the representation. The trajectories of a collection of droplets are plotted, revealing the spray shape. The classical downward motion of droplets on the upper left subfigure is changed in an up-and-down motion, with strong perturbations induced by the carrying flow, on the upper right subfigure where droplets are injected upward. This is expected to induce a longer residence time. Finally, a

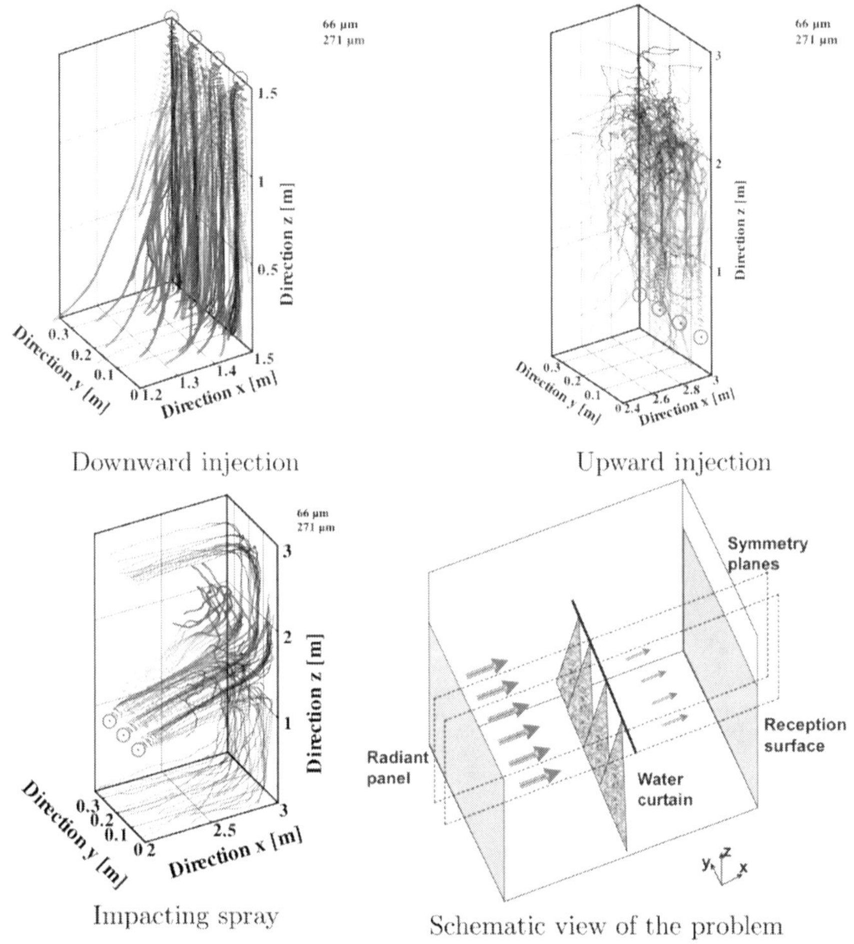

Figure 1. Illustration after post-processing of typical droplet trajectories for the three injection strategies. Red circles indicate some selected nozzle locations.

complex trajectory is also observed for the impacting case where small droplets can even be carried away from the curtain area. For the present illustration a condition of no-rebound has been chosen, meaning that all droplets caught by the surface are trapped and provide a deposit. Direct observations by rapid camera have led us to such hypothesis of droplet deposit without rebound (experiments carried out on droplets with similar size but on a cold wall). Droplets escaping at the top of the picture are simply carried away by the air flow deviated in front of the wall. Rebound conditions could be introduced, with possible splash or split which could provide smaller droplets and would be even better from the radiation attenuation point of view since attenuation increases when droplet size decreases (given the water quantity).

The residence time is reported on table 1 for the two droplet sizes whose trajectories have been presented on Figure 1. These are averages obtained on a huge number of droplets of these given classes followed during their trajectories.

As can be seen through table 1, the classical downward injection penalizes the droplet residence time, more especially in what concerns the small droplets (the most efficient ones for the radiation attenuation). The upward injection appears to be the most efficient solution regarding this time criterion, with a gain highly better than a simple factor 2 due to the up

Table 1. Residence time (in s) depending on the injection direction.

Droplet size	Downward inj.	Upward inj.	Impacting spray
66 μm	0.88	12.19	4.93
271 μm	0.26	2.68	2.03

and down trajectory. This is due to the fact that droplets first decelerate during their upward trajectory, before falling with a zero initial velocity (whereas droplets in the downward case are injected with a high velocity which leads them very quickly to the floor). The impacting spray still allows a better use of the water, even less satisfactory than the upward case. However, the droplets which are reaching the wall may continue to contribute to radiation attenuation with the streaming film. The present times must be considered with criticism as they are clearly conditionned by the exact geometrical configuration which has not been optimized. An injection at a different vertical position or shifted from, or toward, the wall for the impacting case will affect the numerical results. Yet, the present comparison of table 1 still allows a qualitative understanding of the various injection efficiencies.

3. Radiative transfer simulation

3.1. Transfer through the water curtain

Radiation is supposed to come from a radiant panel like it is presented on the schematic view on Figure 1. The case of a panel behaving like a blackbody at 1000K has been treated here.

Radiative properties are simulated as in [6]. Without recalling the details, Mie theory is applied assuming that droplets are spherical and using the optical indices for water given in Hale and Querry reference paper [11]. The radiative properties for gases (water vapor and carbone dioxide) are modelled with a classical C-k model with 43 bands as presented in [12], except for one representation below, where properties will be finally post-processed in the form of averaged values on bands following a 367 band discretisation for a finer description of the properties.

A Monte Carlo method is used for the simulation of radiation propagation, neglecting emission inside the spray still considering that the heat source does not induce a relevant spray heating. A huge number of quanta are followed in the domain, with a numerical process based on a distance of interaction computed with the scattering coefficient. The exact number of quanta of course depends on the information required (spectral or total flux) with convergence criteria based on the variance of the local flux received by the target. Using the droplet distribution, absorption coefficients are computed in the whole domain and the absorption of the quanta is accounted for until their total absorption or their escape from the domain. Results are finally obtained considering the number of quanta received on the target surface. Details on the specific sensitivity study may be found in [10].

Results for the radiative properties are commented first. Based on the distributions predicted with the simulation of the dynamics, the size distribution and the volumetric fractions for droplets and vapor are known. Then, the radiative properties may be computed for each phase. A simple additivity is considered for the prediction of the curtain radiative properties. Typical radiative properties are presented on Figure 2 for the present nozzle under a pressure of 4 bars. For better illustrating the variations a fine representation based on a 367 band formulation has been chosen as explained above. Coefficients are averaged on each band in order to show the attenuation coefficient as a function of the wavelength. For the rest of the simulation a classical

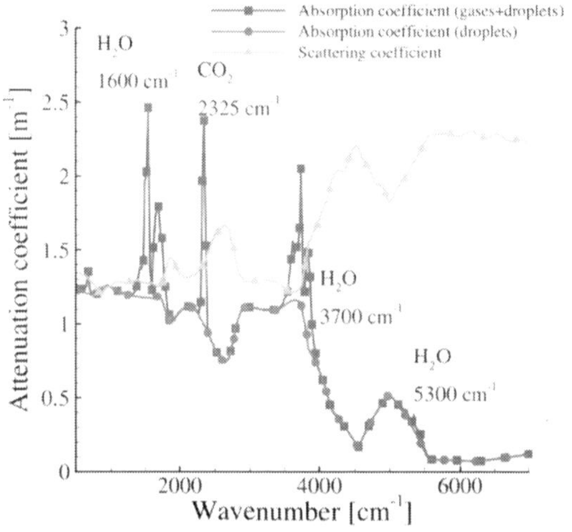

Figure 2. Absorption and scattering coefficients yielded by a single nozzle TP400067 under 4 bars

C-k formulation for the gases with a 43 band resolution has been observed to be sufficient for an accurate description of the spectral attenuation by the spray. As can be seen the attenuation by the droplets is possible in the whole infrared range. Absorption and scattering have a similar influence in the low wavenumber range, then absorption falls to very small levels and scattering is the main effect in the larger wavenumber range. This can be of importance since scattering is highly forward oriented for droplets with size really larger than the wavelength. As a consequence the effective attenuation may be limited despite a high attenuation in terms of coefficients. This is especially the case for applications where sources and targets have a large emission or reception surface, like in the present case. One can also note the contribution of gases (H_2O and CO_2) on the figure. They do not contribute as the main attenuation cause, yet they will affect the curtain efficiency.

Using the Monte Carlo simulation a flux received by the target from the source can be computed in the absence of the spray for a reference value and then with the spray on. The ratio of the two values provides a kind of transmissivity for the curtain :

$$T_{curtain} = \frac{\text{Flux computed through the spray}}{\text{Reference flux without spray}} \tag{1}$$

Attenuation is defined as the complementary value to 1 and finally total values can be computed applying Planck's averaging for example. Following results will make use of these different definitions.

The resulting spectral transmissions through the water curtain are shown on Figure 3 considering the different injection possibilities and at different vertical positions on the target panel. Comparing the three subfigures, the differences between the three injection solutions are obvious.

One can first comment the results for the downward injection as our reference value. The mean transmission level is around 80% with a little decrease when going down in the curtain (because droplets are decelerating and their residence time becomes higher near the bottom). The global evolution as a function of the wavelength is not strong. Yet, there is an increase in the transmission for larger wavenumbers (which is in fact more pronounced for the other injection

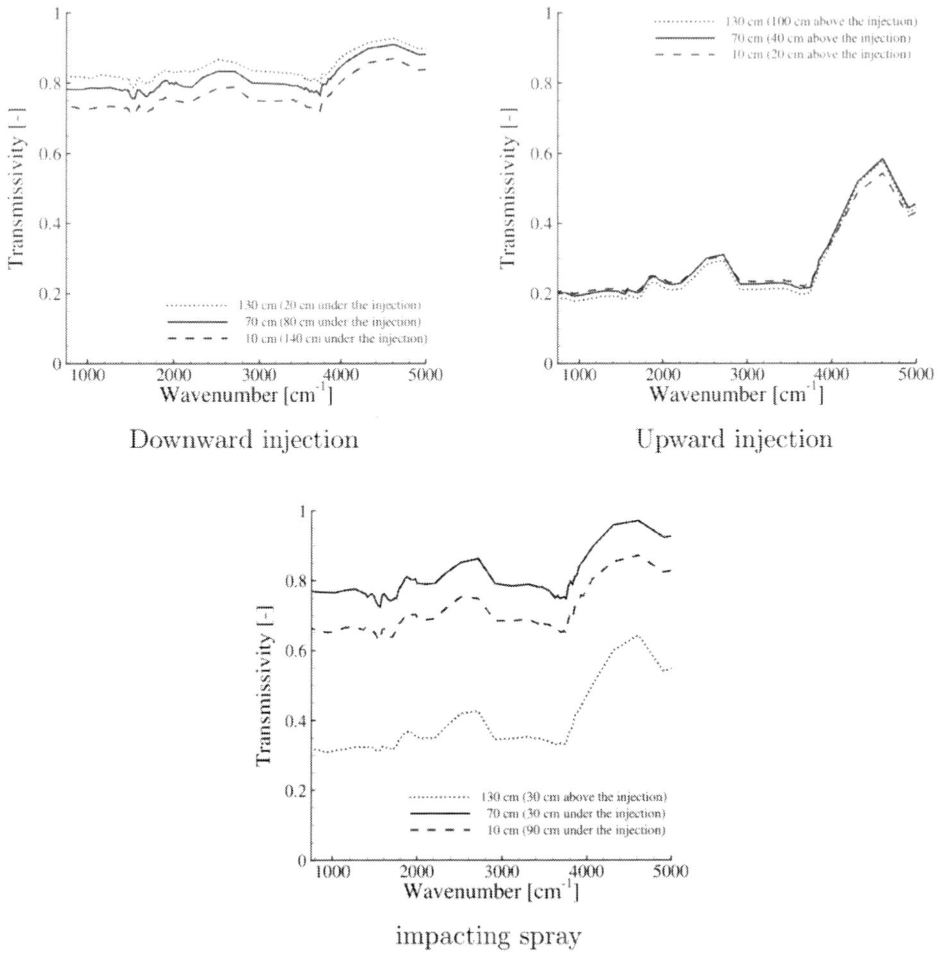

Figure 3. Spectral transmission through the water curtain produced by the droplet injection

directions). As above-explained the extinction coefficient of the droplets is quite constant, but for high wavenumbers absorption falls to a near zero value due to the low absorption index of water. It appears that scattering alone is less efficient for an actual attenuation of radiation in the present scenario where the target is wide. There is a deviation of radiation from initial direction due to scattering, but as scattering is highly forward oriented the deviation is weak and the target still receive the incident flux. It is not really protected and the attenuation ability is finally weak. For the downward injection the attenuation level reaches values close to 25% (transmission equal to 75%) near the bottom of the curtain where droplets have already decelerated. In the upward injection case the attenuation ability is of course better, in relation with a higher residence time for droplets. Transmissions down to 20% are predicted in the best case (low wavenumbers) for the same water consumption. The transmission is higher in the high wavenumber range but the water use remains more efficient than in the downward case. Finally, the impacting spray provides attenuations in an intermediate range. For this particular case, transmission strongly varies with the vertical position, which could be related to the droplet dynamics. The best attenuation occurs near the impacting area where droplet concentration is the highest. Of course the total attenuation in the impacting case is expected to be even better

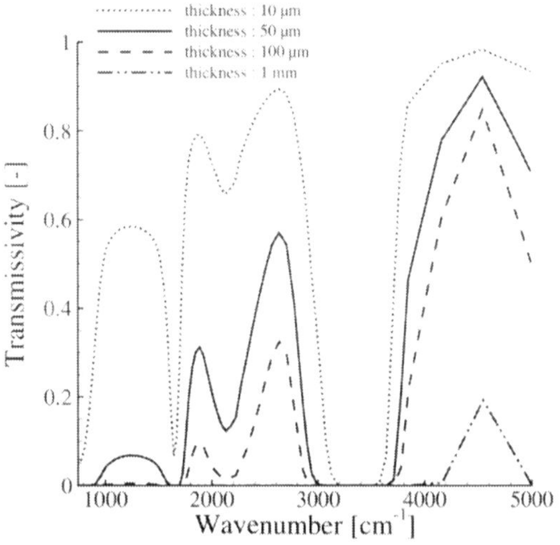

Figure 4. Spectral transmission through a water film with given thickness

whatever the vertical position if the streaming film is also accounted for in the evaluation (as it will be done below).

3.2. Water film description

One major difficulty is the prediction of the actual thickness of the film, which is an exercise of fluid mechanics beyond the scope of the present paper aimed at comparing the different strategies and at defining if a gain can be achieved by using water in a different manner than being simply downward injected. Therefore the film is treated aside from the rest of the spray dynamics, assuming a constant thickness. It is clear that its thickness should vary with position as a consequence of its streaming on the wall. In order to make the evaluation possible the thickness is set to an arbitrary value (which will be varied in order to see the sensitivity of the results to this parameter). Then, the absorption of this homogeneous semitransparent medium is simply defined on the basis of the optical indices of water. The transmissivity of the film is given by :

$$T_{film} = e^{-4\pi k\nu\delta} \tag{2}$$

where k is the absorption index for water (taken from the database by Hale and Querry [11]), ν is the wavenumber and δ is the film thickness.

The complete transmissivity through the spray plus the film will be finally defined as the product :

$$T_{total} = T_{film} \times T_{curtain} \tag{3}$$

Corrections for the reflection at the water film surface are not considered at this stage where the film thickness is already a rough assumption which mainly affects the results. Figure 4 presents the spectral transmission through an idealized water film. Small thicknesses have been considered (between 10 μm and 1 mm) but the potential gain in attenuation is very high as compared to what dispersed water can do. Of course one can ask if a film with thickness as small as 10 μm can be produced in an homogeneous layer but owing to the high attenuation ability of a film, this technique could be considered as an actual way to better use water and to add a supplementary shielding effect to the spray.

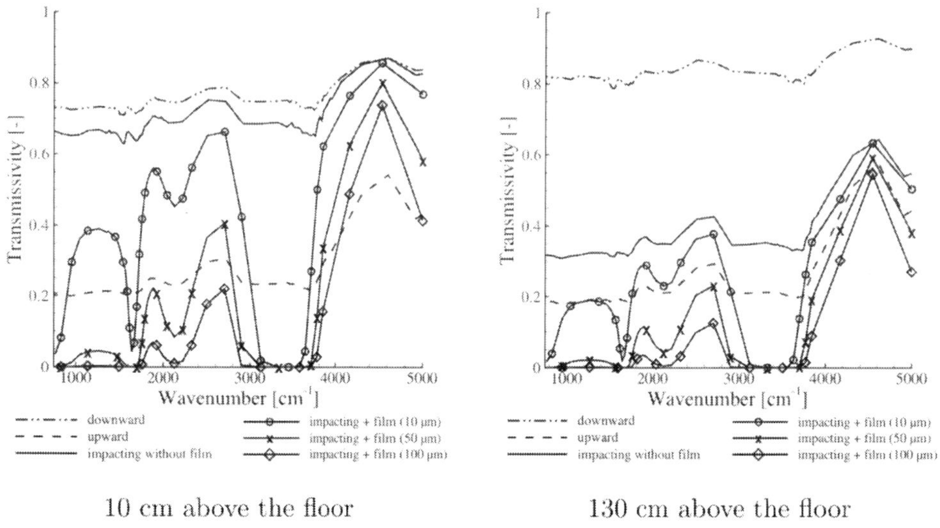

Figure 5. Spectral transmission with the different injection solutions at two vertical positions

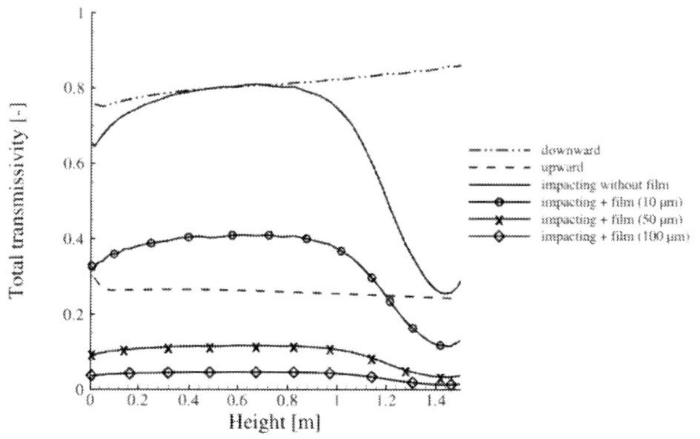

Figure 6. Total transmission through the water curtain as a function of the vertical position

4. Radiation attenuation efficiency

For a complete comparison of the three strategies, efficiency results are now plotted on Figure 5, with the resulting spectral transmission for the downward, the upward and the impacting case in association with an arbitrary film thickness (of course, this assumption of an homogeneous film with given thickness must be kept in mind). The scenario of a radiant panel illuminating a target panel through the spray is kept and the transmission is seen at two different heights on the target : 10 cm and 130 cm above the floor respectively. It is confirmed for both vertical positions that the upward injection highly increases the shielding ability of the curtain, by a factor between 3 and 4, as compared to the downward injection. The efficiency of the impacting curtain is between the two above results, of course potentially highly improved by the film action. The exact ability is dependent on the guessed film thickness but the attractivity of the solution is obvious.

Further analysis can be done on the total transmission which is plotted as a function of the vertical position on Figure 6.

The results have been computed with a Planck's averaging considering a 1000 K blackbody source. As above observed the behavior of the impacting spray is in between the results provided by the downward and the upward injections. Near the bottom of the curtain, the spray solely does not present a clear improvement as compared to the downward case. However, the attenuation efficiency increases with the height in the spray, reaching performances as high as the one of the upward injection. Moreover, even a very thin water film may improve the attenuation ability in a spectacular manner. A theoretical transmission below 10 % is even predicted if a film thickness of 100 μm can be maintained on the surface of the target wall. Of course, this assumes a steady film, with no heterogeneity, but this seems to be a promising and efficient use of water. A possible supplementary advantage is that the stability of the impacting curtain could be better than the one of the upward solution, which can be quickly dispersed by a lateral air flow (like a wind effect or a ventilation flow) as it has been shown in [8].

5. Concluding remarks

A numerical study has been conducted on the ability of a water curtain to build a shield against radiation, considering three different strategies for injecting water with given nozzles: downward for a high vertical position, upward from the bottom, or directly impacting on a target to be protected. Simulations for radiative transfer through the curtain have been carried out on the basis of results obtained for the spray dynamics with an Eulerian-Lagrangian code. Strong assumptions have been considered at this stage for the film created by the impacting spray however, which is still assumed to be homogeneous with given thickness. Comparisons conducted on a scenario of a radiant panel illuminating a target panel through the curtain show that the upward solution should be preferred if the spray solely is considered, as this configuration provides the highest attenuation through the curtain. However, the impacting curtain is also improving the classical use of water as compared to the downward configuration. Moreover, despite the present assumption of a guessed film thickness, it is obvious that even a very thin film (less than 1 mm or even 100 μm thick) highly attenuates radiation. Considering both the impacting spray and the film, the attenuation quickly appears to be even better that in the upward case. As the stability of the impacting curtain would be probably better than the easily dispersed upward spray (which could be confirmed by a further study on the spray dynamics in that configuration when submitted to a lateral air flow), this impacting strategy should be further studied with a true evaluation of the film thickness and an experimental validation in various conditions.

References

[1] Ravigururajan T and Beltran M 1989 *Fire Safety J.* **15** 171
[2] Coppalle A, Nedelka D and Bauer B 1993 *Fire Safety J.* **20** 241
[3] Dembélé S, Delmas A and Sacadura J *J. Heat Transfer* **119**
[4] Hostikka S and McGrattan K *Fire Safety J.* **41**
[5] Nmira F, Kaiss A, Consalvi J and Porterie B 2007 *Numer. Heat Transfer Part A* **52** 1
[6] Collin A, Boulet P, Parent G and Lacroix D 2007 *Int. J. Thermal Sci.* **45** 856
[7] Collin A, Lechêne S, Boulet P and Parent G 2010 *Num. Heat Transfer Part A* **57** 537
[8] Lechêne S, Acem Z, Parent G, Jeandel G and Boulet P 2011 *Int. J. Heat Mass Trans.* **54** 1689
[9] Hald K and Buchlin J 2003 *Proc. Eurotherm 73 on Comput. Thermal Rad. in Part. Media* (Mons, Belgium)
[10] Lechêne S 2010 Ph.D. thesis Nancy Université
[11] Hale G and Querry M *Appl. Optics* **12**
[12] Soufiani A and Taine J *Int. J. Heat Mass Trans.* **40**

Radiative transfer modelling inside thermal protection system using hybrid homogenization method for a backward Monte Carlo method coupled with Mie theory

S Le Foll[1,2,3], F André[2], A Delmas[2], J M Bouilly[1], Y Aspa[1]

[1] ASTRIUM SAS - Space Transportation, Re-entry Systems & Technologies, 33165 St Medard en Jalles, France
[2] Université de Lyon, CNRS, CETHIL, INSA Lyon, UMR5008, F-69621, France

Abstract. A backward Monte Carlo method for modelling the spectral directional emittance of fibrous media has been developed. It uses Mie theory to calculate the radiative properties of single fibres, modelled as infinite cylinders, and the complex refractive index is computed by a Drude-Lorenz model for the dielectric function. The absorption and scattering coefficient are homogenised over several fibres, but the scattering phase function of a single one is used to determine the scattering direction of energy inside the medium. Sensitivity analysis based on several Monte Carlo results has been performed to estimate coefficients for a Multi-Linear Model (MLM) specifically developed for inverse analysis of experimental data. This model concurs with the Monte Carlo method and is highly computationally efficient. In contrast, the surface emissivity model, which assumes an opaque medium, shows poor agreement with the reference Monte Carlo calculations.

1. Introduction

Space vehicles encounter extreme thermal conditions during atmosphere re-entry, as described in Pulci et al. [1], and must be efficiently shielded to protect their contents. One traditional approach to achieve this involves ablative Thermal Protection System (TPS). Furthermore, since weight is a major constraint for space mission, the use of low density fibrous materials is often chosen. These materials are highly porous media and the impact of radiative transfer within them remains an active research area.

Different methods can be used to model radiative transfer inside fibrous media. Farmer and Howell [2] as well as Modest [3] described several Monte Carlo and Backward Monte Carlo methods for the modelling of the transmittance of absorbing and isotropically scattering media that were the starting point of the method presented in this paper. Dombrovsky et al. [4] developed an approximate analytical solution based on the transport approximation and the two-step solution method to compute the normal emittance of an absorbing, scattering, and refracting medium. Results obtained by [4] were reproduced during the early development of our method to ensure its accuracy for gray medium configurations. Milandri et al. [5] used Mie theory (as described for instance by Bohren and Huffman [6]) with the homogenisation method described by Lee [7] to model the directional emittance without axial symmetry for medium made of silica fibres. They computed the spectral bidirectional scattering

[3] Corresponding author, tel. : +33 4 72 43 74 83, E-mail address: Sebastien.le-foll@insa-lyon.fr

coefficient to take into account anisotropic scattering cases. Mie theory and the homogenization method were used in the present work to test our codes against results provided by [5]. Axial symmetry situations were avoided with the use of the phase function of a single fibre and its orientation in space.

The aim of this paper is to study the possibility to derive a simple model for radiative emission to be used in future work to identify complex refractive index required to model the radiative transfer inside fibrous media. The identification process will be based on the analysis of the radiative emission of samples of carbon felt measured by FTIR spectroscopy. Studied sample is a 3 mm thick low density carbon felt (0.18 $g.cm^{-3}$, 89% porosity) with the fibres' mean radius of 5 μm. Fibres are randomly oriented in space with a favoured direction in the felt plane. As a first attempt to determine which model should be used for this purpose, several ones were developed to assess their accuracy. This paper presents a Monte Carlo method for the estimation of spectral directional emittance based on Mie theory and using Lee's method to homogenize the scattering and absorption coefficient of fibres modelled as infinite cylinders. The scattering phase function of a single fibre is used to determine the anisotropic scattering direction. Complex refractive indexes used in Mie theory are obtained with a Drude-Lorenz model for the dielectric function of soot and are assumed in this work to match those of carbon. A surface emissivity model based on the assumption that the fibrous medium emits radiation as an opaque body is also presented. Numerical analyses were also performed on the Monte Carlo model to compute sensitivity coefficients used as building blocks for a multi-linear approximation of the emission factor function.

2. Models for radiative emission

In this section, three different models are presented. The first one is based on the simplifying assumption that the sample emits radiation as an opaque body and only surface emissivity is modelled. This first approximation is motivated by the fact that the medium may be considered as optically thick even for very thin samples. Indeed, this radiative behaviour was observed with preliminary spectroscopic tests performed at room temperature in the 0.2-2.5 μm spectral range. Figure 1 shows an example of the normal hemispherical transmittance and reflectance spectra measured on a 3 mm thick sample with a Perkin Elmer Lambda 900 spectrometer. The second model takes into account the semi-transparent nature of the medium and the apparent directional emission factor is estimated by a Backward Monte Carlo method. The last one is a multi-linear development of the emission factor function based on results from the Monte Carlo model.

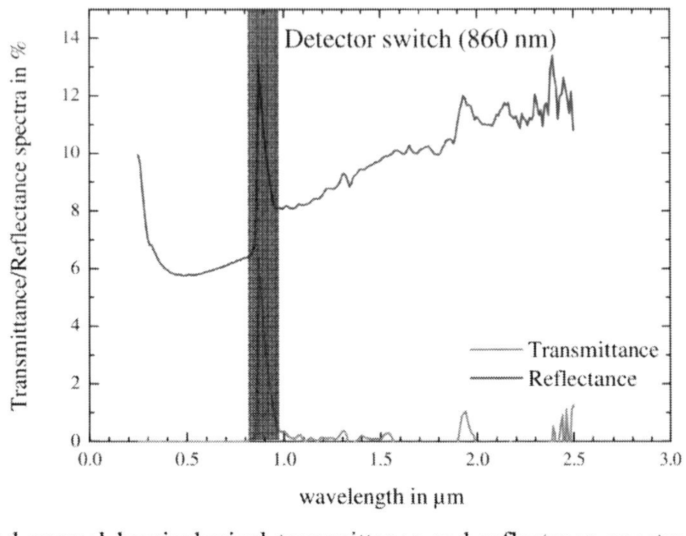

Figure 1. Measured normal hemispherical transmittance and reflectance spectra for a 3 mm thick sample.

2.1. Model for surface emissivity of opaque bodies

The assumption of an opaque body is often chosen in multi-physics code used in the space industry to simplify the already complex and time consuming computation. This model will be used here to show the extent of the inaccuracy brought by such an assumption as it does not suit such materials.

The flux balance for radiation incident in the direction u on an opaque body can be written in spectral form as:

$$\alpha_\nu(\boldsymbol{u}) + \rho_\nu'^{\cap}(\boldsymbol{u}) = 1 \qquad (1)$$

In Eq. (1) $\rho_\nu'^{\cap}(\boldsymbol{u})$ is the spectral directional hemispherical reflectivity. Using Kirchhoff's law, the spectral directional absorptivity $\alpha_\nu(\boldsymbol{u})$ is equal to the spectral directional emissivity $F_\nu(\boldsymbol{u})$ in the same direction. Hence, the spectral directional emissivity is obtained as:

$$F_\nu(\boldsymbol{u}) = 1 - \rho_\nu'^{\cap}(\boldsymbol{u}) \qquad (2)$$

The spectral directional emissivity under unpolarized normal incident radiation for a medium in vacuum can be computed as [8]:

$$F_\nu = 1 - \frac{(n-1)^2+k^2}{(n+1)^2+k^2} \qquad (3)$$

Where n and k are the real and imaginary part of the complex refractive index respectively $(m = n + ik, i^2 = -1)$. Results provided by Eq. (3) will be compared to the Monte Carlo data to assess the relevance of such a model in our problem.

2.2. Backward Monte Carlo model for the radiative emission of fibrous media

A backward Monte Carlo method based on the reciprocity of radiation distribution factor [9] is used to model the spectral directional emittance of the fibrous medium. Using this approach, only rays that reach a target (the detector) are taken into account which reduces computational time considerably compared to forward Monte Carlo for equivalent accuracy [10]. The path of the ray is followed back from the target to the medium layer that emitted it with the scattering or absorption events locations obtained with the extinction length

$$L_{ext} = -\frac{log(Rand)}{\sigma_{\{ext\}}} \qquad (4)$$

In Eq. (4) *Rand* is a number randomly generated between 0 and 1 and $\sigma_{\{ext\}}$ is the extinction coefficients of single fibres. At these locations the ray is absorbed if $Rand > \sigma_{\{sca\}} . \sigma_{\{ext\}}{}^{-1}$ and scattered otherwise. If the ray is scattered, a new direction is computed in accordance with the scattering angle θ, as shown in Figure 2, provided by the scattering phase function of a single fibre. Scattering angles are generated following several successive steps. The first one consists in computing the incident angle Φ based on the former direction of the ray and the orientation of the fibre, randomly chosen in accordance with the orientation distribution function. Then, the scattering angle θ is obtained from the following equation that involves the normalized cumulated phase function defined as:

$$P(\theta, \Phi) = \frac{\int_0^\theta S_{11}(x,\Phi)dx}{\int_0^\pi S_{11}(x,\Phi)dx} = Rand \qquad (5)$$

In Eq. (5) S_{11} is the first element of the amplitude scattering matrix that can be estimated by Mie theory as detailed in Ref. [6]. If the ray is absorbed, the radiative distribution factor $D_\nu^{sj}(\boldsymbol{u})$ is updated. It is defined as the fraction of the total radiation emitted in the emission direction $\boldsymbol{u}$ from the detector and absorbed by a volume element j within the fibrous medium. Then, in the inverse procedure absorption sites j become emission sites and $\boldsymbol{u}$ become the direction in which the emission factor is computed. The radiative spectral intensity in the direction $\boldsymbol{u}$ is the sum of all the radiative distribution factor $D_\nu^{sj}(\boldsymbol{u})$ times the radiative intensity of the black body $I_\nu^b(T_j)$:

$$I_\nu(\boldsymbol{u}) = \sum_{j=0}^{N} D_\nu^{sj}(\boldsymbol{u}) \times I_\nu^b(T_j) \quad (6)$$

The apparent spectral directional emission factor based on a temperature T_0 can then be calculated as

$$F_\nu(\boldsymbol{u}) = \frac{I_\nu(\boldsymbol{u})}{I_\nu^b(T_0)} \quad (7)$$

The Monte Carlo model requires estimating the scattering and extinction coefficients of the medium as well as the scattering phase function of a single fibre. Radiative properties are obtained by homogenisation of those of single fibres assumed to be infinite cylinders and estimated with Mie theory. Homogenisation is based on the method described in Lee [7]. This method uses the orientation of fibres in the medium, as shown in Figure 2.

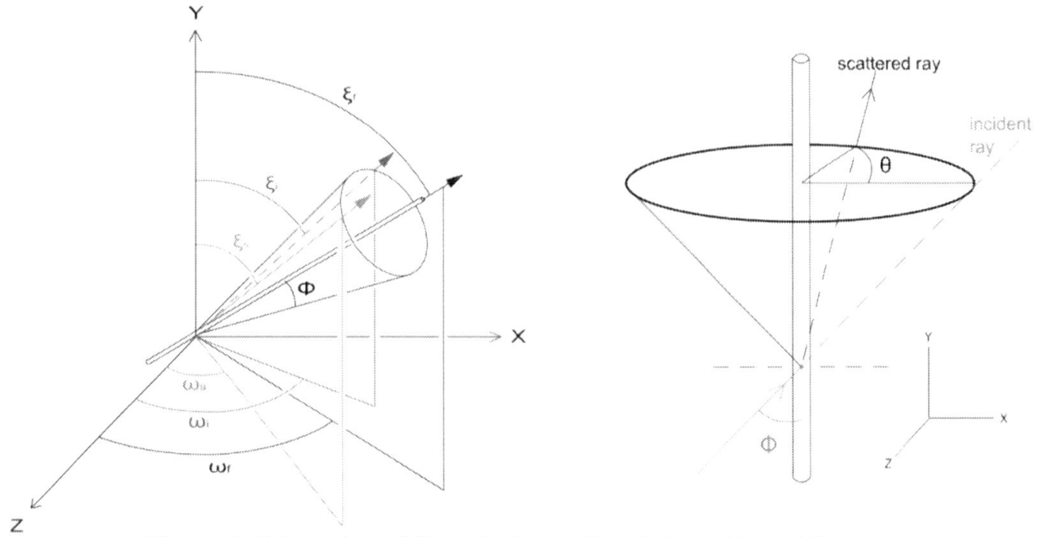

Figure 2. Orientation of fibres in the medium (adapted from [6]).

The scattering/absorption coefficient for fibres randomly oriented in space (see Eq. 8 below) or randomly oriented in a plane (Eq. 9) for media composed of fibres of same diameters are given as

$$\sigma_{\{ext,sca\}} = \frac{f_v}{\pi R} \int_0^{\pi} Q_{\{ext,sca\}}(\Phi)\, sin(\Phi)\, d\Phi \quad (8)$$

$$\sigma_{\{ext,sca\}} = \frac{f_v}{2\pi^2 R} \int_0^{2\pi} \int_0^{\pi} Q_{\{ext,sca\}}(\Phi)\, sin(\xi_f) d\xi_f\, d\omega \quad (9)$$

For configuration involving fibres in specific directions ω_{fi} and ξ_{fi}, we obtain:

$$\sigma_{\{ext,sca\}} = \sum_{i=1}^{n} \sigma_{\{ext,sca\}}(\omega_{fi}, \xi_{fi}) p(\omega_{fi}, \xi_{fi}) \delta(\omega_f - \omega_{fi}) \delta(\xi_f - \xi_{fi}) \quad (10)$$

Where $\sigma_{\{ext,sca\}}$ are the extinction/scattering coefficients of single fibres and $Q_{\{ext,sca\}}$ their corresponding efficiencies f_v is the volume fraction and R is the radius of a fibre assumed to be the same for all fibres. The incident angle is estimated as:

$$cos\,\Phi = sin(\xi_i)\,sin(\xi_f)\,cos(\omega_i - \omega_f) + cos(\xi_i)\,cos(\xi_f) \qquad (11)$$

The main parameters for Mie computations are the size factor $x = 2\pi R\nu$ (ν is the wavenumber) and the complex refractive index of the material fibres are made of. This quantity can be modelled by the Drude-Lorenz formula for the dielectric function with two bound electron $_j$ and one free one $_{fr}$:

$$\epsilon = 1 - \frac{\omega_{p,fr}^2}{\omega_a^2 + i\gamma_{fr}\omega_a} + \sum_{j=1}^{2} \frac{\omega_{p,j}^2}{\omega_{r,j}^2 - \omega_a^2 - i\gamma_j\omega_a} \qquad (12)$$

with $\omega_{p,j}^2 = \frac{e^2 Ne_j}{m_e \varepsilon_0}$, $\omega_a = 2\pi c_0 \nu$ and $m = n + ik = \sqrt{\epsilon}$

In Eq. (12) m_e, $\omega_{r,j}^2$, Ne_j and γ_j are respectively the mass of the electron [9.1096×10^{-31} kg], the resonant frequency (0 for free electrons, $\omega_{r,1} = 1.25 \times 10^{15}$ [rad/s] and $\omega_{r,2} = 7.25 \times 10^{15}$ [rad/s] [10,11]), the electron number density [#/m^3] and the damping constant [rad/s]. e is the charge of an electron [1.6022×10^{-19} C] and ε_0 is the dielectric constant in vacuum [8.8542×10^{-12} F.m^{-1}].

Computation took up to 902 seconds on an Intel Xeon E5620 CPU dual core at 2.40 GHz to calculate the 2.26 milliard bundles required to achieve a standard deviation lower than 0.1% with the Monte Carlo method (see Table 1). The code is parallelized on 8 different processors. Computation made with 113 and 226 wavelengths taken uniformly in the 0.25-2.5 μm spectral range provided the results given in Table 1 in terms of computation time and standard deviation. For each calculation, 4 different samples (4 computation sets with the same parameters) are simulated and the mean value is taken as solution in order to reduce the impact of statistical noise on the results.

Table 1. Monte Carlo method computation times for two different numbers of computed wavelengths.

Number of bundles	Mie theory computation time		Monte Carlo computation time		Standard deviation
	113 wavelengths	226 wavelengths	113 wavelengths	226 wavelengths	
100000	8 sec.	16 sec.	4 sec.	9 sec.	1.18 %
500000	8 sec.	16 sec.	22 sec.	44 sec.	0.74 %
1000000	8 sec.	16 sec.	44 sec.	88 sec.	0.44 %
5000000	8 sec.	16 sec.	224 sec.	443 sec.	0.17 %
10000000	8 sec.	16 sec.	445 sec.	891 sec.	0.10 %

Standard deviation values given in Table 1 are the maximum values of the mean deviation between each sample and their mean value calculated for each wavelength as

$$stddev = max\left(\overline{stddev}(\nu)\right) \qquad (13)$$

With

$$\overline{stddev}(\nu) = \sum_{i=0}^{N_{sample}} \frac{|F_\nu^i - \overline{F_\nu}|}{\overline{F_\nu}} \times N_{sample}^{-1} \qquad (14)$$

The Monte Carlo computation can be quite computationally expensive for identification purpose. As one of the goals of this study is to estimate refractive indexes from experimental emittance data, a simplified model is required. Coupling conduction and radiative heat transfers will be considered as future work to treat non isothermal configurations and may also benefits from such a simplified model.

The modifications made on a standard Backward Monte Carlo method using radiation distribution factor are summarized in the flowcharts shown in Figure 4.

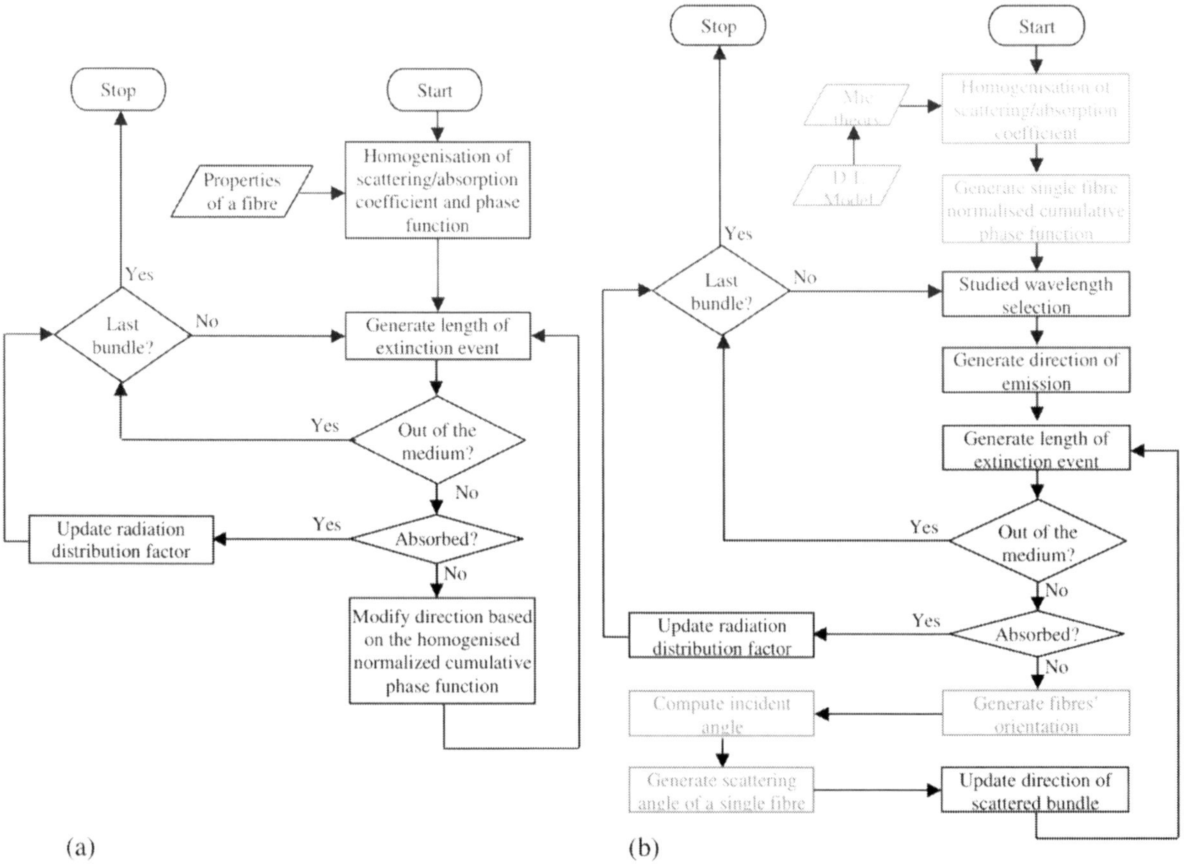

(a) (b)

Figure 4. Flowcharts for usual Backward Monte Carlo method(a) and the one used in this study (b)

2.3. Multi-Linear Model (MLM) from sensitivity analyses

As shown in Eq. (12) six main parameters are used in the Drude-Lorenz model for the dielectric function, Ne_{fr}, Ne_1, Ne_2, γ_{fr}, γ_1 and γ_2 chosen in this work. Using a multi-linear approximation of the emission function we can obtain a simple model for the emission factor that takes into account the complex refractive index of the bulk material:

$$\widetilde{F_\nu} = F_{\nu 0} + \frac{\partial F_\nu}{\partial \log Ne_{fr}}\left(\frac{\widetilde{Ne_{fr}}}{Ne_{fr}} - 1\right) + \frac{\partial F_\nu}{\partial \log Ne_1}\left(\frac{\widetilde{Ne_1}}{Ne_1} - 1\right) + \frac{\partial F_\nu}{\partial \log Ne_2}\left(\frac{\widetilde{Ne_2}}{Ne_2} - 1\right)$$
$$+ \frac{\partial F_\nu}{\partial \log \gamma_{fr}}\left(\frac{\widetilde{\gamma_{fr}}}{\gamma_{fr}} - 1\right) + \frac{\partial F_\nu}{\partial \log \gamma_1}\left(\frac{\widetilde{\gamma_1}}{\gamma_1} - 1\right) + \frac{\partial F_\nu}{\partial \log \gamma_2}\left(\frac{\widetilde{\gamma_2}}{\gamma_2} - 1\right) \quad (15)$$

In Eq. (15) Ne_{fr}, Ne_1, Ne_2, γ_{fr}, γ_1 and γ_2 are the mean value between those given in Table 2 and $\widetilde{Ne_{fr}}, \widetilde{Ne_1}, \widetilde{Ne_2}, \widetilde{\gamma_{fr}}, \widetilde{\gamma_1}$ and $\widetilde{\gamma_2}$ are the estimated parameters. $F_{\nu 0}$ is the emission factor computed with the mean values for all parameters. The influence of each parameter on this model sensitivity analyses have been studied on the Monte Carlo method based on values given in Table 2. These sensitivity analyses have allowed us to calculate sensitivity spectra as:

$$\frac{\partial F_\nu}{\partial \log X} = X\,\frac{F_\nu(X+dX) - F_\nu(X-dX)}{2dX} \quad (16)$$

where X represent either Ne_{fr}, Ne_1, Ne_2, γ_{fr}, γ_1 or γ_2. Those spectra are required to use Eq. (15).

Table 2. Range of the parameters used for the sensitivity analysis.

	Min	Max
Electron number density [#/m^3]		
Ne_{fr}	1.00x10^{24}	6.40x10^{25}
Ne_1	5.00x10^{26}	5.00x10^{27}
Ne_2	5.00x10^{27}	1.00x10^{29}
Damping constant [rad/s]		
γ_{fr}	5.00x10^{14}	3.00x10^{15}
γ_1	1.00x10^{15}	1.00x10^{16}
γ_2	1.00x10^{15}	1.00x10^{16}

Several sensitivity spectra are provided in Figure 5 as examples.

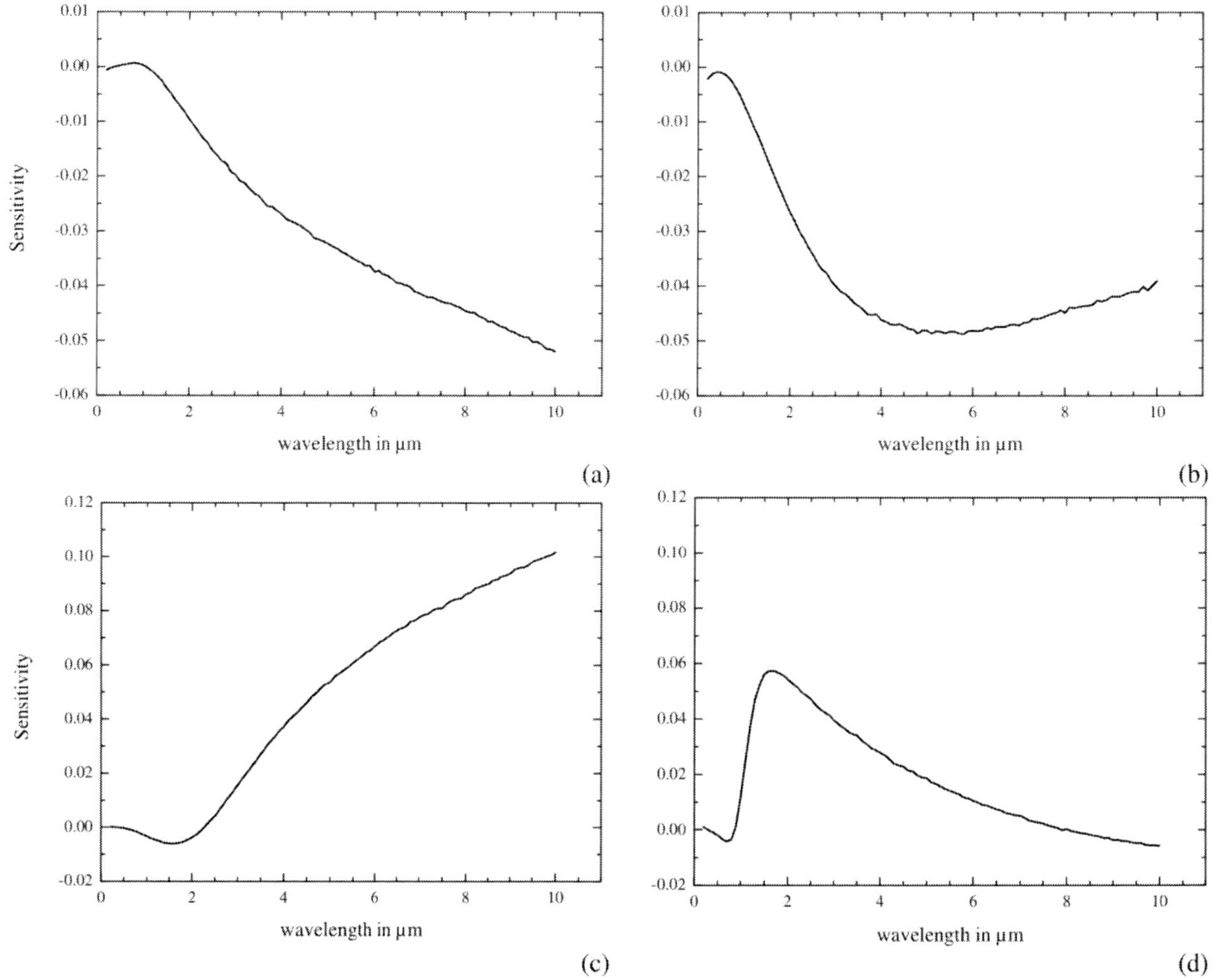

Figure 5. Sensitivity spectra for $\frac{\partial F_\nu}{\partial \log Ne_{fr}}$ (a), $\frac{\partial F_\nu}{\partial \log Ne_1}$ (b) $\frac{\partial F_\nu}{\partial \log \gamma_{fr}}$ (c) and $\frac{\partial F_\nu}{\partial \log \gamma_1}$ (d).

As shown in Figure 5, each parameter has a more or less predominant effect on the calculated emission factor depending on the spectral range that is considered. Consequently an analysis of the sensitivity spectra will enable to define the most relevant spectral regions for identification of given spectral properties. It is important to notice that as a Taylor development was used, this method will only work for properties close to those of the reference state $F_{\nu 0}$. In our case, the MLM model will be used to identify those parameters for our carbon TPS sample. Following results found in the literature for different carbonaceous compounds (soot, amorphous carbon, and pyrolytic graphite) it seems safe to assume that the variation of the parameters will not be too large to use this model.

3. Results and discussion

The Monte Carlo model is used as a reference for the modelling of the spectral directional emission factor of the fibrous medium. For each case, computations were made for a 3 mm thick medium composed of fibres randomly oriented in space. Fibres radius is 5 μm and the volume fraction is 11%. The medium is isothermal. Results presented here are for the direction normal to the surface. Results are shown in Figures 6 to 8 for refractive index estimated with the parameters from Habib and Vervisch [11] for soot obtained from ethylene and acetylene flames or from Charalampopoulos and Chang [12]. Results from the MLM approximation and for surface emissivity assuming an opaque body are also provided.

The MLM is in good agreement with the Monte Carlo method for most of the studied cases, whereas the assumption of surface emissivity is not satisfactory in any of them. The maximal error is 3% for the MLM and reaches 25% for the surface emissivity model. It should be kept in mind that the surface emissivity model is for specular reflection which underestimates the emission factor but the spectral behaviour should not be affected anyhow. Higher order derivation should improve the MLM approximation quite a bit without involving prohibitive computation costs. Other parameters such as the fibre orientation could also be taken into account using other coefficients.

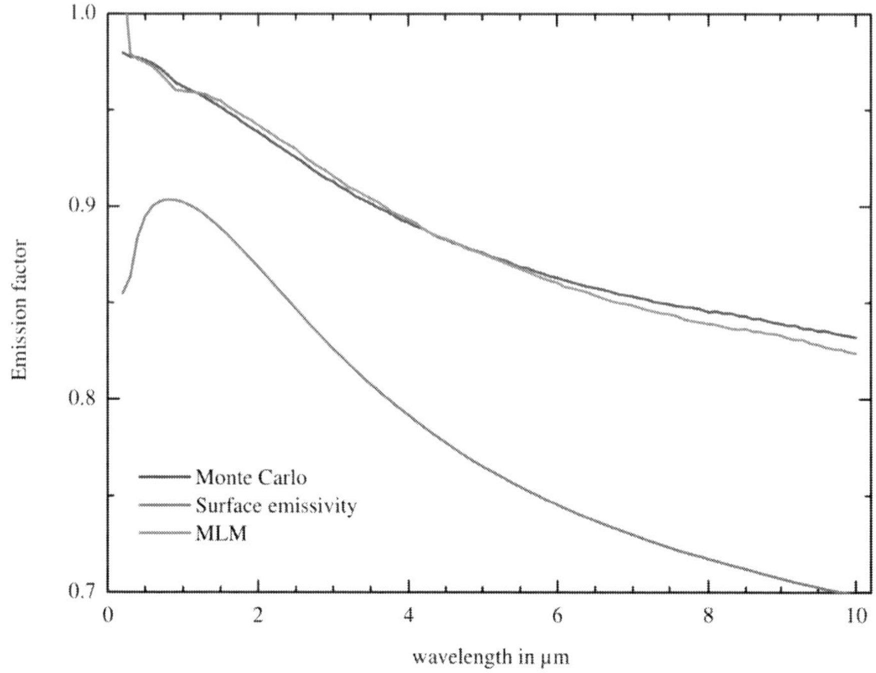

Figure 6. Spectral normal emission factor computed with the Monte Carlo method, the opaque surface model and MLM. Refractive index parameters are from Habib and Vervisch [11] for soot particles produced by ethylene flames.

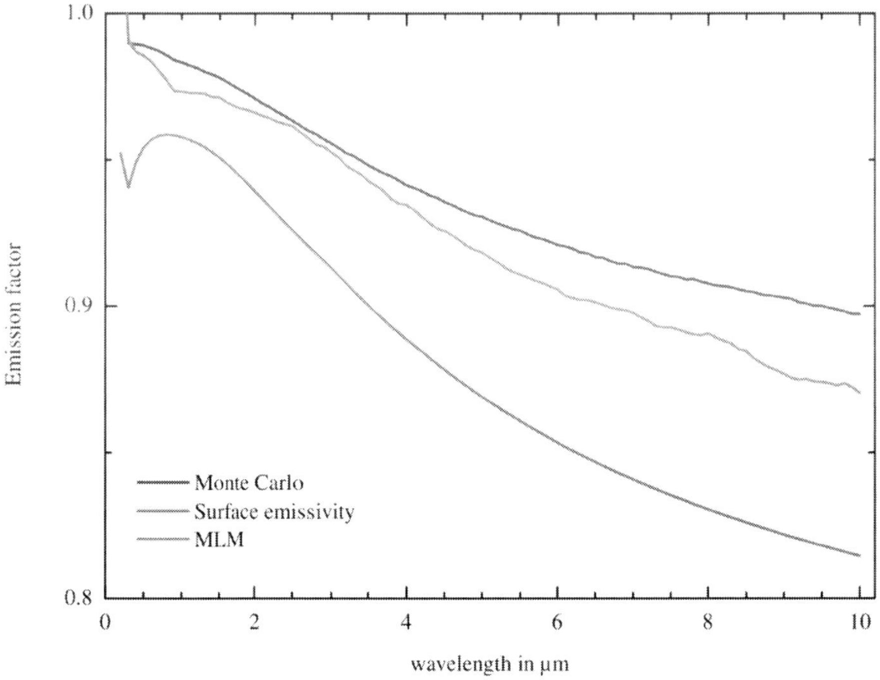

Figure 7. Spectral normal emission factor computed with the Monte Carlo method, the opaque surface model and MLM. Refractive index parameters are from Habib and Vervisch [11] for soot particles produced by acetylene flames.

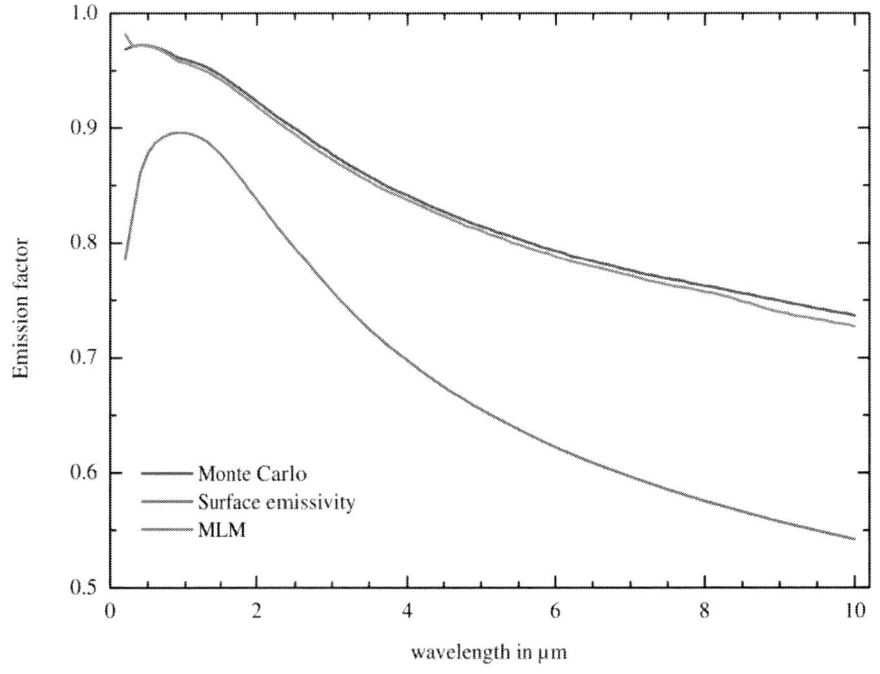

Figure 8. Spectral normal emission factor computed with the Monte Carlo method, the opaque surface model and MLM. Refractive index parameters are from Charalampopoulos and Chang [12].

4. Conclusion

Several models have been developed to study the possibility to derive a simplified model to identify the complex refractive index required to estimate the radiative transfer inside fibrous media at high temperature.

A backward Monte Carlo method for spectral directional emittance of fibrous media has been presented in this paper. Radiative properties of the fibres have been obtained from Mie theory with the complex refractive index computed from the Drude-Lorenz model for the dielectric function. Radiative properties are homogenised over ensembles of fibres whereas the direction of the radiation after scattering events is determined by the phase function of a single one.

This method is computationally expensive and unsuitable for inversion process. For identification purpose, a quicker but accurate model is required. As expected, the simplest model assuming an opaque body is not relevant for the studied material, so the MLM was proposed. Calculations of the normal spectral emission factor performed with the different methods have been compared for four different sets of parameters for Drude-Lorenz model for the dielectric function.

Sensitivity analysis of the refractive index on the spectral directional emission factor was performed on the Monte Carlo model to derive a multi-linear model. This model can be improved furthermore with higher order derivation without significant influence in computation time. Other parameters such as the fibre orientation may also be taken into account using other coefficients. This model will be used for identification purpose of refractive indexes from experimental radiative emission spectra of samples of carbon felt obtained by FTIR spectroscopy. Another interesting point of this Monte Carlo method is the ability to compute the emission factor of a fibrous medium filled with a participating gas. This will later be used to study the impact of the pyrolytic gazes on the radiative transfer inside low density carbon TPS.

References

[1] Pulci G, Tirillò J, Marra F, Fossati F, Bartuli C and Valente T 2010 Carbon–phenolic ablative materials for re-entry space vehicles: Manufacturing and properties *Composites: Part. A* **41** 1483–90

[2] Farmer J T and Howell J R 1998 Comparison of Monte Carlo strategies for radiative transfer in participating media, *Adv. Heat Transfer* **31** 333–429

[3] Modest M F 2003 *Radiative Heat Transfer, Second Edition* (Academic Press)

[4] Dombrovsky L A, Randrianalisoa J H, Lipinski W and Baillis D 2011 Approximate analytical solution to normal emittance of semi-transparent layer of an absorbing, scattering, and refracting medium *J. Quant. Spec. Rad. Transf.* **112** 1987-94

[5] Milandri A, Asllanaj F, Jeandel G and Roche J R 2002 Heat transfer by radiation and conduction in fibrous media without axial symmetry *J.Quant. Spec. Rad. Transf.* **74** 585–603

[6] Bohren C F and Huffman D R 1983 *Absorption and Scattering of Light by Small Particles* (Wiley Interscience)

[7] Lee S C 1986 Radiative transfer through a fibrous medium: allowance for fiber orientation *J.Quant. Spec. Rad. Transf.* **36** 253-63

[8] Siegel R, Howell J and Mengüç M P 2010 *Thermal Radiation Heat Transfer* (New York: CRC Press)

[9] Mahan J R 2002 *Radiation Heat Transfer* (New York: Wiley) 390–408

[10] Lia B X, Yub X J and Liua L H 2005 Backward Monte Carlo simulation for apparent directional emissivity of non-isothermal semitransparent slab *J. Quant. Spec. Rad. Transf.* **91** 173-79

[11] Habib Z G and Vervisch P 1988 On the refractive index of soot at flame temperature *Combustion Science and Technology* **59** 261-74

[12] Charalampopoulos T T and Chang H 1988 In-situ optical properties of soot particles in the wavelength range from 340 nm to 600 nm *Combustion Science and Technology* **59** 401-21

Journal of Physics: Conference Series **369** (2012) 012029 doi:10.1088/1742-6596/369/1/012029

SOLFAST, a Ray-Tracing Monte-Carlo software for solar concentrating facilities

J P Roccia[1], **B Piaud**[1], **C Coustet**[1]
[1]HPC-SA, 3 chemin du Pigeonnier de la Cépière, 31100 Toulouse, Fr.

C Caliot[2], **E Guillot**[2], **G Flamant**[2]
[2]PROMES-CNRS, 7 rue du Four Solaire, 66120 Font Romeu Odeillo, Fr.

J Delatorre[3]
[3]Centre Rapsodee, Ecole des Mines d'Albi, Albi, Fr.

E-mail: benjamin.piaud@hpc-sa.com

Abstract. In this communication, the software SOLFAST is presented. It is a simulation tool based on the Monte-Carlo method and accelerated Ray-Tracing techniques to evaluate efficiently the energy flux in concentrated solar installations.

1. Introduction

According to the International Energy Agency, the world production of electricity by concentrating solar power (CSP) could reach 10% by 2050. That is as much as the solar photovoltaic (PV) or the nuclear industry. Compared to other renewable energy (particularly wind power and PV), the CSP technology has the advantage of adapting electricity production compared to the demand even at night, thanks to the possibility of storage of the heat for several hours before electricity production by thermodynamic cycles. However the cost of installation is still very high compared to other renewable technologies. This high cost is mainly due to the optical concentration system either linear concentrators (parabolic trough, Fresnel) or punctual concentrators (tower, beam-down) compared to the cost of the thermodynamic processes facilities (steam turbines and heat storage). From a technological point of view, the whole thermodynamic process for electricity generation benefits from the experience of existing power plants (thermal or nuclear). Thus reducing the cost of installing a CSP plant will be primarily by optimizing the optical system (receiver, heliostat field, tower). One difficulty is that the design of optical system is unique for a particular facility. Indeed it must be adapted to the latitude, the ground topology and the solar resource. Given the peculiarity of the optical system, CSP plants design and optimization require high-performance computer codes that can efficiently simulate different configurations of systems with complex geometries with thousands of heliostats themselves composed of several mirrors. In terms of modelling to assist in the design of CSP plants there are two types of tools: codes for predicting the radiative flux at the entrance of solar receiver (optical codes) and on the other hand codes to model the whole process to generate electricity ("process" codes). These can be coupled to optical codes or use simple models to predict the radiative flux as boundary condition of the process model. In this paper, we present an algorithm implemented in the Solfast (SOLar FAcilities Simulation Tools) software developed in partnership by HPC-SA and PROMES-CNRS laboratory. Solfast aims to calculate maps of directional solar flux to design or to characterize solar concentration facilities by considering the reflective properties of mirrors. Other Ray-Tracing solvers exist [1] and among them Soltrace [2] and Tonatiuh [3] are freely available and well documented. However these two solvers use collision-based Monte-Carlo algorithms that follow the bundles of ray from the source to the receiver. A first drawback is that the source modelling leads to lose many bundles that do not impact reflective surfaces. A second drawback is that it uses bundles with identical Monte-Carlo weights with russian roulettes causing supplementary bundle losses. These drawbacks slow down Monte-Carlo algorithm convergence. In this article, we present a Monte-Carlo (MC) algorithm which uses an integral formulation with variance reduction techniques to overcome the drawbacks of collision-based Ray-Tracing solvers described previously. Sec. 2 presents the Monte-Carlo algorithm implemented in Solfast. The employed methodology is particularly suitable to solve the radiative transfer equation in complex geometries such as optical system in the concentrated solar processess [4]. Sec. 3 presents the features enabled in Solfast and the graphical interface which allows users to design a solar facility optical system. Sec. 4 presents some academic test cases and a more realistic complex geometry to validate the gain in performance of our Monte-Carlo integral formulation.

2. Monte-Carlo Ray-Tracing model

The Monte-Carlo Ray-Tracing (MCRT) model implemented in the Solfast software includes (1) the reflection events involved in a solar concentrating system having a complex geometry and (2) techniques to increase the speed of the MCRT algorithm. The first point is addressed by using an interface able to manage tessellated complex geometry where reflection events are simulated with a bi-directional reflection distribution function (BRDF). The second point is addressed by reformulating the integral and by using a variance reduction technique and a fast intersector (RayBooster technology [5]).

The aim of the computation is to predict the concentrated solar flux spatial distribution incident on virtual surfaces located anywhere in the scene, e.g. at the receiver entrance as shown in the sketch of a solar concentrating system depicted in Fig.1. Three types of surfaces are considered in Solfast: two reflecting surfaces i.e. heliostats $\mathcal{H}$ and concentrators $\mathcal{C}$, and one virtual surface $\mathcal{V}$ where radiative power density is computed. Each surface has two sides, denoted by a plus or minus sign, for which separate optical properties can be specified. The virtual surface $\mathcal{V}$ is discretized in surface elements that are called hereafter *pixels*. The concentrated radiant solar flux P (in $Watts$) incident on an arbitrary pixel of the virtual surface $\mathcal{V}^+$ is expressed by the following integral:

$$P = \int_{\nu_{min}}^{\nu_{max}} d\nu \int_{\mathcal{H}^+} \mathrm{d}S(\boldsymbol{r_1}) \int_{\Delta\Omega_{\mathrm{sun}}} \mathrm{d}\boldsymbol{\omega_0}\, I_{\mathrm{sun}}(\boldsymbol{\omega_0}, \boldsymbol{r_0}, \nu) \left|\boldsymbol{n}(\boldsymbol{r_1}) \cdot \boldsymbol{\omega_0}\right| \left[H(\boldsymbol{r_0} \notin \{\mathcal{H} \cup \mathcal{C} \cup \mathcal{V}\})\, \mathcal{I}_1 \right] \tag{1}$$

with the sun intensity related to the Direct Normal Irradiance (DNI) by: DNI$= \int_{\nu_{min}}^{\nu_{max}} d\nu \int_{\Delta\Omega_{\mathrm{sun}}} \mathrm{d}\boldsymbol{\omega_0}\ I_{\mathrm{sun}}(\boldsymbol{\omega_0}, \nu) |\boldsymbol{\omega_0} \cdot \boldsymbol{n_0}|$. This integral expression for P assumes a specific formulation of the problem [6, 4] by considering (after the integral over the frequencies) an integral over the surface $\mathcal{H}^+$ which represents the reflecting surface involved in the first reflection. The general principle of the MCRT algorithm highlighted by this formulation is to first sample the $\mathcal{H}^+$ surface for the first reflection location, then the direction of the incident radiation can be sampled and the Ray-Tracing of the reflected bundle of ray can begin. The presented equations include tests whose symbol is $H(x)$, resulting in $H(x) = 1$ if x is true and $H(x) = 0$ if x is false. The test in Eq.(1) is related to the position of the bundle source with location $\boldsymbol{r_0}$ and direction $\boldsymbol{\omega_0}$. The test ensures that there is no shadowing by any geometry element and the Ray-Tracing can begin with the recurrence term $\mathcal{I}_i$ dealing with the multiple reflections:

$$\begin{aligned} \mathcal{I}_i \;=\; & \int_{2\pi} \mathrm{d}\omega_i \, \rho''(\boldsymbol{\omega_i}|\boldsymbol{\omega_{i-1}}, \boldsymbol{r_i}, \nu) \left| \boldsymbol{n}(\boldsymbol{r_i}) \cdot \boldsymbol{\omega_i} \right| \\ & \left[H(\boldsymbol{r_{i+1}} \in \mathcal{V}^+) + H(\boldsymbol{r_{i+1}} \in \{\mathcal{H}^+ \cup \mathcal{C}^+\}) \, \mathcal{I}_{i+1} \right] \end{aligned} \quad (2)$$

The Figure 2 shows a representation of directions and angles. The recurrence term includes the reciprocal and conservative BRDF, ρ'', and represents the Ray-Tracing of the bundle with multiple reflections. The equation Eq.(2) involves two tests:

- $H(\boldsymbol{r_{i+1}} \in \mathcal{V}^+)$ means that if the location of the next intersection is contained in the pixel surface, then $H = 1$ and the realization of the Monte-Carlo weigth is stored.
- $H(\boldsymbol{r_{i+1}} \in \{\mathcal{H}^+ \cup \mathcal{C}^+\})$ means that if the next intersection is located on a reflective surface, another Eq.(2) should be computed to account for another intersection.

It should be noted that if the ray is neither reflected nor impacted on the pixel, the ray is lost.

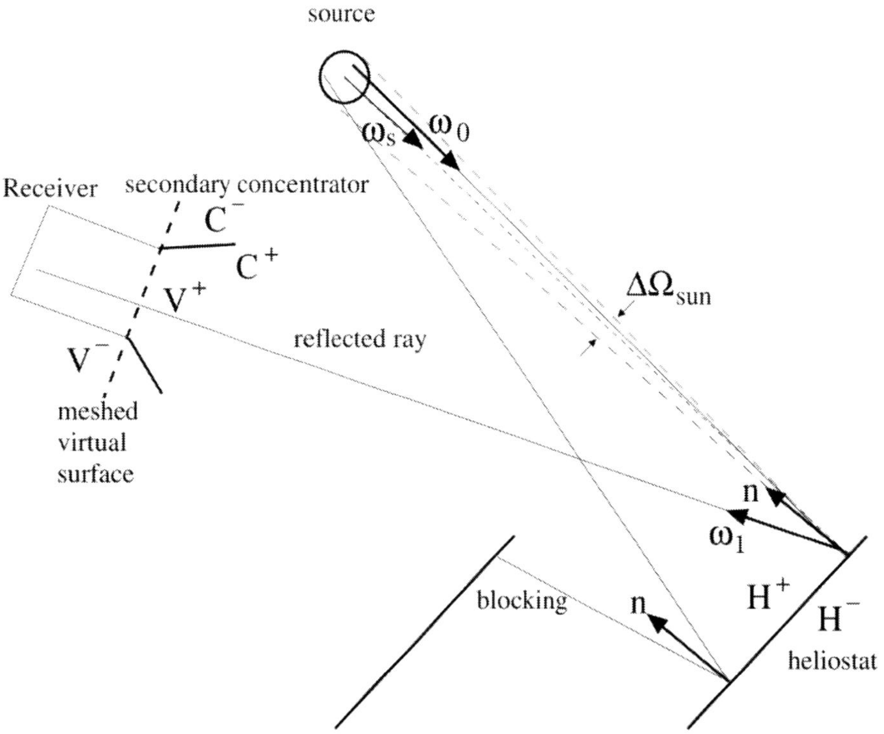

Figure 1. Sketch of a typical simple solar concentration system including the sun source, reflecting mirrors (heliostat and secondary concentrator), a receiver with a secondary concentrator and notations for reflective ($\mathcal{H}^+$) and virtual ($\mathcal{V}^+$) surfaces.

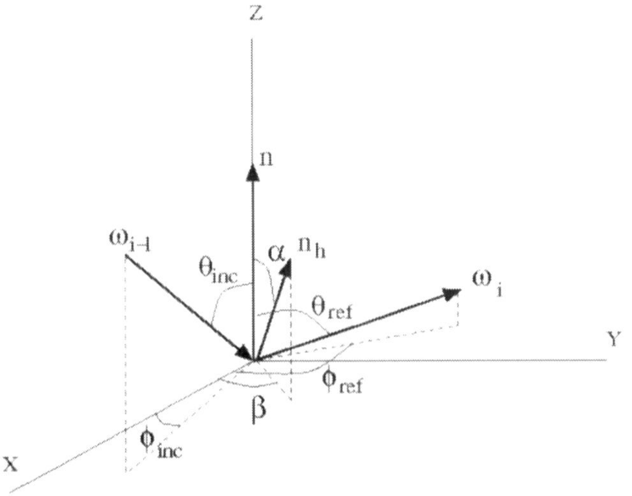

Figure 2. Representation of angles used in Solfast; the Z positive axis is aligned with the theoritical surface normal $\boldsymbol{n}$.

The MCRT model is used to evaluate the integral P (Eq.(1)) and with the Monte-Carlo model the quantity P is considered as the mean of a random variable W_P : $P = E[W_P]$. An approximation is used to compute a finite (but large) number N of realizations $w_{P,k}$ of W_P:

$$P \approx \widetilde{P} = \frac{1}{N} \sum_{k=1}^{N} w_{P,k} \tag{3}$$

This radiative flux is computed for each pixel to obtain a spatial distribution of the radiative flux. Furthermore, with each evaluation of $\widetilde{P}$, a standard deviation is associated $\sigma_{\widetilde{P}}$ to each estimate of $\widetilde{P}$ and can be approximated by a statistical uncertainty $\widetilde{\sigma}_{\widetilde{P}}$:

$$\sigma_{\widetilde{P}} \approx \widetilde{\sigma}_{\widetilde{P}} = \frac{1}{\sqrt{N}} \sqrt{\left(\frac{1}{N} \sum_{k=1}^{N} {w_{P,k}}^2 \right) - \widetilde{P}^2} \tag{4}$$

The MCRT algorithm is used to compute $\widetilde{P}$ and $\widetilde{\sigma}_{\widetilde{P}}$ for each pixel. The Monte-Carlo model allows one to use probability density functions (PDF) in order to improve convergence and reduce the variance. Dropping the frequency integral, assuming a microfacet model (distribution of facet normals) with specular reflection and introducing the PDF and the Monte-Carlo weight, equations Eqs.(1)-(2) become:

$$P = \int_{\mathcal{H}^+} p_S \, \mathrm{d}S(\boldsymbol{r_1}) \int_{\Delta\Omega_{\mathrm{sun}}} p_{\Delta\Omega_{\mathrm{sun}}} \mathrm{d}\boldsymbol{\omega_0} \left[H(\boldsymbol{r_0} \notin \{\mathcal{H} \cup \mathcal{C} \cup \mathcal{V}\})\, \mathcal{I}'_1 \right] \tag{5}$$

$$\mathcal{I}'_i = \int_0^{\Delta\Omega_{\max}} p_{\boldsymbol{N_h}} \, \mathrm{d}\boldsymbol{n_h} \left[H(\boldsymbol{r_{i+1}} \in \mathcal{V}^+)\hat{w}_P + H(\boldsymbol{r_{i+1}} \in \mathcal{H}^+)\mathcal{I}'_{i+1} \right] \tag{6}$$

with the Monte-Carlo weight

$$\hat{w}_P = I_{\mathrm{sun}}(\boldsymbol{\omega_0}, \boldsymbol{r_0}) \left| \boldsymbol{n}(\boldsymbol{r_1}) \cdot \boldsymbol{\omega_0} \right| \frac{1}{p_S} \frac{1}{p_{\Delta\Omega_{\mathrm{sun}}}} \prod_{j=1}^{i} \rho(\boldsymbol{\omega_{\mathrm{spec}}} | \boldsymbol{\omega_{j-1}}, \boldsymbol{n_h}(\boldsymbol{r_j})) \tag{7}$$

Journal of Physics: Conference Series **369** (2012) 012029 doi:10.1088/1742-6596/369/1/012029

with the specular direction defined by $\boldsymbol{\omega_{spec}} = \boldsymbol{\omega_{i-1}} - 2(\boldsymbol{\omega_{i-1}} \cdot \boldsymbol{n})\boldsymbol{n}$, and the following expressions for the PDFs:

$$p_S(S(\boldsymbol{r_1})) = \frac{1}{\mathcal{H}^+} \tag{8}$$

$$p_{\Delta\Omega_{sun}}(\boldsymbol{\omega_0}) = \frac{1}{\int_{\Delta\Omega_{sun}} |\boldsymbol{\omega_0} \cdot \boldsymbol{n_0}| \mathrm{d}\boldsymbol{\omega_0}} = \frac{2}{\pi(1 - \cos 2\theta_{sun})} \tag{9}$$

$$p_{N_h}(\boldsymbol{r_j})\,\mathrm{d}\boldsymbol{n_h} = p_{\boldsymbol{\alpha}}(\alpha)\,\mathrm{d}\alpha\; p_{\boldsymbol{\beta}}(\beta)\,\mathrm{d}\beta$$

$$p_{\boldsymbol{\beta}}(\beta) = \frac{1}{2\pi}, \quad \forall\, \beta \in [0; 2\pi] \tag{10}$$

with θ_{sun} the apparent radius of the sun. Several PDFs can be chosen to sample $\boldsymbol{\alpha}$. In Solfast, three distributions are available i.e. the gaussian Eq.(11), circular normal Eq.(12) and uniform Eq.(13). The $\boldsymbol{\alpha}$ domain is truncated to the value of $\alpha_{max} = \frac{\pi}{4} - \cos^{-1}(|\boldsymbol{\omega_i} \cdot \boldsymbol{n}|)/2$ to prevent any reflection below the mirror surface.

$$p_{\boldsymbol{\alpha}}(\alpha) = \exp\left[-\frac{\alpha^2}{2\sigma^2}\right]\sqrt{2}\left(\sigma\sqrt{\pi}\,\mathrm{Erf}\left[\frac{\alpha_{max}}{\sigma\sqrt{2}}\right]\right)^{-1}, \quad \forall\, \alpha \in [0; \alpha_{max}] \tag{11}$$

$$p_{\boldsymbol{\alpha}}(\alpha) = \frac{\tan\alpha \exp\left[-\frac{(\tan\alpha)^2}{2\sigma^2}\right]\cos^2\alpha}{\sigma^2\left(1 - \exp\left[-\frac{(\tan\alpha_{max})^2}{2\sigma^2}\right]\right)}, \quad \forall\, \alpha \in [0; \alpha_{max}] \tag{12}$$

$$p_{\boldsymbol{\alpha}}(\alpha) = \left(\min[\sigma, \alpha_{max}]\right)^{-1}, \quad \forall\, \alpha \in [0; \min[\sigma, \alpha_{max}]] \tag{13}$$

With the presented expressions for the PDFs, the MCRT algorithm proceeds with the computation of the cumulatives which allow to randomly sample their related variable. The Monte-Carlo integral formulation can be directly interpreted as the algorithm below:

```
P̃ = 0;
foreach event do
    Uniform sampling of r0 on H+;
    Sampling of ω0 on solar disk ;
    if No shadowing between sun and r0 then
        ŵP = DNI |n(r1) · ω0| H+;
        ri = r0;
    else
        ŵP = 0;
        break;
    end
    while ri ∈ H+ ∪ C+ do
        ŵP = ŵP · ρ;
        sampling α and β to generate nh;
        specular reflection ωi = ωi + 2|nh · ωi|nh;
        ri = Intersection of Ray(ri, ωi) with geometry element;
    end
    if ri ∈ V+ then
        P̃ = P̃ + ŵP
    end
end
P̃ = (1/N) P̃;
```

3. Solfast features

In this section, we present briefly the Solfast features and some graphical interface elements which allow user to design the system to simulate. The optical system elements can be classified in four categories : the emitter, the reflectors, the receivers and the occulting objects. These elements except the occulting objects can be chosen in the Solfast library by a simple drag-and-drop action. When an element is placed, a double-click on it brings up a specific graphical user interface to parametrize the object.

3.1. the emitters

Three types of source are available in Solfast : a sun-like emitter, a parallel emitter and a point emitter. The emission intensity of the three emitters is characterized by the Direct Normal Irradiance (DNI) which is defined by the irradiance measured (in $W.m^{-2}$) with a surface element perpendicular to the emitter direction.

3.1.1. sun-like emitter The sun-like emitter is modelized by a pill-box with an apparent diameter (9.3 $mrad$ by default). The emitter direction can be indicated by vector components or by the zenith and azimuth angles. The NREL solar position algorithm [7] is also implemented to get the sun direction in function of the date and the geographical positions.

3.1.2. parallel emitter The parallel emitter is characterized by the direction of incident rays. To measure DNI, the surface element must be perpendicular to this direction.

3.1.3. point emitter The point emitter is characterized by its position. To measure DNI, the surface element normal vector must be pointed toward the emitter position.

3.2. the reflectors

Two types of reflectors are available in the library: the simple reflector and the patched reflector. This last can be considered as a group of simple reflectors. Many menus in the reflector interface allow the user to place the object at a given position, to determine the surface type (the generic quadratic surface is implemented) and the aperture (any polygonal aperture), to choose a pointing direction ('manually' fixed or source tracking) and to parametrize different noises. Among noises we focus on the BRDF model choices which modelizes the specularity defects of the reflecting surface. The defects are represented by a microfacets distribution model [8]. Three distributions have been implemented :

- the gaussian α which consists to sample the angle α between the theoretical normal vector and the microfacet normal vector with a truncated gaussian distribution.
- the circular normal distribution [9] which is implemented in Tonatiuh [3] software for the surface slope error.
- the uniform distribution which also implemented in Tonatiuh.

Another important reflector parameter which is the 'primary' reflector setting. For example, for the Odeillo solar furnace, only the heliostat field mirrors are considered as primary reflectors and not the big parabola. This is an advantage of the Monte-Carlo integral formulation presented in Sec. 2 which allows to get an efficient algorithm because only rays reflected on primary reflectors are generated. No rays which can touch directly the ground are sampled.

3.3. the receivers
The user can chose two types of receivers: simple receivers and directional receivers. The simple receivers are used to get a flux map. They are parametrized by the size and a number of pixels. The directional receivers are similar but with a directional information about the incident energy. Thus the parametrization of directional discretization is also required. Receivers produce the output results of a simulation including the standard deviation. These results can be visualized on a PNG picture and written in a ASCII file.

3.4. the occulting objects
The occulting objects are completely absorbing (no reflection or diffuse reflection). These objects can be imported from a Sketchup model. This is for example the building of the Odeillo solar furnace. The only parameters of an occulting object are its position and orientation.

3.5. the solver
When the optical system is designed, the Ray-Tracing Monte-Carlo simulation can be run from the solver interface. The number of sampled rays must be indicated. The computing performance of Solfast comes from the Monte-Carlo integral formulation and the techniques to accelerate Ray-Tracing (RayBooster technology).

4. Validations and test cases
Two preliminary test cases are presented and compared with two other softwares: Soltrace [2] developed by the NREL and Tonatiuh [3] developed by the CENER. The cases include two simplified geometries: the Eurodish parabolic dish receiver and the LS-2 parabolic trough collector. For the validation cases a DNI of 1000 $W.m^{-2}$ was considered with a pill-box sunshape of 9.3 *mrad* full width. The material properties of reflectors are set with the Gaussian sigma slope error model in Soltrace and the normal sigma slope error model in Tonatiuh. These two BRDF models give similar results [3]. In Solfast the material properties settings are slightly different (cf. 3.2): BRDF models are implemented with different choices of microfacets distributions. To be compatible between the sigma slope error models of Soltrace and Tonatiuh, the chosen microfacets distribution was the circular normal distribution in both cases (Eq.(12)). We present also a more complex system which is a simplified but realistic representation of the THEMIS solar plant. This last example shows the benefits in term of convergence of the Monte-Carlo integral formulation implemented in Solfast compared with the collision-based Monte-Carlo algorithm of both Soltrace and Tonatiuh.

4.1. Eurodish
Figures 3 and 4 represent respectively a screenshot of the Solfast interface showing the simplified Eurodish parabolic receiver geometry and the profiles of the radial distributions of photons obtained by Solfast and Tonatiuh. This study was presented in fig. 5 of [3] to compare Tonatiuh to Soltrace. The present comparison shows a very good agreement between the results of Solfast and Tonatiuh algorithms with a maximum relative difference less than 5%.

4.2. LS-2 parabolic trough
Figure 5 represents a screenshot of the Solfast interface where the simplified LS-2 parabolic trough is designed. It is composed of 20 cylindro-parabolic mirrors. Figure 6 represents the mean flux density profile obtained by Solfast and Soltrace. The flux density was averaged along the main axis of the trough. A good agreement between the results is also observed. Two results are shown for Solfast: a real-time preview simulation obtained with 10^4 rays in less than 0.1 *sec* computational time and an accurate result obtained with 10^7 rays in 23.5 *sec* on an Intel Core

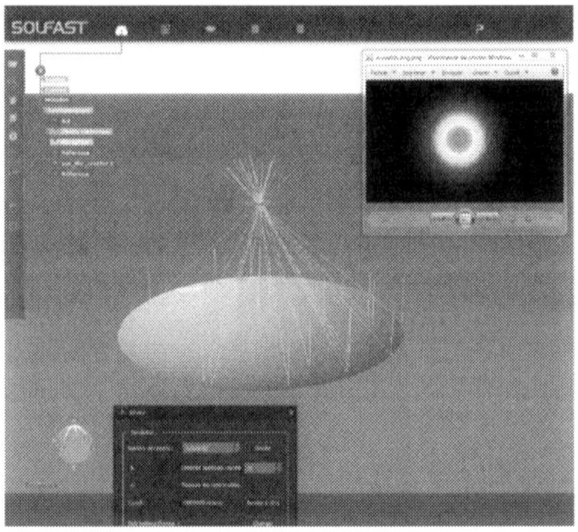

Figure 3. The simplified Eurodish geometry in Solfast interface. The flux map is visualized on an external viewer.

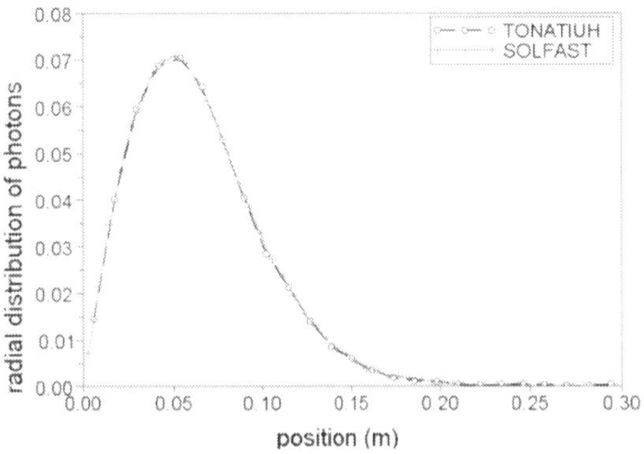

Figure 4. Comparison of the radial distributions of photons obtained by Solfast and Tonatiuh.

i7 Q720. It is worth noting that Solfast gives the possibility to obtain a very accurate result in real-time.

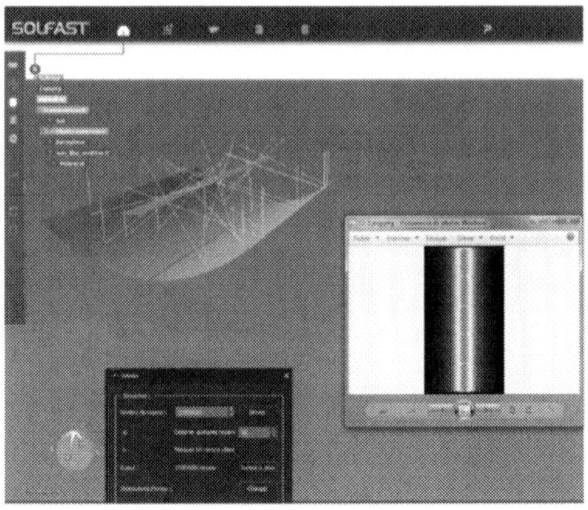

Figure 5. The simplified LS-2 geometry in the Solfast interface.

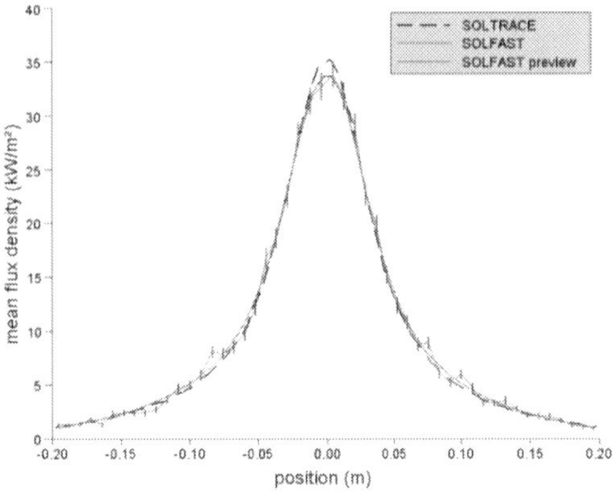

Figure 6. Mean flux density profile obtained by Solfast (10^7 rays), Solfast preview (10^4 rays in 0.1 *sec*) and Soltrace (1.5 10^6 rays). Error bars are only perceptible for the Solfast preview.

4.3. Discussion about efficiency of Monte-Carlo integral formulations

To highlight the importance of the Monte-Carlo integral formulation a comparison was done between Tonatiuh (version 1.2.5), Soltrace (version 2011.7.5) and Solfast for a simplified representation of THEMIS (see Fig. 7). Each of the 115 heliostats is composed with a single 7.5 m $\times$ 7.5 m square mirror. The reflectivity is set to 0.94 and the reflection is specular (no slope or specularity errors). The simulation was done for the March 21th at solar noon. The figures 8 and 9 present respectively the flux received by the target and the Monte-Carlo standard deviation with respect to the number of realizations. These results allow one to compare objectively the performance of the Monte-Carlo algorithm independently of the Ray-Tracing acceleration techniques. These results show for a typical solar facility configuration that the Monte-Carlo integral formulation implemented in Solfast improves standard deviation by a 9.25 factor versus Tonatiuh and a 17 factor versus Soltrace. Regarding the number of realizations required to reach a given standard deviation target, Solfast needs about 85 times less realizations than Tonatiuh and 289 times less realizations than Soltrace.

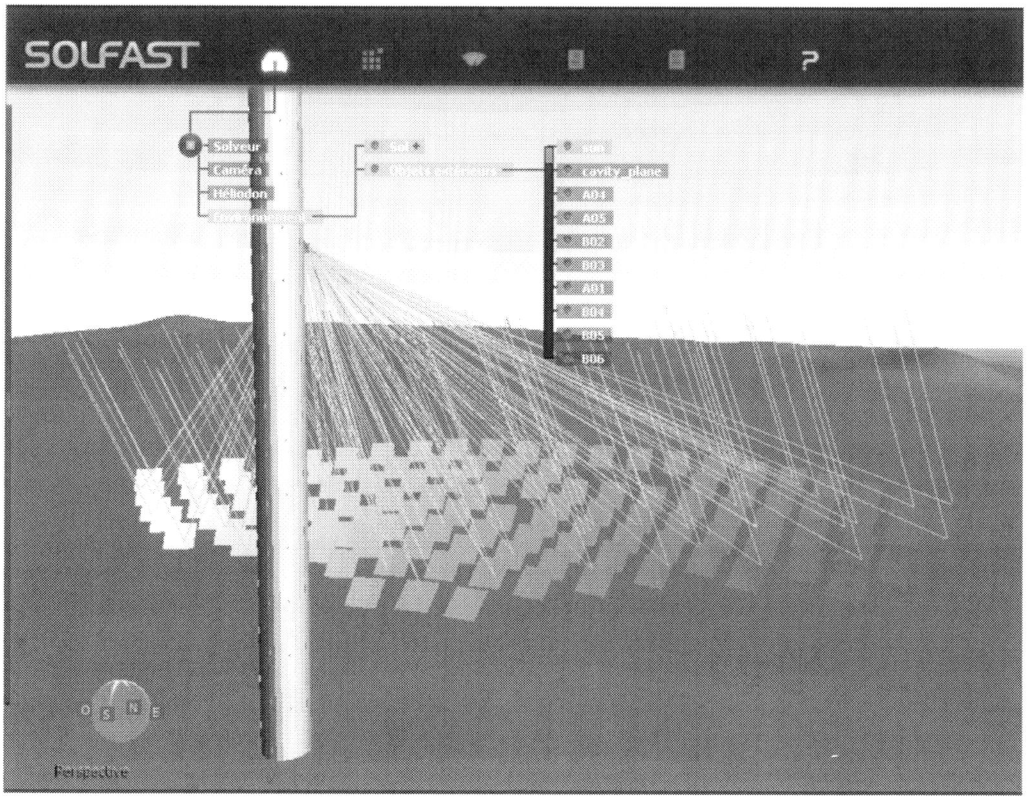

Figure 7. The simplified THEMIS representation in the Solfast main window.

4.4. Computation time
As the ultimate performance measurement is running time, they were measured on an Intel Core i7 Q720 both for a given number of rays (10^6) and for a given standard deviation target (500 W) using the simplified THEMIS model introduced above (Figs. 10 and 11). As a result Solfast takes benefit from using the RayBooster technology for Ray-Tracing computations [5] (Fig. 10), and the gap between Solfast and Tonatiuh or Soltrace is enlarged when the comparison criteria is the running time required to achieve a given accuracy (Fig. 11) .

5. Conclusion and future work
Solfast is a new software for optical simulation of solar concentrating systems with a greater efficiency than the state of the art software Soltrace and Tonatiuh. This is due to the use of Monte-Carlo integral formulation instead of collision based formulation which requires more calculation to provide the same quality of results. Another advantage of the Monte-Carlo integral

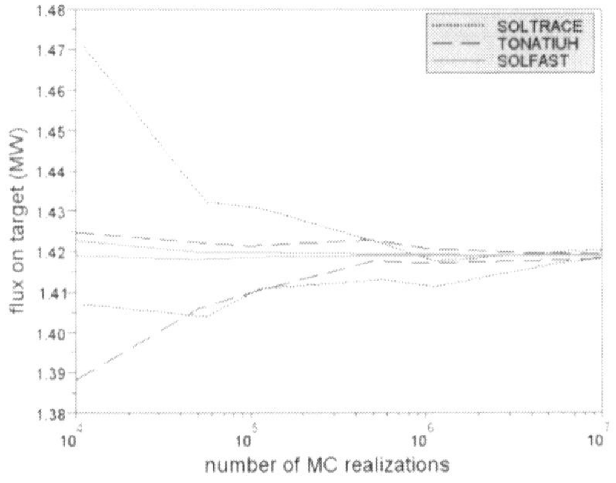

Figure 8. Flux on Themis target in function of the number of MC realizations. Two sets of curves are plotted: flux + standard deviation and flux − standard deviation.

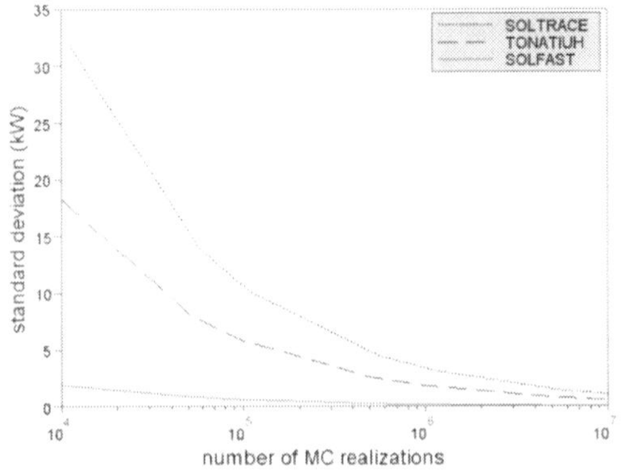

Figure 9. Comparison of the standard deviation obtained by Tonatiuh and Solfast in function of the number of MC realizations.

Solfast	Tonatiuh	Soltrace
3.6	8	11

Figure 10. Computation time in sec. for 1 million MC realizations

Solfast	Tonatiuh	Soltrace
1.2	241	497

Figure 11. Computation time in sec. needed to reach a 500 W standard deviation target

formulation is the ability to estimate systematically for a negligible computational cost the sensitivity of any physical quantity to some parameters reusing the flux ray paths [10, 11]. This, in conjunction with the RayBooster high performance Ray-Tracing library, will enable effective analysis and optimization of complex models on mainstream computers.

References

[1] Garcia P., Ferrière A. and Bézian J-J. 2008 Codes for solar flux calculation dedicated to central receiver system applications : a comparative review. *Solar Energy* **82** 189.

[2] Wendelin T. 2003 Soltrace: A new optical modeling tool for concentrating solar optics *ASME International Solar Energy Conference*, Kohala Coast, Hawaii.

[3] Blanco M.J., Mutuberria A., Garcia P, Gastesi R. and Martin V. 2009 Preliminary validation of Tonatiuh *SOLARPACES Symposium*, Berlin, Germany.

[4] Delatorre J. et al. Monte-Carlo advances and concentrated solar applications, to be submitted.

[5] Roccia J-P. 2012 Hybrid CPU/GPU KD-Tree Construction for Versatile Ray Tracing. *Eurographics Symposium on Rendering*, Cagliari, Italy.

[6] Delatorre J. 2011 *Optimization of concentrating solar power plants using Monte-Carlo methods*, PhD. Thesis, INPT, Toulouse, France (http://ethesis.inp-toulouse.fr/archive/00001811/01/delatorre.pdf).

[7] Reda I. and Andreas A. 2008 Solar position algorithm for solar radiation applications *NREL Report* No. TP-560-34302.

[8] Beckmann P. and Spizzichino A. 1963 *The scattering of electromagnetic waves from rough surface* (Oxford:Pergamon).

[9] Biggs F. and Vittitoe N. 1976 The Helios model for the optical behavior of reflecting solar concentrators *SANDIA Report* No. SAND76-0347.

[10] de Lataillade A, Blanco S., Clergent Y., Dufresne J.L., El Hafi M. and Fournier R. 2002 Monte-Carlo method and sensitivity estimations *JQSRT* **75** 529.

[11] Roger M., Blanco S., El Hafi M. and Fournier R. 2005 Monte-Carlo estimates of domain-deformation sensitivities *Phys. Rev. Lett.* **95** 18.

Radiative heat transfer modelling in a concentrated solar energy bubbling fluidized bed receiver using the Monte Carlo Method

Germain BAUD[1], Jean Jacques BEZIAN[1], Mouna EL HAFI[1] and Gabriel OLALDE[2]

1) Université de Toulouse; CNRS; Centre RAPSODEE, Ecole des Mines d'Albi, Albi, France
2) PROMES, CNRS, Odeillo, France

E-mail: `mouna.elhafi@enstimac.fr`

Abstract. According to recent studies in the solar energy community, a promising ways of solar energy conversion seems to be the beam down concentration technology associated to a fluidized bed receiver. The advantage of this system is its ability to heat air at temperature reaching 1000 K in a receiver directly exposed to a concentrated solar beam and integrated in a thermodynamic cycle. This paper focus on the modelling of radiative heat transfer from the optical concentrator to the receiver by taking into account absorption and multiple scattering of light in the particles bed. To achieve this objective, we develop in this work an efficient tool based on an algorithm solving the integral formulation of the Radiative Transfer Equation by the Monte Carlo Method. This algorithm is implemented in EDStaR environment where computer graphics libraries, parallel computing and specific functionalities to produce statistical quantities and their associated derivatives are available. One of the main advantage of the proposed radiative transfer modelling is the determination of the sensitivities (derivatives) of the physical quantities to any physical or geometrical parameter without significant additional CPU time.

1. Introduction

Fluidized beds are conventionally used as media of thermal exchanges for high temperature processes because of their excellent performance in terms of heat transfer. To improve the performance of thermodynamic cycles for solar electricity production, a technological gap will be the direct gas heating in solar receiver, to obtain at the entrance of a turbine temperatures at a level higher than 1000 K. One of the best solutions would be to use solid particles in a fluidized bed directly submitted to the concentrated solar flux (see fig.1.b). The main advantages of the fluidized bed receiver are its thermal efficiency, an improved resistance to thermal stress compared to metallic and ceramic receivers and its adaptability with beam down concentrators allowing to produce electricity directly on the ground. For these reasons, fluidized bed solar receivers and reactors have been designed and studied by numerous scientists [1, 2, 3, 4].

To evaluate the efficiency of these solar receivers, we need to determine the heat losses caused by radiation which becomes the major mechanism for heat transfer losses from the receiver. The accuracy and the power level of the flux map obtained at the surface of the receiver depends on many optical parameter such as the reflectivity or the quality of the mirrors, the design of the concentrators, etc. The optical and geometrical design of such systems require powerful tools to provide radiative heat transfer exchanges inside the receiver and also to compute the distribution of the irradiation on the receiver.

In this context, several softwares have been developed such as Mirval [5], Soltrace [6], Stral [7] and Tonatiuh [8, 9]. They commonly use the Monte Carlo ray-tracing to predict the concentrated solar flux distribution incident on a receiver [10]. In this method, solar radiation is assumed to be composed of discrete energy bundles and the computation consists of following their probable paths that are generated randomly from the solar source to the optical components. Steinfeld and co-workers have also developed Monte Carlo ray-tracing in-house codes to compute the distribution of solar irradiation on receiver boundaries [11, 12, 13, 14]. Most of the time, these existing tools allow to design the optical part of the concentrated solar system only. What we propose in this study is to model radiation in the whole system: the concentrated solar source from the optical device and the absorption by particles inside the solar receiver. The aim of this study is to calculate the penetration of the incident solar beams through the window (see. fig.1.a) and to estimate the radiative heat loss by reflection. We need for that a radiative model to calculate the absorbed and scattered radiative contribution by particles inside the receiver. Knowing this radiative contribution, the energy balance could be performed and the heat exchange between air and particles could be evaluated.

In this work, the Radiative Transfer Equation (RTE) is solved through its integral formulation. The Monte Carlo Method (MCM) in this case is used as a numerical method to calculate explicitly these integrals. As it is explained in previous studies [16], the main advantage of this integral formulation is to analyse physically the radiative exchanges but also to calculate physical quantities such as radiative fluxes, radiative source terms with their associated sensitivities to any parameter, without additional CPU time.

A first part is dedicated to the description of the radiative properties of particles. Then, the radiative model is presented. The associated algorithm using MCM is detailed determining the radiative heat loss with the multiple scattering by particles and reflection by the walls. An illustration of sensitivity calculation by MCM is shown.

2. The system under investigation

In this work, we model a 45 cm high solar receiver consisting of a diffuser at the bottom of a lower cylindrical column mesuring 25x7 cm, a medium conical part of 7.5 cm high and an upper part of 12.5x12 cm surmounted by a quartz window (see fig.1.a). The concentrated solar radiation enters the top of the receiver and is directly absorpted by the fluidized particles of silicon carbide. The solar receiver allows us to heat the gas sent through the metallic diffuser at the bottom of the column. Air mass flow must reach a threshold to support a good fluidisation of solid particles to maximise heat exchanges and solar radiation depth of penetration. The rise of diameter in the middle of the receiver prevent particles from exiting the fluidisation column or impacting the window.

The present receiver is coupled with a beam-down concentrator available in the laboratory of PROMES, Odeillo, France (see fig.1.b). This system is constituted by a flat heliostat reflecting solar energy to a dish with a diameter of 4.6 m, a focal length of 3.9 m and a focal spot of approximatively 6 cm.

A laboratory scale set-up of this receiver have been studied in the PhD thesis of A.Bounaceur [17]. According to this work, it was possible to heat air from the ambient temperature up to 930 K with a thermal efficiency of 70%. But as discussed in this work, even better results could be reached by reducing radiative losses and avoiding the overheating of the walls of the receiver.

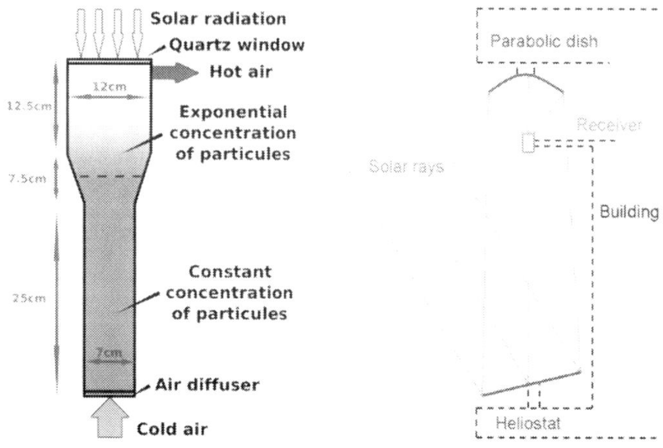

Figure 1. Scheme of the solar fluidized bed receiver on the left (a). Scheme of the Beam down solar concentrator on the right (b).

3. Modelling of radiative heat transfer

3.1. Radiative properties

The stainless steel surface of the receiver ($\mathcal{R}$) is represented by a diffuse-reflective surface with a grey reflectivity $\rho^{\mathcal{R}}$. The quartz window ($\mathcal{G}$) is considered as a specularly diffuse and transmittive surface with a reflectivity $\rho^{\mathcal{G}}(\omega, \mathbf{n})$.

When the particle bed is submitted to radiation, the mechanisms responsible for the distribution of light are absorption and scattering. The spread of radiation occurs through the interstices of the particle fluidized bed and by transmission and scattering through the particles. The air is considered transparent. The length of penetration of radiation depends largely on the receiver geometry, the volumetric fraction of the fluidized bed and the optical properties of particles. Absorption and scattering coefficients in the fluidized bed derive from density and optical properties of particles. The phase function describing scattering in the particle cloud is modelled by the Henyey-Greenstein phase function.

To determine the optical properties of the particles, a calculation over the whole solar spectrum between 0.24 μm and 4 μm is processed using a code developed by Mishchenko based on the Mie theory [19], where the index of refraction is 0.78. Preliminary results [18] have shown that spectral resolution make no significant difference compared to wavelength-integrated calculations when computing the flux in a SiC fluidized bed. Thus, incoming solar radiation is considered monochromatic and we average the values of optical properties Q_s, Q_a and g from the spectral absorption and scattering efficiency factors $Q_{\nu,s}$, $Q_{\nu,a}$ and the asymmetry factor g_ν. The optical properties are described by the following values of the absorption and scattering cross sections :

$$\sigma_a = \frac{Q_a}{\pi d_p^2/4} \quad ; \quad \sigma_s = \frac{Q_s}{\pi d_p^2/4} \tag{1}$$

with $Q_s = 1.51$, $Q_a = 0.5$, $g = 0.7$ and the particle diameter d_p.
Absorption and scattering coefficients are calculated from cross sections and density through :

$$k_a = \eta(\mathbf{x}, p)\sigma_a \quad ; \quad k_s = \eta(\mathbf{x}, p)\sigma_s \tag{2}$$

The determination of the distribution of particles within the fluidized bed has been a critical parameter for the calculation of heat transfer. According to our experimental observations, the receiver can be divided in two parts : (i) the bottom part of the bed within which the particle concentration is constant and high, (ii) the upper part of the bed, within which the particle concentration can be conceptualized by a exponential decay (see fig.1.a). The exponential density of particles from the upper part of the receiver is thus described by :

$$\eta(\mathbf{x}, p) = \eta_0 exp(-\frac{z - z_0}{p}) \quad (3)$$

with η_0 the density in the constant zone, z the height in the fluidized bed, z_0 the height indicating the change from a constant to an exponential density and p a parameter tuning the exponential distribution.

3.2. Method of modelling and algorithm : Calculation of reflection losses

Our concern is to describe in detail the calculation of the fraction of radiative power exiting through the window without being absorbed by the receiver and to compute the sensitivity of this loss to the p parameter of the density of particle. The solar radiation is considered collimated. The algorithm allowing us to calculate this value is the following :

- Uniform drawing of a point x_0 on the window, according to the density probability function $p_{x_0} = 1/S_{\mathcal{G}}$;
- Emission of a photon bundle from the point x_0 in the normal direction to the surface $\omega_0 = n_{x_0}$;
- Drawing of a first optical thickness κ_0 according to the law $p_{\kappa_0}(\kappa_0) = exp(-\kappa_0)$ if the bundle is in the bottom dense zone of the receiver and $p_{\kappa_0}(\kappa_0) = exp(-\kappa_0)/(1 - exp(-\kappa_{max}))$ if the bundle is in the upper zone of the receiver ;
- *Interaction* - If κ_j is smaller than κ_p the optical thickness of the optical path travelled by the photon bundle until its next interaction with the boundaries of the receiver, then the interaction point $x_j(x_{j-1}, \omega_{j-1})$ belongs to the volume of the receiver. If κ_j is bigger than κ_p, then x_j belongs to the surface $(\mathcal{R})$ or the windows $(\mathcal{G})$;
- *reflection on the surface of the receiver* - If $x_j \in \mathcal{R}$ and $\hat{w}_j \geq \epsilon$, then we draw one reflection direction ω_{j+1} according to the lambertian law. We also reduce the energy of the photon bundle by $exp(-\kappa_a)$ the absorption of the cloud of particle among the optical path and $\rho^{\mathcal{R}}$ the reflectivity of the surface ;
- *reflection on the surface of the receiver, Russian roulette mode* - If $x_j \in \mathcal{R}$ and $\hat{w}_j < \epsilon$, then we draw a number $r_{j+1} \in [0, 1]$. If r_{j+1} is smaller than $\rho^{\mathcal{R}}$, then the photon bundle continue its path (drawing of a reflection angle, drawing of a new optical thickness...). Otherwise, the photon bundle stops its path, and its weight is void ;
- *Interaction with the windows* - If $x_j \in \mathcal{G}$, then we draw a number $r_{j+1} \in [0, 1]$. If r_{j+1} is smaller than $\rho^{\mathcal{G}}$, then the photon bundle continue its path (drawing of a reflection angle, drawing of a new optical thickness...). Otherwise, the photon bundle exits the receiver : the algorithm comes to an end and it process to the calculation of the weights of E and its sensitivity;
- *Scattering in the volume* - If $x_j \in \mathcal{V}$, then we draw a direction of scattering ω_{j+1} according to the phase function. Then we reduce the energy of the photon bundle by $exp(-\kappa_j)$ the absorption of the cloud of particle among the optical path and continue the optical path of the photon bundle (drawing of a reflection angle, drawing of a new optical thickness...) ;
- *Calculation of $\hat{w}_j$, the weight of E* - Physically, the weight is the product of the area of the emitting surface $S_{\mathcal{G}}$ by the attenuation of the energy of the photon bundle through its path among the receiver. The extenuation is due to both the absorption from the impacts with the surface of the receiver and the absoprtion of the participating medium ;

- *Calculation of $\hat{w}_{p,j}$, the weight of the sensitivity of E to parameter p* - The expression of the weight of $\frac{\partial E}{\partial p}$ is a little more complicated than the one of the weight of E, but it can be processed at the same time and with the same algorithm.

A "Russian roulette" has been implemented to deal with the reflection at the boundaries of the receiver when the weight become smaller than a certain criterion. Doing so, the computer time necessary for one computation is reduced from 1 hour to a few minutes. The reason why a calculation could take so much time is that the bottom of the bed contains a high particle density and thus the scattering and absorption mean free path are very small in this zone. Consequently, a ray entering this zone will encounters a large number of scattering events before it exits from the receiver by the window. In the meanwhile, the extinction contribution becomes higher and the energy decreases quickly. So, when performing a calculation, the dedicated computing time to multiple scattering of paths with very low energy (ie. low weight) remains significant.

With the "Russian roulette", when the energy of a ray is lower than a defined value, we do not treat the reflection as a ray being reflected and attenuated by the surface, but we test if the interaction point is an absorption or a reflection. When the interaction is a reflection, we process until the next interaction. If the interaction is an absorption, the optical path dies.

This algorithm is implemented in the environment EDStaR which is constituted by an advanced technology allowing the modelling of corpuscular transport and in particular radiative transfer focusing on the Monte Carlo Method. EDStaR is not a code, but rather a development environment providing a library of useful functions to deal with parallel calculation, calculation of sensitivities, ray tracing in complex 3D geometries using computer graphics libraries [21]. More informations about EDStaR are available on line at the following url : http://wiki-energetique.laplace.univ-tlse.fr/wiki/index.php/Edstar.

The integral formulation of the Monte Carlo algorithm described above can be written as follow. Using the same notations as previously, the fraction of the radiative power E exiting through the quartz window is :

$$E = \int_{\mathcal{G}} p_{\mathbf{X}_0}(\mathbf{x}_0) d\mathbf{x}_0 \int_0^{\infty} p_{\kappa_0}(\kappa_0) d\kappa_0 \left\{ \begin{array}{l} H(\mathbf{x}_1 \in \mathcal{R}) \int_{2\pi} p^{\mathcal{R}}_{\mathbf{\Omega}_1}(\omega_1) d\omega_1 \int_0^{\infty} p_{\kappa_1}(\kappa_1) d\kappa_1 \; \mathcal{I}_1 \\ +H(\mathbf{x}_1 \in \mathcal{V}) \int_{4\pi} p^{\mathcal{V}}_{\mathbf{\Omega}_1}(\omega_1|\omega_0) d\omega_1 \int_0^{\infty} p_{\kappa_1}(\kappa_1) d\kappa_1 \; \mathcal{I}_1 \end{array} \right\} \tag{4}$$

where $\mathcal{I}_j, j > 1$ is recursively defined as

$$\begin{array}{ll} \mathcal{I}_j & = H(\mathbf{x}_{j+1} \in \mathcal{G}) \int_0^1 p_{R^{\mathcal{G}}_{j+1}}(r^{\mathcal{G}}_{j+1}) dr^{\mathcal{G}}_{j+1} \times ... \\ & \left\{ \begin{array}{l} H(r^{\mathcal{G}}_{j+1} > \rho^{\mathcal{G}}) \; \hat{w}_{j+1} \\ +H(r^{\mathcal{G}}_{j+1} \leq \rho^{\mathcal{G}}) \int_{2\pi} p^{\mathcal{G}}_{\mathbf{\Omega}_{j+1}}(\omega_{j+1}|\omega_j) d\omega_{j+1} \int_0^{\infty} p_{\kappa_{j+1}}(\kappa_{j+1}) d\kappa_{j+1} \; \mathcal{I}_{j+1} \end{array} \right\} \\ & +H(\mathbf{x}_{j+1} \in \mathcal{R}) \left\{ \begin{array}{l} H(\hat{w}_j \leq \epsilon) \int_0^1 p_{R^{\mathcal{R}}_{j+1}}(r^{\mathcal{R}}_{j+1}) dr^{\mathcal{R}}_{j+1} \times ... \\ \left\{ \begin{array}{l} H(r^{\mathcal{R}}_{j+1} > \rho^{\mathcal{R}}) \times 0 \\ +H(r^{\mathcal{R}}_{j+1} \leq \rho^{\mathcal{R}}) \int_{2\pi} p^{\mathcal{R}}_{\mathbf{\Omega}_{j+1}}(\omega_{j+1}) d\omega_{j+1} \int_0^{\infty} p_{\kappa_{j+1}}(\kappa_{j+1}) d\kappa_{j+1} \; \mathcal{I}_{j+1} \end{array} \right\} \\ +H(\hat{w}_j > \epsilon) \int_{2\pi} p^{\mathcal{R}}_{\mathbf{\Omega}_{j+1}}(\omega_{j+1}) d\omega_{j+1} \int_0^{\infty} p_{\kappa_{j+1}}(\kappa_{j+1}) d\kappa_{j+1} \; \mathcal{I}_{j+1} \end{array} \right\} \\ & +H(\mathbf{x}_{j+1} \in \mathcal{V}) \int_{4\pi} p^{\mathcal{V}}_{\mathbf{\Omega}_{j+1}}(\omega_{j+1}|\omega_j) d\omega_{j+1} \int_0^{\infty} p_{\kappa_{j+1}}(\kappa_{j+1}) d\kappa_{j+1} \; \mathcal{I}_{j+1} \end{array} \tag{5}$$

Where the various angle probability density functions are :

$$\begin{array}{l} p^{\mathcal{R}}_{\mathbf{\Omega}_j}(\omega_j) = \frac{\omega_j . \mathbf{n}_j}{\pi} \\ p^{\mathcal{V}}_{\mathbf{\Omega}_j}(\omega_{j+1}|\omega_j) = \frac{1-g^2}{[1+g^2-2g(\omega_{j-1} . \omega_j)]^{3/2}} \\ p^{\mathcal{G}}_{\mathbf{\Omega}_j}(\omega_{j+1}|\omega_j) = \delta(\omega_{j+1} . t_j + \omega_j . t_j)\delta(\omega_{j+1} . n_j - \omega_j . n_j) \end{array} \tag{6}$$

with n_j and t_j are respectively the normal vector and a tangent vector to the considered surface at the point x_j. And where the optical thickness probability density function is :

$$\kappa_j = \int_0^{||x'-x_j||} k_s(x_j + s\omega_j)ds \tag{7}$$

with $x_j = x'$ if $\kappa_j < \int_0^{||x_j-y_j||} k_s ds$ and $x_j = y_j$ if $\kappa_j \geq \int_0^{||x_j-y_j||} k_s ds$, $\omega_0 = n_{x_0}$ and $x' = x_j + ||x' - x_j||\omega_j$.

The weight is then recursively defined as

$$\hat{w}_{j+1} = \hat{w}_j exp(-\sigma_a \alpha_j) \left\{ \begin{array}{l} H(\mathbf{x}_{j+1} \in \mathcal{G}) \times 1 \\ +H(\mathbf{x}_1 \in \mathcal{R}) \left\{ \begin{array}{l} H(\hat{w}_j > \epsilon)\rho^{\mathcal{R}} \\ +H(\hat{w}_j \leq \epsilon) \times 1 \end{array} \right\} \\ +H(\mathbf{x}_1 \in \mathcal{V}) \times 1 \end{array} \right\} \tag{8}$$

with $\hat{w}_0 = S_{\mathcal{G}}$ and

$$\alpha_j = \int_0^{||\mathbf{x}_{j+1}-\mathbf{x}_j||} \eta(\mathbf{x}_j + s\omega_j, p)ds \tag{9}$$

In the particular case of a uniform particle density $\sigma_a \sum_{q=0}^{j-1} \alpha_q = \sigma_a \eta d_j = k_a d_j$ and the integral formulation corresponding to the sensitivity $\frac{\partial E}{\partial \eta}$ is identical to that of E in Eqs. 4 and 5, replacing only $\hat{w}_j$ with

$$\hat{w}_{\eta,j} = \frac{\partial \hat{w}_j}{\partial \eta} + \hat{w}_j \sum_{q=0}^{j-1} \frac{\frac{\partial p_{\lambda_q}}{\partial \eta}}{p_{\lambda_q}} = \hat{w}_j \left[-\sigma_a d_j + \sum_{q=0}^{j-1} \frac{1-\kappa_q}{\eta} \right] \tag{10}$$

In the general case, the integral formulation corresponding to the sensitivity $\frac{\partial E}{\partial p}$ is identical to that of E, replacing only $\hat{w}_j$ with

$$\hat{w}_{p,j} = \hat{w}_j \sum_{q=0}^{j-1} \left[-\sigma_a \frac{\partial \alpha_q}{\partial p} - \sigma_s \frac{\int_0^{||\mathbf{x}'-\mathbf{x}_q||} \eta(\mathbf{x}_q + s\omega_q, p)ds}{\partial p} + \frac{\partial_2 \eta(\mathbf{x}_{q+1}, p)}{\eta(\mathbf{x}_{q+1}, p)} \right] \tag{11}$$

where $\partial_2 \eta(\mathbf{x}, p)$ is the partial derivative with p of η.

As the expression of the sensitivity is not correlated with the density profile outside of the receiver, we can choose by convention that κ_j does not depend on $\eta(\mathbf{x}, p)$ and that $\partial_2 \eta(\mathbf{x}_{q+1}, p) = 0$ when $\mathbf{x}_{q+1} \notin \mathcal{V}$ (when reflection occurs before scattering). As a result, we have :

$$\begin{aligned} \int_0^{||\mathbf{x}'-\mathbf{x}_q||} \eta(\mathbf{x}_q + s\omega_q, p)ds &= \int_0^{||\mathbf{x_{q+1}}-\mathbf{x}_q||} \eta(\mathbf{x}_q + s\omega_q, p)ds + \int_0^{||\mathbf{x}'-\mathbf{x_{q+1}}||} \eta(\mathbf{x}_q + s\omega_q, p)ds \\ &= \int_0^{||\mathbf{x_{q+1}}-\mathbf{x}_q||} \eta(\mathbf{x}_q + s\omega_q, p)ds \end{aligned}$$

and the sensitivity becomes :

$$\hat{w}_{p,j} = \hat{w}_j \sum_{q=0}^{j-1} \left[-(\sigma_a + \sigma_s) \frac{\partial \alpha_q}{\partial p} + H(x_{q+1} \in \mathcal{V}) \frac{\partial_2 \eta(\mathbf{x}_{q+1}, p)}{\eta(\mathbf{x}_{q+1}, p)} \right] \tag{12}$$

3.3. Modelling of the concentrator and optical errors

In the description above, for didactic purpose, the solar radiation have been considered collimated. In our case study, the incoming concentrated solar radiation reaching the receiver is focalized. To model such systems, a beam-down like concentrator as been implemented in the previous code featuring a dish focalizing solar energy from infinity to the receiver.

As no concentrator is perfect, we take into account the misalignment of the heliostat, the surface structural defects of the mirrors, and the sunspot effect, by including a statistical error when tracing the reflection angle on the surface of the dish. Our model is inspired by Blinn [21] in order to render rough shiny materials in physically based picture rendering.

As a theoretical reflected ray is reflected specularly with the normal vector to the surface n_j as a reference, the imperfect reflected ray is traced using a modified normal vector n_h. In practical terms, the modified normal vector is drawn inside of a cone centered on the normal vector to the surface, following the probability density function :

$$P_{n_h}(n_h|\omega_j;p) = \frac{2+1/e}{2\pi(1-cos(\frac{\pi}{4}-\frac{1}{2}arccos(\omega_j.n_{j+1})))^{2+1/e}}(n_h.n_{j+1})^{1+1/e} \quad (13)$$

Using the experimental map flux taken at the focus point of a real concentrator, the parameter e of the model is adjusted. Figure 2.b shows the comparison between experimental [20] and adjusted numerical map flux taken at the focus point of the concentrator.

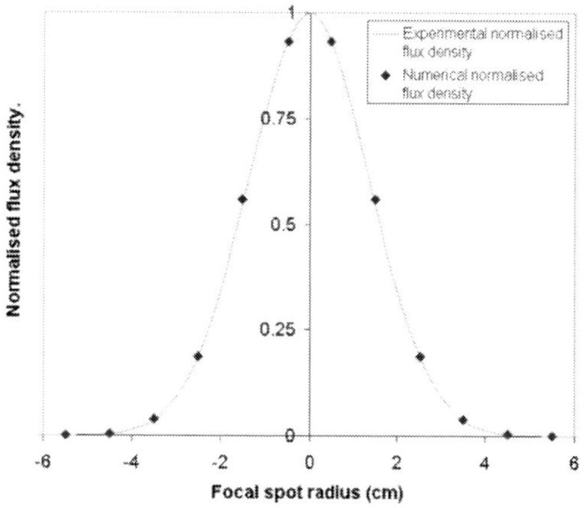

Figure 2. Scheme of the beam-down solar concentrator of Odeillo, France (a). Experimental [20] and adjusted numerical map flux taken at the focus point of the concentrator (b).

4. Results : Calculation of reflection losses and its sensitivity analysis

4.1. Validation of the sensitivity calculation in the case of collimated light

The radiative power exiting the receiver without being absorbed (see fig.3) and its sensitivity to the parameter p specifying the exponential decrease of the density of particles (see fig.4) are computed using a reflectivity of the stainless steel surface of the receiver of $\rho^{\mathcal{R}}$=0.4 and a reflectivity of the quartz window ($\mathcal{G}$) of $\rho^{\mathcal{G}}$=0.1. The diameter of particule and the density of particles in the constant zone are chosen such as d_p =280 μm and $\eta_0 = 26.10^3$ cm^{-3}.

The average computer time for a run of 10^6 realizations as used in the results shown in figure 3 is about a few minutes on a AMD Opteron Processor 246 with 2Gb ram.

Comparing the value of the fraction of losses and its sensitivity to p, we can minimize the energy losses by choosing p around 0.014. As p parameter increases over 0.014, the concentration of particles in the upper part of the fluidized bed decreases and the solar radiation have a higher probability to reach the stainless steel surface of the choke point and to be reflected outside of

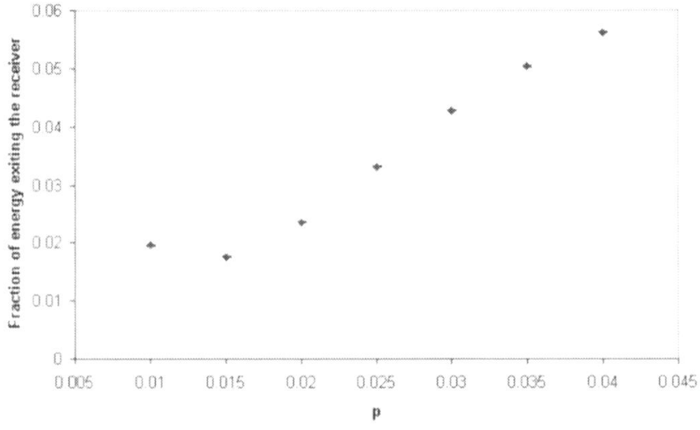

Figure 3. Fraction of incoming energy that exit the receiver without being absorbed, the p parameter of the density of particle is fixed. Error bars corresponding to the estimated standard deviation of the Monte Carlo Method are less than 1%.

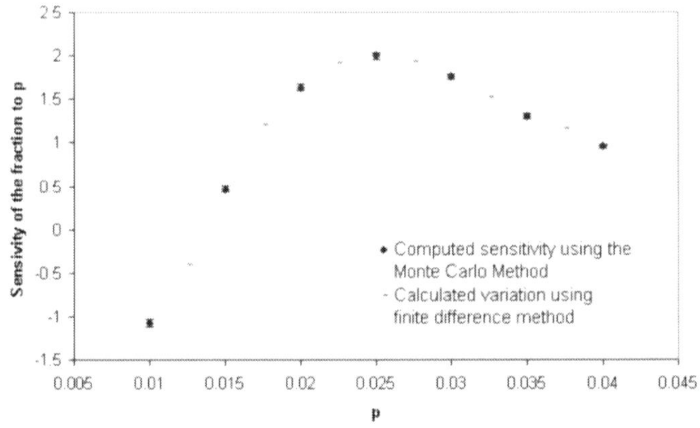

Figure 4. Sensitivity of the exiting fraction to p calculated by both the MCM and the finite difference method as reference, p is fixed. Error bars corresponding to the estimated standard deviation of the Monte Carlo Method are negligible.

the receiver. On the contrary, when p decreases below the value of 0.014, we notice that the losses increase again. Indeed since the energy refracted by the particles becomes predominant when high particle density move closer to the top of the receiver.

4.2. Fraction of losses and its sensitivity calculation to p

For different concentrator / receiver layouts, the data of table 1 represent incoming solar radiation losses. The different layouts of the receiver that have been moved in a range of 100 mm to allow the focus plan of the concentrator to move in a range of 50 mm at the front and rear of the quartz window of the fluidized bed receiver (see fig.5). For each test, the losses are obtained for the optimal value of parameter p, found using the sensitivity calculations.

Table 1. Visible radiative losses for different concentrator/receiver layouts

	Case 1	**Case 2**	**Case 3**	**Case 4**
Incoming radiation	collimated	focused	focused	focused
Focal point position/window (mm)	x	0	-50	+50
Optimal p parameter	0.014	0.018	0.015	0.027
Aiming losses (%)	0	0	1.17	1.16
Window reflection losses (%)	3.47	5.35	5.31	5.24
Bed reflection losses (%)	1.7	3.08	1.89	5.24
Total losses (%)	5.17	8.43	8.37	12.09

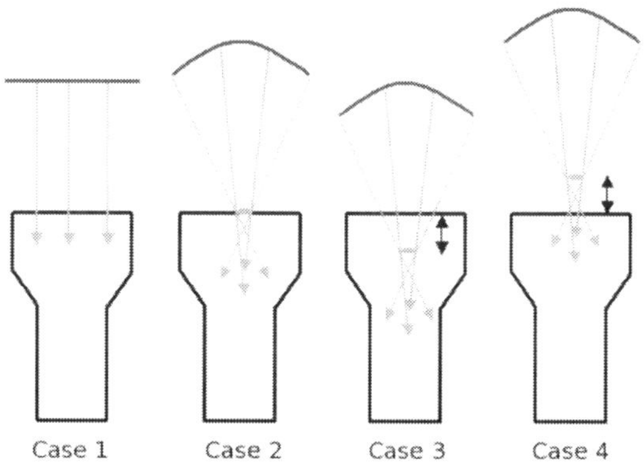

Figure 5. The four cases studied corresponding to different concentrator / receiver layouts

According to the results, we notice that reflection of the window is the main source of radiative heat loss. Even if the total losses reach the maximum when the focus point of the concentrator is located outside the receiver, we point out that for each receiver position, a specific value of parameter p can be found in order to reduce the bed reflection losses. In other words, it implies that the control of the particle distribution produced by the fluidisation inside the receiver must be adjusted according to the incoming solar radiation.

5. Conclusion

We presented in this study an original radiative transfer model that allows us to calculate the propagation of light from a beam-down concentrator to a fluidized bed receiver, taking into account the multiple scattering of light in the particulate medium and the directionality of the concentrated solar radiation. Our approach is based on the integral formulation of the RTE solved by the MCM. The algorithm is implemented in a new environment EDStaR, allowing the convenient implementation of an in house radiative model and the calculation of parametric sensitivities with nearly no additional CPU cost. To illustrate this approach, we studied the effect of the concentration of particles of a fluidized bed receiver on radiative loss by reflection. This information can be useful in order to maximise the efficiency of the receiver. As infra red radiative losses can reach 30% of the global losses of the receiver, the next step is to apply this method to model infra red radiative transfer in the fluidized bed receiver.

Nomenclature

Greek letters

α, β	limits of integration
η	concentration of particles, m^{-3}
κ	optical thickness, m^{-1}
ω, Ω	angle of emission
ϕ	flux density, W/m^2
ρ	Reflectivity
σ	cross section, m^2
σ^2	variance

Subscripts

0	(subscript) initial
$\mathcal{G}$	(subscript) window
$\mathcal{R}$	(subscript) wall
$\mathcal{V}$	(subscript) volume
a	(subscript) absorption
i, j, q	(subscript) indexes
p	(subscript) particle
s	(subscript) scattering

Symbols

d	diameter, m
E	radiative power exiting fraction, %
e	parameter describing the modified normal vector drawing probability density function
g	asymmetry factor

H	Heavy side function	Q	efficiency factor
I	integral, recursive term	r	random number in [0,1]
l	length of the path between two interactions	S	area, m^2
N	number of realizations	t	tangent vector to the surface
n	normal vector to the surface	x	variable, position, interaction position
p	parameter describing the exponential decay of particles	y	interaction point with a boundary
$P(x)$	density probability function	z	height, m
		$\hat{w}$	weight

References

[1] Flamant G. and Menigault T., 1986. Combined wall-to-fluidized bed heat transfer. Bubbles and emulsion contributions at high temperature, Int. J. Heat Mass Transfer, 30, 9, 1803-1812.
[2] Haddad I. M. and Elsayed M. M., 1988. Transient performance fluidised bed solar receiver at various parametric conditions, Solar and Wind Technology Journal., Vol. 5, No. 6, pp 653-659.
[3] Muller. R., Zedtwitz P. V., Wokaun A. and Steinfeld A., 2003. Kinetic investigation on steam gasification of charcoal under direct high-flux irradiation, Chem. Engin. Sci. No. 58, pp. 5111-5119.
[4] Trommer D. and al., 2005. Hydrogen production by steam-gasification of petroleum coke using concentrated solar power: Thermodynamic and Kinetic analyses, Int. Journal of Hydrogen Energy. No. 30, pp. 605-618.
[5] Leary, P.L., Hankins, J.D., 1979. Users guide for MIRVAL: a computer code for comparing designs of heliostat-receiver optics for central receiver solar power plants, Sandia Report SAND-77-8280.
[6] Wendelin, T., 2003. SolTRACE: a new optical modeling tool for concentrating solar optics, in: Proceedings of the ISEC 2003: International Solar Energy Conference, 15-18 March 2003, Kohala Coast, Hawaii, New York, American Society of Mechanical Engineers, pp. 253-260.
[7] Belhomme B., Pitz-Paal R., Schwarzbzl P., Ulmer S., 2009. A new fast ray tracing tool for high-precision simulation of heliostat fields. ASME Journal of Solar Energy Engineering 131(3), 031002.
[8] Blanco M.J., Mutuberria A., Garcia P., Gastesi R., Martin V., 2009. Preliminary validation of Tonatiuh, in: Proceedings of the International Energy Agency's SolarPACES Symposium 2009,
[9] Blanco M.J., Amieva J., Mancilla A., 2005. The Tonatiuh Software Development Project: An Open Source Approach to the Simulation of Solar Concentrating Systems, in: Proceedings of 2005 ASME IMEC: 2005 ASME International Mechanical Engineering Congress & Exposition, Orlando, Florida.
[10] Garcia P., Ferriere A., Bezian J.-J., 2007. Codes for solar flux calculation dedicated to central receiver system application: a comparative review, Solar Energy 82(3), 189-197.
[11] Lipinski W., Steinfeld A., 2004. Heterogeneous thermochemical decomposition under direct irradiation. International Journal of Heat and Mass Transfer 47, 1907-1916.
[12] Muller R., Steinfeld A., 2007. Band-approximated radiative heat transfer analysis of a solar chemical reactor for the thermal dissociation of zinc oxide. Solar Energy 81(10), 1285-1294.
[13] Petrasch J., Steinfeld A., 2007. Dynamics of a solar thermochemical reactor for steam-reforming of methane. Chemical Engineering Science 62(16), 4214-4228.
[14] Muller R., Lipinski W., Steinfeld A., 2008. Transient heat transfer in a directly-irradiated solar chemical reactor for the thermal dissociation of ZnO. Applied Thermal Engineering 28, 524-531.
[15] Roger, M. 2006. Modeles de sensibilite dans le cadre de la methode de Monte-Carlo : Illustrations en transfert radiatif, 24 mai 2006, Institut National Polytechnique de Toulouse, PhD thesis.
[16] Bounaceur A., 2008. Interaction lit fluidise de particules solides-rayonnement solaire concentre pour la mise au point d'un procede de chauffage de gaz a plus de 1000K, PhD Thesis, Ecole des Mines de Paris.
[17] Baud G., 2011. Conception de recepteurs solaires a lit fluidise sous flux radiatif concentre, PhD Thesis, Institut National Polytechnique de Toulouse.
[18] Mie G., 1908. Beitrage zur optik truber medien spiezell kolloidallen metallosungen, Ann. d. physik, 25, 377-445.
[19] Flamant G., 1982. Theoretical and experimental study of radiant heat transfer in a solar fluidized bed receiver, AIChE Journal, Vol. 28, No. 4, pp. 529-535
[20] Pharr M., Humphreys G., 2004. Physically based rendering : from theory to implementation, Elvesier.
[21] Bezian JJ., Bounaceur A., De Ryck A., El Hafi M., 2007. Un nouveau concept de centrale solaire thermodynamique base sur un recepteur a lit fluidise, In : Actes des Journees Internationales de Thermique, Albi, FRANCE, ISBN 978-2-95-11591-6-7, p. 346-350

AUTHOR INDEX

Institute of Physics
Dirac House, Temple Back
Bristol BS1 6BE UK

proceedings.com